Lecture Notes in Computer Science 15726

Founding Editors

Gerhard Goos
Juris Hartmanis

The series Lecture Notes in Computer Science (LNCS), including its subseries Lecture Notes in Artificial Intelligence (LNAI) and Lecture Notes in Bioinformatics (LNBI), has established itself as a medium for the publication of new developments in computer science and information technology research, teaching, and education.

LNCS enjoys close cooperation with the computer science R & D community, the series counts many renowned academics among its volume editors and paper authors, and collaborates with prestigious societies. Its mission is to serve this international community by providing an invaluable service, mainly focused on the publication of conference and workshop proceedings and postproceedings. LNCS commenced publication in 1973.

Jens Petersen · Vedrana Andersen Dahl
Editors

Image Analysis

23rd Scandinavian Conference, SCIA 2025
Reykjavik, Iceland, June 23–25, 2025
Proceedings, Part II

Editors
Jens Petersen
University of Copenhagen
Copenhagen, Denmark

Vedrana Andersen Dahl
Technical University of Denmark
Kongens Lyngby, Denmark

ISSN 0302-9743 ISSN 1611-3349 (electronic)
Lecture Notes in Computer Science
ISBN 978-3-031-95917-2 ISBN 978-3-031-95918-9 (eBook)
https://doi.org/10.1007/978-3-031-95918-9

This Springer imprint is published by the registered company Springer Nature Switzerland AG
The registered company address is: Gewerbestrasse 11, 6330 Cham, Switzerland

Preface

This book constitutes the refereed proceedings of the 23rd Scandinavian Conference on Image Analysis (SCIA 2025), held in Reykjavík, Iceland, in June 2025. The 60 papers presented in these proceedings were selected from 99 submissions, resulting in an acceptance rate of 61%.

The reviewing process for SCIA 2025 was conducted in a double-blind manner, with Microsoft CMT used to manage the reviewing workflow. One submission was desk-rejected, while the remaining 98 submissions were each assigned to three reviewers. The majority of the submissions–68 papers–received three reviews, while 30 papers received two reviews. To ensure that each paper had at least two assessments, a small number of emergency reviews were commissioned as needed.

The Program Chairs made the final acceptance decision based on the review grades and review comments. Initially, 61 papers were accepted; however, one was subsequently withdrawn, resulting in a total of 60 papers included in these proceedings

The most popular topics among the accepted papers were:

- Machine learning and deep learning (10 papers),
- Detection, recognition, classification, and localization in 2D and/or 3D (7 papers),
- 3D vision from multiview and other sensors (7 papers),
- Medical, biological, and cell microscopy (6 papers),
- Vision applications and systems (5 papers),
- Dataset and evaluation (5 papers),
- Segmentation, grouping, and shape (4 papers), and
- Image and video processing, analysis, and understanding (3 papers).

The scientific program of SCIA 2025 featured two invited talks and presentations of the accepted papers. The invited talks were delivered by Pascal Fua from EPFL and Dima Damen from the University of Bristol and Google DeepMind. Accepted papers were presented either in oral sessions (12 papers) or in poster sessions (all papers). In addition, all poster presenters gave accompanying lightning talks. SCIA 2025 also welcomed abstract submissions, with 21 accepted abstracts presented as posters.

The vast majority of paper and abstract submissions to SCIA 2025 came from Scandinavian countries: 35 from Denmark, 21 from Finland, 16 from Sweden, and 4 from Norway. Additionally, we received 16 submissions from other European countries, with Czechia particularly well-represented, contributing 6 papers. The remaining 7 submissions came from outside Europe.

We would like to thank all authors for their contributions, the reviewers for their careful and constructive evaluations, and the members of the organizing committee for their dedicated efforts.

We gratefully acknowledge all authors for submitting their work, the reviewers for providing thorough and thoughtful evaluations, and the organizing committee for their hard work and dedication. We would also like to thank our sponsors for their generous support, which was invaluable in making SCIA 2025 possible.

June 2025

Jens Petersen
Vedrana Andersen Dahl

Organization

General Chairs

Rasmus Larsen	Technical University of Denmark, Denmark
Thomas B. Moeslund	Aalborg University, Denmark

Program Committee Chairs

Jens Petersen	University of Copenhagen, Denmark
Vedrana A. Dahl	Technical University of Denmark, Denmark

Local Organizing Committee

Lotta María Ellingsen	University of Iceland, Iceland
Jakob Sigurdsson	University of Iceland, Iceland

Organizing Committee

Alejandro Cortina Uribe	University of Copenhagen, Denmark
Juan Miguel Valverde	Technical University of Denmark, Denmark
Pablo Delgado-Rodriguez	Technical University of Denmark, Denmark

Program Committee

Andreas Aakerberg	Aalborg University, Denmark
Marco Acerbis	Uppsala University, Sweden
Shuichi Akizuki	Chukyo University, Japan
Ameen Ali	Tel Aviv University, Israel
Constantino Álvarez Casado	University of Oulu, Finland
Akmaral Amanturdieva	Norwegian University of Science and Technology, Norway
Valery N. Ashu	LUT University, Finland
Kalle Åström	Lund University, Sweden
Ewert Bengtsson	Uppsala University, Sweden

Amanda Berg	Linköping University, Sweden
Miguel Bordallo Lopez	University of Oulu, Finland
Gunilla Borgefors	Uppsala University, Sweden
Jonathan Bøss	University of Southern Denmark, Denmark
Dimitri Bulatov	Fraunhofer IOSB, Germany
Hakan Cevikalp	Eskisehir Osmangazi University, Turkey
Denis Chaplygin	Tampere University, Finland
Tong Chen	University of Copenhagen, Denmark
Anthony Cioppa	University of Liège, Belgium
Alejandro Cortina Uribe	University of Copenhagen, Denmark
Anders B. Dahl	Technical University of Denmark, Denmark
Vedrana A. Dahl	Technical University of Denmark, Denmark
Sneha Das	Technical University of Denmark, Denmark
Tareen Dawood	Technical University of Denmark, Denmark
Felipe Delestro	Technical University of Denmark, Denmark
Yifan Ding	Linköping University, Sweden
Barbara Dzaja	University of Split, Croatia
Johan Edstedt	Linköping University, Sweden
Tuomas Eerola	LUT University, Finland
Gabriel Eilertsen	Linköping University, Sweden
Junyuan Fang	Aalto University, Finland
Aasa Feragen	Technical University of Denmark, Denmark
Axel Flinth	Umeå University, Sweden
Jan Flusser	UTIA, Czech Academy of Sciences, Czech Republic
Per-Erik Forssén	Linköping University, Sweden
Rikke Gade	Aalborg University, Denmark
Gaël Gendron	University of Auckland, New Zealand
Prashant Goswami	Blekinge Institute of Technology, Sweden
Anna Gummeson	Lund University, Sweden
Ulas Gunes	Tampere University, Finland
Gustav Hanning	Lund University, Sweden
Andreas Lau Hansen	Technical University of Denmark, Denmark
Manabu Hashimoto	Chukyo University, Japan
Soren Hauberg	Technical University of Denmark, Denmark
Joakim Bruslund Haurum	Aalborg University, Denmark
Jan Held	University of Liège, Belgium
Emily Honey Andersen Hoffmann	University of Copenhagen, Denmark
Bulat Ibragimov	University of Copenhagen, Denmark
Christian Igel	University of Copenhagen, Denmark
Nikolai Ilinykh	University of Gothenburg, Sweden
Xingxun Jiang	Southeast University, China

Sheng Joevenller	Luleå University of Technology, Sweden
Linnea H. Juul	Technical University of Denmark, Denmark
Sander Riisøen Jyhne	Norwegian Mapping Authority, Norway
Fredrik Kahl	Chalmers University of Technology, Sweden
Heikki Kälviäinen	LUT University, Finland
Michael C. Kampffmeyer	UiT The Arctic University of Norway, Norway
Justus Karlsson	Linköping University, Sweden
Mehmet Onurcan Kaya	Technical University of Denmark, Denmark
Ammar Kheder	LUT University, Finland
Huai Qian Khor	University of Oulu, Finland
Hans Martin Kjer	Technical University of Denmark, Denmark
Václav Košík	UTIA, Czech Academy of Sciences, Czech Republic
Arjan Kuijper	Fraunhofer Institute for Computer Graphics Research IGD and TU Darmstadt, Germany
Jorma Laaksonen	Aalto University, Finland
Manuel Lage Cañellas	University of Oulu, Finland
Erik Landolsi	Chalmers University of Technology, Sweden
Viktor Larsson	Lund University, Sweden
Olivier Lézoray	University of Caen Normandy, France
Deng Li	LUT University, Finland
Yante Li	University of Oulu, Finland
Manxi Lin	Technical University of Denmark, Denmark
Tony Lindeberg	KTH Royal Institute of Technology, Sweden
Shipeng Liu	Linköping University, Sweden
Xin Liu	LUT University, Finland
Zhi-Song Liu	LUT University, Finland
Miguel López-Pérez	Universitat Politècnica de València, Spain
Zhijin Lyu	Technical University of Denmark, Denmark
Jyri Maanpää	National Land Survey of Finland, Finland
Julia Machnio	University of Copenhagen, Denmark
Neelu Madan	Aalborg University, Denmark
Baptiste Magnier	IMT Mines Alès, France
Christian Mai	Aalborg University, Denmark
Filip Malmberg	Uppsala University, Sweden
Gustav E. Mark-Hansen	University of Copenhagen, Denmark
Jiří Matas	Czech Technical University in Prague, Czech Republic
Mostafa Mehdipour Ghazi	University of Copenhagen, Denmark
Satya Mahesh M. Muddamsetty	Aalborg University, Denmark
Lazaros Nalpantidis	Technical University of Denmark, Denmark
Costanza Navarretta	University of Copenhagen, Denmark

Nhi Nguyen	University of Oulu, Finland
Allan A. Nielsen	Technical University of Denmark, Denmark
Mads Nielsen	University of Copenhagen, Denmark
Mikael Nilsson	Lund University, Sweden
Silas N. Ørting	University of Copenhagen, Denmark
Magnus Oskarsson	Lund University, Sweden
Sumit Pandey	University of Copenhagen, Denmark
Dim P. Papadopoulos	Technical University of Denmark, Denmark
Marco Parola	University of Pisa, Italy
Rasmus R. Paulsen	Technical University of Denmark, Denmark
Malte Pedersen	Aalborg University, Denmark
Marius Pedersen	Norwegian University of Science and Technology, Norway
Rasmus J. Pedersen	Technical University of Denmark, Denmark
Ivar Persson	Lund University, Sweden
Jens Petersen	University of Copenhagen, Denmark
Mark Philip Philipsen	Aalborg University, Denmark
Lidia M. Pivovarova	University of Helsinki, Finland
Kumari Pooja	Technical University of Denmark, Denmark
Xuqian Ren	Tampere University, Finland
Robert Sablatnig	TU Wien, Austria
Suaiba A. Salahuddin	UiT The Arctic University of Norway, Norway
Helene Schulerud	SINTEF, Norway
Raghavendra Selvan	University of Copenhagen, Denmark
Anders Skaarup Johansen	Aalborg University, Denmark
Stefan Sommer	University of Copenhagen, Denmark
Kristine Aa. Sørensen	Novo Nordisk A/S, Technical University of Denmark, Denmark
Radim Spetlik	Czech Technical University in Prague, Czech Republic
Jon Sporring	University of Copenhagen, Denmark
Nataliya Strokina	Tampere University, Finland
Josefine Vilsbøll Sundgaard	Novo Nordisk A/S, Technical University of Denmark, Denmark
Jaakko Suutala	University of Oulu, Finland
Jussi Tohka	University of Eastern Finland, Finland
Juan Miguel Valverde	Technical University of Denmark, Denmark
Marc Van Droogenbroeck	University of Liège, Belgium
Mårten Wadenbäck	Linköping University, Sweden
Haishan Wang	Aalto University, Finland
Ruilin Wang	University of Helsinki, Finland
Zihan Wang	Aalto University, Finland

Elisabeth Wetzer	UiT The Arctic University of Norway, Norway
Kristoffer K. Wickstrøm	UiT The Arctic University of Norway, Norway
Tristan Wirth	TU Darmstadt, Germany
Bohao Xing	LUT University, Finland
Yonghao Xu	Linköping University, Sweden
Christopher Zach	Chalmers Institute of Technology, Sweden
Fereidoon Zangeneh	KTH Royal Institute of Technology, Sweden
Yejun Zhang	Aalto University, Finland
Xiaoyan Zhou	National University of Defense Technology, China

Sponsors

3Shape
Chemometec
Foss Analytical
Gubra
IHFood
JLIVision
Trackman
Videometer
Visiopharm
SCIA 2025 is endorsed by IAPR

3shape

FOSS

Endorsed by IAPR

Contents – Part II

Detection, Recognition, Classification, and Localization in 2D and/or 3D

Medical, Biological, and Cell Microscopy

Explainable AI for CV

Vision + Language (+ Other Modalities)

Computational Photography

Fairness

Low-Level and Physics-Based Vision

Contents – Part I

Segmentation, Grouping, and Shape

Vision for Robotics and Autonomous Vehicles

Biometrics, Faces, Body Gestures and Pose

3D Vision from Multiview and Other Sensors

Vision Applications and Systems

Datasets and Evaluation

FIORD: A Fisheye Indoor-Outdoor Dataset with LIDAR Ground Truth for 3D Scene Reconstruction and Benchmarking

Ulas Gunes[1(✉)], Matias Turkulainen[2], Xuqian Ren[1], Arno Solin[2], Juho Kannala[2,3], and Esa Rahtu[1]

[1] Tampere University, Tampere, Finland
{ulas.gunes,xuqian.ren,esa.rahtu}@tuni.fi
[2] Aalto University, Espoo, Finland
{matias.turkulainen,arno.solin,juho.kannala}@aalto.fi
[3] University of Oulu, Oulu, Finland

Abstract. The development of large-scale 3D scene reconstruction and novel view synthesis methods mostly rely on datasets comprising perspective images with narrow fields of view (FoV). While effective for small-scale scenes, these datasets require large image sets and extensive structure-from-motion (SfM) processing, limiting scalability. To address this, we introduce a fisheye image dataset tailored for scene reconstruction tasks. Using dual 200-degree fisheye lenses, our dataset provides full 360-degree coverage of 5 indoor and 5 outdoor scenes. Each scene has sparse SfM point clouds and precise LIDAR-derived dense point clouds that can be used as geometric ground-truth, enabling robust benchmarking under challenging conditions such as occlusions and reflections. While the baseline experiments focus on vanilla Gaussian Splatting and NeRF based Nerfacto methods, the dataset supports diverse approaches for scene reconstruction, novel view synthesis, and image-based rendering. The dataset is available here.

Keywords: fisheye image dataset · 3D scene reconstruction · novel view synthesis · Gaussian splatting · image-based rendering

1 Introduction

Recent advances in computer vision and graphics have revolutionized 3D scene reconstruction, novel view synthesis, and image-based rendering. Techniques such as 3D Gaussian splatting (3DGS, [14]), with its explicit point-based representation, and Neural Radiance Fields (NeRFs, [24]) have demonstrated remarkable results in these tasks. Gaussian splatting, in particular, offers faster rendering speeds and scalability for real-time applications. Unlike NeRFs, which encode

Supplementary Information The online version contains supplementary material available at https://doi.org/10.1007/978-3-031-95918-9_1.

J. Petersen and V. A. Dahl (Eds.): SCIA 2025, LNCS 15726, pp. 3–17, 2025.
https://doi.org/10.1007/978-3-031-95918-9_1

Fig. 1. Example data capture setup. The top image shows an example placement of camera in one of our scenes (depicted as its dense point cloud from Faro scanner), while the bottom image illustrates wide-angle 360° photo captured with a single shot of the camera.

volumetric data in neural networks, Gaussian splatting directly models scene geometry and appearance using Gaussian primitives, making it highly efficient for large-scale reconstruction.

Existing datasets commonly used for developing these techniques are predominantly composed of perspective images with narrow fields of view (FoV, typically <120°) and they are tailored for specific applications such as object-centric scenes or urban driving scenarios. Additionally, these datasets are often captured as videos during movement in the environment, introducing motion blur and compromising image quality. These limitations make them less suitable for broader applications in 3D reconstruction, novel view synthesis, and image-based rendering, particularly in large-scale, non-object-centric environments.

To address these challenges, we introduce a high-resolution, ultra-wide-angle fisheye dataset designed to support a wide range of applications. Additionally, we include precise LIDAR-derived ground truth point clouds of the captured environments using a Faro Focus 3D scanner [8]. Baseline evaluation results for vanilla Gaussian splatting and Nerfacto on this dataset are also provided, serving as a reference for future methods.

Our contributions are as follows:

- **A high-resolution, wide-angle fisheye still image collection:** The wide-angle images provide comprehensive scene coverage with fewer captures compared to narrow-field-of-view perspective images. The use of static photographs avoids motion blur associated with video frame extraction, ensuring sharp feature matching and improved reconstruction accuracy in Structure-from-Motion (SfM) pipelines.
- **Dense LIDAR-derived ground truth:** Dense point clouds generated with a Faro Focus 3D laser scanner serve as authoritative references for evaluating and improving reconstruction pipelines, particularly for alignment-sensitive techniques like Gaussian Splatting and NeRF.
- **SfM-compatible sparse point clouds:** Sparse point clouds generated with the Structure-from-Motion (SfM) tool COLMAP [27–29] are included in the dataset.
- **Baseline evaluations and benchmarks:** We provide baseline results for vanilla Gaussian splatting and Nerfacto methods, which offer insight into the potential of the dataset for novel view synthesis and scene reconstruction tasks. These benchmarks can guide future research and serve as references for evaluating new rendering, reconstruction, and depth-based methods.

By addressing the limitations of existing datasets, our work enables the development of novel techniques for diverse real-world scenarios in 3D reconstruction, image-based rendering, and novel view synthesis.

2 Related Work

Out of the numerous datasets available for 3D reconstruction and novel view synthesis tasks, we present a curated selection here, chosen for their diversity, scene coverage complexity and modality, while noting that many can also be repurposed for broader computer vision applications.

Datasets for 3D Scene Reconstruction and Novel View Synthesis. The Tanks and Temples [16] dataset provides high-quality ground truth data derived from an industrial laser scanner and high-resolution video input for both indoor and outdoor settings, and serves as a benchmark for static scene reconstruction. Waymo Open Dataset [31], KITTI-360 [19], and nuScenes [3] similarly provide extensive multi-modal sensor data, including LIDAR, RGB imagery, and trajectory information, making them suitable for scene reconstruction tasks.

The ScanNet++ [34] dataset extends the original ScanNet [7] dataset by adding object-level semantics and refining camera pose alignments, making it particularly suitable for semantic and geometric indoor reconstructions.

MuSHRoom [26] emphasizes diverse indoor environments, captured using high-precision and consumer-grade sensors. Replica [30] offers photorealistic reconstructions of indoor spaces, widely used in visual SLAM and neural rendering. The dataset used in Mip-NeRF360 [1] is another well-known one, which focuses on unbounded, object-centric scenes and small-scale, bounded indoor environments, accompanied by estimated camera poses from COLMAP.

The Aerial Coastline Imagery Dataset (ACID, [22]) captures natural coastal scenes using aerial drone footage, which allows long-range trajectory synthesis in scenes. UrbanScene3D [21] similarly facilitates bird-eye view urban reconstructions and was pivotal in an early large-scale 3DGS-based work [20]. MatrixCity [18] also delivers synthetic data for controlled Gaussian splatting experiments across ground-level and aerial scenes, which have been used in another early work on the 3DGS-based large-scale scene reconstruction [23]. Similar to our work, the dataset used in the hierarchical Gaussian splatting [15], collected with a multi-camera GoPro rig that captures time-elapsed narrow FoV images in motion (walking on foot or moving by bicycle), focuses on large-scale scene modeling.

The 360Roam dataset [9] provides full 360° imagery optimized for Gaussian splatting, while EgoNeRF [4] dataset focuses on omnidirectional modeling for large-scale indoor reconstructions. OmniGS [17] work leverages the panoramic datasets 360Roam [9] and EgoNeRF [4] for indoor reconstructions, highlighting the utility of omnidirectional data for large-scale modeling in panoramic format. The LetsGo project [6] uses a commercial LIDAR and fisheye imaging device in the same coordinate system for garage-scale environments, which have bounded (indoor) and semi-bounded (garage with outdoor openings) scene components. Unlike these works, our dataset captures images in raw fisheye format rather than the equirectangular format, which is commonly used to stitch the two fisheye views from a 360° camera. This approach eliminates potential stitching artifacts that may arise during the transformation process. Additionally, our dataset is captured in a completely motion-free setting, ensuring that each frame remains still and unaffected by movement, in contrast to datasets that rely on moving systems that record videos or capture time-elapsed images.

Fisheye Image Based Rendering. Recent works addressing the challenges of rendering wide-angle fisheye images (>180° FoV) include On the Error Analysis of 3D Gaussian Splatting [10], which introduces a rasterizer for fisheye rendering without rectification, and 3DGUT [33], which extends 3D Gaussian splatting to support non-linear camera projections and secondary rays for simulating effects like reflections and refractions. Both methods demonstrate these capabilities using indoor and unbounded fisheye images.

3 The FIORD Dataset

The FIORD comprises of still fisheye images captured from ten distinct scenes, provided in their stitched (two fisheyes side to side) and split (single fisheye) formats. Additionally, the dataset includes two types of point clouds for each scene: the sparse point cloud generated via Structure-from-Motion (SfM) and the dense point cloud captured with a Faro LIDAR scanner.

In this section, we first explain the camera calibration and data collection procedures performed using the Insta360 RS One-Inch camera [11] and the Faro Focus LIDAR Scanner. Then, we explain the post-data collection processing steps to create our dataset, which yields the generation of sparse and dense point clouds for the scenes. Finally, we describe the sparse and dense point cloud alignment process and the image rectification step for our experiments mentioned in the next section. The overall process is visualized in Fig. 2.

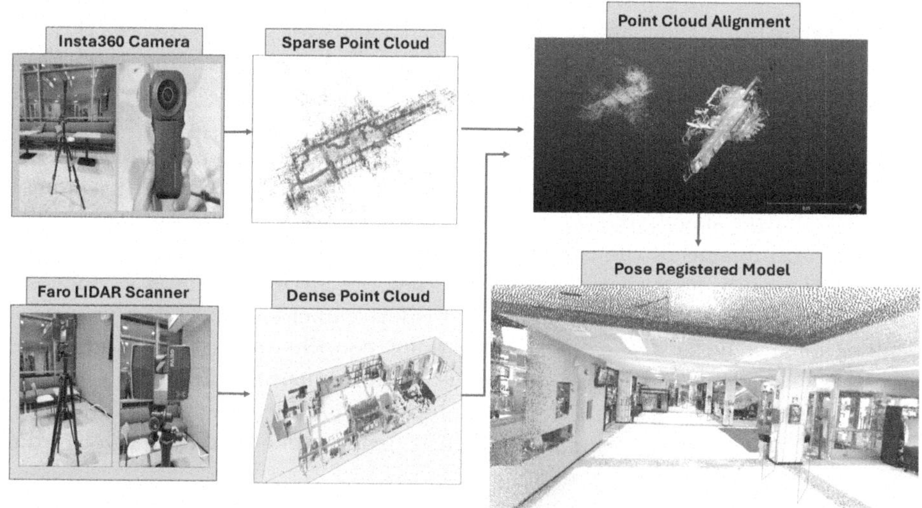

Fig. 2. Sparse and dense point cloud alignment. Fisheye images and LIDAR scans are used to generate and align sparse and dense point clouds. Camera poses can be registered to the aligned model in real-world or COLMAP coordinate system scale.

3.1 Camera Calibration

To prepare this data set, we used two wide-angle (200° FoV per lens) fisheye cameras of the Insta360 One RS One-Inch Sensor camera [11], mounted on a tripod. In other words, all the images we captured, both for scene data collection and calibration purposes, are still (motion-free). Although the Insta360 camera can stitch the images from its two fisheye lenses to generate full 360° equirectangular panoramic images, this stitching process often introduces alignment errors, such as ghosting artifacts along stitching lines, particularly in scenes with complex geometry or significant depth variation. To avoid these issues and preserve geometric accuracy, we opted to work directly with the raw fisheye images.

The raw fisheye images were initially stored in the camera manufacturers' .insp file format, and we converted them to JPEG. JPEG was chosen because it is supported by both the Camera Calibration Toolbox and SfM software, and it offers a practical balance of compatibility, quality, and file size. Higher-fidelity formats like PNG could be used for tasks requiring lossless quality, albeit with increased computational cost. Each capture results in a single image that contains two side-by-side fisheye views (Fig. 1). These images were split into two separate 3264 × 3264 images, each corresponding to one fisheye lens.

Fisheye lenses produce heavily distorted images, particularly toward the edges, where the distortion effect becomes most visible (Fig. 1). We performed camera calibration to correct these distortions. We extracted precisely estimated intrinsic camera parameters for each fisheye lens separately.

Using the estimated intrinsic camera parameters, we performed camera calibration. We utilized the Camera Calibration Toolbox for Generic Lenses [12,13], which provides calibration support for wide-angle lenses exceeding 180° FoV. We used the radial camera model option from the toolbox for our lenses. The calibration pattern displayed on a flat screen was captured from both the heavily distorted edges and the less distorted central region of each lens by placing the tripod-mounted camera in different locations in front of the pattern. Camera exposure, shutter speed, and white-balancing settings were kept constant while capturing still images of the calibration target.

After the calibration process, we converted the intrinsic camera parameters we obtained from the toolbox to OpenCV-Fisheye camera format [2]. This format includes focal lengths (f_x, f_y), principal point coordinates (c_x, c_y), and radial distortion coefficients $(k_1, ..., k_4)$, and is accepted by the SfM pipeline COLMAP [27,29], which will be used in the next steps. We verified the accuracy of the calibration by using the estimated intrinsic parameters to remove the radial distortion from our fisheye images.

3.2 Data Collection

To create a diverse dataset, we selected ten distinct scenes (five indoor environments and five outdoor environments) from the Tampere University Hervanta Campus, with varying color, lighting characteristics, scale, and geometric complexity. Depending on the scene size and complexity, a total of 500–1300 images (single fisheye) per scene are captured. Our scenes were captured during winter conditions, which introduced unique realistic challenges for our outdoor environments, such as snow or ice-induced glare, foggy conditions, occlusions, complex lighting conditions and repetitive structures that might be challenging for the SfM pipeline based sparse point cloud reconstruction. We believe that these challenges will push forward the research in 3D scene reconstruction techniques. The scenes in our dataset and their brief descriptions are given in Table 1.

The Insta360 camera [11] was mounted on a tripod, and placed at a fixed location to take a single shot for the scene image capturing process (an example camera placement in a scene is shown in Fig. 1). Afterwards, it was repositioned with minimal rotation and movement (less than 10 cm between each camera positioning and less than 60° of rotation to either side on the lateral axis of the camera) to another location within the scene, and another image was taken. This

Table 1. Descriptions of Indoor and Outdoor Scenes in the Dataset

Name	Description
Indoor Scenes	
Kitchen_In	A 12 m^2 kitchen featuring repetitive, detailed objects (chairs), appliances, and reflective countertops, providing moderate geometric complexity
MeetingRoom_In	A 15 m^2 room with simple geometry, flat walls, and minimal objects, with heavy ceiling light exposure
Building_In	A hallway-like 62 m^2 indoor environment with uniform light distribution, repetitive tiles and reflective materials
Hall_In	A large 80 m^2 hallway with tall ceiling, nonuniform light distribution, repetitive tiles, shiny and or highly detailed objects
Upstairs_In	A large 260 m^2 hall area with irregular shapes, textured surfaces, and reflective materials such as glass
Outdoor Scenes	
Bridge_Out	A 125 m^2 outdoor walkway with snow, reflective glasses and repetitive texture buildings.
Night_Out	A 125 m^2 outdoor garden area with trees, buildings with reflective surfaces, repetitive window patterns and non-uniform light. Captured during evening conditions
Corridor_Out	A 207 m^2 outdoor walkway with snow, repetitive structured stairs, glasses and non-homogenous (pepper-salt style small rocks) floor structure
Building_Out	A 305 m^2 outdoor space, includes a couple of moving objects such as people or cars
Road_Out	A 930 m^2 large unbounded outdoor space with occlusions, non-uniform light conditions, non-homogenous (pepper-salt style small rocks) floor structure, reflective surfaces and fog

process was repeated systematically until the entire scene was covered, with each lens consistently covering the same side of the scene at all times. This consistency, combined with the minimal rotation or movement between each capture ensured sufficient overlap between the images, which played a key role in the subsequent accurate SfM based sparse point cloud generation step. The short focal length of the camera also enabled sharp capture of wide areas, even from long distances. Similar to the calibration step, consistent settings for shutter speed, exposure, and white balance during image captures are used to capture true lighting and color in scenes.

To avoid disruptions from moving objects, such as people in indoor environments or cars in outdoor settings, images were captured at times and locations with minimal activity. The photographer ensured they remained out of the frame by strategically positioning themselves in occluded areas or sequentially capturing the two fisheye images from the same fixed position—first taking a shot while remaining outside the field of view (FoV) of one lens, then repositioning to avoid the FoV of the second lens before capturing the next image.

To obtain the geometry ground truth of each scene, we used the Faro Focus 3D LIDAR scanner fixed on a tripod, to capture high-resolution XYZRGB point clouds. The Faro scanner covers a 360° horizontal and 170° vertical FoV (−60° to 90°) with a data capture range of 0.6–200 m depending on the indoor or outdoor capture modes. Similar to the camera setup, the Faro scanner was placed at fixed locations in the scene and multiple scans, each corresponding to a different location in the scene, are taken at 1/4 or 1/5 resolution and 4× quality. Each scan lasted around 11 min and depending on the scale of the scene, 5–25 scans at different fixed positions were necessary to fully cover each scene. Minor artifacts from moving objects were negligible relative to scene scale and were not visible in the dense point clouds created. An example dense point cloud captured with the Faro Scanner for the Kitchen_In scene is given in Fig. 1.

3.3 Formation of Sparse and Dense Point Clouds

The Structure-from-Motion (SfM) is a fundamental step in many 3D scene reconstruction methods, as it allows for the recovery of camera poses and the forming of sparse 3D structures from multiple overlapping images. In our dataset, we utilized COLMAP [27,29] version 3.9.1, an incremental SfM pipeline, to generate sparse point clouds of the scenes using the raw fisheye images.

As the first step of SfM, feature extraction, in COLMAP, we employed the OpenCV Fisheye camera model and supplied the original calibration parameters. To accommodate the high-resolution fisheye images, we increased the maximum image size allowed and the number of features extracted. After extracting feature points from the images, feature matching was performed using the vocabulary tree matcher. In COLMAP, images are registered into the scene representation, followed by triangulation of 3D points and a global bundle adjustment to simultaneously optimize camera poses and the 3D structure. This reconstruction process results in a sparse point cloud.

The COLMAP SfM pipeline outputs the sparse point cloud representing the 3D structure of each scene in binary (.bin) format. The binary format is included in each scenes' sparse model in our dataset, and these can be converted to the text format if necessary. The COLMAP format is also supported by Nerfstudio [32], a common 3D scene reconstruction framework that can enable real-time rendering in navigable environments for a number of NeRF and 3DGS-based scene reconstruction methods.

The dense, ground-truth point cloud is obtained from the Faro scanner software SCENE [8], after the software processes the scans we have taken for each scene. Minimal cropping operations are performed to remove redundant points.

3.4 Point Cloud Alignment and COLMAP Model Modification

Aligning sparse point clouds generated by COLMAP with dense ground truth data from the Faro scanner establishes a shared coordinate system and enables direct comparison and evaluation for downstream applications.

We performed an alignment between the two point clouds, using the CloudCompare [5] software. First we selected 7–10 easily identifiable points (e.g., corners, edges, and structural features) from the COLMAP point cloud and then have marked their correspondences in the FARO Scanner generated point cloud. Based on these correspondences, we estimated a transformation matrix defining the rotation, translation, and scaling required to map the sparse COLMAP point cloud to the Faro scan's coordinate system.

A key challenge in this process is the significant difference in the density of the point clouds. For example, in the largest indoor scene, the sparse point cloud produced by COLMAP contains about 400,000 points, while the dense Faro scan for the same scene includes nearly 500 million points. This disparity complicates the task of mapping corresponding points between the two datasets.

The alignment accuracy was validated using the Root Mean Square Error (RMSE) metric calculated in CloudCompare. RMSE quantifies the mean distance between these corresponding points in the two ppoint clouds, indicating how closely the two clouds overlap after alignment. For instance, an RMSE of 25 cm in our largest indoor scene suggests that, on average, the corresponding points in the sparse and dense clouds differ by 25 cm, which is acceptable for scenes of this scale (e.g., a 220 m^2 room). In addition to the RMSE metric, the alignment was verified visually. The final transformation matrix was applied to the COLMAP model, updating all reconstructed points and camera poses. A simplistic illustration of the alignment process is given in Fig. 2.

3.5 Image Rectification

For compatibility with our experiments presented in the next section, we performed image rectification (undistortion) on the raw fisheye images using the estimated intrinsic camera calibration parameters we obtained in Sect. 3.1. This step transformed the fisheye images into a pinhole camera model-compatible format, and facilitated our baseline experiments using Gaussian Splatting [14] and Nerfstudio's Nerfacto [32] methods, which lack native support for fisheye rendering for wide-angle lenses ($>180°$ FoV).

The rectification process was implemented for convenience and allowed the dataset to be easily integrated into the novel 3D scene reconstruction pipelines. However, this approach is not optimal, as the rectified images may lose scene information from the heavily distorted parts of the fisheye images. Despite this, the processed dataset provides a practical solution for our experiments and works sufficiently well, based on the quantitative results.

4 Experiments

In this section, we present two experiments that demonstrate the capabilities of our dataset, highlighting the usage of both the sparse COLMAP data and the dense Faro scanner data. For our experiments, we applied a standard 90%-10% train-test split to the images for each scene. The visual results and evaluation metrics presented correspond to the rendered test images, which were randomly selected from the complete set of images for each scene.

4.1 Novel View Synthesis with Vanilla Gaussian Splatting (3DGS) and Nerfacto Using SfM (Sparse) Point Cloud

In the first experiment, we establish a baseline for rendering quality and performance using the Gaussian splatting (3DGS) [14] and Nerfstudio (v1.1.4)'s NeRF-based scene reconstruction method Nerfacto [32]. For both of these models we use the sparse point clouds generated by the SfM software COLMAP. These point clouds originate from fisheye images, which we rectify (undistort) using camera calibration parameters to be compatible with the Gaussian splatting implementation's rasterization requirements for pinhole camera models [14], and with the Nerfacto models' supported camera types.

For training the Gaussian splatting and Nerfacto models, we down-sample the high-resolution images by 4 (800×800 pixels per image), because processing them at full resolution exceeds available RAM capacity. All tunable parameters for these models remained at their default values. Training and evaluation were carried out on an NVIDIA RTX 4090 GPU, which has 24 GB VRAM. The highest VRAM consumption was recorded as 13 GB for training the Nerfacto model for our largest outdoor scene, Road_Out. The training and evaluation

Fig. 3. Compact comparison of Ground Truth vs. Gaussian Splatting renders for four representative scenes.

pipeline took between 20 to 30 min for Gaussian splatting model, and between 7–15 min for the Nerfacto model depending on scene complexity. We trained each model for 30k iterations for each scene.

Figure 3 presents example image-based rendering results obtained with the Gaussian splatting method for four scenes in our dataset. Each scene is illustrated by a single example to highlight the versatility of the data set across various environments. Example video renders of two of our scenes created from the Nerfacto model are also provided in the Supplementary Materials. Meanwhile, Table 2 provides standard image quantitative metrics (PSNR, SSIM, and LPIPS [35]) for all scenes in the dataset, averaged over the test images of each scene for both models.

The baseline results from the vanilla Gaussian Splatting (3DGS) and Nerfacto-based scene reconstruction methods demonstrate the dataset's immediate applicability for novel view synthesis and 3D reconstruction tasks. As illustrated in Fig. 3, the Gaussian Splatting method effectively handles varying lighting conditions and complex reflections, such as glare from glass surfaces and produces high-quality renders. Quantitative metrics in Table 2 and Table 3 further validate the robustness of these two models, with 3DGS providing better quantitative results than Nerfacto for our scenes.

Table 2. Quantitative Metrics (PSNR, SSIM, LPIPS) of 3DGS Method

Indoor Scenes				Outdoor Scenes			
Scene	PSNR	SSIM	LPIPS	Scene	PSNR	SSIM	LPIPS
Upstairs_In	23.33	.8187	.3693	Bridge_Out	27.58	.8426	.2544
Hall_In	25.47	.8354	.1961	Corridor_Out	28.06	.8507	.2331
Building_In	26.28	.8076	.3017	Building_Out	24.44	.7525	.3000
MeetingRoom_In	27.41	.8628	.2133	Road_Out	25.67	.8109	.2882
Kitchen_In	27.15	.8705	.2200	Night_Out	26.12	.8328	.3437

Table 3. Quantitative Metrics (PSNR, SSIM, LPIPS) of Nerfacto Model

Indoor Scenes				Outdoor Scenes			
Scene	PSNR	SSIM	LPIPS	Scene	PSNR	SSIM	LPIPS
Upstairs_In	18.64	.7768	.5384	Bridge_Out	22.07	.7799	.3362
Hall_In	21.30	.7039	.4590	Corridor_Out	18.91	.5576	.5630
Building_In	24.49	.7936	.4878	Building_Out	20.31	.4549	.4081
MeetingRoom_In	23.90	.8428	.2622	Road_Out	19.15	.6402	.5086
Kitchen_In	21.28	.7922	.3386	Night_Out	18.19	.6327	.4979

4.2 Using Dense LIDAR Data for Gaussian Scene Initialization

Table 4. COLMAP vs. COLMAP+LIDAR Point Clouds for Gaussian Initialization on the Building_In Scene (averaged over all test images)

Gaussian Initialization	Number of Points (M)	PSNR	SSIM	LPIPS
COLMAP	~0.40	25.31	0.802	0.302
COLMAP+LIDAR	~2.6	26.24	0.811	0.269

In the second experiment, we incorporate the dense point cloud we obtained from the Faro scanner to enrich the sparse point cloud provided by COLMAP. We begin by uniformly sub-sampling the Faro data to keep it computationally manageable, limiting the maximum available points in the dense point cloud to 2–3 times the points available in the sparse point cloud for easy and sufficiently accurate alignment, and then scale this sub-sampled cloud to match the scale of the COLMAP point cloud. After a rough manual alignment of rotation and translation, we register the sampled dense point cloud with Iterative Closest Point (ICP) [25], yielding a fused point cloud. This LIDAR-aided point cloud serves as initialization for Gaussian splatting. Compared to the sparse point cloud, this fused point cloud enables the adaptive Gaussian training process to start from a better representative set of points. An example image-based

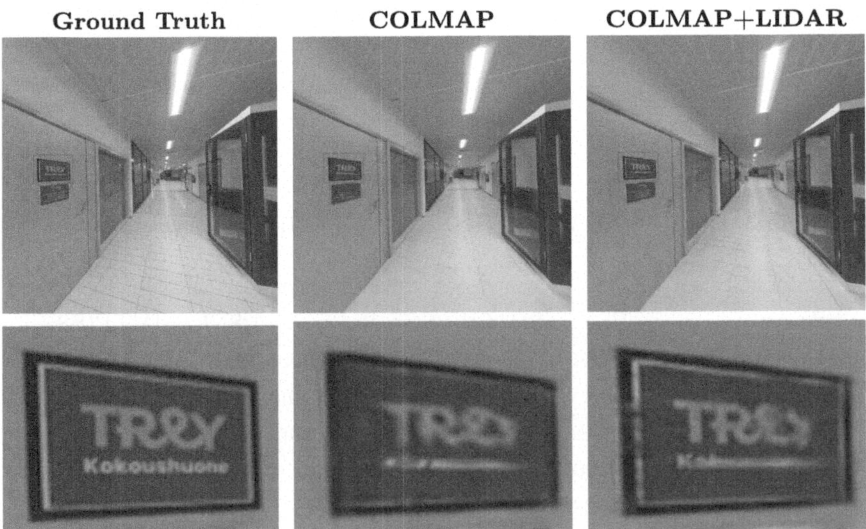

Fig. 4. Comparison of an image rendering result for the Building_In scene. The ground truth image is compared against the rendering initialized with only COLMAP data versus the fused COLMAP+LIDAR point cloud.

rendering result of this densification process is shown from our "Building_In" scene, in Fig. 4. For a true comparison, the test image set remains the same for both the sparse and fused approaches during rendering (Table 4).

This experiment demonstrates that incorporating dense Faro LIDAR point clouds could improve both quantitative metrics and visual reconstruction quality, particularly in complex scenes. Beyond our experiment demonstrating a use case for the Faro scanner LIDAR data, there are many other potential uses for researchers, including but not limited to geometric accuracy evaluation, depth-based benchmarking, environment mapping and scene understanding.

5 Conclusion

We introduced a high-quality dataset to address limitations in existing resources for large-scale 3D scene reconstruction, novel view synthesis, and image-based rendering. Captured using an Insta360 camera with dual 200-degree fisheye lenses, the dataset provides comprehensive 360-degree coverage while compensating for heavy lens distortion through calibration and maintaining high scene detail. Complemented by dense ground truth point clouds from a Faro Focus 3D LiDAR scanner, it enables robust geometric evaluation and alignment benchmarking. The dataset presents unique challenges, which makes it well suited to 3D scene reconstruction under real-world complexities. By relying on still images, it avoids motion blur, maintains high detail, and provides a solid basis for advancing 3D reconstruction and novel view synthesis in complex environments.

Acknowledgements. We acknowledge the financial support of the Intelligent Work Machines Doctoral Education Pilot Program (IWM VN/3137/2024-OKM-4) and funding from the Research Council of Finland (grants 352788, 353138, 362407, 362408, 339730, 353139, 362409) and the Finnish Center for Artificial Intelligence. We also acknowledge the Centre for Immersive Visual Technologies for the equipment used.

References

1. Barron, J.T., Mildenhall, B., Verbin, D., Srinivasan, P.P., Hedman, P.: Mip-NeRF 360: unbounded anti-aliased neural radiance fields. In: CVPR (2022)
2. Bradski, G.: The OpenCV library. Dr. Dobb's J. Softw. Tools (2000)
3. Caesar, H., et al.: nuScenes: a multimodal dataset for autonomous driving. In: CVPR (2020)
4. Choi, C., Kim, S.M., Kim, Y.M.: Balanced spherical grid for egocentric view synthesis. In: Proceedings of the IEEE/CVF Conference on Computer Vision and Pattern Recognition (CVPR), pp. 16590–16599 (2023)
5. CloudCompare: CloudCompare (version 2.13.2) [GPL software] (2024). http://www.cloudcompare.org/
6. Cui, J., et al.: LetsGo: large-scale garage modeling and rendering via lidar-assisted gaussian primitives. ACM Trans. Graph. **43**(6) (2024)

7. Dai, A., Chang, A.X., Savva, M., Halber, M., Funkhouser, T., Nießner, M.: ScanNet: richly-annotated 3D reconstructions of indoor scenes. In: Proceedings of the Computer Vision and Pattern Recognition (CVPR). IEEE (2017)
8. Faro Technologies, Inc.: Faro Focus 3D LiDAR Scanner (2025). https://www.faro.com/en/Products/
9. Huang, H., Chen, Y., Zhang, T., Yeung, S.K.: 360Roam: real-time indoor roaming using geometry-aware 360° radiance fields. arXiv preprint arXiv:2208.02705 (2022)
10. Huang, L., Bai, J., Guo, J., Li, Y., Guo, Y.: On the error analysis of 3D gaussian splatting and an optimal projection strategy. In: Computer Vision – ECCV 2024, pp. 247–263. Springer, Cham (2025)
11. Insta360: Insta360 One RS 1-Inch 360 Edition Specifications (2024). https://www.insta360.com/product/insta360-oners/1inch-360
12. Kannala, J., Brandt, S.S.: A generic camera model and calibration method for conventional, wide-angle, and fish-eye lenses. IEEE Trans. Pattern Anal. Mach. Intell. **28**(8), 1335–1340 (2006)
13. Kannala, J., Heikkilä, J., Brandt, S.S.: Geometric camera calibration. Wiley Encycl. Comput. Sci. Eng. **13**(6), 1–20 (2008)
14. Kerbl, B., Kopanas, G., Leimkühler, T., Drettakis, G.: 3D gaussian splatting for real-time radiance field rendering. ACM Trans. Graph. **42**(4) (2023)
15. Kerbl, B., Meuleman, A., Kopanas, G., Wimmer, M., Lanvin, A., Drettakis, G.: A hierarchical 3D gaussian representation for real-time rendering of very large datasets. ACM Trans. Graph. **43**(4) (2024)
16. Knapitsch, A., Park, J., Zhou, Q.Y., Koltun, V.: Tanks and temples: benchmarking large-scale scene reconstruction. ACM Trans. Graph. **36**(4) (2017)
17. Li, L., Huang, H., Yeung, S.K., Cheng, H.: OmniGS: fast radiance field reconstruction using omnidirectional gaussian splatting (2024)
18. Li, Y., et al.: MatrixCity: a large-scale city dataset for city-scale neural rendering and beyond. In: Proceedings of the IEEE/CVF International Conference on Computer Vision, pp. 3205–3215 (2023)
19. Liao, Y., Xie, J., Geiger, A.: KITTI-360: a novel dataset and benchmarks for urban scene understanding in 2D and 3D. Pattern Analysis and Machine Intelligence (PAMI) (2022)
20. Lin, J., et al.: VastGaussian: vast 3D gaussians for large scene reconstruction (2024)
21. Lin, L., Liu, Y., Hu, Y., Yan, X., Xie, K., Huang, H.: Capturing, reconstructing, and simulating: the urbanscene3d dataset. In: ECCV, pp. 93–109 (2022)
22. Liu, A., Tucker, R., Jampani, V., Makadia, A., Snavely, N., Kanazawa, A.: Infinite nature: perpetual view generation of natural scenes from a single image. In: Proceedings of the IEEE/CVF International Conference on Computer Vision (ICCV) (2021)
23. Liu, Y., Guan, H., Luo, C., Fan, L., Peng, J., Zhang, Z.: CityGaussian: real-time high-quality large-scale scene rendering with gaussians (2024)
24. Mildenhall, B., Srinivasan, P.P., Tancik, M., Barron, J.T., Ramamoorthi, R., Ng, R.: NeRF: representing scenes as neural radiance fields for view synthesis. In: ECCV (2020)
25. Park, J., Zhou, Q.Y., Koltun, V.: Colored point cloud registration revisited. In: Proceedings of the IEEE International Conference on Computer Vision (ICCV) (2017)
26. Ren, X., Wang, W., Cai, D., Tuominen, T., Kannala, J., Rahtu, E.: Mushroom: multi-sensor hybrid room dataset for joint 3D reconstruction and novel view synthesis (2023)

27. Schönberger, J.L., Frahm, J.M.: Structure-from-motion revisited. In: Conference on Computer Vision and Pattern Recognition (CVPR) (2016)
28. Schönberger, J.L., Price, T., Sattler, T., Frahm, J.M., Pollefeys, M.: A vote-and-verify strategy for fast spatial verification in image retrieval. In: Asian Conference on Computer Vision (ACCV) (2016)
29. Schönberger, J.L., Zheng, E., Pollefeys, M., Frahm, J.M.: Pixelwise view selection for unstructured multi-view stereo. In: European Conference on Computer Vision (ECCV) (2016)
30. Straub, J., et al.: The replica dataset: a digital replica of indoor spaces. arXiv preprint arXiv:1906.05797 (2019)
31. Sun, P., et al.: Scalability in perception for autonomous driving: waymo open dataset. In: Proceedings of the IEEE/CVF Conference on Computer Vision and Pattern Recognition (CVPR) (June 2020)
32. Tancik, M., et al.: NeRFStudio: a modular framework for neural radiance field development. In: ACM SIGGRAPH 2023 Conference Proceedings, SIGGRAPH 2023 (2023)
33. Wu, Q., Esturo, J.M., Mirzaei, A., Moenne-Loccoz, N., Gojcic, Z.: 3DGUT: enabling distorted cameras and secondary rays in gaussian splatting (2024)
34. Yeshwanth, C., Liu, Y.C., Nießner, M., Dai, A.: ScanNet++: a high-fidelity dataset of 3D indoor scenes. In: Proceedings of the International Conference on Computer Vision (ICCV) (2023)
35. Zhang, R., Isola, P., Efros, A.A., Shechtman, E., Wang, O.: The unreasonable effectiveness of deep features as a perceptual metric. In: CVPR (2018)

SunVid: A Curated Online Video Database for Sundown Syndrome Research

Qianru Xu[1,2], Mengting Wei[1], Huai-Qian Khor[1], Feng Vankee Lin[2], and Guoying Zhao[1(✉)]

[1] Center for Machine Vision and Signal Analysis, University of Oulu, Oulu, Finland
{qianru.xu,mengting.wei,huai.khor,guoying.zhao}@oulu.fi

[2] Cogtlab, Department of Psychiatry and Behavioral Sciences, Stanford University, Stanford, USA
vankee_lin@stanford.edu
https://cogtlab.stanford.edu/

Abstract. Sundown Syndrome (SS) is a condition characterized by cognitive, emotional, and behavioral disturbances, primarily affecting individuals with dementia in the late afternoon and evening. Despite being long recognized, research on SS remains limited due to the lack of standardized diagnostic criteria and the limited availability of structured multimodal datasets. To address this gap, we present SunVid, a curated online video database specifically designed for SS research. This dataset includes 74 SS-related videos from online platforms and 405 annotated segments labeled with SS states, affective states, symptoms, and context. To assess the effectiveness of SunVid, we perform three evaluation experiments using facial features, Emonet-extracted emotional features, and body gesture features, employing state-of-the-art benchmark models to establish baseline performance. For facial feature analysis, the Swin Transformer model achieves the highest accuracy of 69.03% in SS vs. non-SS classification. Emonet-extracted emotional features, including emotion category, valence, and arousal, produce slightly better results with an accuracy of 72.85%. Body gesture analysis across different models further confirms that movement patterns contribute to SS identification, although with lower predictive accuracy than facial features. Despite limitations such as class imbalance and variations in video quality, our findings demonstrate the feasibility of AI-assisted SS detection. This study also highlights the potential of video data as a novel avenue for SS analysis and underscores the promise of AI-driven methods in supporting SS detection and intervention.

Keywords: Sundown Syndrome · Video Database · Facial Expression Recognition · Body Gesture Analysis · AI-Assisted Diagnosis

1 Introduction

Sundown syndrome (SS), or sundowning, refers to a set of cognitive, emotional, and behavioral changes that typically occur during the late afternoon or evening

J. Petersen and V. A. Dahl (Eds.): SCIA 2025, LNCS 15726, pp. 18–31, 2025.
https://doi.org/10.1007/978-3-031-95918-9_2

hours. It predominantly affects older adults, particularly those with dementia [3, 13]. Despite being recognized for decades [2,7], awareness of SS remains limited, which presents significant challenges in identifying and addressing the condition. This lack of understanding often leads to misconceptions about its symptoms and their effects on behavior, resulting in considerable distress for both patients and caregivers [3].

Traditionally, SS has been studied primarily through questionnaires and observational methods. Caregivers or trained researchers are often asked to observe and code SS symptoms during the sundown period or compare symptom frequency and severity at different times of the day to study the condition (e.g., [7,17]). While these studies have provided valuable initial insights, they are labor-intensive and prone to human error, which may lead to missed observations or biased assessments. To introduce more objective measurements, later studies incorporated motion sensors and found that movement activity tended to increase during sundowning(e.g., [8,29]). While these methods effectively capture pronounced behavioral symptoms such as wandering and agitation, they are less effective in detecting subtle movement changes or tracking emotional and cognitive fluctuations associated with SS. Moreover, distinguishing SS-specific patterns from normal activity variations remains challenging without additional contextual information. Given the complexity and individual variability of SS, there is a growing need for multi-modal approaches that can capture a broader spectrum of symptoms beyond agitation and increased movement [31].

Video technology enables effective tracking of behavioral and emotional changes over time. With advances in related technologies, it offers a valuable tool for SS research by reducing manual effort and providing more detailed information. Currently, video-based AI has been successfully applied to detecting and managing conditions such as depression and agitation, where emotional and behavioral changes serve as key indicators (for reviews, see [10,14]). However, as far as we know, these methods have not yet been applied to SS detection, likely due to the limited attention this condition has received and the lack of suitable datasets for this purpose.

Although existing evidence suggests that colder seasons and higher latitudes may increase the likelihood of SS occurrence [19], making SS potentially more severe in Scandinavia, practical constraints, due to the complexity of SS and its relatively low prevalence, pose significant challenges to collecting a large-scale dataset in real-world settings. To address this challenge, we collect and curate online videos that explain SS and share personal experiences for educational purposes. Building on these videos, we aim to provide deeper insights into SS and explore the potential of video-based sensing for assisting in SS detection and management.

In this context, the paper aims to: (1) present SunVid, a dataset of SS-related online videos with emotion-based representations to support future research; and (2) provide baseline SS/non-SS classification results to demonstrate the potential of AI-based approaches for SS detection and intervention. This paper is organized as follows: Sect. 2 describes the process of data collection, annotation, and basic feature analysis, along with key insights obtained from the dataset. Section 3 presents a baseline analysis for SS detection, focusing on facial expressions and body movements. Facial expressions are primary indicators of emotional changes, while body gestures are valuable for identifying conditions such as agitation and wandering, which are commonly associated with SS. Finally, Sect. 4 discusses the findings and their potential future applications.

2 Database Framework

2.1 Video Collection Process

To collect related videos, We first conduct separate searches on public video-sharing platforms using the keywords *sundown syndrome*, *sundowning*, and *sundown syndrome + dementia* to find relevant content. Additionally, since many previous studies (for reviews, see [1,13]) have primarily focused on agitation as one of the most significant symptoms of SS, we also search using the keywords *agitation + sundown syndrome* to include as many videos as possible. We review all the videos and select those featuring humans who clearly exhibit or demonstrate relevant symptoms, excluding lectures without sample cases, as well as cartoons and other unrelated content, such as music videos. As a result, we curated 74 publicly available videos, which were then segmented for annotation and analysis. The overall data construction framework is illustrated in Fig. 1.

2.2 Video Segmentation and Annotation

Five annotators were recruited for the annotation process, with each video reviewed by three annotators to determine the final label. Prior to annotation, all annotators underwent thorough training and were provided with detailed instructions. This training included familiarization with each symptom, completing the online NPI-Q Interviewer Training Module, and obtaining the NPI-Q Interviewer Certification issued by the National Alzheimer's Coordinating Center[1].

During the annotation process, annotators carefully reviewed each video to identify relevant events involving the patient, whether during symptom exhibition or in a normal state. They were instructed to mark the start and end times of each event and assign appropriate labels based on their observations. The labeling framework included SS-related categories such as Conditions, States, Symptoms, and Time. Additionally, contextual information, including Persons, Positions, and Cues, was recorded to provide a comprehensive understanding of each event. A detailed breakdown of all categories and their descriptions is provided in Table 1. Please note that since different studies employ various questionnaires to assess SS, establishing a common or gold-standard set of symptoms for annotation remains challenging. In this sense, we refer to a recent study that developed the Sundown Syndrome Questionnaire (SSQ) to operationally measure SS based on previous literature and clinical experience [21]. The selected symptoms and the detailed explanations provided as annotation instructions can also be found in Table 1.

[1] https://naccdata.org.

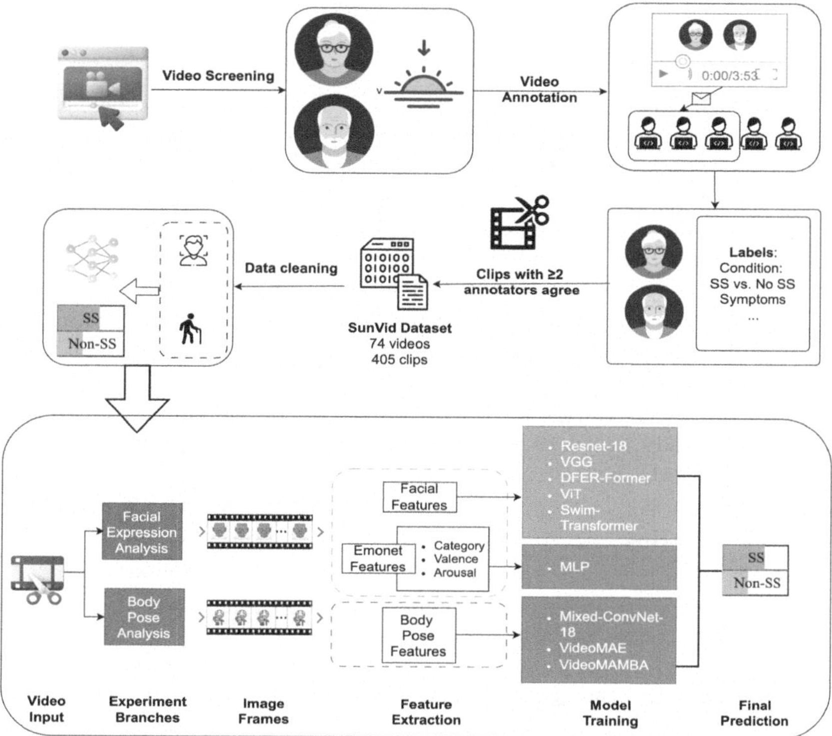

Fig. 1. Pipeline for SunVid Database Construction and Model Validation. Note: Icons used in the figure is created by Freepik - Flaticon (https://www.flaticon.com)

To isolate meaningful instances within complex video data, each event was segmented based on observable changes or transitions. The start time was defined as the first frame where a symptom, behavior, or interaction became noticeable, while the end time marked the frame where the event concluded or transitioned into a different state. Examples included changes in posture (e.g., sitting to standing), mood (e.g., neutral to aggressive), or interaction context (e.g., patient alone to patient with caregiver). Irrelevant content, such as caregiver-only segments or static video clips lacking meaningful context, was excluded from the segmentation process.

To determine the final dataset segments and their corresponding labels, we retained only those segments that were annotated by at least two annotators and whose annotations reached agreement. The start and end times of each segment were defined as the overlapping time range shared by the selected annotations.

Since our labeling system includes both single-choice and multiple-choice items, we established the following consensus rules:

Table 1. Annotation Categories and Descriptions

Category	Description
1-Conditions	The patient's state is categorized as either **1A Normal**, where no observable SS symptoms are present, or **1B Abnormal**, where unusual or concerning SS behaviors are displayed
2-States	Affective states fall into three categories: **2A Positive**, involving behaviors such as smiling or engaging socially; **2B Negative**, characterized by signs of distress or other negative moods; and **2C Neutral**, where no significant emotional changes are observed
3-Persons	This category identifies whether the video features only the patient (**3A Only patient**) or includes interactions with caregivers (**3B Patient and caregivers**)
4-Cues	Observable indicators fall into four categories: **4A Facial expressions**, **4B Body gestures**, **4C Audio** signals, and textual elements within the video (**4D Text**)
5-Position	The patient's posture is classified as **5A Lying down**, **5B Standing**, or **5C Sitting**
6-Symptoms	Specific symptoms are identified and categorized as follows: – **6A Disorientation**: Difficulty identifying time, place, or familiar people, often appearing confused or lost – **6B Anxiety:** Excessive nervousness or tension, potentially accompanied by physical signs such as shaking or rapid speech – **6C Depression:** Persistent sadness or lack of interest in previously enjoyable activities, often observed through slow speech or a slumped posture – **6D Wandering:** Aimless walking or pacing, typically without a clear purpose, sometimes reflecting restlessness or heightened energy – **6E Aggression:** Hostile or violent behavior, including physical resistance, shouting, or threatening gestures – **6F Emotional lability:** Rapid, exaggerated mood changes, such as sudden laughter or crying without clear reason – **6G Hoarding/Obsessive behavior:** Excessive collection or retention of items with little value, or refusal to discard unnecessary objects – **6H Delusion:** False beliefs, such as paranoia about theft or harm, often accompanied by suspicion or fear – **6I Hallucination:** Perception of non-existent stimuli, such as seeing or hearing things that are not present – **6J Other:** Any symptoms not covered above, including unique physical actions, expressions, or vocalizations relevant to SS – **6K No symptoms:** No symptoms are observed
7-Time	The temporal context is categorized as **7A Daytime**, **7B Sundowning time**, **7C Midnight**, or **7D Unknown**

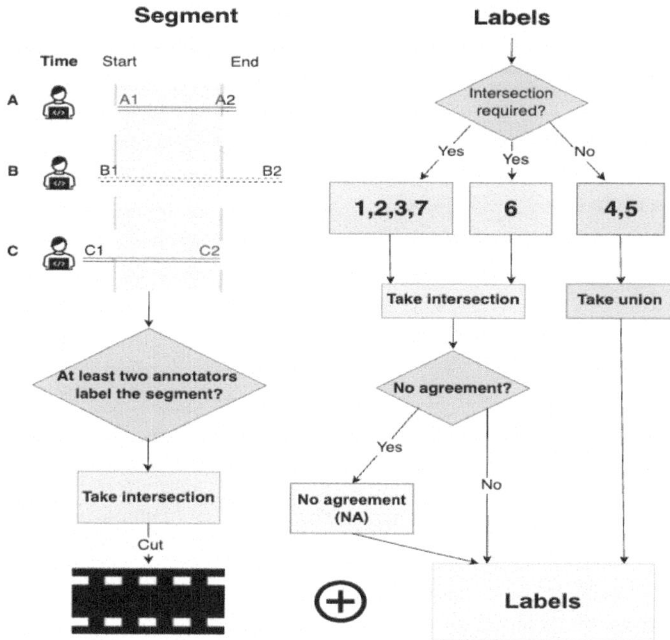

Fig. 2. Segmentation and Label Merging Process. Note: Categories are labeled as follows: 1-Conditions, 2-States, 3-Persons, 4-Cues, 5-Position, 6-Symptoms, and 7-Time.

- **Single-choice categories** (1–Conditions, 2–States, 3–Persons, and 7–Time): Labels were retained only if selected by at least two annotators to ensure accuracy.
- **Multiple-choice categories** (6–Symptoms): Labels appearing in at least two annotations were kept, while unique selections were excluded.
- **Union-based categories** (4–Cues and 5–Position): As these serve as additional contextual information for future modality selection, all labels noted by annotators were retained.

If no consensus was reached, the segment was marked as "No agreement (NA)". Finally, the consensus segments and labels were compiled, and the videos were segmented accordingly, resulting in 405 video clips in our final dataset. The segmentation and label assignment process is visualized in Fig. 2.

2.3 Database Statistics

Among the annotated events, 284 cases are labeled as "abnormal", indicating that individuals exhibit certain symptoms, which might be associated with SS. In contrast, 91 cases are labeled as "normal", suggesting no observable symptoms during these events. The remaining 30 cases show no agreement among annotators and are therefore excluded from further analysis.

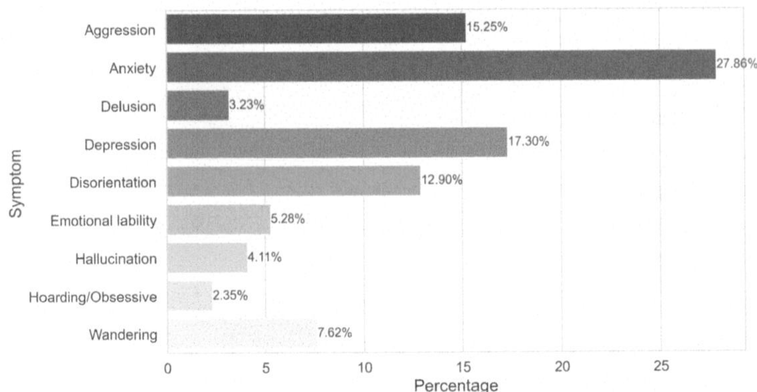

Fig. 3. Symptom Distribution in Sundown Syndrome Cases

For the cases where SS is recognized, the majority (195 cases) exhibit negative affective states, indicating a predominant presence of distress-related emotions. Fifty-two cases are classified as neutral, and 11 cases display a positive mood, representing a relatively small proportion of the dataset. This distribution suggests that emotional changes can serve as an important indicator for detecting SS, highlighting the potential role of affective state monitoring in SS evaluation and intervention.

Since time is a significant indicator and contributing factor to SS, we analyze the temporal distribution of cases where SS is recognized. Based on our annotations, 129 cases occur during daytime, 66 during sundowning time, 28 at midnight, 23 are classified as unknown, and 38 cases fall under non-agreement. Although some videos are recorded for scientific demonstrations, and the indoor setting of most cases may make precise time determination challenging, these findings align with previous studies, indicating that SS can manifest across a broad timeframe, from afternoon to midnight.

In cases where SS is identified, anxiety is the most frequently observed symptom, followed by depression and aggression. Many cases also exhibit emotional lability, highlighting the strong link between SS and emotional disturbances. Beyond these, disorientation and wandering are common, reflecting cognitive and spatial confusion. Additionally, less frequent but notable symptoms include hoarding/obsessive behaviors, delusions, and hallucinations, which may require individualized assessment and tailored interventions. See Fig. 3 for a detailed distribution of symptoms.

3 Methods for Sundown Syndrome Identification

Dataset statistics indicate that emotional cues and movement patterns play a key role in determining SS. We therefore conduct separate experiments on facial and body pose features. Although a combined analysis would offer a more comprehensive view, low video resolution limits its feasibility. In most cases, the face and body cannot be reliably detected at the same time, and facial details are often unclear, affecting feature quality and performance. Given these constraints, we report only separate results for face and body analyses. In addition to providing readers with more information, we also extract emotion-related representations using Emonet [26] for facial features and conduct further analyses based on these representations. The following sections provide a detailed examination of all these aspects.

3.1 Facial Expression Analysis

Data Preprocessing. For facial data preprocessing, we first split each video clip into frames, detect facial landmarks [32], and crop the face regions from the images. However, some events inevitably include both caregivers and patients, or the patient's face is not visible. To address this, we conduct an additional manual cleaning step, removing caregivers and incorrectly detected faces to ensure that only relevant samples are retained. Furthermore, we exclude some segments due to excessive facial blurriness or cases where SS detection relies primarily on non-visual cues (e.g., dialogue).

Following this process, 382 video segments (88 normal/non-SS, 267 abnormal/with SS) were selected as a baseline subset for experiments focusing on facial expression recognition. These segments ensure consistency and data quality for benchmarking. The remaining segments, though excluded from the current analysis, remain available for future studies targeting other aspects of SS. To ensure uniform input, we randomly sample 16 frames per clip, resize each frame to 48 × 48 pixels, normalize and convert them into tensors before further processing.

Experiment 1: Video-Based Baseline for SS Classification. For facial feature-based SS classification, we evaluate five benchmark models: ResNet-18 [9], VGG [25], DFER-Former [33], ViT [6], and Swin Transformer [18]. Among these models, we fine-tune Swin Transformer, VGG and DFER-Former using our data, with Swin Transformer and VGG pre-trained on ImageNet-1K [5] and DFER-Former pre-trained on DFEW [11], respectively. ViT and ResNet-18 are trained from scratch using our data. Each model is trained using a stochastic gradient descent (SGD) optimizer [22] with a constant learning rate of 0.005. We follow 5-fold cross-validation for all models. Specifically, in our case, the folds are separated based on videos, ensuring that all data from a single video is confined to one fold. During the process, one fold is used for testing while the other four are used for training, and this is repeated five times, with each fold serving as the test set once. This approach ensures that the model is evaluated on unseen videos, robustly measuring its generalization ability. The model's task is a binary classification problem (normal/non-SS vs. abnormal/with SS), so we use the cross-entropy loss function. All implementations are conducted using PyTorch [20] on an NVIDIA GeForce RTX 4090. As summarized in Table 2, Swin Transformer achieves the highest accuracy of 69.03%, and DFER-Former achieves the best macro-averaged F1-score (UF1) of 50.03%.

Table 2. Baseline Performance of SS Classification Using Facial Features

Method	Accuracy (%)	UF1 (%)
VGG [25]	62.08	47.25
DFER-Former [33]	63.26	**50.03**
ViT [6]	65.13	49.23
Swin Transformer [18]	**69.03**	46.09
ResNet-18 [9]	67.72	46.92

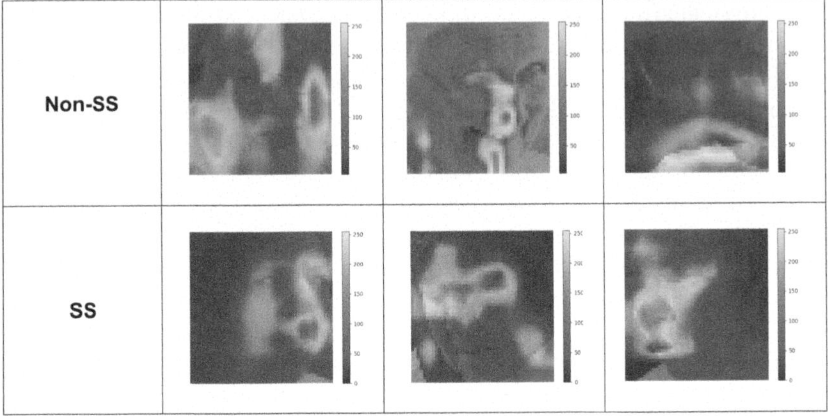

Fig. 4. Grad-CAM-based saliency maps for SS classification.

Additionally, we employ Grad-CAM [24] to visualize the saliency regions, identifying the most informative facial features contributing to SS classification. As shown in Fig. 4, the model's correct predictions of SS and Non-SS cases primarily rely on its ability to capture key emotional and facial movement characteristics. The heatmaps indicate that the model predominantly focuses on critical facial regions, such as the eyes, mouth, and forehead, which align with known behavioural manifestations of typical facial expressions. The variation in activation patterns across images suggests that different facial features contribute to the model's decision-making process. Higher-intensity activation areas (yellow to red regions) likely correspond to salient features, such as frowning, eye gaze shifts, or mouth movements, often associated with emotional expressions potentially linked to agitation, confusion, or emotional distress in SS patients. These results support the hypothesis that facial behaviours could serve as key indicators of SS and highlight their potential effectiveness in assisting early detection and monitoring of SS-related behavioural changes.

Experiment 2: SS Classification Using Emonet-Derived Emotion Features. Since emotional fluctuation is a key characteristic of SS, we employ

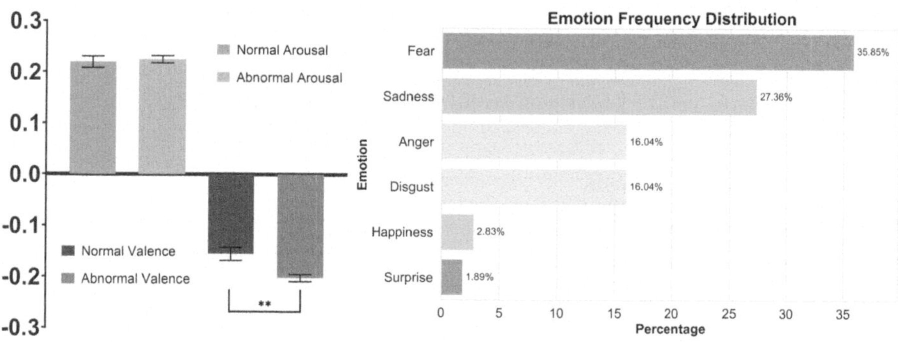

Fig. 5. Overview of emotion features predicted by Emonet. The left panel shows the distribution of valence and arousal values across all frames. The right panel summarizes the frequency of predicted facial expression categories, based on the dominant label in each video. (Error bars indicate SEM; ** represent p < 0.01.)

Emonet [26] to predict each video's facial expression category, valence, and arousal levels, two continuous dimensions commonly used to describe emotional experience. Specifically, valence reflects the positivity or negativity of the emotion (ranging from pleasant to unpleasant), while arousal measures the intensity or activation level of the emotional state (from calm to excited) [23]. Emonet is a pre-trained model that has been shown to effectively recognize emotions in naturalistic settings [4,15,30]. Based on its predictions, we compute the average valence and arousal values for each video and the category with the highest number of corresponding frames.

As shown in Fig. 5, independent samples t-tests were conducted to compare the affective states between video segments labeled with and without SS symptoms. Although there is no significant difference in arousal levels between individuals exhibiting SS symptoms (abnormal) and those without SS symptoms (normal) ($t(353) = 0.373, p = .710$), cases with SS symptoms tend to exhibit a more negative affective representation compared to the normal group. Specifically, valence scores are lower in the abnormal SS group ($M = -0.204, SD = 0.115$) than in the normal group($M = -0.154, SD = 0.123$), with a statistically significant difference($t(353) = -3.343, p = .001$). This suggests that while arousal levels remain comparable, SS cases are characterized by a more negative emotional state, reinforcing the association between SS and emotional disturbances. Regarding specific facial expressions, since our results are determined by a voting mechanism, neutral facial expressions constitute the majority (161 cases). However, among the remaining cases, negative emotions, particularly fear, are predominant and may be associated with symptoms such as anxiety and aggression (see Fig. 5).

To assess the predictive power of Emonet-extracted features in SS classification, we trained a multi-layer perceptron (MLP) model using these features as input. The model was evaluated using five-fold cross-validation, with accuracy and macro F1-score as performance metrics. The results show an average accuracy of 72.85% and an average macro F1-score of 46.35%, suggesting that the extracted emotion-related features could benefit SS classification.

3.2 Body Pose Analysis

Data Preprocessing. We perform a semi-automatic data cleaning for body pose analysis to obtain clean and trainable data. We first employ Mask R-CNN (ResNet-50 backbone) to segment individuals from the background and utilize OpenCV contours to remove black-background regions, ensuring the isolation of the subject in each frame. A subsequent manual cleaning step is performed to refine the dataset by removing caregivers, incorrectly detected objects, and cases where the camera angle is too close to the patients (e.g., only capturing the upper body without observable gestures). After this process, 109 video segments (91 normal/non-SS, 283 abnormal/with SS) remain for further analysis. For temporal consistency, 16 uniformly sampled frames are extracted from each clip, with each frame resized to 112 × 112 pixels before further processing.

Experiment: SS Classification Using Body Pose Features. For our experiments, we select Mixed-ConvNet-18 [28], VideoMAE [27], and VideoMAMBA [16], all pre-trained on body pose-related datasets (Kinetic400 [12]), as they may share similar features to our task. From there, we extract spatiotemporal features of the uniformly sampled frames to analyze the changes in body pose spatially and temporally. In experimental setting-wise, we use Adam optimizer with a learning rate of 0.001 and ExponentialLR decay with a gamma of 0.9. To standardize evaluation, we use the same subject-based 5-fold cross-validation protocol as in facial analysis because we deduce that body pose also contains its identity uniqueness, similar to facial identity.

Table 3. Baseline Performance of SS Classification With Body Pose Features

Method	Accuracy (%)	UF1 (%)
Mixed-ConvNet-18 [28]	60.51	**46.84**
VideoMAE [27]	**63.82**	43.49
VideoMAMBA [16]	61.68	45.01

Table 3 presents the baseline results for SS classification using body pose features. VideoMAE achieves the highest accuracy (63.82%), while Mixed-ConvNet-18 attains the highest UF1 score (46.84%). These results establish a baseline for SS classification using body pose features, serving as a reference for both comparisons with facial features and future studies. From the results, we observe that while transformer-based models such as VideoMAE and VideoMAMBA achieve higher accuracy, they may be more sensitive to variations in temporal continuity due to their reliance on token-based receptive fields for modeling long-range relationships. In contrast, Mixed-ConvNet-18 may benefit from its ability to capture local spatial relationships, leading to better performance in terms of F1 scores. Due to class imbalance, the macro-averaged F1-score is recommended as a more reliable metric for evaluating model performance. Future work may also investigate how different model architectures respond to temporal variations to optimize feature extraction in body pose analysis.

4 Conclusion and Future Directions

In this study, we introduce SunVid, a newly curated online video database designed to support future SS research using video-based sources. Despite challenges such as varying video quality and symptom imbalance, which are common in online-sourced data, our study demonstrates the feasibility of video-based SS detection. Building on this, SunVid serves as the first dataset to integrate online video data with automated analysis methods for SS research.

Our experimental results confirm that video-based SS detection, particularly using emotion-related features, can be effective, achieving reasonable accuracy even with limited data. Although some videos are demonstrations or educational materials rather than real patient recordings, they often depict prototypical SS-like symptoms. This makes them valuable for exploring behavioral modeling, which is especially important given the complexity and behavioral nature of SS. These findings provide initial proof of concept that automated video analysis is both feasible and promising for SS research. Future clinical studies may benefit from moving beyond manual symptom recording toward video-based monitoring with post-episode annotation and AI-driven analysis.

To further advance this approach, addressing practical challenges in real-world applications remains crucial. As our dataset was collected from existing online videos in naturalistic settings, standardizing camera conditions and video quality was not possible. While we validated both facial and body features, data limitations prevented the integration of these cues into a unified multimodal framework. Future work should consider incorporating additional modalities, such as audio, to better capture language-dependent symptoms like verbal agitation or delusions. Combining video with motion or biological sensors may also offer deeper insights.

Ultimately, we hope that SunVid serves as a foundation for future research, fostering the development of AI-driven tools to enhance the early detection, monitoring, and management of SS.

Acknowledgements. This work was supported by the Finnish-American Research Innovation Accelerator (FARIA) project (GP FARIA CC4 Menot), the Research Council of Finland (former Academy of Finland) Academy Professor project EmotionAI (No. 336116, 345122, 359854), the University of Oulu and the Research Council of Finland Profi 7 (No. 352788), the EU HORIZON-MSCA-SE-2022 project ACMod (No. 101130271), and Infotech Oulu. The work of Dr. Qianru Xu was supported by Finnish

Cultural Foundation for Säätiöiden post doc-pooli Grant (No. 00240135). The authors also acknowledge CSC–IT Center for Science, Finland, for computational resources. We also thank Xingxun Jiang, Yujie Liu, and Lili Liu for their valuable efforts in assisting with the annotation work during the development of our database.

Disclosure of Interests. The authors have no competing interests to declare that are relevant to the content of this article.

References

1. Boronat, A.C., Ferreira-Maia, A.P., Wang, Y.P.: Sundown syndrome in older persons: a scoping review. J. Am. Med. Dir. Assoc. **20**(6), 664–671 (2019)
2. Cameron, D.E.: Studies in senile nocturnal delirium. Psychiatric Q. (1941)
3. Cipriani, G., Lucetti, C., Carlesi, C., Danti, S., Nuti, A.: Sundown syndrome and dementia. Eur. Geriatric Med. **6**(4), 375–380 (2015)
4. Daněček, R., Black, M.J., Bolkart, T.: EMOCA: emotion driven monocular face capture and animation. In: Proceedings of the IEEE/CVF Conference on Computer Vision and Pattern Recognition, pp. 20311–20322 (2022)
5. Deng, J., Dong, W., Socher, R., Li, L.J., Li, K., Fei-Fei, L.: ImageNet: a large-scale hierarchical image database. In: 2009 IEEE Conference on Computer Vision and Pattern Recognition, pp. 248–255. IEEE (2009)
6. Dosovitskiy, A.: An image is worth 16×16 words: transformers for image recognition at scale. arXiv preprint arXiv:2010.11929 (2020)
7. Evans, L.K.: Sundown syndrome in institutionalized elderly. J. Am. Geriatr. Soc. **35**(2), 101–108 (1987)
8. Ghali, L.M., Hopkins, R.W., Rindlisbacher, P.: Temporal shifts in peak daily activity in Alzheimer's disease. Int. J. Geriatr. Psychiatry **10**(6), 517–521 (1995)
9. He, K., Zhang, X., Ren, S., Sun, J.: Deep residual learning for image recognition. In: Proceedings of the IEEE Conference on Computer Vision and Pattern Recognition, pp. 770–778 (2016)
10. He, L., et al.: Deep learning for depression recognition with audiovisual cues: a review. Inf. Fusion **80**, 56–86 (2022)
11. Jiang, X., et al.: DFEW: a large-scale database for recognizing dynamic facial expressions in the wild. In: Proceedings of the 28th ACM International Conference on Multimedia, pp. 2881–2889 (2020)
12. Kay, W., et al.: The kinetics human action video dataset. arXiv preprint arXiv:1705.06950 (2017)
13. Khachiyants, N., Trinkle, D., Son, S.J., Kim, K.Y.: Sundown syndrome in persons with dementia: an update. Psychiatry Investig. **8**(4), 275 (2011)
14. Khan, S.S., Ye, B., Taati, B., Mihailidis, A.: Detecting agitation and aggression in people with dementia using sensors-a systematic review. Alzheimer's Dementia **14**(6), 824–832 (2018)
15. Le, N., Nguyen, K., Tran, Q., Tjiputra, E., Le, B., Nguyen, A.: Uncertainty-aware label distribution learning for facial expression recognition. In: Proceedings of the IEEE/CVF Winter Conference on Applications of Computer Vision, pp. 6088–6097 (2023)
16. Li, K., Li, X., Wang, Y., He, Y., Wang, Y., Wang, L., Qiao, Y.: VideoMamba: state space model for efficient video understanding. In: European Conference on Computer Vision, pp. 237–255. Springer (2025)

17. Little, J.T., Satlin, A., Sunderland, T., Volicer, L.: Sundown syndrome in severely demented patients with probable Alzheimer's disease. J. Geriatr. Psychiatry Neurol. **8**(2), 103–106 (1995)
18. Liu, Z., et al.: Swin transformer: hierarchical vision transformer using shifted windows. In: Proceedings of the IEEE/CVF International Conference on Computer Vision, pp. 10012–10022 (2021)
19. Madden, K.M., Feldman, B.: Weekly, seasonal, and geographic patterns in health contemplations about sundown syndrome: an ecological correlational study. JMIR Aging **2**(1), e13302 (2019)
20. Paszke, A., et al.: PyTorch: an imperative style, high-performance deep learning library. In: Advances in Neural Information Processing Systems, vol. 32 (2019)
21. Pyun, J.M., Kang, M.J., Yun, Y., Park, Y.H., Kim, S.: APOE ε_4 and REM sleep behavior disorder as risk factors for sundown syndrome in Alzheimer's disease. J. Alzheimers Dis. **69**(2), 521–528 (2019)
22. Ruder, S.: An overview of gradient descent optimization algorithms. arXiv preprint arXiv:1609.04747 (2016)
23. Russell, J.A.: A circumplex model of affect. J. Pers. Soc. Psychol. **39**(6), 1161 (1980)
24. Selvaraju, R.R., Cogswell, M., Das, A., Vedantam, R., Parikh, D., Batra, D.: Grad-CAM: visual explanations from deep networks via gradient-based localization. In: Proceedings of the IEEE International Conference on Computer Vision, pp. 618–626 (2017)
25. Simonyan, K., Zisserman, A.: Very deep convolutional networks for large-scale image recognition. arXiv preprint arXiv:1409.1556 (2014)
26. Toisoul, A., Kossaifi, J., Bulat, A., Tzimiropoulos, G., Pantic, M.: Estimation of continuous valence and arousal levels from faces in naturalistic conditions. Nat. Mach. Intell. **3**(1), 42–50 (2021)
27. Tong, Z., Song, Y., Wang, J., Wang, L.: VideoMAE: masked autoencoders are data-efficient learners for self-supervised video pre-training. In: Advances in Neural Information Processing Systems, vol. 35, pp. 10078–10093 (2022)
28. Tran, D., Wang, H., Torresani, L., Ray, J., LeCun, Y., Paluri, M.: A closer look at spatiotemporal convolutions for action recognition. In: Proceedings of the IEEE Conference on Computer Vision and Pattern Recognition, pp. 6450–6459 (2018)
29. Trumpf, R., Haussermann, P., Zijlstra, W., Fleiner, T.: Circadian aspects of mobility-related behavior in patients with dementia: an exploratory analysis in acute geriatric psychiatry. Int. J. Geriatr. Psychiatry **38**(6), e5957 (2023)
30. Xu, J., Li, Y., Yang, G., He, L., Luo, K.: Multiscale facial expression recognition based on dynamic global and static local attention. IEEE Trans. Affect. Comput. (2024)
31. Xu, Q., Lin, F.V., Liu, Y., Zhao, G.: Bridging gaps in sundown syndrome research: a roadmap for future multimodal approaches. Archives of Clinical Neuropsychology (2025). In press
32. Yang, J., Bulat, A., Tzimiropoulos, G.: Fan-face: a simple orthogonal improvement to deep face recognition. In: Proceedings of the AAAI Conference on Artificial Intelligence, vol. 34, pp. 12621–12628 (2020)
33. Zhao, Z., Liu, Q.: Former-DFER: dynamic facial expression recognition transformer. In: Proceedings of the 29th ACM International Conference on Multimedia, pp. 1553–1561 (2021)

Comparative Analysis of rPPG and Motion-Based Approaches for Heart and Respiration Rate Estimation from Videos

Nhi Nguyen(✉), Constantino Álvarez Casado, Le Nguyen, Manuel Lage Cañellas, and Miguel Bordallo López

Center for Machine Vision and Signal Analysis, University of Oulu, Oulu, Finland
{thi.tn.nguyen,constantino.alvarezcasado,le.nguyen, manuel.lage,miguel.bordallo}@oulu.fi

Abstract. Reliable estimation of physiological indicators such as heart rate (HR) and respiratory rate (RR) from camera-based systems is important for enabling non-invasive contactless monitoring in healthcare. This work evaluates recent remote photoplethysmography (rPPG) and motion-based methods for physiological signal measurement using the OMuSense-23 database. The dataset comprises recordings from 50 participants and includes synchronized RGB-D video and ground truth photoplethysmography (GT PPG) and respiratory signals collected with medical-grade sensors. The analysis assesses traditional signal processing and deep learning-based methods in controlled breathing patterns (normal respiration, guided respiration, reading, and apnea). HR estimation is evaluated through multiple rPPG methods, while RR estimation includes two approaches: the first based on body motion tracking and the second based on derived rPPG signals. The study highlights the challenges encountered by signal processing methods under certain breathing patterns and identifies factors that affect deep learning performance, such as motion artifacts and respiratory-induced variations. These findings provide practical insights for improving non-contact physiological monitoring and enhancing real-world applications in remote healthcare.

Keywords: Non-contact physiological measurement · rPPG · Respiratory rate estimation · Telemedicine · Biosignals · Vital signs · Contactless

1 Introduction

Remote photoplethysmography (rPPG) enables HR estimation by analyzing subtle skin color variations or imperceptible motion from facial videos [4,25]. Advances in computer vision and deep learning have improved rPPG-based HR estimation, leading to extensive research and benchmarking efforts. On the other hand, non-contact vision-based RR estimation has received considerably less attention, despite its importance in monitoring respiratory health, stress levels,

J. Petersen and V. A. Dahl (Eds.): SCIA 2025, LNCS 15726, pp. 32–46, 2025.
https://doi.org/10.1007/978-3-031-95918-9_3

and sleep quality. Unlike HR, which can be inferred from color fluctuations in the skin, RR estimation involves detecting periodic chest or facial movements, which are influenced by posture, clothing, and voluntary control over breathing patterns. A systematic evaluation of existing methods is needed to determine their effectiveness under different conditions.

This study presents a comprehensive evaluation of rPPG and motion-based methods for HR and RR estimation using the OMuSense-23 database, which includes 50 participants with synchronized medical-grade PPG and respiratory signals. The dataset captures multiple postures (sitting, standing, lying down) and breathing patterns (normal breathing, guided breathing, reading, and apnea). We assess several recent rPPG algorithms for HR estimation and different RR estimation approaches. The results highlight the strengths and limitations of each method, explaining why certain signal processing techniques struggle under specific conditions and where deep learning models may improve performance or face challenges. The main contributions of this work are:

- A standardized evaluation of vision-based HR and RR estimation using a controlled dataset with synchronized GT signals.
- A comparative analysis of rPPG-based and motion-based methods for RR estimation, addressing an area with limited prior research.
- An investigation into the impact of different breathing patterns on HR and RR estimation accuracy.
- Insights into the limitations of traditional signal processing methods and the potential of deep learning models for non-contact physiological measurement.

By systematically assessing these approaches, this work provides a foundation for improving non-contact vital sign monitoring and advancing real-world applications in telemedicine, clinical settings, and continuous health monitoring.

2 Related Work

rPPG is a technique for estimating physiological signals using video-based analysis of skin color variations. It detects changes in light absorption due to blood volume pulsations and extracts these signals from facial regions. Unlike contact-based PPG, rPPG does not require direct skin contact, which makes it relevant for applications in telemedicine and remote health monitoring [4,13].

2.1 rPPG Methods for HR Estimation

Early rPPG methods extracted blood volume pulse (BVP) signals from facial videos using signal processing techniques. The GREEN method showed that the green channel carries the strongest pulsatile information due to hemoglobin absorption [32]. CHROM reduced specular reflections by leveraging chrominance ratios in the RGB space [12]. OMIT applied orthogonal matrix decomposition to separate uncorrelated RGB components, reducing motion artifacts [4]. POS projected the RGB color space onto a plane orthogonal to the skin tones, enhancing the extraction of pulsatile signals [33]. LGI mapped signals into an invariant feature space, improving robustness to motion and illumination changes [27]. Despite these improvements, these methods remained sensitive to variations in skin tone, ambient lighting, and occlusions, leading to signal distortions [4,18,20,25]. To address these limitations, deep learning methods have been introduced. HR-CNN used a two-stage pipeline in which a convolutional neural network optimized the signal-to-noise ratio before estimating HR [28]. DeepPhys used attention mechanisms to focus on relevant facial regions, reducing motion artifacts [7]. RhythmNet applied spatial-temporal mappings to track HR in dynamic conditions [26]. AutoHR leveraged neural architecture search to optimize deep learning pipelines for rPPG extraction [34]. PhysFormer introduced a transformer-based model to capture temporal dependencies in skin color variations, improving performance over traditional convolutional methods [35]. Contrast-Phys [30] enabled unsupervised training by leveraging spatio-temporal contrast, improving noise robustness. These deep learning approaches improved accuracy but required large annotated datasets and lacked interpretability [8,36].

2.2 rPPG Methods for RR Estimation

Respiration modulates the blood volume pulse (BVP) signal due to cyclic changes in venous return and intrathoracic pressure, allowing RR estimation from rPPG signals. This is typically achieved by filtering the BVP signal within the respiratory frequency range, applying signal decomposition techniques, or using learning-based methods [6,29]. Chen et al. [6] proposed a signal processing framework incorporating motion compensation, temporal filtering, and signal pruning to enhance respiration-related oscillations in rPPG signals, improving robustness over motion-based respiration estimation. Ghodratigohar et al. [23] applied Independent Component Analysis (ICA) and Complete Ensemble Empirical Mode Decomposition with Adaptive Noise (CEEMDAN) to isolate respiration-induced variations from noise, achieving better performance in non-stationary conditions. Machine learning techniques have further improved estimation accuracy. Vatanparvar et al. [17] developed a learning-based approach for artifact removal, validated on a diverse dataset, showing improved RR estimation. Du et al. [14] introduced a weakly supervised learning method that pre-estimates rPPG signals using chrominance-based techniques modulated by breathing waves, enhancing generalization across datasets. Multimodal strategies have also been explored. Gwak et al. [22] integrated rPPG with facial motion analysis, demonstrating higher accuracy compared to rPPG alone. Alnaggar et al. [24] applied Eulerian Video Magnification (EVM) to amplify respiration-induced fluctuations, improving signal visibility. Luguern et al. [9] evaluated contact-based PPG algorithms on rPPG signals, showing their applicability to non-contact respiration monitoring. A hospital-based study by Allado et al. [1] validated rPPG-based RR estimation in 963 patients, confirming strong agreement with clinical measurements. While vision-based methods have improved, challenges such as motion artifacts, skin tone variations, and the lower signal-to-noise ratio of respiration-induced fluctuations compared to cardiac signals remain.

2.3 Motion-Based Methods for RR Estimation via Cameras

Unlike rPPG-based methods that infer respiration-induced fluctuations from BVP signals, motion-based approaches via cameras estimate RR by tracking physical movements of the head, face, or chest [2,10,11,21]. These methods rely on detecting periodic motion patterns associated with breathing, offering an alternative when rPPG signals are weak or unreliable due to low skin reflectance, poor lighting, or significant motion artifacts. Early approaches used optical flow analysis to track thoracoabdominal motion [2] and structured light projection to measure size variations in a projected dot on the chest and abdomen [21]. Makkapati et al. [21] applied ICA and motion filtering in the Clifford Fourier Transform domain to estimate neonatal RR from chest and abdomen movements. Schrumpf et al. [10] demonstrated that weak head movements, induced by respiration, can be used for estimation when direct chest motion detection is not feasible. More recent studies have improved robustness through motion artifact removal and adaptive feature selection. Gwak et al. [11] introduced a method that adaptively selects between head and chest movements to minimize artifacts, achieving mean absolute errors of 1.95 breaths per minute (BrPM) using head motion and 1.28 BrPM using chest motion. Cheng et al. [15] proposed a two-level fusion framework combining pixel intensity variations with movement-based signals, reducing errors across different motion conditions. Depth cameras have also been explored for respiration tracking, with Kempfle et al. [16] showing that accuracy decreases beyond four meters and improves when monitoring a larger chest region. Multimodal approaches further enhance accuracy by combining motion-based respiration tracking with rPPG. Gwak et al. [22] demonstrated that the combination of facial motion and rPPG signals improves the estimation performance, achieving an error of 1.33 BrPM. Despite these advances, motion artifacts, variations in camera distance, and environmental lighting conditions remain challenges, requiring further research to enhance real-world applicability.

3 Evaluation Methodology

The evaluation framework follows a structured pipeline for HR and RR estimation from video signals (Fig. 1). It begins with an input dataset containing GT physiological signals and synchronized video frames. GT signals are used to compute reference HR and RR values for performance evaluation. Facial landmarks and pose keypoints are extracted from video frames using Mediapipe landmark detection. From these, rPPG signals are extracted using both signal processing and deep learning methods. In addition, respiratory signals are extracted using vision-based motion analysis methods by measuring distance variations between nose-shoulder and nose-lip keypoints or by filtering to isolate respiration-induced components in rPPG signals. The extracted physiological signals are then evaluated against GT measurements to quantify the accuracy of the estimate.

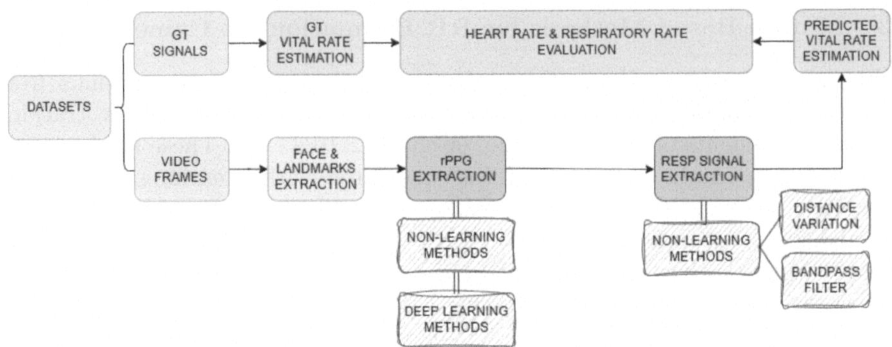

Fig. 1. Evaluation pipeline for HR and RR estimation.

3.1 Datasets

The experiments are conducted on an extended version of the OMuSense-23 dataset [5], which incorporates synchronized depth and RGB video recordings in addition to the original CSV-based physiological and metadata annotations. Although the original data set included the Texas Instruments IWR1443 FMCW mmWave radar signals, this study focuses exclusively on the newly added video recordings and their corresponding GT physiological signals for evaluating rPPG and RR estimation methods.

The dataset includes RGB and depth video recordings captured with an Intel RealSense D435 RGB-D camera at 1280 $\times$ 720 resolution and 30 FPS. Depth is obtained through active stereo vision, where an infrared projector emits a structured pattern detected by two IR sensors to compute the disparity. Depth values are encoded as 8-bit pixel intensities using the JET colormap. RGB and depth sequences are stored as MP4 files, with timestamps in CSV format for precise synchronization with GT physiological data. GT signals include PPG and respiratory waveforms recorded using the CONTEC CMS50E pulse oximeter (60 Hz) and Venier Go Direct® Respiration Belt (20 Hz). The dataset consists of 50 participants recorded in three postures: standing, sitting, and lying down, as shown in Fig. 2. Each session follows a structured protocol [5] with nine phases: normal breathing (NB), reading (R), guided breathing (GB), holding breath (HB), four transition gaps (GAP 1–4), and an apnea release (AR) phase. Figure 3 illustrates an example of aligned GT PPG, respiratory signals, frame number, and phase transitions.

3.2 Methodology for Physiological Signals Extraction

This study extracts physiological signals by analyzing facial landmarks and upper-body motion from RGB-D video data. Facial landmarks and body keypoints are detected in RGB images using a modern deep-learning model, enabling ROI selection. The extracted features include facial landmarks for rPPG-based HR estimation and body keypoints (nose, lip, and shoulders) for motion-based

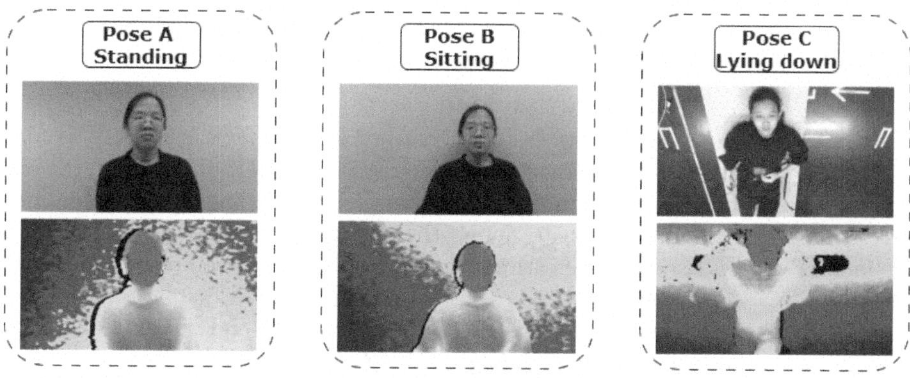

Fig. 2. Examples of three poses captured by an RGB-D camera.

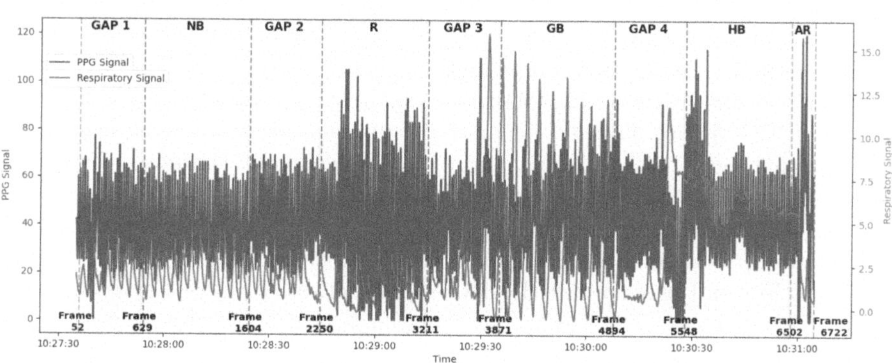

Fig. 3. Visualization of aligned GT PPG, respiratory signals, and phase transitions.

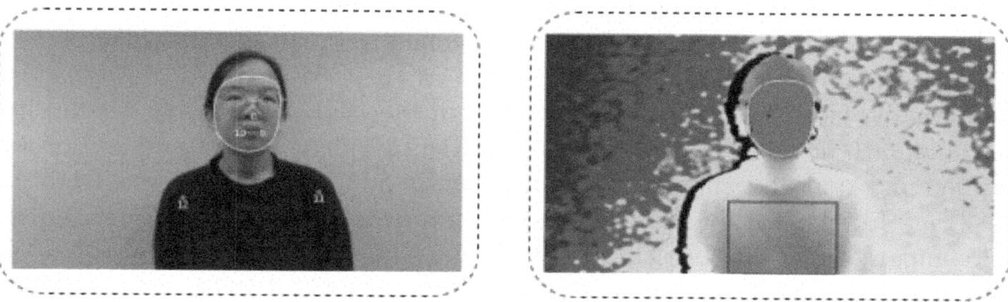

Fig. 4. Face ROI selection and pose keypoints for rPPG and motion-based respiration (left), and face ROI selection with a fixed chest bounding box for depth-based respiration tracking (right).

RR estimation. Additionally, depth data is processed by selecting a fixed bounding box in the chest area to analyze respiratory-induced displacement. Figure 4 illustrates the extracted features, displaying Mediapipe keypoints overlaid on the RGB image and depth image, along with the selected ROI for chest motion analysis in the depth image.

RGB Videos: HR and RR are estimated from RGB videos using rPPG and motion-based methods. To extract these signals, facial and body landmarks are detected using the Google MediaPipe framework [3], allowing facial and upper body geometrical segmentation to define the regions of interest (left of Fig. 4). The facial region is used for the estimation of HR and RR based on rPPG, while upper body motion is analyzed to track respiratory-induced displacement.

HR is derived from rPPG signals extracted from the facial region using non-learning methods, such as POS [33], CHROM [12], and OMIT [4], as well as deep learning approaches based on ContrastPhys in unsupervised, semi-supervised, and supervised configurations [30,31]. The extracted BVP signal is bandpass filtered between 0.75 and 4 Hz (45–240 BPM) to isolate cardiac oscillations, and the HR is estimated from the dominant frequency component.

RR is estimated using two complementary methods. The first applies a band-pass filter (0.1–0.5 Hz, 6–30 BrPM) to extract respiration-induced modulations from the rPPG signal. The second analyzes upper body motion by computing Euclidean distances between pose landmarks. Specifically, the nose-to-shoulder distance is measured between the nose (landmark 0) and the shoulder midpoint (landmarks 11 and 12), while the nose-to-lip distance is derived from the mid-point of the lip corners (landmarks 9 and 10). Both methods track breathing-related motion and are evaluated on the OMuSense-23 dataset across various breathing conditions.

Depth Videos: The depth stream captures respiratory motion by analyzing a manually selected chest region (fixed bounding box), to ensure consistent positioning across subjects given the fixed recording setup [16]. The RGB values of this region are averaged per frame, producing a scalar signal, which is read from the publicly available depth CSV file in the OMuSense-23 dataset [19]. Additionally, facial landmarks are detected in the RGB video using MediaPipe face landmark detector and mapped to the depth video, as both modalities are rectified and spatially aligned. The mean depth values of both regions (right of Fig. 4) are filtered within the RR bandpass range to extract respiratory signals and compared with GT breathing waveforms. The same values are also filtered in the HR bandpass range and evaluated against the GT PPG heart waveform.

3.3 Protocol and Metrics

Due to privacy restrictions, 11 users did not consent to share their facial data, so they were assigned to testing, and the remaining 39 to training. The training process for one model took approximately 8 h on a PC equipped with an Nvidia GeForce RTX 4070 GPU, a 10-core CPU, and 32 GB of RAM. For training, two 10-second face ROI clips (72×72) were sampled from different videos. The model runs for 40 epochs with a learning rate of 1×10^{-7}, using model parameters

$K = 4$ and $S = 2$. The semi-supervised ContrastPhys model is trained using 50% labeled and 50% unlabeled data. After extracting signals using various methods, the extracted and GT signals are segmented with a 10-second sliding window (1-second step, 9-second overlap). Each segment is then bandpass filtered within the appropriate range for the task, using Welch's method with a Hamming window to estimate the vital rate. An example of the signal is shown in Fig. 5.

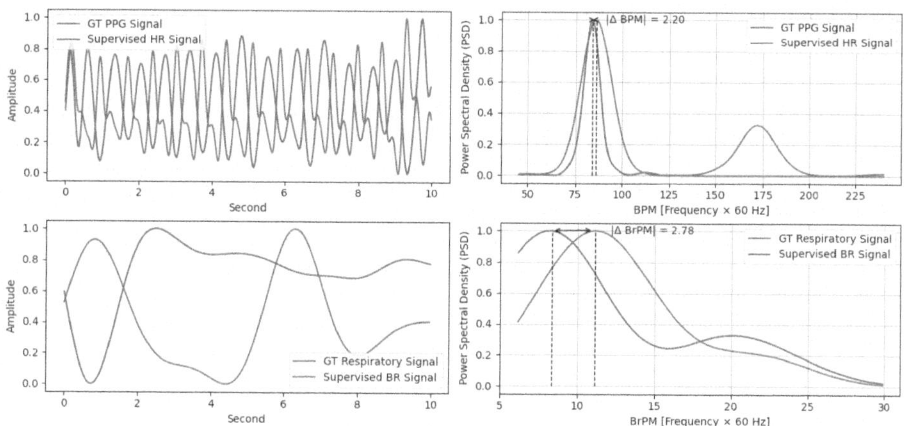

Fig. 5. ContrastPhys supervised signal vs GT ppg, respiratory signal.

Afterward, the vital rate from each rPPG window is compared to the vital rate from reference signals recorded simultaneously. This is done using the mean absolute error (MAE) and the Pearson correlation coefficient (PCC). MAE measures the average absolute deviation between predicted and GT vital rate values while PCC assesses the linear correlation between them:

$$\text{PCC} = \frac{\sum_i (\hat{y}_i - \hat{\mu})(y_i - \mu)}{\sqrt{\sum_i (\hat{y}_i - \hat{\mu})^2}\sqrt{\sum_i (y_i - \mu)^2}} \tag{1}$$

where y_i represents the GT HR values, $\hat{y}_i$ denotes the predicted HR values, and μ and $\hat{\mu}$ are their respective mean values.

4 Experimental Analysis and Results

This section describes our experiments with learning-based models and non-learning methods. We assess the performance of motion tracking and rPPG techniques to capture respiratory movements in RGB and depth videos. Furthermore, we investigate the impact of motion artifacts on HR and RR estimation.

4.1 RGB Video Analysis

The results in Table 1 indicate that deep learning-based rPPG models achieve lower errors and higher correlation than non-learning-based methods. Among them, the fully supervised ContrastPhys demonstrates the best performance, followed by the semi-supervised and unsupervised configurations. The relatively small performance gap between the fully supervised and semi-supervised variants suggests that semi-supervised learning is a viable alternative when labeled data is limited. The unsupervised model also maintains competitive accuracy, highlighting the potential of contrastive learning to extract physiological signals without explicit supervision.

Task-specific trends reveal that NB, GB, and AR produce the most accurate HR estimates across all methods. These breathing activities involve relatively stable facial appearance and controlled respiration, leading to a stronger and less noisy pulsatile signal. In contrast, the R task appears to be the most challenging, followed by the gap and HB phases. The reading activity involves more facial movements (e.g., talking and reading aloud) that could introduce motion artifacts and affect the rPPG signal quality. The gap phases, which are transitions between different breathing patterns, might introduce instability in physiological signals. Similarly, the HB phase may reduce rPPG accuracy due to changes in blood perfusion and skin color, affecting the consistency of the extracted pulse signal. Among the non-learning methods, POS achieves the best results, outperforming CHROM and OMIT. This suggests that the projection-

Table 1. HR errors of rPPG methods for each phase

rPPG Methods	Dataset									
	Full Test		NB Test		R Test		GB Test		HB Test	
	MAE	PCC	MAE	PCC	MAE	PCC	MAE	PCC	MAE	PCC
CHROM	15.89	0.20	9.31	0.37	18.76	0.13	10.67	0.19	14.56	0.20
OMIT	13.43	0.24	8.61	0.37	16.71	0.14	9.32	0.28	12.82	0.26
POS	11.88	0.31	7.43	0.46	14.36	0.21	8.10	0.32	10.71	0.35
Unsupervised	9.04	0.44	3.03	0.91	13.47	0.10	4.85	0.72	10.20	0.44
Semi-supervised	7.87	0.54	3.61	0.86	11.80	0.28	4.05	0.80	8.66	0.62
Fully-supervised	7.14	0.61	3.06	0.91	10.19	0.38	3.97	0.83	8.50	0.60
	GAP 1 Test		GAP 2 Test		GAP 3 Test		GAP 4 Test		AR Test	
	MAE	PCC	MAE	PCC	MAE	PCC	MAE	PCC	MAE	PCC
CHROM	23.15	0.03	17.52	0.18	14.15	0.33	14.73	0.22	9.76	0.46
OMIT	16.36	0.10	13.46	0.30	14.59	0.32	14.99	0.25	7.04	0.67
POS	15.73	0.16	13.18	0.23	13.43	0.31	16.01	0.21	7.17	0.71
Unsupervised	8.80	0.36	10.15	0.29	13.36	0.13	12.61	0.28	8.01	0.53
Semi-supervised	7.31	0.52	9.79	0.34	8.68	0.56	8.91	0.56	7.07	0.67
Fully-supervised	6.34	0.64	8.84	0.43	8.04	0.61	8.07	0.64	6.33	0.74

based approach of POS is more resilient to motion and illumination variations compared to chrominance-based or orthogonal decomposition methods. However, an exception is observed in the AR phase, where OMIT unexpectedly outperforms most methods, achieving a high PCC and lower MAE only behind the fully supervised ContrastPhys. This indicates that its decomposition strategy may be more robust to abnormal physiological responses that occur after holding breath, potentially capturing subtle variations in BVP that other methods struggle to extract. However, all non-learning methods remain sensitive to artifacts, reinforcing the advantages of deep learning models for robust HR estimation.

From Table 2, while ContrastPhys models demonstrate strong performance in HR estimation, their rPPG signals exhibit a weaker respiratory component. All methods achieve their best performance during the GB task, though results vary across other phases. The nose-to-lip and nose-to-shoulder distance approaches do not perform as expected, showing similar outcomes even in the R task, suggesting limitations in their ability to capture respiratory motion effectively. Interestingly, respiratory signals extracted from PPG GT hold results comparable to certain non-learning methods in specific phases, emphasizing the continued relevance of traditional methods for RR estimation, particularly under controlled conditions.

Table 2. RR errors of rPPG-based methods for each phase

rPPG Methods	Dataset									
	Full Test		NB Test		R Test		GB Test		HB Test	
	MAE	PCC	MAE	PCC	MAE	PCC	MAE	PCC	MAE	PCC
PPG	4.68	0.11	5.48	0.29	5.15	−0.07	2.82	−0.07	4.95	0.14
CHROM	4.54	0.09	5.24	0.20	4.43	−0.01	2.54	0.05	4.52	0.07
OMIT	4.49	0.07	5.42	0.18	4.32	0.04	2.83	0.02	4.36	0.08
POS	5.92	0.09	5.30	0.27	6.52	−0.06	4.82	0.01	6.18	0.08
Unsupervised	6.03	0.03	6.64	0.12	6.98	−0.04	4.18	−0.04	6.15	−0.03
Semi-supervised	6.03	0.03	7.23	0.10	5.82	0.05	4.69	−0.04	5.45	0.06
Fully-supervised	5.97	0.03	6.75	0.10	5.99	0.02	4.72	−0.07	5.91	−0.02
Nose-Lip	5.00	0.04	5.14	0.15	5.53	−0.03	3.44	−0.06	5.55	−0.01
Nose-Shoulder	4.59	0.12	4.51	0.26	5.24	0.06	2.57	−0.01	5.87	−0.02
	GAP 1 Test		GAP 2 Test		GAP 3 Test		GAP 4 Test		AR Test	
	MAE	PCC	MAE	PCC	MAE	PCC	MAE	PCC	MAE	PCC
PPG	5.75	0.27	6.16	0.07	5.61	0.05	4.91	−0.05	4.53	0.13
CHROM	7.12	−0.02	5.93	0.13	5.57	0.09	4.47	0.07	5.10	0.05
OMIT	6.83	0.01	5.83	0.12	5.67	−0.02	4.07	0.17	4.93	0.10
POS	7.27	−0.02	7.04	0.11	6.71	0.06	6.54	0.04	6.25	0.16
Unsupervised	7.48	−0.09	7.51	0.01	7.05	0.03	6.27	0.08	5.93	0.12
Semi-supervised	7.46	0.05	7.29	−0.01	6.77	0.11	5.62	0.04	5.20	0.17
Fully-supervised	7.57	0.04	7.20	−0.01	6.76	0.06	5.93	0.03	5.50	0.20
Nose-Lip	6.34	−0.02	6.55	−0.10	5.52	0.07	4.60	−0.02	5.17	0.01
Nose-Shoulder	5.63	0.11	6.42	−0.06	5.55	0.02	4.93	0.07	5.42	−0.06

4.2 Depth Video Analysis

Table 3 shows that respiratory motion extracted from the depth video, particularly from the chest region, provides reliable RR estimates. The best performance is observed during NB and GB tasks, where the breathing patterns remain regular. The R task also yields reasonable accuracy, suggesting that depth-based motion tracking can be robust to mild facial and upper-body movement. However, accuracy declines during the HB and GAP 1–4 phases, likely due to irregular breathing, posture changes, and transient motion artifacts that weaken the periodicity of the signal.

Table 3. RR errors of depth data for each phase

Depth Regions	Dataset									
	Full Test		NB Test		R Test		GB Test		HB Test	
	MAE	PCC	MAE	PCC	MAE	PCC	MAE	PCC	MAE	PCC
Depth Face	4.05	0.17	3.70	0.30	4.65	0.16	1.75	0.05	5.51	−0.01
Depth Chest	3.45	0.25	2.83	0.36	3.71	0.24	0.99	0.11	5.53	0.02
	GAP 1 Test		GAP 2 Test		GAP 3 Test		GAP 4 Test		AR Test	
	MAE	PCC	MAE	PCC	MAE	PCC	MAE	PCC	MAE	PCC
Depth Face	5.27	0.12	5.04	0.28	5.25	0.13	4.78	0.03	4.39	0.24
Depth Chest	4.64	0.24	5.52	0.07	4.54	0.22	4.09	0.19	4.20	0.10

Table 4 confirms that depth-based signals lack a strong cardiac component, as expected. The extracted waveforms do not exhibit a clear HR-related periodicity, reinforcing that depth imaging is more suited for capturing respiratory motion rather than BVP variations.

Table 4. HR errors of depth data for each phase

Depth Regions	Dataset									
	Full Test		NB Test		R Test		GB Test		HB Test	
	MAE	PCC	MAE	PCC	MAE	PCC	MAE	PCC	MAE	PCC
Depth Face	31.51	−0.11	32.59	−0.16	33.80	−0.20	27.15	−0.14	38.30	0.03
Depth Chest	35.02	-0.18	36.80	−0.24	35.82	−0.23	32.77	−0.26	39.28	−0.05
	GAP 1 Test		GAP 2 Test		GAP 3 Test		GAP 4 Test		AR Test	
	MAE	PCC	MAE	PCC	MAE	PCC	MAE	PCC	MAE	PCC
Depth Face	27.71	−0.17	31.14	−0.10	25.24	0.01	31.63	−0.18	30.03	−0.34
Depth Chest	32.76	−0.28	32.70	−0.19	28.91	−0.18	37.28	−0.21	36.47	−0.30

5 Discussion and Conclusion

The comparative evaluation of vision-based HR and RR estimation methods, challenges several conventional assumptions on the strengths and limitations of different approaches. A key finding is that motion-based RR estimation, often assumed to be the most reliable non-contact approach, does not always outperform rPPG-based methods. Although motion tracking can effectively capture respiratory movements, rPPG-derived RR estimates performed comparably in many scenarios, highlighting the potential of physiological signal extraction from subtle skin color variations. However, both approaches struggled in the presence of significant motion artifacts. Unexpectedly, the traditional OMIT method demonstrated strong performance in the apnea release phase, outperforming deep learning models in certain conditions. This suggests that classical signal decomposition techniques may still be valuable, particularly to detect subtle physiological changes following disrupted breathing. While deep learning-based HR estimation showed improved accuracy in general, its advantage diminished in more dynamic scenarios, such as reading aloud or transitioning between breathing patterns. Notably, semi-supervised learning achieved results close to fully supervised models, indicating that large-scale labeled datasets may not be strictly necessary for effective training.

The evaluation of depth-based RR estimation from RGB-D cameras provides further insights. While chest motion extracted from depth data yielded reliable RR estimates under controlled breathing conditions, performance deteriorated when subjects moved, emphasizing the need for improved motion compensation techniques. These findings suggest that depth-based tracking alone may not be sufficient for real-world applications. Our analysis highlighted critical challenges in non-contact physiological monitoring, particularly the need to mitigate motion artifacts and account for varying breathing patterns. Future research should explore multimodal fusion strategies that integrate rPPG, motion tracking, and depth-based signals to enhance robustness, developing models that explicitly account for task-dependent noise and variability in breathing dynamics.

Acknowledgements. The research was supported by the Research Council of Finland 6G Flagship programme (Grant Number: 369116) and PROFI5 HiDyn programme (Grant Number: 326291).

References

1. Allado, E., et al.: Remote photoplethysmography is an accurate method to remotely measure respiratory rate: a hospital-based trial. J. Clin. Med. (2022). https://doi.org/10.3390/jcm11133647
2. Avishek, C., A., P., Pragathi, P.: Real-time respiration rate measurement from thoracoabdominal movement with a consumer grade camera. In: Annual International Conference of the IEEE Engineering in Medicine and Biology Society (2016). https://doi.org/10.1109/EMBC.2016.7591289

3. Bazarevsky, V., Grishchenko, I., Raveendran, K., Zhu, T.L., Zhang, F., Grundmann, M.: BlazePose: on-device real-time body pose tracking. In: CVPR Workshop on Computer Vision for Augmented and Virtual Reality (2020). https://arxiv.org/abs/2006.10204
4. Casado, C.A., López, M.B.: Face2 PPG: an unsupervised pipeline for blood volume pulse extraction from faces. IEEE J. Biomed. Health Inform. (2023). https://doi.org/10.1109/JBHI.2023.3307942
5. Cañellas, M.L., et al.: OMuSense-23: a multimodal dataset for contactless breathing pattern recognition and biometric analysis (2024). https://arxiv.org/abs/2407.06137
6. Chen, M., Zhu, Q., Zhang, H., Wu, M., Wang, Q.: Respiratory rate estimation from face videos. In: IEEE EMBS International Conference on Biomedical and Health Informatics (2019). https://doi.org/10.1109/bhi.2019.8834499
7. Chen, W., McDuff, D.: DeepPhys: video-based physiological measurement using convolutional attention networks. In: Ferrari, V., Hebert, M., Sminchisescu, C., Weiss, Y. (eds.) ECCV 2018. LNCS, vol. 11206, pp. 356–373. Springer, Cham (2018). https://doi.org/10.1007/978-3-030-01216-8_22
8. Cheng, C.H., Wong, K.L., Chin, J.W., Chan, T.T., So, R.: Deep learning methods for remote heart rate measurement: a review and future research agenda. Sensors (2021). https://doi.org/10.3390/s21186296
9. Duncan, L., et al.: An assessment of algorithms to estimate respiratory rate from the remote photoplethysmogram. In: IEEE/CVF Conference on Computer Vision and Pattern Recognition Workshops (2020). https://doi.org/10.1109/CVPRW50498.2020.00160
10. Fabian, S., Christoph, M., Gerold, B., M., F.: Exploiting weak head movements for camera-based respiration detection. In: International Conference of the IEEE Engineering in Medicine and Biology Society (2019). https://doi.org/10.1109/EMBC.2019.8856387
11. Gwak, M., Vatanparvar, K., Kuang, J., Gao, A.: Motion- based respiratory rate estimation with motion artifact removal using video of face and upper body. In: International Conference of the IEEE Engineering in Medicine and Biology Society (2022). https://doi.org/10.1109/EMBC48229.2022.9871231
12. de Haan, G., Jeanne, V.: Robust pulse rate from chrominance-based rPPG. IEEE Trans. Biomed. Eng. **60**(10), 2878–2886 (2013). https://doi.org/10.1109/TBME.2013.2266196
13. Huang, B., et al.: Challenges and prospects of visual contactless physiological monitoring in clinical study. NPJ Digit. Med. **6**(1), 231 (2023). https://doi.org/10.1038/s41746-023-00973-x
14. Jingda, D., Siqi, L., Bochao, Z., P., Y.: Weakly supervised rPPG estimation for respiratory rate estimation. In: IEEE/CVF International Conference on Computer Vision Workshops (2021). https://doi.org/10.1109/ICCVW54120.2021.00271
15. Juan, C., Runqing, L., Jiajie, L., Rencheng, S., Yu, L., Xun, C.: Motion-robust respiratory rate estimation from camera videos via fusing pixel movement and pixel intensity information. IEEE Trans. Instrum. Meas. (2023). https://doi.org/10.1109/TIM.2023.3291770
16. Kempfle, J., Van Laerhoven, K.: Respiration rate estimation with depth cameras: Dn evaluation of parameters. In: International Workshop on Sensor-Based Activity Recognition and Interaction (2018). https://doi.org/10.1145/3266157.3266208
17. Korosh, V., Migyeong, G., Li, Z., Jilong, K., Gao, A.: Respiration rate estimation from remote PPG via camera in presence of non-voluntary artifacts. In: Interna-

tional Conference on Wearable and Implantable Body Sensor Networks (2022). https://doi.org/10.1109/BSN56160.2022.9928485

18. Kumar, M., Veeraraghavan, A., Sabharwal, A.: DistancePPG: robust non-contact vital signs monitoring using a camera. Biomed. Opt. Express **6** (2015). https://doi.org/10.1364/BOE.6.001565
19. Lage Cañellas, M., et al.: OMuSense-23: a multimodal dataset for contactless breathing pattern recognition and biometric analysis (2024). https://doi.org/10.5281/zenodo.12705176. [Data set]
20. Lin, Y., Lin, Y.H.: A study of color illumination effect on the snr of rPPG signals. In: International Conference of the IEEE Engineering in Medicine and Biology Society, vol. 2017 (2017). https://doi.org/10.1109/EMBC.2017.8037807
21. Makkapati, V.V., Rambhatla, S.S.: Camera based estimation of respiration rate by analyzing shape and size variation of structured light. In: IEEE International Conference on Acoustics, Speech and Signal Processing (2016). https://doi.org/10.1109/icassp.2016.7472071
22. Migyeong, G., et al.: Multimodal breathing rate estimation using facial motion and RPPG from RGB camera. In: IEEE International Conference on Acoustics, Speech, and Signal Processing (2024). https://doi.org/10.1109/ICASSP48485.2024.10446086
23. Ghodratigohar, M., Ghanadian, H., Al Osman, H.: A remote respiration rate measurement method for non-stationary subjects using CEEMDAN and machine learning. IEEE Sens. J. (2020). https://doi.org/10.1109/JSEN.2019.2946132
24. Alnaggar, M., Siam, A.I., Handosa, M., Medhat, T., Rashad, M.Z.: Video-based real-time monitoring for heart rate and respiration rate. Expert Syst. Appl. (2023). https://doi.org/10.1016/j.eswa.2023.120135
25. Nguyen, N., Nguyen, L., Li, H., Bordallo López, M., Álvarez Casado, C.: Evaluation of video-based rPPG in challenging environments: artifact mitigation and network resilience. Comput. Biol. Med. **179**, 108873 (2024). https://doi.org/10.1016/j.compbiomed.2024.108873
26. Niu, X., Shan, S., Han, H., Chen, X.: RhythmNet: end-to-end heart rate estimation from face via spatial-temporal representation. IEEE Trans. Image Process. **29**, 2409–2423 (2019). https://doi.org/10.1109/TIP.2019.2947204
27. Pilz, C.S., Zaunseder, S., Krajewski, J., Blazek, V.: Local group invariance for heart rate estimation from face videos in the wild. In: IEEE/CVF Conference on Computer Vision and Pattern Recognition Workshops (2018). https://doi.org/10.1109/CVPRW.2018.00172
28. Špetlík, R., Franc, V., Matas, J.: Visual heart rate estimation with convolutional neural network. In: British Machine Vision Conference (2018)
29. Sun, Y., Thakor, N.: Photoplethysmography revisited: from contact to noncontact, from point to imaging. IEEE Trans. Biomed. Eng. **63**(3), 463–477 (2016). https://doi.org/10.1109/TBME.2015.2476337
30. Sun, Z., Li, X.: Contrast-phys: unsupervised video-based remote physiological measurement via spatiotemporal contrast. In: Avidan, S., Brostow, G., Cissé, M., Farinella, G.M., Hassner, T. (eds.) ECCV 2022. LNCS, vol. 13672, pp. 492–510. Springer, Cham (2022). https://doi.org/10.1007/978-3-031-19775-8_29
31. Sun, Z., Li, X.: Contrast-phys+: unsupervised and weakly-supervised video-based remote physiological measurement via spatiotemporal contrast. IEEE Trans. Pattern Anal. Mach. Intell. **46**(08), 5835–5851 (2024). https://doi.org/10.1109/TPAMI.2024.3367910

32. Verkruysse, W., Svaasand, L.O., Nelson, J.S.: Remote plethysmographic imaging using ambient light. Opt. Express **16**(26), 21434–21445 (2008). https://doi.org/10.1364/oe.16.021434
33. Wang, W., den Brinker, A.C., Stuijk, S., de Haan, G.: Algorithmic principles of remote PPG. IEEE Trans. Biomed. Eng. **64**(7), 1479–1491 (2017). https://doi.org/10.1109/TBME.2016.2609282
34. Yu, Z., Li, X., Niu, X., Shi, J., Zhao, G.: AutoHR: a strong end-to-end baseline for remote heart rate measurement with neural searching. IEEE Signal Process. Lett. **27**, 1245–1249 (2020). https://doi.org/10.1109/LSP.2020.3007086
35. Yu, Z., Shen, Y., Shi, J., Zhao, H., Torr, P., Zhao, G.: PhysFormer: facial video-based physiological measurement with temporal difference transformer. In: IEEE/CVF Computer Vision and Pattern Recognition Conference (2022). https://doi.org/10.1109/CVPR52688.2022.00415
36. Zhan, Q., Wang, W., de Haan, G.: Analysis of CNN-based remote-PPG to understand limitations and sensitivities. Biomed. Opt. Express **11**(3), 1268–1283 (2020). https://doi.org/10.1364/BOE.382637

Aligning Subjective and Objective Assessments in Super-Resolution Models

Muhammad Hamza Zafar(✉) and Jon Y. Hardeberg

Colourlab, Department of Computer Science, NTNU – Norwegian University of Science and Technology, Gjøvik, Norway
{muhamhz,jon.hardeberg}@ntnu.no

Abstract. The evaluation of super-resolution (SR) models remains a challenge due to discrepancies between objective metrics and human perception. This study assesses four state-of-the-art SR models, including ResShift, Real-ESRGAN, BSRGAN, and SwinIR, through both computational metrics and human-centered evaluations. Standard metrics such as PSNR, SSIM, LPIPS, and CLIPIQA were used to quantify fidelity and perceptual alignment, while subjective assessments were conducted through online and controlled lab-based experiments. Results indicate that ResShift consistently achieves superior perceptual quality, while BSRGAN, despite competitive objective scores, ranks lower in human evaluations. To the best of our knowledge, this is the first study to provide an in-depth subjective evaluation of ResShift alongside a systematic comparison with other SR models. These findings underscore the limitations of traditional metrics and emphasize the need for hybrid evaluation approaches that integrate subjective assessments. Code and Data Available at https://github.com/hamzafer/super-resolution-color.

Keywords: Super-resolution · Subjective Assessment · Image Quality

1 Introduction

Super-resolution (SR) is a key task in image processing that reconstructs high-resolution (HR) images from low-resolution (LR) inputs, with applications in medical imaging, satellite analysis, video enhancement, and restoration of degraded images [1,2]. While SR models have traditionally been evaluated using objective metrics such as Peak Signal-to-Noise Ratio (PSNR), Structural Similarity Index Measure (SSIM) [5], and Learned Perceptual Image Patch Similarity (LPIPS) [3], these metrics often fail to align with human perception [4]. A model with a high PSNR may still produce artifacts, oversmoothing, or unnatural colors, limiting the reliability of objective evaluations.

Recent advancements have introduced SR models leveraging architectures such as transformers, generative adversarial networks (GANs), and diffusion-based approaches. Among these, ResShift [7], Real-ESRGAN [6], BSRGAN [9],

Supplementary Information The online version contains supplementary material available at https://doi.org/10.1007/978-3-031-95918-9_4.

J. Petersen and V. A. Dahl (Eds.): SCIA 2025, LNCS 15726, pp. 47–60, 2025.
https://doi.org/10.1007/978-3-031-95918-9_4

and SwinIR [8] have demonstrated strong performance. ResShift, in particular, claims state-of-the-art (SOTA) status [7]. However, like many SR models, it has not undergone rigorous subjective evaluation, which is essential for validating perceptual quality.

Subjective evaluation captures human perception by assessing factors such as sharpness, color fidelity, and artifact presence [10,11]. Despite its importance, subjective assessment remains underexplored in SR research due to the complexity of psychophysical experiments and challenges in obtaining consistent human judgments [12]. To address this gap, we conduct two psychophysical experiments to evaluate the perceptual quality of HR images generated by ResShift, Real-ESRGAN, BSRGAN, and SwinIR. The first experiment gathers user preferences through an online platform, considering diverse viewing conditions, while the second is a controlled lab study using pairwise comparisons to validate preference consistency.

This paper makes the following contributions: (1) We conduct two comprehensive psychophysical experiments to evaluate the perceptual quality of SR outputs from four leading models, under both uncontrolled (online) and controlled (lab) settings. (2) We benchmark these results against widely used objective metrics, including PSNR, SSIM, LPIPS [3], and CLIP-IQA [13], providing insight into their correlation with human perception. (3) We perform a systematic failure case analysis to uncover perceptual artifacts and limitations common across models. Together, these contributions form a unified evaluation framework that connects algorithmic performance with human visual quality, advancing the perceptual assessment of super-resolution models.

To the best of our knowledge, this is the first study to conduct an in-depth subjective quality assessment of ResShift [7]. The findings contribute to the development of SR models that excel not only in standard benchmarks but also in delivering visually appealing results across varied viewing conditions, enhancing their real-world applicability.

2 Related Work

Super-resolution (SR) has evolved from traditional interpolation-based methods such as bilinear and bicubic interpolation [14] to deep learning-driven approaches. While early interpolation techniques were computationally efficient, they often resulted in oversmoothed images with limited detail recovery and poor adaptability to real-world scenarios.

The advent of deep learning enabled SR models to learn end-to-end mappings from low-resolution (LR) to high-resolution (HR) images. CNN-based models like SRCNN and VDSR demonstrated significant improvements by extracting hierarchical features [17], while later architectures incorporated generative adversarial networks (GANs) [18], transformers, and diffusion-based approaches to further enhance detail restoration and perceptual quality. Among these, ResShift [7], Real-ESRGAN [6], BSRGAN [9], and SwinIR [8] have achieved state-of-the-art (SOTA) performance.

ResShift introduces residual shifting for efficient reconstruction and has demonstrated strong performance on real-world datasets. Real-ESRGAN extends ESRGAN by modeling real-world degradations, making it more practical for diverse applications. BSRGAN is designed for robustness against various degradation types, while SwinIR leverages transformers to capture both local and global dependencies for superior restoration. However, these models rely primarily on objective evaluation metrics such as PSNR, SSIM, and LPIPS [3,5], which often fail to reflect human perceptual preferences regarding sharpness, color fidelity, and artifact presence.

To address these limitations, subjective evaluation studies have emerged. Latent Diffusion Models (LDM) integrate subjective assessments by comparing HR images to ground truth or other model outputs, but they lack dataset diversity and systematic failure case analysis [15]. VQD-SR explores subjective evaluation for animation datasets using pairwise comparisons across video frames, yet its applicability to real-world imagery is limited, and it does not validate findings under controlled conditions [16].

These gaps underscore the need for a more comprehensive approach to SR evaluation. This study builds on prior work by integrating two psychophysical experiments: an online study capturing diverse viewing conditions and a controlled lab-based study validating consistency in perceptual assessments.

3 Dataset and Preparation

This study utilizes the DIV2K dataset [19,20], a widely recognized benchmark for super-resolution tasks in NTIRE (CVPR 2017, 2018) and PIRM (ECCV 2018) [22]. DIV2K consists of diverse high-quality 2K resolution images, divided into training, validation, and test subsets, each containing high-resolution (HR) images and their corresponding low-resolution (LR) versions generated with different downscaling factors.

To ensure consistency with benchmarked setups, this study uses the validation data from Track 1 (bicubic downscaling). LR images at 255 $\times$ 169 pixels were selected from the bicubic x8 dataset and upsampled by a factor of 4$\times$ to generate HR images of 1020 $\times$ 676 pixels. The HR ground truth images were taken from the bicubic downscaling x2 dataset, avoiding additional preprocessing and ensuring comparability with existing work. The selected resolution and scaling factor provide a balance between computational efficiency and perceptual quality.

For structured evaluation, a subset of images was chosen (see Fig. 1. Experiment 1 used 30 images based on pilot studies, ensuring a manageable evaluation time of approximately 15 min per participant. Each participant selected the best HR image from four options for each LR reference. Experiment 2 used a refined subset of 10 images, resulting in 60 unique pairwise comparisons. This configuration balanced evaluation depth with participant engagement.

The DIV2K dataset was selected for its extensive use in previous SR studies. Further details about the dataset and its usage are discussed in [19–22].

(a) Selected 30 images for Experiment 1 (single-choice evaluation).

(b) Selected 10 images for Experiment 2 (pairwise comparison).

Fig. 1. Selected images used in the experiments.

4 Methodology

4.1 Inference

After dataset preparation, inference was performed on four selected SR models: ResShift, Real-ESRGAN, BSRGAN, and SwinIR. Official repositories for each model were cloned, and dependencies were set up following provided instructions to ensure reproducibility.

Inference was conducted using an NVIDIA RTX 4090 GPU, enabling efficient computation. The LR images from the DIV2K dataset were upscaled to a resolution of 1020 $\times$ 676 pixels, and the generated HR images were collected for further evaluation. This step ensured consistency across models and allowed a structured comparison of outputs.

4.2 Objective Metrics

The performance of super-resolution models was evaluated using four objective metrics: Peak Signal-to-Noise Ratio (PSNR), Structural Similarity Index Measure (SSIM), Learned Perceptual Image Patch Similarity (LPIPS), and CLIP-based Image Quality Assessment (CLIPIQA). These metrics quantify fidelity, perceptual quality, and semantic alignment of high-resolution (HR) images.

PSNR and SSIM are commonly used to assess pixel-level fidelity. PSNR measures absolute pixel intensity differences, while SSIM accounts for structural and perceptual similarities [5]. LPIPS evaluates perceptual quality by computing feature-level differences via deep neural networks [3], whereas CLIPIQA leverages CLIP embeddings to assess semantic consistency [13].

Objective metrics were computed for ResShift, Real-ESRGAN, BSRGAN, and SwinIR using corresponding low-resolution (LR) inputs from the DIV2K dataset. Table 1 summarizes the results, showing that while PSNR and SSIM values were comparable across models, perceptual metrics like LPIPS and CLIPIQA indicated a stronger alignment with human perception, with ResShift achieving the highest perceptual quality.

A closer analysis revealed that models with high PSNR and SSIM scores did not always yield perceptually superior results, reinforcing the limitations of traditional metrics. Figure 2 shows the distribution of metric values, highlighting variability across models.

Table 1. Objective metrics for evaluated SR models.

Method	PSNR↑	SSIM↑	LPIPS↓	CLIPIQA↑
BSRGAN [9]	24.35	0.703	0.172	0.976
RealESRGAN [6]	23.73	0.696	0.168	0.971
ResShift [7]	24.65	0.723	0.136	0.986
SwinIR [8]	23.85	0.705	0.163	0.968

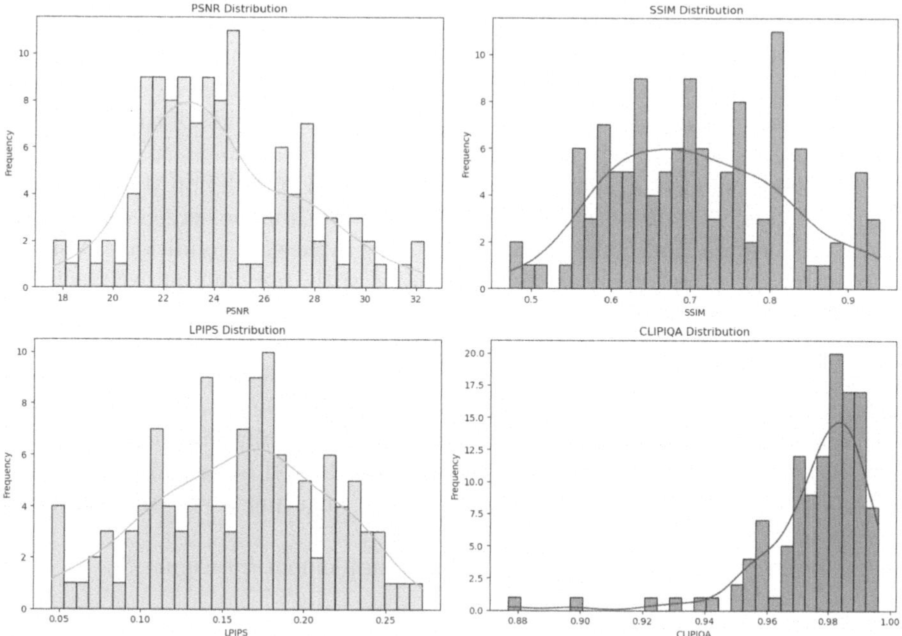

Fig. 2. Distribution of metric values across all images, highlighting variability and outliers.

Identifying Hard Negatives with Metric Thresholds. To detect poor-quality outputs, an empirical threshold-based approach was applied across multiple metrics. PSNR was set at 24.0, SSIM at 0.7 [5], LPIPS at 0.15 [3], and CLIPIQA at 0.96 [13]. This method successfully flagged visually unpleasant or hallucinated images more reliably than any single metric.

This analysis underscores the importance of integrating multiple evaluation methods. While objective metrics provide quantitative comparisons, their limitations in perceptual alignment highlight the necessity of subjective assessments to capture human visual preferences [10,11].

4.3 Subjective Assessment

To complement objective evaluations, two psychophysical experiments were conducted to assess human perception of the super-resolution (SR) models.

Experiment 1: Online Assessment. The first experiment was conducted online through a custom-built web platform designed for user-friendly evaluation of HR images generated by different SR models. Participants were recruited randomly to ensure diverse representation. Each participant was shown a low-resolution (LR) reference image at the center of the screen, with four high-resolution (HR) images surrounding it. These HR images, upscaled by different SR models, had a resolution of 1020 × 676 pixels. Participants selected the image they perceived as the best based on visual quality. The position of the HR images was randomized to eliminate bias.

The study included 30 images, with an estimated completion time of 15 min per participant. The platform ensured images fit within the screen dimensions, preventing scrolling and maintaining consistency in viewing conditions. A single-choice evaluation method was used, aligning with previous super-resolution research methodologies.

To enhance clarity, the platform provided structured instructions, a neutral background to minimize distractions, and recorded participant choices, response times, and demographic data. At the end of the experiment, participants provided qualitative feedback on their selection criteria, highlighting aspects such as color, detail, and artifacts. Pilot studies were conducted beforehand to validate the experiment design and estimated duration, ensuring a smooth evaluation process.

The online assessment setup is shown in Supplementary Figure S1.

Experiment 2: On-Site Controlled Assessment. The second experiment was conducted in a controlled laboratory setting to validate findings from the online study. Participants, primarily students and researchers from our research group with expertise in image quality, were shown a low-resolution (LR) reference image (255 × 169 pixels) at the center of the screen, with two high-resolution (HR) images (1020 × 676 pixels) displayed side by side. They selected the image they perceived as superior based on visual quality. Image positions were randomized to eliminate bias, and a total of 10 images were used, resulting in 60 unique pairwise comparisons. The estimated completion time per participant was 15 min.

The experiment followed a pairwise comparison methodology, a well-established approach for evaluating perceptual preferences. To ensure consistent viewing conditions, a BenQ SW321C monitor was calibrated to sRGB with a gamma of 2.2, a white point of D65, and a maximum luminance of $80\,\mathrm{cd/m^2}$. Calibration was performed using the i1 Display Pro instrument and iProfiler software.

Pilot studies refined the setup and validated the estimated evaluation time. Participant choices, response times, and qualitative feedback were recorded. To maintain engagement, participants were briefed beforehand.

Participants could zoom into HR images for detailed inspection, ensuring thorough evaluation. Although the Quick Eval platform used for the experiment cropped images by default, participants were instructed on how to explore all areas, minimizing any potential limitations.

The pairwise comparison interface is illustrated in Supplementary Figure S2.

5 Results

Here we present the findings from the objective analysis and subjective assessments of the evaluated super-resolution (SR) models. The results are structured into three parts: objective metric evaluation, findings from Experiment 1 (online assessment), and findings from Experiment 2 (on-site controlled assessment).

5.1 Objective Analysis

The performance of SR models was evaluated using four objective metrics: Peak Signal-to-Noise Ratio (PSNR), Structural Similarity Index Measure (SSIM), Learned Perceptual Image Patch Similarity (LPIPS), and CLIP-based Image Quality Assessment (CLIPIQA). These metrics assess fidelity, perceptual quality, and semantic alignment [3,5,13].

ResShift consistently achieved the highest scores for LPIPS (lower is better) and CLIPIQA (higher is better), indicating strong perceptual quality and semantic alignment. These results support findings in prior work emphasizing the effectiveness of perceptual metrics in evaluating image quality [3,13]. In contrast, PSNR and SSIM, which measure pixel-level fidelity, showed comparable results across models. BSRGAN achieved slightly higher PSNR values than Real-ESRGAN, while ResShift outperformed all models in SSIM. However, these traditional metrics do not always align with human perception, as also noted in previous studies [4,10].

A more detailed analysis of these discrepancies and their implications for SR model evaluation is provided in the Discussion section.

5.2 Experiment 1: Online Assessment

The first psychophysical experiment assessed human preferences for HR images in an online setting. A total of 30 images were evaluated, with participants selecting the best HR image from four randomized options for each LR reference image. While 134 sessions were initiated, only 54 were completed, reflecting the voluntary nature of participation and challenges in retention.

Preferences for the SR models showed ResShift as the most favored with 624 selections, followed by SwinIR (377), Real-ESRGAN (351), and BSRGAN (268), aligning with ResShift's strong performance in perceptual metrics like LPIPS and CLIPIQA. While ResShift's dominance across most images highlights its perceptual appeal, other models showed strengths in specific cases, suggesting that effectiveness may depend on dataset characteristics and use cases. Age-based trends further revealed that younger participants preferred ResShift more strongly, whereas older participants exhibited a more balanced distribution of preferences, as illustrated in Fig. 3.

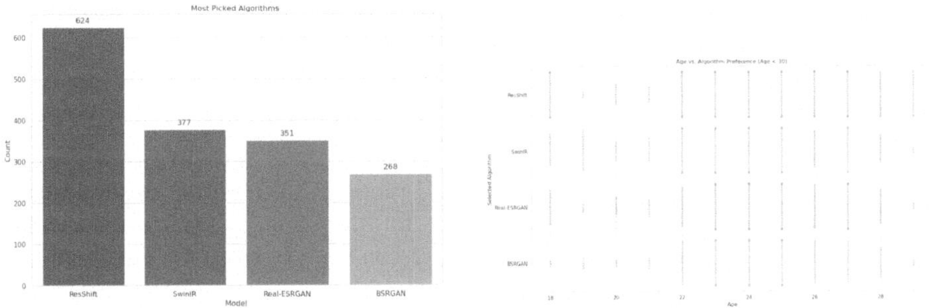

(a) SR model preferences across participants.

(b) Model preferences across different age groups.

Fig. 3. Participant preferences for SR models.

Moreover, analysis of response times and qualitative feedback further supported these trends, showing that images with subtle perceptual differences required longer decision times. Also, participants consistently cited sharpness, vibrant colors, and reduced artifacts as reasons for their selections. These patterns reinforce the alignment between participant engagement and the perceptual strengths identified through subjective assessment, highlighting the limitations of relying solely on traditional quantitative metrics.

5.3 Experiment 2: On-Site Controlled Assessment

The second experiment validated the findings from Experiment 1 under controlled conditions. Participants evaluated 60 image pairs, selecting their preferred HR image from each pair. A total of 15 observers participated, resulting in 900 individual comparisons. This controlled setup eliminated potential distractions and inconsistencies present in the online experiment. ResShift emerged as the most preferred model with 309 selections, consistent with its performance in Experiment 1. RealESRGAN (228) and SwinIR (220) followed, while BSRGAN (143) received the fewest selections, reinforcing its lower perceptual quality.

Variability in decision times across participants indicated differences in perceptual difficulty. Some tasks required longer decision times, suggesting challenges in differentiating image quality between models. Individual engagement also varied; some participants took significantly longer to complete selections.

To further analyze model preferences, Borda count rankings confirmed the dominance of ResShift in terms of perceptual quality. Additionally, the Bradley-Terry model was applied to estimate the relative ability of each model to be preferred [23]. The results reinforced the raw selection counts, ranking ResShift highest, followed by RealESRGAN and SwinIR, while BSRGAN consistently ranked lowest.

A Chi-Square test was conducted to evaluate statistical significance, yielding a Chi-Square statistic of 61.40 and a p-value of 0.0000, confirming that model preferences deviated significantly from a uniform selection pattern [24]. Pairwise confidence intervals for win ratios (Table 2) highlight the likelihood of one model being preferred over another. ResShift consistently outperformed other models, while SwinIR and RealESRGAN remained competitive in some cases.

Details on the decision time distribution can be found in Supplementary Figure S3.

Experiment 2 corroborates findings from Experiment 1, with ResShift consistently outperforming other models. Statistical analyses validate these results, offering a robust framework for evaluating subjective image quality for SR.

Table 2. Pairwise Confidence Intervals for Win Ratios.

Model	ResShift	SwinIR	RealESRGAN	BSRGAN
ResShift	–	(0.61, 0.76)	(0.55, 0.71)	(0.66, 0.80)
SwinIR	(0.24, 0.39)	–	(0.43, 0.59)	(0.57, 0.72)
RealESRGAN	(0.29, 0.45)	(0.41, 0.57)	–	(0.58, 0.73)
BSRGAN	(0.20, 0.34)	(0.28, 0.43)	(0.27, 0.42)	–

6 Discussion

This study aimed to evaluate super-resolution (SR) models using both objective metrics and subjective assessments, providing a comprehensive understanding of their performance. The findings underscore the importance of combining quantitative metrics with human-centric evaluations to capture perceptual quality effectively.

6.1 Objective Metrics Analysis

The objective evaluation leveraged metrics such as PSNR, SSIM, LPIPS, and CLIPIQA. As shown in Table 3, ResShift consistently outperformed other models across most metrics, including achieving the highest PSNR (25.01) and LPIPS (0.231). These results align with findings reported in prior literature [6–8], reaffirming ResShift's technical superiority.

However, BSRGAN, despite achieving competitive scores in PSNR and SSIM, performed poorly in subjective evaluations, as detailed in Experiment 1 and Experiment 2. This discrepancy highlights the limitations of traditional metrics, which often fail to account for perceptual quality attributes like sharpness and artifact suppression [3,4]. Figure 4a provides a visualization of these results on the DIV2K dataset, while Fig. 4b compares them with metrics reported in the original papers.

Table 3. Performance Metrics of Selected Models. Metrics derived from [7].

Method	PSNR↑	SSIM↑	LPIPS↓	CLIPIQA↑
BSRGAN [9]	24.42	0.659	0.259	0.581
SwinIR [8]	23.99	0.667	0.238	0.564
RealESRGAN [6]	24.04	0.665	0.254	0.523
ResShift [7]	25.01	0.677	0.231	0.592

6.2 Experiment 1: Online Assessment

The first psychophysical experiment, conducted with 54 observers, revealed clear preferences for ResShift, which was chosen 624 times, far surpassing the counts for SwinIR (377), RealESRGAN (351), and BSRGAN (268). Figure 4c illustrates the aggregated preferences.

Participants frequently cited sharpness, absence of artifacts, and color fidelity as reasons for their choices, aligning with ResShift's superior LPIPS and CLIPIQA scores. This experiment highlights the divergence between subjective preferences and traditional metrics, as BSRGAN's lower subjective scores contrast with its relatively high PSNR and SSIM values. These findings underscore the necessity of incorporating human-centric evaluations in SR research [10,11].

6.3 Experiment 2: On-Site Controlled Assessment

The second experiment validated the findings from Experiment 1 in a controlled environment, involving 15 observers and 900 pairwise comparisons. ResShift was the most preferred model with 309 selections, followed by RealESRGAN (228), SwinIR (220), and BSRGAN (143), reinforcing the model's lower perceptual quality.

Statistical analyses further supported these findings. The Bradley-Terry [23] model provided ability estimates for each model, confirming ResShift's dominance. A Chi-Square test [24] indicated statistical significance with a Chi-Square statistic of 61.40 and a p-value of 0.0000, demonstrating that model preferences deviated significantly from a uniform distribution. Additionally, pairwise confidence intervals highlighted ResShift's consistent preference over other models.

Supplementary Figure S4 provides an overview of the preference results from Experiment 2.

6.4 Key Insights and Implications

These findings lead to several important insights:

1. **ResShift's Robustness:** The model's consistent performance across objective metrics and subjective evaluations underscores its suitability for diverse real-world applications, aligning with prior research [7].
2. **Limitations of Traditional Metrics:** BSRGAN's low subjective scores despite competitive PSNR and SSIM values highlight the need for perceptual metrics and subjective assessments [3].
3. **Alignment with Prior Studies:** RealESRGAN and SwinIR exhibited similar trends to those reported in their original papers [6,8], further validating their relative performance.

Figure 4 provides a unified visualization of model performance across objective metrics, Experiment 1, and Experiment 2.

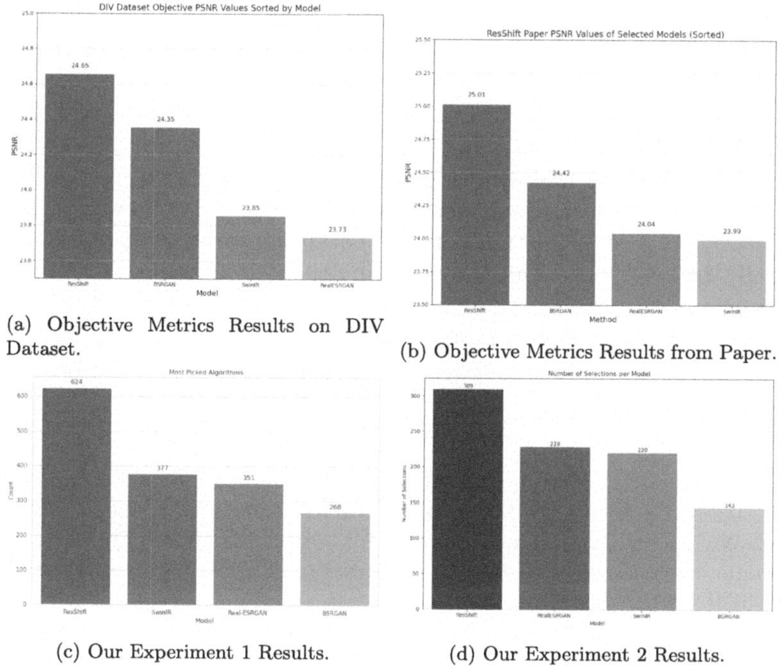

(a) Objective Metrics Results on DIV Dataset.

(b) Objective Metrics Results from Paper.

(c) Our Experiment 1 Results.

(d) Our Experiment 2 Results.

Fig. 4. Summary of SR model performance across objective metrics (a, b), Experiment 1 (c), and Experiment 2 (d).

7 Conclusion

This study provides a comprehensive evaluation of state-of-the-art super-resolution (SR) models, combining objective metrics and subjective assessments. The findings highlight ResShift's superior performance across both quantitative benchmarks and human preferences, making it a robust choice for real-world applications. However, some limitations were observed consistently across all evaluated models, including ResShift. These issues appeared particularly in challenging regions such as faces and fine textures, where hallucinated details or unnatural artifacts occasionally emerged. The results also emphasize the limitations of traditional metrics like PSNR and SSIM, underscoring the importance of incorporating subjective evaluations to better capture perceptual quality.

While this study offers valuable insights into the perceptual quality of SR models, it has several limitations. First, the evaluation relies on commonly used objective metrics, without introducing novel metrics. Future work could explore the development or adoption of more perceptually aligned evaluation techniques to better reflect human preferences. Second, the analysis focuses on only four state-of-the-art SR models. Including a broader range of methods would provide a more comprehensive benchmark and uncover additional strengths and weaknesses across model architectures. Third, the dataset used, particularly in the controlled lab experiment, was relatively small, comprising only 10 images. Expanding the dataset in both scale and diversity would improve the robustness and generalizability of the findings. More diverse datasets (e.g., RealSR [25], DRealSR [26]) could provide a broader evaluation and better reflect real-world conditions. Nevertheless, the evaluation framework presented in this work is modular and scalable, allowing future studies to extend it to additional SR methods with minimal adaptation.

In addition to these limitations, future work could also focus on analyzing failure cases more deeply and exploring hybrid evaluation frameworks that combine subjective and objective measures to better capture perceptual quality.

Overall, this work contributes to bridging the gap between algorithmic evaluation and human perception, advancing the development of SR models that perform well in both benchmarks and practical applications.

References

1. Wang, Z., Chen, J., Hoi, S.: Deep learning for image super-resolution: a survey. IEEE Trans. Pattern Anal. Mach. Intell. **43**(10), 3365–3387 (2020)
2. Yang, W., Zhang, X., Tian, Y., Wang, W., Xue, J.H., Liao, Q.: Deep learning for single image super-resolution: a brief review. IEEE Trans. Multimedia **21**(12), 3106–3121 (2019)
3. Zhang, R., Isola, P., Efros, A.A., Shechtman, E., Wang, O.: The unreasonable effectiveness of deep features as a perceptual metric. In: Proceedings of the IEEE Conference on Computer Vision and Pattern Recognition (CVPR), pp. 586–595 (2018)

4. Blau, Y., Michaeli, T.: The perception-distortion tradeoff. In: Proceedings of the IEEE Conference on Computer Vision and Pattern Recognition (CVPR), pp. 6228–6237 (2018)
5. Wang, Z., Bovik, A.C., Sheikh, H.R., Simoncelli, E.P.: Image quality assessment: from error visibility to structural similarity. IEEE Trans. Image Process. **13**(4), 600–612 (2004)
6. Wang, X., Xie, L., Dong, C., Shan, Y.: Real-ESRGAN: training real-world blind super-resolution with pure synthetic data. In: Proceedings of the IEEE/CVF International Conference on Computer Vision Workshops (ICCVW) (2021)
7. Yue, Z., Wang, J., Loy, C.C.: ResShift: efficient diffusion model for image super-resolution by residual shifting. In: Advances in Neural Information Processing Systems (NeurIPS) (2023)
8. Liang, J., Cao, J., Sun, G., Zhang, K., Van Gool, L., Timofte, R.: SwinIR: image restoration using swin transformer. In: Proceedings of the IEEE/CVF International Conference on Computer Vision Workshops (ICCVW) (2021)
9. Zhang, K., Liang, J., Van Gool, L., Timofte, R.: Designing a practical degradation model for deep blind image super-resolution. In: Proceedings of the IEEE/CVF International Conference on Computer Vision (ICCV) (2021)
10. Sheikh, H.R., Sabir, M.F., Bovik, A.C.: A statistical evaluation of recent full reference image quality assessment algorithms. IEEE Trans. Image Process. **15**(11), 3440–3451 (2006)
11. Liu, X., Zhang, L., Li, C.: Visual quality assessment of super-resolution images. IEEE Trans. Image Process. **20**(12), 3442–3452 (2011)
12. Lugmayr, A., Danelljan, M., Timofte, R.: NTIRE 2020 challenge on real-world image super-resolution: Methods and results. In: Proceedings of the IEEE/CVF Conference on Computer Vision and Pattern Recognition Workshops (CVPRW), pp. 494–495 (2020)
13. Wu, J., Wu, T., Wu, Y., Huang, J., Zhai, G.: CLIP-IQA: Better text-image agreement evaluation via CLIP. In: Proceedings of the 30th ACM International Conference on Multimedia (2022)
14. Han, D.: Comparison of commonly used image interpolation methods. In: Proceedings of the 2nd International Conferences on Computer Science and Electronic Engineering (ICCSEE) (2013)
15. Rombach, R., Blattmann, A., Lorenz, D., Esser, P., Ommer, B.: High-resolution image synthesis with latent diffusion models. In: Proceedings of the IEEE/CVF Conference on Computer Vision and Pattern Recognition (CVPR), pp. 10366–10376 (2022)
16. Tuo, Z., Zhao, H., Huang, H., Liu, J.: Learning data-driven vector-quantized degradation model for animation video super-resolution. In: Proceedings of the IEEE/CVF International Conference on Computer Vision (ICCV) (2023)
17. Kappeler, A., Yoo, S., Dai, Q., Katsaggelos, A.K.: Video super-resolution with convolutional neural networks. IEEE Trans. Comput. Imaging **2**(2), 109–122 (2016)
18. Xu, X., Sun, D., Pan, J., Zhang, H., Pfister, H., Yang, M.: Learning to super-resolve blurry face and text images. In: Proceedings of the IEEE/CVF International Conference on Computer Vision (ICCV) (2017)
19. Agustsson, E., Timofte, R.: NTIRE 2017 challenge on single image super-resolution: dataset and study. In: Proceedings of the IEEE Conference on Computer Vision and Pattern Recognition Workshops (CVPRW) (2017)
20. Timofte, R., et al.: NTIRE 2017 challenge on single image super-resolution: methods and results. In: Proceedings of the IEEE Conference on Computer Vision and Pattern Recognition Workshops (CVPRW) (2017)

21. Timofte, R., et al.: NTIRE 2018 challenge on single image super-resolution: methods and results. In: Proceedings of the IEEE Conference on Computer Vision and Pattern Recognition Workshops (CVPRW) (2018)
22. Ignatov, A., Timofte, R., et al.: PIRM challenge on perceptual image enhancement on smartphones: report. In: Proceedings of the European Conference on Computer Vision Workshops (ECCVW), January (2019)
23. Bradley, R.A., Terry, M.E.: The rank analysis of incomplete block designs: I. The method of paired comparisons. Biometrika **39**(3/4), 324–345 (1952)
24. Pearson, K.: X. On the criterion that a given system of deviations from the probable in the case of a correlated system of variables is such that it can be reasonably supposed to have arisen from random sampling. Philos. Mag. J. Sci. **50**(302), 157–175 (1900)
25. Cai, J., Zeng, H., Yong, H., Cao, Z., Zhang, L.: Toward real-world single image super-resolution: a new benchmark and a new model. In: Proceedings of the IEEE/CVF International Conference on Computer Vision (ICCV), pp. 3086–3095 (2019)
26. Wei, P., et al.: Component divide-and-conquer for real-world image super-resolution. In: Computer Vision-ECCV 2020. Lecture Notes in Computer Science, vol. 12353, pp. 101–117. Springer, Cham (2020)

Comparing Next-Day Wildfire Predictability of MODIS and VIIRS Satellite Data

Justus Karlsson[1], Yonghao Xu[1(✉)], Amanda Berg[1,2], and Leif Haglund[1,2]

[1] Computer Vision Laboratory, Linköping University, Linköping, Sweden
{justus.karlsson,yonghao.xu,amanda.berg,leif.haglund}@liu.se
[2] Maxar Intelligence, Linköping, Sweden

Abstract. Multiple studies have performed next-day fire prediction using satellite imagery. Two main satellites are used to detect wildfires: MODIS and VIIRS. Both satellites provide fire mask products, called MOD14 and VNP14, respectively. Studies have used one or the other, but there has been no comparison between them to determine which might be more suitable for next-day fire prediction.

In this paper, we first evaluate how well VIIRS and MODIS data can be used to forecast wildfire spread one day ahead. We find that the model using VIIRS as input and VNP14 as target achieves the best results. Interestingly, the model using MODIS as input and VNP14 as target performs significantly better than using VNP14 as input and MOD14 as target. Next, we discuss why MOD14 might be harder to use for predicting next-day fires. We find that the MOD14 fire mask is highly stochastic and does not correlate with reasonable fire spread patterns. This is detrimental for machine learning tasks, as the model learns irrational patterns. Therefore, we conclude that MOD14 is unsuitable for next-day fire prediction and that VNP14 is a much better option. However, using MODIS input and VNP14 as target, we achieve a significant improvement in predictability. This indicates that an improved fire detection model is possible for MODIS. The full code and dataset are available at https://github.com/justuskarlsson/wildfire-mod14-vnp14.

Keywords: Wildfire · Deep Learning · Remote Sensing

1 Introduction

Wildfires are getting more frequent and severe due to climate change [20,27]. Being able to more accurately predict wildfire behavior can help with wildfire management and response [25,26]. One option for next-day fire prediction is to use satellite imagery. MODIS and VIIRS [11,23] satellites provide twice-daily global coverage and are equipped with both infrared and visible sensors suitable

---

Supplementary Information The online version contains supplementary material available at https://doi.org/10.1007/978-3-031-95918-9_5.

J. Petersen and V. A. Dahl (Eds.): SCIA 2025, LNCS 15726, pp. 61–74, 2025.
https://doi.org/10.1007/978-3-031-95918-9_5

for detecting wildfires. These factors make satellite imagery a very powerful tool for wildfire detection and prediction, compared with, for example, aerial imagery, since the coverage is consistent both spatially and temporally. Furthermore, the wealth of data makes machine learning a good candidate for learning wildfire behavior.

Both MODIS and VIIRS provide products for active fire detection, called MOD14 [11] and VNP14 [23], respectively. VNP14 is based on the same contextual algorithm for detecting fires as MOD14. Several studies have performed next-day fire prediction using either MOD14 [9,10,12,19] or VNP14 [1,2,5,31]. However, to the best of our knowledge, there has not yet been a comparison between the two data sources for the task of next-day fire prediction. Having a better understanding of the pros and cons of the two fire mask products would be very beneficial for future research on wildfire detection.

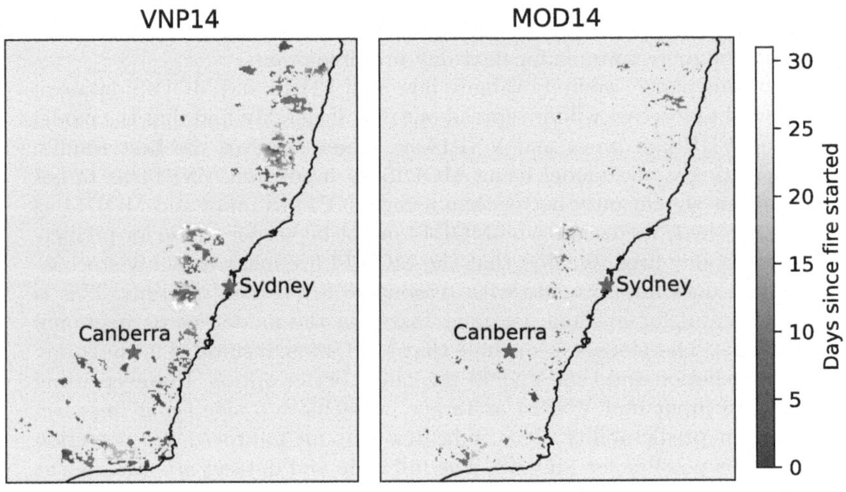

Fig. 1. Fire progression during the wildfire period 2019-10-01 to 2020-01-31 according to VNP14 (left) and MOD14 (right). The color scale shows progression of each individual fire measured from the ignition date of said fire. The area shown, southeast Australia, is a subsection of the study area. Each pixel in the image represents about 1 km.

In this paper, we conduct a two-step study to assess which data source is better suited for next-day fire prediction. First, we compare the predictability of VIIRS and MODIS for next-day fire prediction. To restrict the amount of data for this study, we only gather data from mainland Australia. Data from two time periods are collected. For the training and validation set, the period from 2019-10-01 to 2020-01-31 was used. For the test set, the period from 2023-10-01 to 2024-01-31 was used. Both these periods contained widespread wildfires in Australia. A visualization of fires from the training and validation set is shown in

Fig. 1. We frame the task as a binary-classification image segmentation problem. The comparison is done by training identical deep learning models, where the only difference is the input and target data. Predictability is not conclusive evidence for real-world accuracy. For example, a trivial case where the product only outputs zeros is 100% predictable, but not very useful. Therefore, the second step is to assess why one of them might be easier for a deep learning model to predict. The goal of such an assessment is to reach a conclusion on which data source leads to next-day fire prediction that correlates best with real-world fire spread. To facilitate further research and evaluation on MODIS and VIIRS for next-day fire prediction, the code and data are made public[1]. To summarize, the main contributions of this paper are:

1. A quantitative analysis of the predictability of MOD14 and VNP14 for next-day fire prediction using deep learning models. We then try to explain the results by looking at the fire progression according to the two products.
2. A public dataset and codebase enabling further research comparing MODIS and VIIRS for wildfire prediction.
3. A framework for constructing a global dataset for next-day fire prediction. Even though this study is restricted to Australia, the code and data retrieval is compatible with all regions globally.

2 Related Work

First, we describe the two fire mask products, MOD14 and VNP14. Then we discuss previous work using MOD14 and VNP14 for next-day fire prediction.

2.1 MOD14 and VNP14 Fire Mask Products

In 2006, Justice et al. [11] described their MODIS fire products (including MOD14). The 3.929–3.989 µm band at 1 km spatial resolution was the primary band used for fire detection. A number of other bands at different wavelengths and spatial resolutions were used to detect false alarms, such as sun glint. To perform fire detection, a contextual algorithm was used that first rejected obvious non-fire pixels, then analyzed neighboring pixels to estimate background signal, and finally applied threshold tests comparing brightness temperatures to detect fires while eliminating false detections.

In 2013, Schroeder et al. [23] presented their VNP14 active fire product based on VIIRS data. The algorithm builds on the MOD14 algorithm by Justice et al. [11]. A major difference compared with MOD14 is that the VNP14 fire detection only uses bands at a spatial resolution of 375 m. Schroeder et al. assert that this leads to "improved consistency of fire perimeter delineation for biomass burning lasting multiple days", compared with MOD14. The 3.550–3.930 µm band was the primary band used for fire detection in VNP14. The contextual algorithm

[1] https://github.com/justuskarlsson/wildfire-mod14-vnp14.

used in VNP14 is similar to the one used in MOD14 but adapted to the different band characteristics of the VIIRS sensor.

In addition to differences in spatial resolution and algorithm design, the observation times of the platforms also differ, which can influence fire detection capabilities. Specifically, the Terra platform used for MOD14 captures images at 10:30 and 22:30 local time. The S-NPP/VIIRS platform, used for VNP14, captures images at 13:30 and 01:30 local time.

2.2 Previous Work Using MOD14

Huot et al. [10] used Google Earth Engine (GEE) [7] to retrieve MOD14 fire masks and auxiliary data over the United States. They make no mention of VNP14, however, at the time of writing their publication, VNP14 was not yet available on GEE. They evaluate multiple deep learning models on the next-day fire prediction task and frame it as an image segmentation problem. In a subsequent publication, Huot et al. [9] used a similar approach, but again, they make no mention of potentially using VNP14. Their analysis was limited to data accessible through Google Earth Engine. Prapas and Kondylatos et al. [19] used MOD14 data together with historical burned areas from the European Forest Fire Information System (EFFIS). Their study is restricted to the Eastern Mediterranean region. In contrast with [9,10], they focused specifically on predicting fire *ignition*. The problem was also set up as image classification as opposed to image segmentation, since only the pixel in the center of each image was evaluated. In a subsequent publication, Prapas and Kondylatos et al. [12] used a similar approach as [19]. VNP14 is not mentioned as a potential fire mask source here either.

In summary, the papers using MOD14 make no mention of VNP14.

2.3 Previous Work Using VNP14

Ali et al. [1] present a dataset for next-day fire spread prediction. The dataset uses VNP14 for their fire masks. The two main reasons they mention for using VNP14 instead of MOD14 are that VNP14 has a higher spatial resolution and that it has a higher number of fire detections. They make no direct evaluation of the quality between MOD14 and VNP14. Gerard et al. [5] present the dataset *WildfireSpreadTS* for next-day fire spread prediction. No thorough discussion of the choice between MOD14 and VNP14 is made, but it seems that increased spatial resolution was the main reason for using VNP14. Zhao et al. [31] present a multi-task dataset called *TS-SatFire*. They label one of the tasks as "wildfire progression prediction", which fits into our label of next-day fire prediction. The fire masks used for the next-day fire prediction tasks are from VNP14. The increased spatial resolution is cited as the reason for choosing VNP14 over MOD14. Ali et al. [2] present a dataset for next-day wildfire spread prediction. The fire masks used are from VNP14, but no discussion of the choice between MOD14 and VNP14 is made.

In summary, the papers using VNP14 primarily mention the increased spatial resolution as the reason for using VNP14 over MOD14.

2.4 Comparison of MOD14 and VNP14

We can see a mix of MOD14 and VNP14 in the literature, but there has not been a definitive comparison of which data source is better for predicting fire spread. The motivation for using one or the other is typically not well formulated. One stated argument for using VIIRS is the higher spatial resolution. This argument makes no judgment of the quality of the data, even though it might be natural to assume that higher spatial resolution correlates with better quality. Studies that use MODIS typically do not motivate its use in comparison with VIIRS. One reason might be data availability. In remote sensing, it is common to use Google Earth Engine (GEE) [7] for data retrieval. Although both VNP14 and MOD14 are available on GEE, VNP14 is only available from 2023-09-03, whereas MOD14 is available from 2000-02-24. Using NASA's own data portal (LP-DAAC), VNP14 can be retrieved back to 2012-01-17.

3 Method

First, the method of selecting, retrieving, and processing the data is described. We then list some choices and their motivations in relation to the predictability experiments. The underpinning motivation is to maximize the fairness of the comparison between VIIRS and MODIS. Finally, we briefly describe how the qualitative assessment was set up.

Table 1. List of data products retrieved via *earthaccess*. The products are divided by their use, sensor name, and source. LAADS and LP-DAAC are NASA data distribution services.

Data Products		
Use	Product	Data portal
Geolocation	MOD03 [18]	LAADS
Geolocation	VNP03IMG [13,14,28,29]	LAADS
Geolocation	VNP03MOD [13,14,28,29]	LAADS
1 km bands	MOD021KM [15]	LAADS
500 m bands	MOD02HKM [17]	LAADS
250 m bands	MOD02QKM [16]	LAADS
375 m bands	VNP02IMG [13,14,28,29]	LAADS
750 m bands	VNP02MOD [13,14,28,29]	LAADS
Fire Mask	MOD14 [6]	LP-DAAC
Fire Mask	VNP14IMG [22]	LP-DAAC

3.1 Data Retrieval, Selection and Processing

We focus our study on Australia, a country that has experienced severe wildfires [30], while also limiting the volume of data to be retrieved. For the training and validation dataset, we use the period from 2019-10-01 to 2020-01-31. The test dataset is from 2023-10-01 to 2024-01-31. These two periods contained a large number of fire detections in both MOD14 and VNP14. The period from October to January is part of the eastern Australian wildfire season [20], which also motivated the specific months used (Fig. 2).

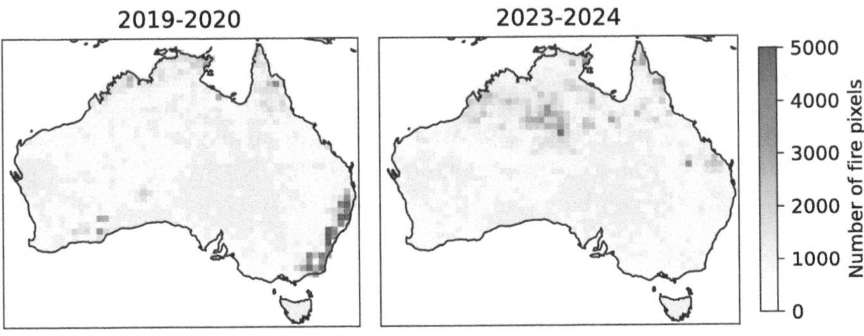

Fig. 2. Fire detections (VNP14) in Australia during the two study periods. The left image shows fire detections from 2019-10-01 to 2020-01-31, which is used for training and validation. The right image shows fire detections from 2023-10-01 to 2024-01-31, which is used for testing. Both periods show significant wildfire activity, albeit in slightly different areas.

The training and validation dataset is split uniformly based on x and y grid cells by 75% and 25% respectively. In practice, this is done by putting grid cells where both x and y grid indices are even numbers into the validation set, and the rest in the training set. This is to ensure no overlap between the training and validation set that the model might overfit to. The test dataset is instead separated from the training/validation dataset temporally. This split results in 2,654 training samples, 753 validation samples, and 4,710 test samples.

As input data, we use the raw bands from MODIS and VIIRS. Furthermore, we also include weather and drought data from the input day. This data is from ERA5 [4] and KBDI [24] respectively. The raw bands data (Level 1 products) for MODIS and VIIRS is available from LAADS. The fire mask products are available from LP-DAAC. The Python package *earthaccess* [3] is used to gather data from both data portals. The package allows for retrieving images based on the desired date range and area of interest. The data products retrieved via earthaccess are shown in Table 1. The weather[2] and drought[3] data are retrieved

[2] ECMWF/ERA5_LAND/DAILY_AGGR.
[3] UTOKYO/WTLAB/KBDI/v1.

from Google Earth Engine (GEE). Similarly to *earthaccess*, this service also allows for easy filtering according to the desired date range and area of interest.

Table 2. Overview of MODIS bands and their resolutions. The bands are divided into reflective and emissive types.

MODIS Bands		
Resolution	Band	Type
250 m	B1–B2	Reflective
500 m	B3–B7	Reflective
1 km	B8–B19, B26	Reflective
1 km	B20–B25, B27–B36	Emissive

Table 3. Overview of VIIRS bands and their resolutions. The bands are divided into reflective and emissive types.

VIIRS Bands		
Resolution	Band	Type
375 m	I1–I3	Reflective
375 m	I4–I5	Emissive
750 m	M01–M06, M09	Reflective
750 m	M07–M08, M10–M16	Emissive

MODIS and VIIRS images have different projections that vary from day to day. For the prediction task, we want each data sample to relate to the same area from the current day to the next day. For this purpose, we need a common grid from which to retrieve samples. A simple geodetic grid was chosen, where each cell corresponds to $0.75 \times 0.75°$ longitude and latitude. To avoid storing an excessive amount of data, the images (shown in Table 2 and 3) are resampled to a patch size that roughly matches their corresponding geodetic resolution at the equator. This corresponds to 64 pixels for 1 km data, 96 pixels for 750 m, and 192 pixels for 375 m. The geolocation data (Table 1) provides geodetic coordinates for each pixel, which are converted to floating-point coordinates within each patch. Integer pixel values are avoided at this stage to preserve information for the final resampling step. For raw sensor bands, we use an inverse distance resampling with a kernel size of 3×3. For the fire mask data, we use nearest neighbor resampling with a kernel size of 3×3. The kernel size in this case refers to the maximum distance we allow a pixel to be. If no pixel is found within the region, the pixel is set to no data. The weather and drought data were already in a geodetic projection. Therefore, we simply patchify them and insert each patch into the corresponding cell (in the geodetic grid).

3.2 Experiments

The focus of the predictability study is wildfire spread, and not wildfire ignition. Therefore, we only select cells and dates where there is at least one fire pixel in both the MODIS and VIIRS data for the current day. The requirement that both datasets have a fire pixel ensures that the two models are trained on the same set of samples.

The input data is upsampled into a shared resolution corresponding to 192 × 192 pixels per patch for both MODIS and VIIRS models. This ensures that the VIIRS model can benefit from its higher resolution, while still allowing the MODIS model to potentially benefit from increased model capacity through upsampling. Both MODIS and VIIRS sensors capture data during both day and night, with only emissive bands included at night. The input features are then concatenated along the channel dimension. This results in 65 input channels for the MODIS model and 43 input channels for the VIIRS model.

The architecture used for both models is a U-Net [21] with a ResNet-50 backbone [8]. This choice was made to keep the architecture simple and use proven technologies widely used in the field. The output is of the same spatial dimensions as the input, 192 × 192 pixels. However, to allow a fair comparison between the two fire products, the target fire masks are downsampled to the resolution of the MOD14 data (64 pixels). The loss and metrics are, therefore, calculated on the 64 × 64 resolution. The target fire mask is a max aggregation of the daytime and nighttime fire mask for the next day. The loss function is binary cross entropy with a positive weight w to account for the imbalance in the labels:

$$L = -\frac{1}{N}\sum_{i=1}^{N}\left(wy_i \log(\hat{y}_i) + (1 - y_i)\log(1 - \hat{y}_i)\right), \tag{1}$$

where y_i is the true label, $\hat{y}_i$ is the predicted probability, and N is the number of pixels. The positive weight, $w = 3$, was found through an ablation study (see supplementary material). The testing is done with the checkpoint that achieves the highest intersection over union (IoU) on the validation set. The IoU is performed pixel-wise across all samples.

To provide further basis for the suitability of one fire product for the prediction task, we also include a qualitative assessment of the MOD14 and VNP14 fire masks. The purpose of this exercise is to reason about why one of the datasets might perform better than the other for the prediction task. To do this, we pick some of the larger fires that occurred in the dataset. Then, for each of the fires, we plot the progression of the fire according to both VNP14 and MOD14. We then reason about the relative realism of the fire progression of MOD14 versus VNP14.

4 Results and Discussion

We first present results for the persistence baselines, representing a lower bound for the expected performance of the models. We then present results for the two models trained exclusively on either MODIS+MOD14 or VIIRS+VNP14 data. To investigate the underlying reasons for the performance difference, we also present results of models that use a mix of MODIS, MOD14, VIIRS, and VNP14 data for the three different data slots in the experiment: input, training, and evaluation. To increase the confidence of the results, 5 independent runs are done and all metrics are presented with their means and standard deviations. We then compare the fire progression between MOD14 and VNP14 for a select number of individual fires and find large differences in the apparent quality of the fire masks. Finally, we discuss the results and possible implications for the use of MODIS and VIIRS data for wildfire prediction.

4.1 Predictability

We establish a *persistence baseline* for the prediction task by measuring the performance of assuming the next-day fires will equal the current day fires. The results for this baseline are reported in Table 4 for both MOD14 and VNP14. This baseline is also a measurement of the temporal consistency of the products. We can clearly see that VNP14 is more consistent than MOD14.

Table 4. Evaluation of the persistence baselines on the test set.

Product	F1 Score (%)	IoU (%)	Precision (%)	Recall (%)
MOD14	6.56	3.39	5.75	7.64
VNP14	19.62	10.87	18.45	20.94

We train two models that are fed with two different data sources: MODIS (with MOD14 fire masks) and VIIRS (with VNP14 fire masks). The results are shown in Table 5.

Table 5. Test results for the two models trained exclusively on either MODIS/MOD14 or VIIRS/VNP14 data. VIIRS/VNP14 is clearly superior.

Input	Training Target	Evaluation Target	F1 Score (%)	IoU (%)	Precision (%)	Recall (%)
MODIS	MOD14	MOD14	11.96 ± 1.52	6.37 ± 0.86	13.89 ± 0.74	10.82 ± 2.53
VIIRS	VNP14	VNP14	28.57 ± 0.62	16.67 ± 0.42	28.02 ± 1.30	29.28 ± 1.68

In Fig. 3, we can see some sample predictions for the two models. Both of these models beat their respective persistence baselines established in Table 4 by a large margin. We can see that the model trained with VIIRS data performs much better than the MODIS model. There could be multiple reasons for this:

1. The high-resolution VIIRS raw band data as input to the model gives superior prediction performance.
2. The VNP14 fire masks are more consistent and easier to predict.

We, therefore, also train two models where input (MODIS or VIIRS bands) and target (MOD14 or VNP14 fire masks) are swapped. The results are shown in Table 6. We can see that the main problem with the MODIS model was in fact the MOD14 fire masks, and not the MODIS input data. Furthermore, predicting VNP14 from MODIS data is a harder task than the other way around, since VIIRS captures data approximately 2 h later. Using VIIRS input data helps with predicting both VNP14 and MOD14, although the increase in performance is not as significant.

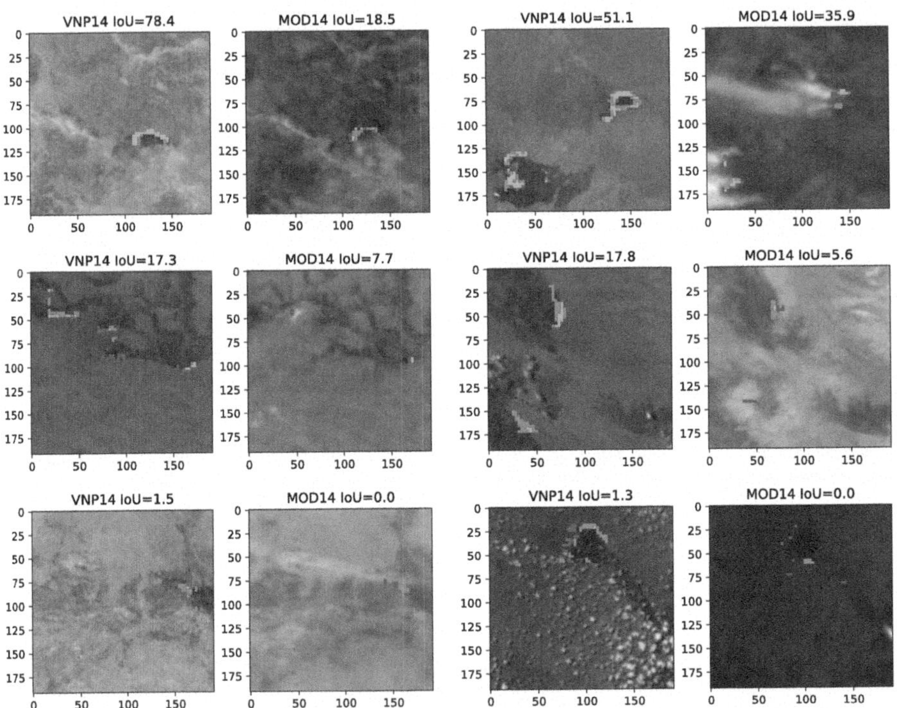

Fig. 3. Sample predictions from the two models in Table 5. Areas filled in green indicate true positives (correctly predicted fires), red indicates false positives (incorrectly predicted fires), and blue indicates false negatives (missed fires). The first row shows good predictions, the second row shows median predictions, and the last row shows bad predictions.

We have thus narrowed down that the biggest detractor from performance is the MOD14 fire masks. Finally, we want to evaluate how much MOD14 as a training target hinders learning. We perform this comparison by training four additional models where the training and evaluation targets are different. The results are shown in Table 7. If we compare row 1 in Table 5 with row 2 in Table 7, we can see that the performance is lower when training on MOD14 instead of VNP14, *even when testing* on MOD14 data. By training on VNP14 instead of MOD14, we get a mean IoU of 8.17% when testing on MOD14, compared with 6.37% when training *and testing* on MOD14 data. Put another way, for the machine learning model evaluated, VNP14 is a better prior for new MOD14 data than MOD14 data itself.

Table 6. Test results where we swap the input used for each model. We can see that the problem lies in the MOD14 fire masks, and not the MODIS data.

Input	Training Target	Evaluation Target	F1 Score (%)	IoU (%)	Precision (%)	Recall (%)
MODIS	VNP14	VNP14	27.16 ± 0.63	15.72 ± 0.42	25.73 ± 0.93	28.86 ± 1.53
VIIRS	MOD14	MOD14	16.09 ± 0.59	8.75 ± 0.35	17.53 ± 0.92	15.05 ± 1.74

Table 7. Measuring the impact on learning by swapping training and testing data sources.

Input	Training Target	Evaluation Target	F1 Score (%)	IoU (%)	Precision (%)	Recall (%)
MODIS	MOD14	VNP14	17.67 ± 2.81	9.72 ± 1.70	23.03 ± 1.26	14.98 ± 4.62
MODIS	VNP14	MOD14	15.11 ± 0.91	8.17 ± 0.53	11.96 ± 0.62	21.40 ± 5.10
VIIRS	VNP14	MOD14	17.03 ± 1.18	9.31 ± 0.71	16.79 ± 0.73	17.56 ± 2.91
VIIRS	MOD14	VNP14	16.66 ± 3.13	9.12 ± 1.86	27.62 ± 1.19	12.24 ± 3.27

4.2 Qualitative Assessment

In Fig. 4, we can see a comparison between the progression of VNP14 and MOD14 fires. A higher-level comparison can be seen in Fig. 1. It is clear that VNP14 produces a more cohesive and uniform fire spread. MOD14 seems to contain stochastic detections of fire. It is reasonable to assume that the MOD14 models struggle with this stochastic nature of detections. The input data does not support why the fire spreads in seemingly random and erratic patterns, and so the learned patterns will not generalize to the test set.

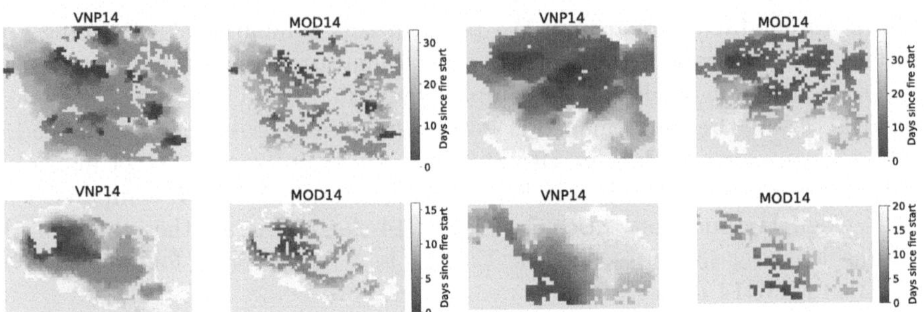

Fig. 4. Qualitative assessment of the wildfire spread for both products. The color scale shows progression of each individual fire measured from the ignition date of said fire.

4.3 Discussion

The results in Table 6 show that the problem does not lie in the quality of the raw bands. When using MODIS input bands but VIIRS target, the model performs much better than the other way around. This supports the conclusion that MOD14 provides an unreliable target particularly ill-suited for the task of next-day prediction using machine learning. A further indicator of this is the qualitative results in Fig. 4. The target data forces the model to learn irrational patterns, and the model is unable to generalize well to the test set. While stochastic detections might not be detrimental for all applications, they are a poor match for this particular task.

As stated previously, MOD14 and VNP14 rely on the same algorithmic basis. Why do we then see such a difference in performance? There is a significant difference in spatial resolution for the infrared bands used to detect temperature anomalies. VNP14 has a spatial resolution of 375 m, while MOD14 has a resolution of 1 km. However, as we saw in Table 6, we were able to achieve good results using MODIS input and VNP14 target. The increased capability of a deep learning model might be able to compensate for the lower resolution of the MODIS data.

5 Conclusion

In this study, we systematically compare the predictability of MOD14 and VNP14 for next-day fire prediction using deep learning models. Our experiments show that the MOD14 fire mask is highly unpredictable, while VNP14 proves to be better suited for this task. The unreliability of MOD14 is not explained by the quality of the raw bands of MODIS, since using them as input and trying to predict VIIRS had much better results. The relative success of using MODIS input and VNP14 target indicates that an improved fire detection model using deep learning is possible for MODIS, which we plan to explore in future work.

Acknowledgment. This work was supported in part by the Excellence Center at Linköping-Lund in Information Technology (ELLIIT) Researcher Funding, the Zenith Research Program, and the Wallenberg Artificial Intelligence, Autonomous Systems and Software Program (WASP), funded by Knut and Alice Wallenberg Foundation. The computational resources were provided by the National Academic Infrastructure for Supercomputing in Sweden (NAISS) at C3SE, and by the Berzelius resource, provided by the Knut and Alice Wallenberg Foundation at the National Supercomputer Centre.

References

1. Ali, S.H., Goel, A., Singirikonda, A., Khan, A.Y., Xiao, T.: Towards a comprehensive dataset for next-day wildfire prediction. In: Proceedings - 2022 IEEE 22nd International Conference on Software Quality, Reliability and Security Companion, QRS-C 2022, pp. 593–598. Institute of Electrical and Electronics Engineers Inc. (2022)
2. Ali, S.H., Zhang, S., Mhatre, S., Xiao, T.: Advancing wildfire predictive models: a novel dataset for next-day wildfire spread. In: 2024 Conference on AI, Science, Engineering, and Technology (AIxSET), pp. 69–76 (2024). https://doi.org/10.1109/AIxSET62544.2024.00015
3. Barrett, A., et al.: earthaccess: python library for NASA earthdata APIs (2023). https://github.com/nsidc/earthaccess
4. (C3S), C.C.C.S.: Era5: fifth generation of ecmwf atmospheric reanalyses of the global climate (2017). https://cds.climate.copernicus.eu/cdsapp#!/home
5. Gerard, S., Zhao, Y., Sullivan, J.: WildfireSpreadTS: a dataset of multi-modal time series for wildfire spread prediction. In: Advances in Neural Information Processing Systems 36 - 37th Conference on Neural Information Processing Systems, NeurIPS 2023, New Orleans, United States of America (2023)
6. Giglio, L., Justice, C.: MOD14 MODIS/Terra thermal anomalies/fire 5-min L2 swath 1km V006 [data set] (2015). https://doi.org/10.5067/MODIS/MOD14.006. Accessed 17 Feb 2025
7. Gorelick, N., Hancher, M., Dixon, M., Ilyushchenko, S., Thau, D., Moore, R.: Google earth engine: planetary-scale geospatial analysis for everyone. Remote Sens. Environ. **202**, 18–27 (2017)
8. He, K., Zhang, X., Ren, S., Sun, J.: Deep residual learning for image recognition. Proceedings of the IEEE Computer Society Conference on Computer Vision and Pattern Recognition, vol. 2016-December, pp. 770–778 (2016)
9. Huot, F., Hu, R.L., Goyal, N., Sankar, T., Ihme, M., Chen, Y.F.: Next day wildfire spread: a machine learning dataset to predict wildfire spreading from remote-sensing data. IEEE Trans. Geosci. Remote Sens. **60**, 1–13 (2022). https://doi.org/10.1109/TGRS.2022.3192974
10. Huot, F., et al.: Deep learning models for predicting wildfires from historical remote-sensing data (2021). https://arxiv.org/abs/2010.07445
11. Justice, C., et al.: Modis fire products algorithm technical background document. Technical report, MODIS Science Team (2006). https://lpdaac.usgs.gov/documents/87/MOD14_ATBD.pdf. eOS ID# 2741
12. Kondylatos, S., et al.: Wildfire danger prediction and understanding with deep learning. Geophys. Res. Lett. **49** (2022). https://doi.org/10.1029/2022GL099368

13. Lin, G., Wolfe, R.E., Dellomo, J.J., Tan, B., Zhang, P.: SNPP and NOAA-20 VIIRS on-orbit geolocation trending and improvements. In: Earth Observing Systems XXV, vol. 11501, pp. 215–225. SPIE (2020)https://doi.org/10.1117/12.2569148
14. Lin, G., Wolfe, R.E., Zhang, P., Dellomo, J.J., Tan, B.: Ten years of VIIRS on-orbit geolocation calibration and performance. Remote Sens. **14**, 4212 (2022). https://doi.org/10.3390/rs14174212
15. MODIS Characterization Support Team (MCST): Modis 1km calibrated radiances product (2017). https://doi.org/10.5067/MODIS/MOD021KM.061
16. MODIS Characterization Support Team (MCST): Modis 250m calibrated radiances product (2017). https://doi.org/10.5067/MODIS/MOD02QKM.061
17. MODIS Characterization Support Team (MCST): Modis 500m calibrated radiance product (2017). https://doi.org/10.5067/MODIS/MOD02HKM.061
18. MODIS Characterization Support Team (MCST): Modis geolocation fields product (2017). https://doi.org/10.5067/MODIS/MOD03.061
19. Prapas, I., et al.: Deep learning methods for daily wildfire danger forecasting (2021). https://arxiv.org/abs/2111.02736
20. Richardson, D., et al.: Global increase in wildfire potential from compound fire weather and drought. NPJ Clim. Atmos. Sci. **5**(1), 23 (2022)
21. Ronneberger, O., Fischer, P., Brox, T.: U-net: convolutional networks for biomedical image segmentation. In: Lecture Notes in Computer Science (including subseries Lecture Notes in Artificial Intelligence and Lecture Notes in Bioinformatics), vol. 9351, pp. 234–241 (2015)
22. Schroeder, W., Giglio, L.: VIIRS/NPP thermal anomalies/fire 6-min L2 swath 750m V001 [data set] (2017). https://doi.org/10.5067/VIIRS/VNP14.001. Accessed 17 Feb 2025
23. Schroeder, W., Oliva, P., Giglio, L., Csiszar, I.A.: The new VIIRS 375m active fire detection data product: algorithm description and initial assessment. Remote Sens. Environ. **143**, 85–96 (2014)
24. Takeuchi, W., et al.: Near-real time meteorological drought monitoring and early warning system for croplands in Asia. In: 36th Asian conference on remote sensing (ACRS), Manila, Philippines (2015)
25. Thompson, M.P., et al.: Application of wildfire risk assessment results to wildfire response planning in the southern sierra Nevada, California, USA. Forests **7**(3), 64 (2016)
26. Thompson, M.P., et al.: Risk management and analytics in wildfire response. Curr. For. Rep. **5**, 226–239 (2019)
27. Westerling, A.L., Hidalgo, H.G., Cayan, D.R., Swetnam, T.W.: Warming and earlier spring increase western U.S. forest wildfire activity. Science (New York N.Y.) **313**, 940–943 (2006)
28. Wolfe, R.E., Lin, G., Nishihama, M., Tewari, K.P., Montano, E.: NPP VIIRS early on-orbit geometric performance. In: Proceedings of the Earth Observing Systems XVII, vol. 8510, p. 851013. SPIE, Bellingham (2013)
29. Wolfe, R.E., Lin, G., Nishihama, M., Tewari, K.P., Tilton, J.C., Isaacman, A.R.: Suomi NPP VIIRS prelaunch and on-orbit geometric calibration and characterization. J. Geophys. Res.: Atmos. **118**, 11508–11521 (2013)
30. Xu, Y., Berg, A., Haglund, L.: Sen2Fire: a challenging benchmark dataset for wildfire detection using sentinel data. In: IGARSS (2024)
31. Zhao, Y., Gerard, S., Ban, Y.: Ts-SatFire: a multi-task satellite image time-series dataset for wildfire detection and prediction (2024). https://arxiv.org/abs/2412.11555

Image and Video Processing, Analysis, and Understanding

A Spectral-Preserving Zero-Shot Technique for Hyperspectral Pansharpening

Giuseppe Guarino[1], Matteo Ciotola[1], and Giuseppe Scarpa[2(✉)]

[1] DIETI Department, Federico II University, 80125 Naples, Italy
[2] Engineering Department, Parthenope University, 80143 Naples, Italy
giuseppe.scarpa@uniparthenope.it

Abstract. Hyperspectral pansharpening is a key fusion process aimed to enhance the spatial resolution of a hyperspectral image benefiting of a simultaneously acquired higher resolution panchromatic band. Promising approaches, which leverage on deep learning models, are optimized in unsupervised manners using full-resolution real datasets with no reference. The lack of ground-truths imposes the use of a two-fold composite loss to ensure both spectral and spatial quality according to some trade-off criterion. The common solution, however, is agnostic with respect to the spectral-spatial balance, giving the same relevance to both properties. Indeed, the spectral information is the most distinctive feature to preserve in any application based on hyperspectral data. Motivated by this last observation, in this work we propose a spectral-preserving fusion approach that puts the spectral quality ahead thanks to a suitably defined unsupervised loss. Our experiments show very promising results with really low spectral distortions, on average and uniformly along the spectral axis, ensuring good sharpening levels too.

Keywords: Convolutional neural networks · image fusion · multiresolution · hyperspectral images · remote sensing

1 Introduction

Remotely sensed hyperspectral (HS) images are key products to enable applications in diverse domains such as geology [20], forestry [9], agriculture [17], snow and ice [23], atmosphere [11], to mention a few. Given their extremely large number of spectral bands, technological constraints impose a relatively lower spatial resolution to the HS acquisitions compared to other products, *e.g.*, multispectral (MS) images. An option to overcome resolution limits at a reasonable cost is to equip the flying system with an additional camera for sensing, synchronously with the HS image, a single higher resolution panchromatic (PAN) band. Then, leveraging a fusion process known as pansharpening, applied by either data providers or end users, the PAN-HS pair is merged giving raise to a single high resolution HS datacube. Examples of satellite missions that feature

J. Petersen and V. A. Dahl (Eds.): SCIA 2025, LNCS 15726, pp. 77–91, 2025.
https://doi.org/10.1007/978-3-031-95918-9_6

the joint acquisition of PAN and HS images are PRISMA (*PRecursore IperSpettrale della Missione Applicativa*), by the Italian Space Agency, ASI, or EO-1/ALI (Earth Observing-1/Advanced Land Imaging) by NASA. In both cases the spatial resolution on the ground is 30 m for the HS component. On the contrary, while the former provides a PAN image at 5 m resolution (resolution ratio $R = 6$), the latter features an halved resolution ratio ($R = 3$) with 10 m resolution PAN images. Needless to say, the PRISMA case is certainly more challenging due to the higher PAN-HS resolution gap.

Pansharpening has been object of intense research in the last decades, especially for the PAN-MS fusion case, where a limited number of spectral bands (from 4 to 8) in the visible to near-infrared spectrum are concerned. Extensive surveys concerning the MS case, in fact, can be found in the literature [10,30]. Recently, the attention has moved toward the more complex HS case as testified by a challenge on HS pansharpening [31] and by a large number of new research works that have been carefully reviewed in [7]. On one side, [7] reviews model-based solutions suitably adapted to the HS case, *e.g.* Component Substitution (CS) [1,5], Multiresolution Analysis (MRA) [3,24], Model-Based Optimization (MBO) [25,29], to mention a few. Besides, data-driven solutions, *e.g.* Deep Learning (DL) ones such as [4,12] are also discussed and analysed in detail. Needless to say, the last decade has registered a sharp turn toward machine/deep learning solutions for pansharpening since the pioneering work of Masi *et al.* [21] in 2016. The early works referred to the MS case [34] [19,27] have been soon followed by many HS-specific ones [4,12–16,35]. The interested Reader is referred to the recent work [7], for a critical review of the state-of-the-art methods and for accessing to tools (implementations of methods and accuracy routines) and reference datasets[1] for a deeper study. Most of the current DL approaches use supervised learning and perform training at reduced resolution using a subsampled version of the original data so that the full-resolution HS datacube can play as ground truth (GT). Recently, to avoid the quality loss induced by subsampling, a paradigm shift towards full-resolution training is taking place for both MS [6,8,28] and HS [12,13,22] pansharpening, with novel self-supervised/unsupervised learning strategies and suitably defined loss functions.

By definition, pansharpening is a mixing of spectral (coming form the MS or HS component) and spatial (coming form the PAN image) features that unavoidably give raise to a trade-off between the two. In the MS case, spectral and spatial qualities are usually equally weighted, In contrast, also in light of the typical applications based on HS imagery, in the HS case the spectral one is certainly the most distinctive and precious feature that should be guaranteed, while sharpness (spatial feature), although being an important characteristic, remains subordinated to spectral integrity. Unfortunately, the current literature on HS pansharpening shows that most solutions are agnostic with respect to the spectral-spatial quality balance. Moreover, the figures employed to push down

[1] As a by-product, [7] provides a toolbox with methods, accuracy indexes and datasets at https://github.com/matciotola/hyperspectral_pansharpening_toolbox.

the spectral distortion are averages along the wavelength axis. As a consequence, large differences in spectral quality hidden into an average over hundreds of bands can be obtained. Such non-uniform responses are also favored by non-uniform signal-to-noise ratios in the HS stack.

Among the many solutions appeared in the literature, the recently proposed R-PNN [13] (Rolling Pansharpening Neural Network) method seems to be particularly suited to face the above raised issue. R-PNN performs a sequential band-wise pansharpening with a full-resolution unsupervised tuning at inference time, where at each step/band the model tuned on the preceding band is further tuned on the current one. In different words, this process can be meant as a cross-band transfer learning. In particular, its band-wise characteristic allows, in principle, to act individually on each different pansharpened spectral band to push its spectral loss to a desired target level common to all bands. In this work, we therefore propose an upgraded version of R-PNN where, thanks to a suitable optimization scheme, we ensure uniform and robust spectral accuracy levels.

In the reminder of the paper, after reviewing the baseline approach and discussing the limitations of the state-of-the-art methods in Sect. 2, we describe the proposed solution in Sect. 3 and then present the experimental results in Sect. 4, before concluding with final remarks and future work in Sect. 5

2 Related Work

Let us begin with the problem statement. Pansharpening is a process that takes two input images, say $\mathbf{H} \in \mathbb{R}^{w \times h \times B}$, and $\mathbf{P} \in \mathbb{R}^{W \times H}$, providing the pansharpened version $\widehat{\mathbf{H}} \in \mathbb{R}^{W \times H \times B}$ of $\mathbf{H}$. The spatial sizes of $\mathbf{H}$ and $\mathbf{P}$ are in a fixed resolution ratio: $R = W/w = H/h$. The output image $\widehat{\mathbf{H}}$ is expected to be consistent in terms of spectral signature with the input component $\mathbf{H}$, or with a resized version $\widetilde{\mathbf{H}} \in \mathbb{R}^{W \times H \times B}$ of it, obtained with some approximation of the ideal interpolator. Due to the lack of details in $\widetilde{\mathbf{H}}$ or $\mathbf{H}$, with respect to $\widehat{\mathbf{H}}$, such consistency has to be meant as limited to the low spatial frequency range. Besides, the additional details conveyed into $\widehat{\mathbf{H}}$ have to be coherent with the structural component (high frequency) of $\mathbf{P}$. The definition of these properties in a unique and universally accepted way is not obvious at all and it is actually a quite debated problem [2,7,26]. The core of the problem is the lack of real datasets with associated ground-truths (GTs), *i.e.*, ideal pansharpened images.

A widespread trick to circumvent this limitation, introduced the first time in 2016 by Masi *et al.* [21] in the context of MS pansharpening based on convolutional neural networks (CNN), is to derive synthetic reduced-resolution datasets from full-resolution real ones, thanks to a suitably defined resolution downgrade process already used in the past for quality assessment purposes (Wald's protocol [32]). Such process consists in decimating (carefully selecting the cutoff frequency) both $\mathbf{H}$ and $\mathbf{P}$ at rate R, so that the real $\mathbf{H}$ can play as GT while the decimated images are used as (synthetic) inputs. By doing so, any CNN or other trainable network suitably shaped for pansharpening can be trained in a supervised manner using any distance (*e.g.*, ℓ_1- or ℓ_2-norm) between the pansharpened images and the corresponding GT as loss function.

Unfortunately, due to an unavoidable mismatch between reduced-resolution datasets and real datasets, the networks trained on the former can reach excellent scores in the same reduced-resolution context but registering a consistent performance drop when tested on the latter. This has motivated a paradigm shift in recent years [8,13,22,28] that has moved the attention toward unsupervised training schemes so that no resolution downgrade is required. Unsupervised training is achieved by means of suitable loss functions capable to account for the consistency between the pansharpened images and the two input components to be fused [6].

2.1 About Spectral Consistency in Hypespectral Pansharpening

The spectral signature of ground elements is the most distinctive feature that HS data can provide and that HS users look for. In light of this, super-resolution and sharpening are certainly welcome as long as spectral fidelity is guaranteed. The critical review proposed in [7] provides insights about current trends in HS pansharpening. In particular, the related toolbox allows us to carry on a deeper analysis of the spectral behavior of the state-of-the-art methods. In Fig. 1(a) we show the spectral distortion of the best ranking methods of the benchmark [7] for a sample test dataset as function of the bands wavelength (indexed). For each fixed band, it is reported the spatial average $\mu_{\mathbf{E}_b}$ of the absolute difference $\mathbf{E}_b$ between the subsampled (low-pass filtered and decimated) version $\widehat{\mathbf{H}}_{b,\downarrow}$ of the pansharpened band $\widehat{\mathbf{H}}_b$ and the corresponding input HS band $\mathbf{H}_b$:

$$\mu_{\mathbf{E}_b} = \langle \mathbf{E}_b \rangle, \quad \text{with} \quad \mathbf{E}_b \triangleq |\widehat{\mathbf{H}}_{b,\downarrow} - \mathbf{H}_b| \in \mathbb{R}^{w \times h}, \tag{1}$$

being $\langle \cdot \rangle$ the spatial average.

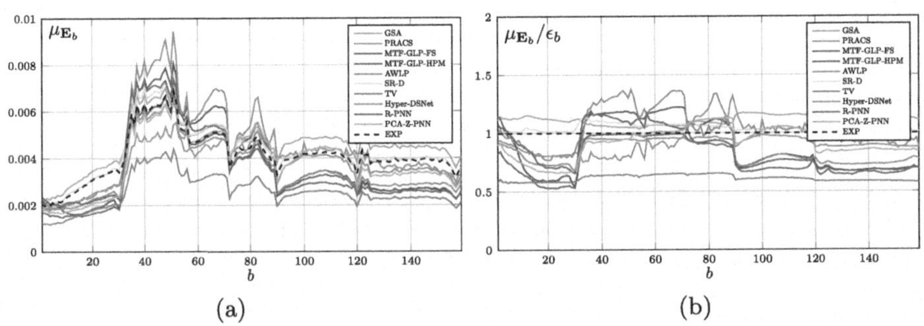

Fig. 1. Band-wise spectral distortion $\mu_{\mathbf{E}_b}$ for the PRISMA image of Macuspana (MEX) [7] (a) and corresponding normalized curve $\mu_{\mathbf{E}_b}/\epsilon_b$ (b).

The first due remark is that these curves only account for spectral accuracy but not for the sharpening level (spatial quality), which require other measurements, therefore they are only partial quality indicators. However, the visual inspection reveals a strong relationship between distortion and wavelength. For example, on the first ~30 bands (visible spectrum) all methods present a relatively low distortion, contrarily to the next ~20 ones (infrared) where the distortion grows remarkably. These differences are in part due to intrinsic features of the spectral dynamics of the dataset (*e.g.*, wavelength-dependent signal-to-noise ratios) and in part due to the specific action of the pansharpening solutions that may force sharpening in different ways along the spectral axis. For example, forcing too much the sharpening of bands that are far from the visible spectrum (the range where the PAN image is acquired). To better appreciate the differences between intrinsic and extrinsic distortion contributions, we can normalize the distortion with respect to the distortion ϵ_b reported by the approximation $\widetilde{\mathbf{H}}_b$ (method EXP in the toolbox) of the "ideal" interpolation:

$$\epsilon_b \triangleq \left\langle \left| \widetilde{\mathbf{H}}_{b,\downarrow} - \mathbf{H}_b \right| \right\rangle \tag{2}$$

As it can be seen from (a), the interpolator distortion (dashed line) seems to follow that of the pansharpening methods although no sharpening is performed, confirming that the above discussed spectral disparities are partly due to the HS imaging characteristics themselves. Therefore, on the basis of this last observation, we can normalize the distortion with respect to the interpolator (b) to roughly highlight the spectral unbalance strictly due to the methods. In this new perspective, the plots in (b) highlight that most methods suffer a non-uniform distortion in the spectral dimension. As "virtuous" example we can mention TV whose (normalized) spectral response look quite uniform and low on average, although its spatial behavior (sharpness) is not equally satisfying [7].

2.2 Rolling Pansharpening Neural Network

The Rolling Pansharpening Neural Network method proposed in [13] for the fusion of HS and PAN images works band-wise leveraging a single lightweight CNN designed for a single band pansharpening task, whose weights are continuously tuned moving from one band to the next one. In other words, the generic HS b-th band $\mathbf{H}_b \in \mathbb{R}^{w \times h}$ is fused with $\mathbf{P}$ to provide the corresponding pansharpened image $\widehat{\mathbf{H}}_b$ by means of a network f_ϕ with weights ϕ, *i.e.*,

$$f_\phi : \ (\mathbf{H}_b, \mathbf{P}) \in \mathbb{R}^{w \times h} \times \mathbb{R}^{W \times H} \longrightarrow \widehat{\mathbf{H}}_b = f_\phi(\mathbf{H}_b, \mathbf{P}) \in \mathbb{R}^{W \times H}. \tag{3}$$

Moving sequentially from the first band to the last one, at each step the parameters, inherited from the previous band, are tuned on the current one before performing the inference. This is achieved thanks to a suitably defined unsupervised loss, applied to the same test image, which is responsible for the consistency of the pansharpening result with both input components. By doing so, we get a cross-band transfer learning which makes perfectly sense given the very high correlation degree between adjacent spectral bands in HS stacks, with the exception of few bands with specific wavelengths, for example on the visible to near infrared edge. The tuning time, however, is set separately for each band to account for such discontinuity, allowing more iterations for bands with lesser correlation with the preceding one (see [13] for further details).

The key for the tuning is an unsupervised loss defined as follows:

$$\mathcal{L} = \mathcal{L}_\lambda\left(\widehat{\mathbf{H}}_b, \mathbf{H}_b\right) + \beta\mathcal{L}_S\left(\widehat{\mathbf{H}}_b, \mathbf{P}\right), \tag{4}$$

being β a weighting hyperparameter. The first term of Eq. 4 accounts for the spectral consistency and is given by

$$\mathcal{L}_\lambda\left(\widehat{\mathbf{H}}_b, \mathbf{H}_b\right) = \left\langle\left|\widehat{\mathbf{H}}_{b,\downarrow} - \mathbf{H}_b\right|\right\rangle. \tag{5}$$

Observe that this loss term equals the spectral distortion $\mu_{\mathbf{E}_b}$ as defined in Eq. 1.

Besides, the second term measures the structural (or spatial) consistency of the pansharpened image with the PAN, being defined as

$$\mathcal{L}_S\left(\widehat{\mathbf{H}}_b, \mathbf{P}\right) = \langle[\rho^{\max}(s) - \rho(s)]_+\rangle \tag{6}$$

where $\rho(s)$ is the correlation coefficient between $\widehat{\mathbf{H}}_b$ and $\mathbf{P}$ in a $R\times R$ window around the point s, and $\rho^{\max}(s)$ a desired value of $\rho(s)$ estimated at a reduced-scale correlating the smoothed version of $\mathbf{P}$ with the interpolated version $\widetilde{\mathbf{H}}_b$ of $\mathbf{H}_b$. The operator $[\cdot]_+$ returns its argument when non-negative or zero otherwise. Therefore, the structural loss weights the correlation gap $\rho^{\max}(s) - \rho(s)$ only for those locations where it is positive.

3 Proposed Solution

The above described R-PNN method seems to be particularly suited to face the above discussed spectral unbalance issue (Sect. 2.1), as it allows to fit the model separately on each individual band. Specifically, one can force each pansharpened band to have a spectral distortion below a prefixed level. This can be done by giving a specific form to the loss which gives a higher priority to the spectral term. In the original R-PNN the balance between the spectral and spatial loss terms is left to a single, constant, hyperparameter β (Eqs. 4–6), with no actual guarantee about which of the two terms is better optimized on each individual band.

A starting keypoint is to define a what is a reasonable spectral distortion level to achieve for each band. A partial answer can be found in with the help of Fig. 1 (b). Considering that the spectral distortion cannot be completely eliminated, because of a intrinsic trade-off with the sharpening goal, and in light of the band-dependent characteristics of noise and, more in general, image content, we can define distortion target in relative terms with respect to the score of the EXP interpolator. In other words, we introduce a rate $\gamma < 1$ of the EXP distortion ϵ_b (Eq. 2) we would like to guarantee, *i.e.*,

$$\widehat{\mathbf{H}}: \quad \mu_{\mathbf{E}_b} = \left\langle|\widehat{\mathbf{H}}_{b,\downarrow} - \mathbf{H}_b|\right\rangle \leq \gamma\epsilon_b = \gamma\left\langle|\widetilde{\mathbf{H}}_{b,\downarrow} - \mathbf{H}_b|\right\rangle, \quad \forall b \tag{7}$$

To this goal, we define the mask

$$\mathbf{M}_b = (\mathbf{E}_b > \gamma\epsilon_b) \in \{0,1\}^{w\times h}, \tag{8}$$

that maps the pixel locations where the distortion exceeds the target threshold $\gamma\epsilon_b$. The mask $\mathbf{M}_b$ can be used to perform a spatially disjoint optimization of the spectral and spatial loss terms (Eqs. 5 and 6, respectively), limiting the former to the regions where Eq. 8 is true (exceeding spectral distortion) and the latter to its complementary domain. However, since $\mathcal{L}_S$ is computed at the full spatial scale $[W\times H]$ whereas $\mathcal{L}_\lambda$ and $\mathbf{M}_b$ are define at the coarser scale $[w\times h]$, we resize $\mathbf{M}_b$ using a nearest neighbor interpolator getting a full-scale mask $\overline{\mathbf{M}}_b$ suited to $\mathcal{L}_S$. With these premises we can now provide the proposed loss

$$\mathcal{L} = \mathcal{L}_\lambda^{(\mathrm{M})} + \beta\mathcal{L}_S^{(\mathrm{M})}, \tag{9}$$

still in the form of a linear combination as in Eq. 4, which involves the following "masked" loss terms:

$$\mathcal{L}_\lambda^{(\mathrm{M})} \triangleq \frac{wh}{\sum_s \mathbf{M}_b(s)} \left\langle \mathbf{M}_b \cdot \left| \widehat{\mathbf{H}}_{b,\downarrow} - \mathbf{H}_b \right| \right\rangle \tag{10}$$

and

$$\mathcal{L}_S^{(\mathrm{M})} \triangleq \frac{WH}{\sum_s (1-\overline{\mathbf{M}}_b(s))} \left\langle (1-\overline{\mathbf{M}}_b) \cdot [\rho^{\max}(s) - \rho(s)]_+ \right\rangle, \tag{11}$$

where $\cdot$ denotes the element-wise multiplication and s is the spatial location. The scaling factors outside the spatial averages renormalize them to the actual number of elements filtered by the mask $\mathbf{M}_b$ or by its complement $1-\overline{\mathbf{M}}_b$.

Thanks to this bi-partition of the loss, we ensure that, for a fixed training iteration, each spatial location is considered exclusively either to reduce the spectral distortion (when it exceed the target) or to improve the spatial loss (once the spectral distortion falls below the limit). This establishes a clear hierarchy between spectral and spatial loss terms, conditioning the activation of the latter to the minimization of the former.

In Fig. 2 we show an example of the loss evolution with or without masking during the tuning performing training of 1000 iterations from scratch on a sample HS band. The loss evolution is drawn in the $\mathcal{L}_\lambda/\epsilon_b$–$\mathcal{L}_S$ plane (hence computed globally without masks) for both solutions to ensure a proper comparison, although the masked solution optimizes modified terms. Both models start from the same (random) weights configuration. Therefore at the first iteration they are in the same point in the loss space (top-right). In the early iterations, the masked solution is likely driven only by its spectral loss term with a mask covering nearly all the spatial domain. This explains the more "horizontal" loss trajectory which points straight to the bound γ. Once approached that limit the spatial loss term clearly activates determining an abrupt 90°C turn aimed to its minimization keeping the spectral loss close to the target. On the contrary, removing the masking the spectral loss term appear more unstable and the overall minimization is driven mainly by the spatial term.

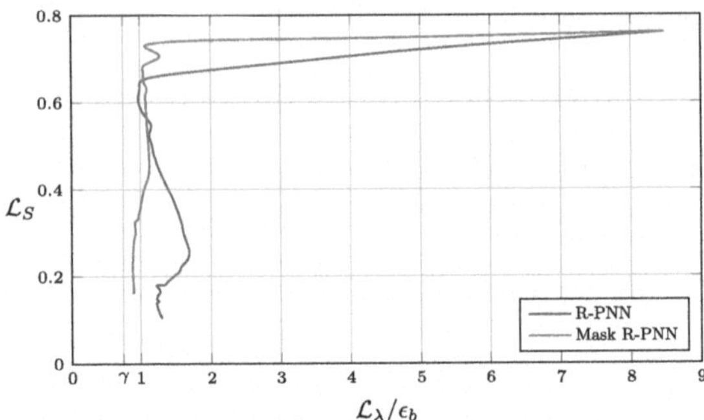

Fig. 2. Loss trajectory (1000 iterations from random weights) in the $\mathcal{L}_\lambda/\epsilon_b$–$\mathcal{L}_S$ plane for a sample HS band, with (blue) and without (red) masking. The $\mathcal{L}_\lambda$ target is fixed to $\gamma\epsilon_b = 0.75\epsilon_b$. (Color figure online)

The above experiment has been conceived to give the sense of the proposal and of its effectiveness. However, in the actual call on a full HS stack, both R-PNN and the proposed variant, when pansharpening the current band, involve initial network weights inherited from the model applied to the preceding band (not random ones). This determines a starting point in the loss space much closer to the origin and, as a consequence, a much smaller number of tuning iterations is required.

4 Experimental Results

To assess the performances of the proposed solution we leverage the benchmark toolbox [7] which provides implementations of state-of-the-art HS pansharpening methods, routines for quality assessment and datasets (PRISMA mission) already split for training, validation and test purposes. In the following, after a brief description of methods and assessment we will provide our comparative results with related discussion.

4.1 Dataset

The benchmark dataset is composed of images acquired through the PRISMA mission. The PRISMA imaging system provides a 5 m resolution PAN band covering the visible spectrum (400–700 nm), coupled with a 30 m resolution HS stack (aligned with the PAN) composed of 239 spectral bands acquired with two spectrometers and distributed from the visible to the short-wave infrared spectrum (400–2505 nm). As it frequently occurs that several bands are affected by severe acquisition errors, it is a common practice to get rid of too noisy/damaged bands. In the specific case of the benchmark dataset the actual number of bands kept is 159 (same selection for all acquisitions).

Specifically, the dataset comprises 163 large tiles of size 1152 × 1152 (PAN resolution) for training (plus 24 for validation) extracted from 12 different acquisitions of places distributed over the globe. Since the resolution ratio for PRISMA is $R = 6$, the HS component of the tiles are 192×192 large. Reduced-resolution dataset (with GT) are also provided for methods designed for supervised training. All the methods provided in the toolbox were trained on such dataset.

In the specific case of R-PNN and of its variant proposed here, the model requires only a pretraining involving exclusively the first band of the HS stack, as from the second band on the model is transferred form the previous band and tuned on the fly on the same test image. Moreover, since they use an unsupervised loss the reference dataset for their training is the full-resolution one not needing any GT. Indeed, with a little additional computational effort (more tuning iterations on the first band), these solutions can work without pretraining using random initial weights.

The test set is derived from four large images, 3600 × 3600 PAN size, acquired in places not covered by training and validation sets: Cagliari (I), Udine (I), Ford Country (Kansas, USA), Macuspana (Tabasco, MEX). Reduced-resolution test images are obtained through the usual 6× resolution downgrade process which provides 600 × 600 images. Instead, the actual full-resolution test images are 1200 × 1200 crops from the entire images.

4.2 Quality Assessment

A numerical quality assessment of pansharpening methods can be obtained using synthetic and/or real data. The use of reduced-resolution synthetic data allows to measure objectively the difference between the pansharpened images and the GT. For real data, instead, there is no GT, therefore only quality indexes such as indicators of the consistency between the pansharpened image and the two input components can be employed. Of course, one must be aware that the characteristics of the synthesized reduced-resolution cannot perfectly match that of the real data. This is particularly true in the case of remote sensing acquisitions where the resolution of the ground elements is fixed, *e.g.*, 5 m (PAN) in our case. In fact, if we move to the synthetic PRISMA, the resolution becomes 30 m (PAN) and categories made of relatively small objects may never occur (when their size falls below 30 m) or lose their geometric properties (when large but not enough). From the deep learning perspective this poses a clear generalization issue when training on synthetic data and test on real data and *vice versa*. In our case, the baseline approach R-PNN is unsupervised and the parameters fitting is run on real data, therefore we expect a performance drop when moving to synthetic images compared to networks trained on synthetic image (supervised approaches). The results discussed in [7] actually confirm this expectation for R-PNN, although the performance drop is rather limited: it ranks on top in the full-resolution framework while in the reduced-resolution one it performs marginally well. In any case, what really count is the generalization in the other direction (toward real data). For this reason, and for the sake of brevity, we will focus here on the quality assessment in the more interesting full-resolution framework, for which the benchmark [7] provides two indexes: the spectral distortion D_λ and the spatial distortion D_S.

The spectral distortion index [18] is defined as $D_\lambda = 1 - Q2^n\left(\widehat{\mathbf{H}}_\downarrow, \mathbf{H}\right)$, where $Q2^n(\cdot,\cdot)$ is a multiband extension of the Universal Image Quality Index $Q(\mathbf{I},\mathbf{J})$ [33] defined for pairs of scalar images $\mathbf{I}$ and $\mathbf{J}$ as

$$Q(\mathbf{I},\mathbf{J}) = \left\langle \frac{\sigma_{\mathbf{IJ}}}{\sigma_{\mathbf{I}}\sigma_{\mathbf{J}}} \cdot \frac{2\sigma_{\mathbf{I}}\sigma_{\mathbf{J}}}{\sigma_{\mathbf{I}}^2 + \sigma_{\mathbf{J}}^2} \cdot \frac{2\mu_{\mathbf{I}}\mu_{\mathbf{J}}}{\mu_{\mathbf{I}}^2 + \mu_{\mathbf{J}}^2} \right\rangle \tag{12}$$

where the statistics $\mu_\cdot$ (image mean), $\sigma_\cdot$ (image standard variation) and $\sigma_{\mathbf{IJ}}$ (covariance) are computed on a 32×32 sliding window before to be average over the whole image domain. The factorization proposed in Eq. 12 highlights the meaning of this index, that is to account for the average differences between $\mathbf{I}$ and $\mathbf{J}$ in terms of intensity, contrast and correlation.

Contrary to the no-reference spectral quality assessment, for which D_λ provides robust indications, the no-reference spatial evaluation is quite debated and there is no clear consensus about the best index to use [26,30]. The index D_S proposed in the toolbox [7] is probably the most popular one and it is defined as the variance of a residue image $\mathbf{I}$ normalized to the variance of the PAN, *i.e.*, $D_S = \sigma_{\mathbf{I}}^2/\sigma_{\mathbf{P}}^2$, where the residue image is the difference between the PAN and a reconstruction of it from the pansharpened image $\mathbf{H}$ via linear combination of its bands,

$$\mathbf{I} = \mathbf{P} - \sum_b w_b \widehat{\mathbf{H}}_b, \tag{13}$$

with weighs w_b optimized to minimize $\sigma_{\mathbf{I}}^2$. Although this index can provide good indications, there are cases where it can fail. To make an extreme example, an optimization of the weights w_b may converge toward degenerate configurations if, for example, one (among hundreds) pansharpened band is just a replacement of the $\mathbf{P}$. The impact on the spectral distortion could be minimal, especially if such band comes from the visible spectrum, while D_S would get its optimal value (0). For this reason we found it useful to also consider a different index for a more comprehensive assessment of the spatial quality. In particular, we will consider the correlation distortion index D_ρ recently proposed in [26], that is the spatial-spectral average of the local ($R \times R = 6 \times 6$ sliding window) correlation coefficient between the PAN and each of the pansharpened bands, complemented to 1:

$$D_\rho = 1 - \left\langle \frac{\sigma_{\mathbf{P}(s)\widehat{\mathbf{H}}_b(s)}}{\sigma_{\mathbf{P}(s)}\sigma_{\widehat{\mathbf{H}}_b(s)}} \right\rangle_{s,b} \tag{14}$$

where $\mathbf{P}(s)$ and $\widehat{\mathbf{H}}_b(s)$ indicate $R \times R$ patches of $\mathbf{P}$ and band $\widehat{\mathbf{H}}_b$, respectively, around the pixel location s.

4.3 Results and Discussion

The numerical results obtained on the four test images are gathered in Tab. 1. Top-5 and worst-5 are highlighted in green and red, respectively, for each index. As expected, due to the importance given to spectral consistency in the proposed solution, it can be observed a consistent improvement over the baseline R-PNN, with D_λ dropping from 0.0090 to 0.0069. This is achieved at a relatively low cost in term of spatial quality according to D_S and D_ρ. The best method according to D_λ remains TV that, however, does not provide good spatial scores (relatively high D_S and D_ρ). The same consideration partially applies to MTF-GLP-FS and MTF-GLP-HPM-R for which the correlation distortion D_ρ is relatively high. If we change perspective and start from the best spatial scores, we can observe a dual problem with several methods that are good according to one or both the spatial indexes whose spectral distortion D_λ is relatively high. Overall, the proposed solution offer the most balanced behavior, providing an excellent spectral score coupled with good spatial figures.

Table 1. Average numerical results on the test datasets. D_λ assesses the spectral distortion. D_S and D_ρ are two alternative indicators of the spatial quality. The first block of model-based methods is followed by deep-learning ones. The top-5 and worst-5 solutions are shown in green and red, respectively. The **best** is in bold green.

Method (Ideal)	D_λ (0)	D_S (0)	D_ρ (0)
GSA	0.0140	0.0098	0.2367
BT-H	0.0513	**0.0001**	0.0688
BDSD-PC	0.0351	0.0007	0.2019
PRACS	0.0102	0.0151	0.2585
MTF-GLP-FS	0.0063	0.0247	0.2393
MTF-GLP-HPM	0.0098	0.0267	0.0619
MTF-GLP-HPM-R	0.0066	0.0263	0.2458
AWLP	0.0094	0.0295	0.1040
MF	0.0442	0.0393	**0.0589**
HySURE	0.0853	0.0018	0.1842
SR-D	0.0096	0.1327	0.8596
TV	**0.0032**	0.0570	0.2669
HyperPNN	0.0264	0.0051	0.2859
HSpeNet	0.0163	0.0159	0.2921
DHP-DARN	0.0607	0.0061	0.2657
DIP-HyperKite	0.0157	0.0048	0.3301
Hyper-DSNet	0.0080	0.0247	0.3230
PCA-Z-PNN	0.0074	0.0153	0.4128
R-PNN	0.0090	0.0195	0.0728
Proposed	0.0069	0.0222	0.0870

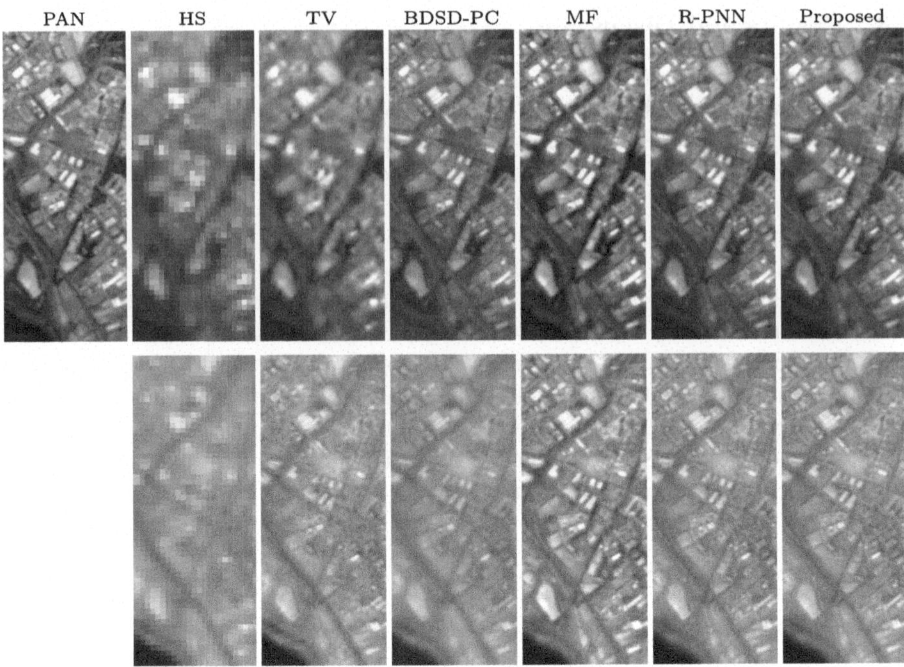

Fig. 3. Cropped (160 × 360) pansharpening results on Cagliari image (full resolution). From left to right: PAN, HS, then five pansharpening results. The RGB channels of the color images are filled with bands from the visible (top) and SWIR (bottom) spectrum.

To gain insight the behavior of the compared methods, beside numerical assessment, the visual inspection can provide further indications. Figure 3 shows zoomed crops from some pansharpening results obtained for the Cagliari test image. For the sake of brevity only a subsets of meaningful solutions are shown. At a first glance the spectral adherence of the pansharpened images to the HS is not immediate to recognize. The differences among TV, R-PNN and proposed in this regard are hard to get. The three solutions are actually quite good D_λ scores, meaning that their downscaled versions should return a very good approximation of the HS image shown on the left. Visible spectral aberrations, instead, are visible for both BDSD-PC and (much more) MF. From the spatial point of view, TV lacks sharpness, same for BDSD-PC but limited to the SWIR bands. MF provides very sharp results that is the consequence of a too strong injection of PAN details into the output that is also cause of the above mentioned spectral aberrations. Proposed and R-PNN seem to provide a similar level of (good) sharpness. Eventually, the differences between proposed and R-PNN are not easy to perceive by visual inspection but only numbers (Table 1) can reveal them. This is also due to the fact that the six displayed bands cannot really summarize the overall pansharpening of 159 bands.

5 Conclusions

In this work we have introduced an improved version of the recently proposed R-PNN method for HS pansharpening. The proposed solution leverages the use of a novel spatially adaptive composite unsupervised loss comprising two competing terms, one to reduce spectral distortion, the other to force spatial sharpness. The loss involves the computation of a spatial mask at each tuning iteration and for each spectral band indicating the regions where the spectral loss is sufficiently small so that the spatial loss term con be activated switching off the spectral one. By doing so, the spectral distortion can be controlled forced to the desired level, uniformly over the whole spectral range of interest. Very encouraging results on four benchmark datasets confirm the expected improvement providing an overall performance at the state of the art. Our future research will focus on new spatial loss functions that could better fit to the actual spatial quality.

Disclosure of Interests. The authors have no competing interests to declare that are relevant to the content of this article.

References

1. Aiazzi, B., Baronti, S., Selva, M.: Improving component substitution pansharpening through multivariate regression of MS+Pan data. IEEE Trans. Geosci. Remote Sens. **45**(10), 3230–3239 (2007)
2. Alparone, L., Aiazzi, B., Baronti, S., Garzelli, A., Nencini, F., Selva, M.: Multispectral and panchromatic data fusion assessment without reference. Photog. Eng. Remote Sens. **74**(2), 193–200 (2008)
3. Alparone, L., Garzelli, A., Vivone, G.: Intersensor statistical matching for pansharpening: theoretical issues and practical solutions. IEEE Trans. Geosci. Remote Sens. **55**(8), 4682–4695 (2017)
4. Bandara, W., Valanarasu, J., Patel, V.M.: Hyperspectral pansharpening based on improved deep image prior and residual reconstruction. IEEE Trans. Geosci. Remote Sens. **60**, 1–16 (2022)
5. Choi, J., Yu, K., Kim, Y.: A new adaptive component-substitution-based satellite image fusion by using partial replacement. IEEE Trans. Geosci. Remote Sens. **49**(1), 295–309 (2011)
6. Ciotola, M., Poggi, G., Scarpa, G.: Unsupervised deep learning-based pansharpening with jointly enhanced spectral and spatial fidelity. IEEE Trans. Geosci. Remote Sens. **61**, 1–17 (2023)
7. Ciotola, M., et al.: Hyperspectral pansharpening: critical review, tools, and future perspectives. IEEE Geosci. Remote Sens. Mag. 2–29 (2024)
8. Ciotola, M., Vitale, S., Mazza, A., Poggi, G., Scarpa, G.: Pansharpening by convolutional neural networks in the full resolution framework. IEEE Trans. Geosci. Remote Sens. **60**, 1–17 (2022)
9. Dalponte, M., Ørka, H.O., Gobakken, T., Gianelle, D., Næsset, E.: Tree species classification in boreal forests with hyperspectral data. IEEE Trans. Geosci. Remote Sens. **51**(5), 2632–2645 (2013)
10. Deng, L.J., et al.: Machine learning in pansharpening: a benchmark, from shallow to deep networks. IEEE Geosci. Remote Sens. Mag. **10**(3), 279–315 (2022)

11. Di, D., Li, J., Han, W., Yin, R.: Geostationary hyperspectral infrared sounder channel selection for capturing fast-changing atmospheric information. IEEE Trans. Geosci. Remote Sens. **60**, 1–10 (2022)
12. Guarino, G., Ciotola, M., Vivone, G., Poggi, G., Scarpa, G.: PCA-CNN hybrid approach for hyperspectral pansharpening. IEEE Geosci. Remote Sens. Lett. **20**, 1–5 (2023)
13. Guarino, G., Ciotola, M., Vivone, G., Scarpa, G.: Band-wise hyperspectral image pansharpening using CNN model propagation. IEEE Trans. Geosci. Remote Sens. **62**, 1–18 (2024)
14. He, L., Zhu, J., Li, J., Plaza, A., Chanussot, J., Li, B.: HyperPNN: hyperspectral pansharpening via spectrally predictive convolutional neural networks. IEEE J. Sel. Top. Appl. Earth Obs. Remote Sens. **12**(8), 3092–3100 (2019)
15. He, L., Xie, J., Li, J., Plaza, A., Chanussot, J., Zhu, J.: Variable subpixel convolution based arbitrary-resolution hyperspectral pansharpening. IEEE Trans. Geosci. Remote Sens. **60**, 1–19 (2022)
16. He, L., Zhu, J., Li, J., Plaza, A., Chanussot, J., Yu, Z.: CNN-based hyperspectral pansharpening with arbitrary resolution. IEEE Trans. Geosci. Remote Sens. **60**, 1–21 (2022)
17. Jia, J., et al.: Tradeoffs in the spatial and spectral resolution of airborne hyperspectral imaging systems: a crop identification case study. IEEE Trans. Geosci. Remote Sens. **60**, 1–18 (2022)
18. Khan, M.M., Alparone, L., Chanussot, J.: Pansharpening quality assessment using the modulation transfer functions of instruments. IEEE Trans. Geosci. Remote Sens. **47**(11), 3880–3891 (2009)
19. Liu, C., et al.: Band-independent encoder-decoder network for pan-sharpening of remote sensing images. IEEE Trans. Geosci. Remote Sens. **58**(7), 5208–5223 (2020)
20. Lorenz, S., Ghamisi, P., Kirsch, M., Jackisch, R., Rasti, B., Gloaguen, R.: Feature extraction for hyperspectral mineral domain mapping: a test of conventional and innovative methods. Remote Sens. Environ. **252**, 112129 (2021)
21. Masi, G., Cozzolino, D., Verdoliva, L., Scarpa, G.: Pansharpening by convolutional neural networks. Remote Sens. **8**(7), 594 (2016)
22. Nie, J., Xu, Q., Pan, J.: Unsupervised hyperspectral pansharpening by ratio estimation and residual attention network. IEEE Geosci. Remote Sens. Lett. **19**, 1–5 (2022). https://ieeexplore.ieee.org/document/9703344/
23. Nolin, A.W., Dozier, J.: A hyperspectral method for remotely sensing the grain size of snow. Remote Sens. Environ. **74**(2), 207–216 (2000)
24. Otazu, X., González-Audícana, M., Fors, O., Núñez, J.: Introduction of sensor spectral response into image fusion methods. Application to wavelet-based methods. IEEE Trans. Geosci. Remote Sens. **43**(10), 2376–2385 (2005)
25. Palsson, F., Sveinsson, J.R., Ulfarsson, M.O.: A new pansharpening algorithm based on total variation. IEEE Geosci. Remote Sens. Lett. **11**(1), 318–322 (2014)
26. Scarpa, G., Ciotola, M.: Full-resolution quality assessment for pansharpening. Remote Sens. **14**(8) (2022)
27. Scarpa, G., Vitale, S., Cozzolino, D.: Target-adaptive CNN-based pansharpening. IEEE Trans. Geosci. Remote Sens. **56**(9), 5443–5457 (2018)
28. Seo, S., et al.: UPSNet: unsupervised pan-sharpening network with registration learning between panchromatic and multi-spectral images. IEEE Access **8**, 201199–201217 (2020)
29. Simoes, M., Bioucas-Dias, J., Almeida, L., Chanussot, J.: A convex formulation for hyperspectral image superresolution via subspace-based regularization. IEEE Trans. Geosci. Remote Sens. **53**(6), 3373–3388 (2015)

30. Vivone, G., et al.: A new benchmark based on recent advances in multispectral pansharpening: revisiting pansharpening with classical and emerging pansharpening methods. IEEE Geosci. Remote Sens. Mag. **9**(1), 53–81 (2021)
31. Vivone, G., Garzelli, A., Xu, Y., Liao, W., Chanussot, J.: Panchromatic and hyperspectral image fusion: outcome of the 2022 whispers hyperspectral pansharpening challenge. IEEE J. Sel. Top. Appl. Earth Obs. Remote Sens. **16**, 166–179 (2023)
32. Wald, L., Ranchin, T., Mangolini, M.: Fusion of satellite images of different spatial resolutions: assessing the quality of resulting images. Photog. Eng. Remote Sens. **63**(6), 691–699 (1997)
33. Wang, Z., Bovik, A.C.: A universal image quality index. IEEE Sig. Proc. Lett. **9**(3), 81–84 (2002)
34. Yang, J., Fu, X., Hu, Y., Huang, Y., Ding, X., Paisley, J.: PanNet: a deep network architecture for pan-sharpening. In: ICCV (2017)
35. Zheng, Y., Li, J., Li, Y., Guo, J., Wu, X., Chanussot, J.: Hyperspectral pansharpening using deep prior and dual attention residual network. IEEE Trans. Geosci. Remote Sens. **58**(11), 8059–8076 (2020)

Characteristic-Driven Deep Learning in Synthetic Aperture Radar Target Recognition

Xiaoyan Zhou[1,2], Zhuo Su[2], Tao Tang[1], Li Liu[1](✉), and Gangyao Kuang[1]

[1] National University of Defense Technology, Changsha 410073, China
liuli_nudt@nudt.edu.cn
[2] University of Oulu, 90014 Oulu, Finland

Abstract. Synthetic Aperture Radar (SAR) automatic target recognition is an important technology in remote sensing, and deep learning has greatly improved its performance. However, SAR images, which belong to the microwave vision spectrum and reflect the target's backscattering, differ fundamentally from optical images. This difference creates challenges when directly applying optical-based deep learning techniques to SAR data, such as poor interpretability and a tendency to overfit. To overcome these issues, integrating SAR-specific characteristics into deep learning methods has become a key research focus. However, there is a lack of clear guidance on how to effectively consider these characteristics, limiting the effectiveness of recognition systems. This review answers three important questions: (1) What are the unique characteristics of SAR images? (2) How can these characteristics be effectively modeled? (3) How can SAR target features be integrated into deep learning models? By addressing these questions, this review aims to support the development of more reliable and interpretable SAR target recognition systems.

Keywords: SAR ATR · Deep Learning · Physical Characteristic

1 Introduction

Synthetic Aperture Radar (SAR) is a high-resolution imaging radar system that emits microwaves and captures the echoes reflected from ground targets, enabling imaging capabilities under all-weather and all-time conditions [1]. Due to its ability, SAR has become a critical tool in earth observation [2–4], as illustrated in Fig. 1(a). The advancements in synthetic large-aperture antennas have significantly improved SAR image resolution, elevating its importance in applications such as ground surveillance and reconnaissance [5,6], air traffic control [7,8], and port surveillance [9,10]. Among these, SAR automatic target recognition (ATR) has emerged as a key technology for the intelligent interpretation of SAR data.

The rapid development of deep learning, especially deep neural networks (DNNs), has significantly advanced SAR ATR and demonstrated superior performance [6,11,12]. However, directly applying deep learning methods designed for

J. Petersen and V. A. Dahl (Eds.): SCIA 2025, LNCS 15726, pp. 92–105, 2025.
https://doi.org/10.1007/978-3-031-95918-9_7

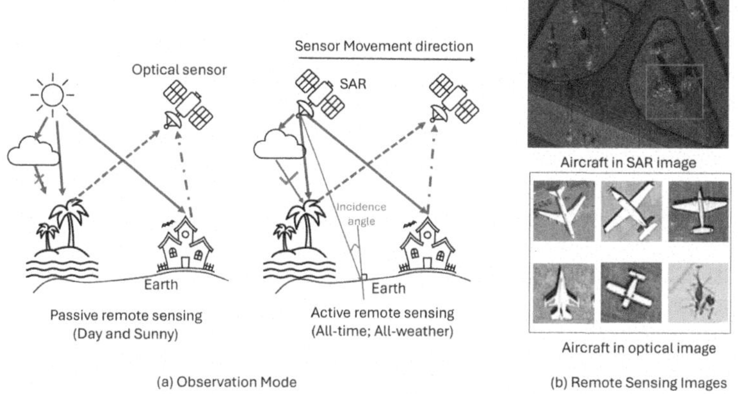

Fig. 1. Difference in optical and SAR: observation mode and images. Radar signals will attenuate in extreme weather conditions but usually not to the extent of becoming undetectable. In contrast, optical satellites, which are passive sensors, cannot work in extreme weather conditions.

optical images to SAR data has proven challenging, leading to issues such as overfitting [13] and poor interpretability [14,15]. SAR images, known as microwave radar image, reflect backscattering characteristics of targets that are fundamentally different from those of optical images, as illustrated in Fig. 1(b). Moreover, SAR targets are strongly influenced by radar and target-specific parameters, which further complicates the adaptation of optical-domain deep learning techniques to SAR imagery.

In recent years, researchers have increasingly focused on integrating the unique target characteristics of SAR with DNNs to enhance robustness [16,17], recognition accuracy [18], and interpretability [19,20]. By leveraging SAR-specific features, such as the attribute scattering center (ASC), and exploring innovative strategies for feature fusion, the community has made significant strides in developing tailored solutions for SAR ATR. However, a review of characteristic-driven deep learning approaches for SAR ATR is still lacking. For instance, [2] reviews SAR ATR methods according to template-matching, machine learning, and model-based techniques, while [21] focuses on few-shot application scenarios. Despite these efforts, there remains a pressing need for further exploration of methodologies that balance SAR-specific knowledge with advanced machine learning techniques, aiming for both robustness and explainability in SAR target recognition.

Inspired by [22], this review aims to provide a comprehensive analysis of SAR ATR methodologies, emphasizing the intersection of physical characteristics and deep learning techniques. By synthesizing the latest advancements in this field, we aim to highlight what SAR imagery characteristics are (Sect. 2), explore how to model (Sect. 3) and apply them (Sect. 4), and propose directions for future research to advance SAR ATR (Sect. 5).

2 SAR Imagery Characteristics

In complex SAR imagery, every pixel, $z(m, n)$, consists of an imaginary (B) and a real (A) part [23], which is formulated as:

$$z(m, n) = A + jB, \tag{1}$$

where m means pixel row index, n is pixel column index. In SAR data processing, we typically use $R = \sqrt{A^2 + B^2}$ and $\angle R = arctan(B, A)$ to denote the amplitude and phase, respectively. Therefore, another description would be:

- The **amplitude** R of each pixel represents the strength of the radar signal, which correlates with the radar backscatter coefficient.
- The **phase** $\angle R$ of each pixel records the change in phase of the radar wave from its emission to its reception.

The radar backscatter coefficient is influenced by various factors, including SAR system parameters (such as the radar's frequencies, polarization modes, and observation angles) and ground (target) parameters (such as materials, shapes, and surface roughness levels). Thus, each element in the complex matrix z results from these complex interactions, and understanding these parameters is crucial for interpreting SAR data.

2.1 Radar Parameters on SAR Imagery

From the perspective of a sensor, factors like frequencies (wavebands) and polarization methods significantly influence SAR imagery. The specific influences are as follows:

- **Wavebands:** The choice of wavebands (e.g., X-band, C-band, or L-band) affects the radar's capability to penetrate different materials and resolve fine-grained details. Different wavebands interact uniquely with surface and subsurface features, influencing the microwave reflection and absorption properties. For example, the L-band has a stronger penetration ability than the X-band due to the longer wavelength.
- **Polarization Methods:** The polarization of the radar signal contains electric and magnetic fields. "Horizontal polarization" is denoted as H, and "Vertical polarization" is denoted as V. The four polarization methods—HH, HV, VH, and VV-refer to the combinations of horizontal or vertical polarization used for transmitting and receiving radar waves, with HH and VV being same-polarization and HV and VH being cross-polarization. The polarization methods determine how the emitted microwaves interact with target materials. This affects the degree of backscatter received by the sensor, which is crucial for identifying material properties and structural details.
- **Observation angles** The observation angles (incidence angle, shown in Fig. 1(a) and target azimuth angle, shown in Fig. 4(a)) significantly affect the radar echo characteristics of the targets. Different viewing angles can

result in different shadow effects and reflective characteristics. This is due to the fact that when the angle at which microwaves strike a target changes, the reflection paths and intensities are altered.

2.2 Target Parameters on SAR Imagery

For fixed sensors used in Earth observation, the characteristics of the observed targets are profoundly affected by materials, shapes, and surface roughness levels. They are summarized as follows:

- **Target materials.** Metallic objects strongly reflect microwaves and typically appear as brighter pixel values in SAR images, whereas vegetation and water bodies, which have different microwave reflective properties, usually appear as darker pixel values.
- **Target shapes.** The shape of a target alters the directions and intensities of microwave reflections. Complex shapes, such as buildings or mountains, with their multifaceted structures, can reflect radar waves in multiple directions, resulting in more complex textures or brightness variations in SAR images.
- **Target surface roughness levels.** Rougher surfaces increase scattering effects, making the reflected echoes more dispersed and uniform in SAR images; conversely, smoother surfaces may cause specular reflections, leading to a lack of reflected waves when observed from certain angles. For instance, smooth surfaces like lake surfaces can reflect radar waves away like mirrors, making these areas appear darker in SAR images.

3 SAR Target Characteristics Representation

Unlike optical images, SAR images represent the backscatter of targets and are encoded as complex data. As such, SAR target characteristic representations vary depending on the data type (complex data), physical properties, and imaging mechanisms. In this context, we will describe **complex-valued neural networks** for processing complex data, the **attribute scattering center model** for physical property modeling, and **azimuth sensitivity prior embedding**, which considers the influence of imaging mechanisms such as the variation of target characteristics with different observation angles.

3.1 Complex-Valued Neural Networks

From a data perspective, SAR data is complex-valued, containing both phase and magnitude information. There exists a statistical correlation between the phase and magnitude, which real-valued neural networks (RVNN) typically overlook. This oversight often leads to poorer performance compared to complex-valued neural networks (CVNN) [24]. A CVNN extends an RVNN by incorporating complex-valued operations, such as complex-valued convolution, batch normalization, non-linear activation functions, and weight initialization [25]. These

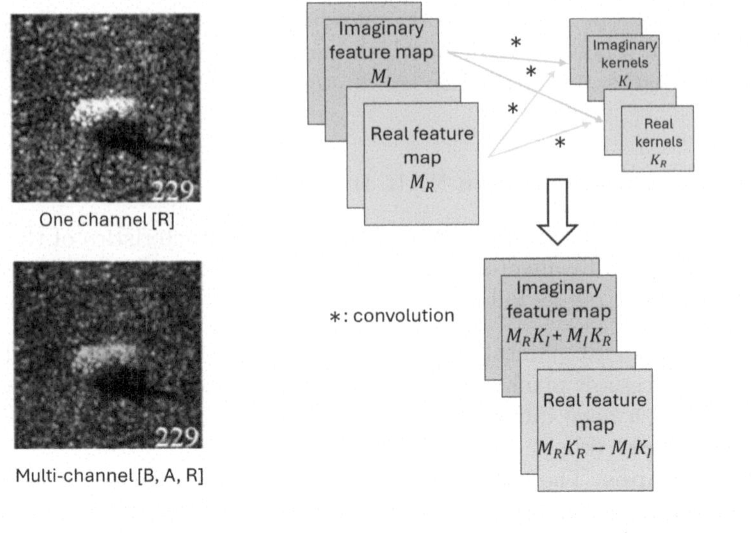

Fig. 2. Representation method of complex data

adaptations enable the network to process and learn from complex data more effectively, leveraging both the phase and magnitude information for improved performance [26].

For the input of CVNNs, a multi-channel approach is commonly used, where data is represented as $[B, A, R]$ defined in Eq. (1) to account for different components of the complex data, as proposed in [27], and illustrated in Fig. 2(a). Regarding the CVNN design, complex-valued convolution, introduced in [25], is illustrated in Fig. 2(b). In complex-valued convolution, both the real and imaginary parts are considered during multiplication, making the operation four times more computationally expensive than in RVNN.

For the activation function, the ReLU used in RVNNs is defined as:

$$ReLU(z) = \max(z, 0), \tag{2}$$

where z is the complex data in Eq. (1). However, in CVNNs, an activation function related to the phase, such as $zReLU$ [28], is formulated as follows:

$$z\,\text{ReLU}(z) = \begin{cases} z \text{ if } \angle R \in [0, \pi/2] \\ 0 \text{ otherwise} \end{cases}, \tag{3}$$

where $\angle R$ represents the phase, as discussed in Sect. 2. For more details on CVNNs in SAR ATR, please refer to [29,30], and [31].

3.2 Attribute Scattering Center Model

When the size of a radar target is much larger than the wavelength of the electromagnetic waves, its total scattering response, $E(f, \phi; \Theta)$, can be viewed as the coherent superposition of the responses from multiple local equivalent scattering sources, $E_i(f, \phi; \theta_i)$, which is expressed as:

$$E(f, \phi; \Theta) = \sum_{i=1}^{K} E_i(f, \phi; \theta_i), \tag{4}$$

where K means the number of scattering sources, f means frequency, and ϕ is the angle of radar wave. These equivalent scattering sources are referred to as the scattering centers of the target, and $\Theta = \{\theta_i\}$ means the attribute set of the scattering centers, which relates to physical properties. The parameterized model of scattering centers aims to describe the relationship between parameters such as frequencies, angles, and polarization methods with the scattering centers. The Attributed Scattering Center (ASC) model, based on the Geometric Theory of Diffraction (GTD [32]), is one of the most widely applied parameterized scattering center models today [33]. In ASC [34], $E_i(f, \phi; \theta_i)$ can be expressed as:

$$\begin{aligned} E_i(f, \phi; \theta_i) = A_i \cdot \left(\mathrm{j}\frac{f}{f_c}\right)^{\alpha_i} \cdot \exp\left(\frac{-\mathrm{j}4\pi f}{c}(x_i \cos\phi + y_i \sin\phi)\right) \\ \cdot \operatorname{sinc}\left(\frac{2\pi f}{c} L_i \sin(\phi - \bar{\phi}_i)\right) \cdot \exp(-2\pi f \gamma_i \sin\phi) \end{aligned}, \tag{5}$$

where $\theta_i = [R_i, \alpha_i, x_i, y_i, L_i, \phi_i, \gamma_i]$. Among these, (x_i, y_i) describes the position of the scattering center within the local coordinate system. The amplitude R_i means the intensity of the scattering. L_i means the length. If $L_i = 0$, the scattering center is localized, and $\gamma_i = 0$ if the scatterer is distributed. $\bar{\phi}_i$ means the orientation angle of the distributed scatterer. The combination of α_i and L_i presents different scattering structures. For example, $\alpha = -1, L = 0$ corresponds to corner diffraction. ASC is used in SAR ATR as physical-related target characteristics [35] (Fig. 3).

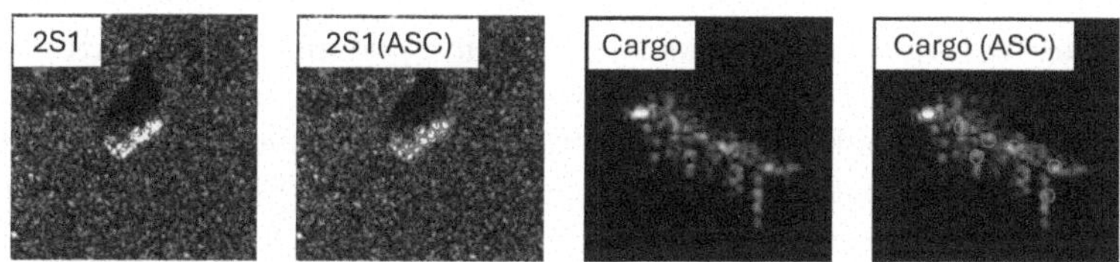

Fig. 3. ASC of SAR target. Red circle means scattering center, and its feature can be presented by physical-related parameters Θ. (Color figure online)

The θ_i can be estimated based on image segmentation [36] or sparse theory [37], and θ_i can be used to reconstruct SAR images, which can serve as a template for traditional SAR ATR methods [38]. Alternatively, the reconstructed image and the original SAR image can be treated as complementary inputs, allowing a deep neural network to learn better feature representations based on these inputs, such as in the PAN network described in [19].

3.3 Target Azimuth Sensitivity Prior Embedding

Target azimuth represents the angle between the target, the north, and the direction of the SAR sensor, which is shown in Fig. 4(a). The multi-azimuth SAR images of vehicle target ZIL131 are shown in Fig. 4(b). It can be found that (1) The smaller the difference between target azimuth angles, the higher the similarity between images:

$$sim(I_\theta, I_{\theta+\Delta\theta_{AZ1}}) > sim(I_\theta, I_{\theta+\Delta\theta_{AZ2}}), \Delta\theta_{AZ1} < \Delta\theta_{AZ2}, \tag{6}$$

where the image I can be $z(m,n)$, R or A in Sect. 2. The θ_{AZ} is the target azimuth angle, $\Delta\theta_{AZ1}$ and $\Delta\theta_{AZ2}$ mean the azimuth angle difference between two SAR targets, and $sim(\cdot)$ represents the similarity. (2) Even for the same target, the electromagnetic scattering patterns at different azimuth angles show large differences (e.g., 0° and 269°).

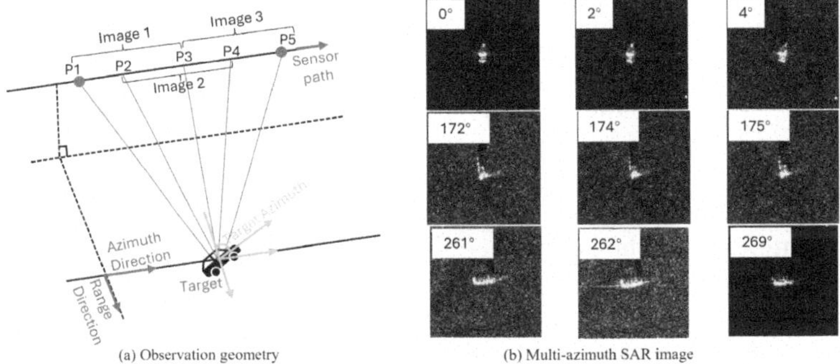

Fig. 4. SAR Observation geometry and multi-azimuth SAR image.

Based on the target azimuth sensitivity prior, the multi-view SAR data have similar characteristics to videos. To fully exploit the relationship between multi-view SAR images, [39] adopts LSTM in SAR ATR, [40] and [41] design a parallel network which fuses features progressively, shown in Fig. 5. Furthermore, [27] rotates SAR images according to their azimuth angle for data normalization during data pre-processing, [42] introduces contrastive loss functions in SAR ATR to make different classes have lower similarity in the feature space.

3.4 Summary

In summary, from the perspectives of data representation, feature representation, and image relationships, the characteristics of SAR can be summarized as follows:

1. Complex data;
2. The ASC features reflect physical properties;
3. Image correlation is closely related to the target azimuth.

To accommodate these characteristics, CVNN is employed to model the complex data. The ASC model is used to represent physically relevant features, while the target azimuth sensitivity prior is embedded into the framework through structural design [40,43] or loss function modifications [42,44].

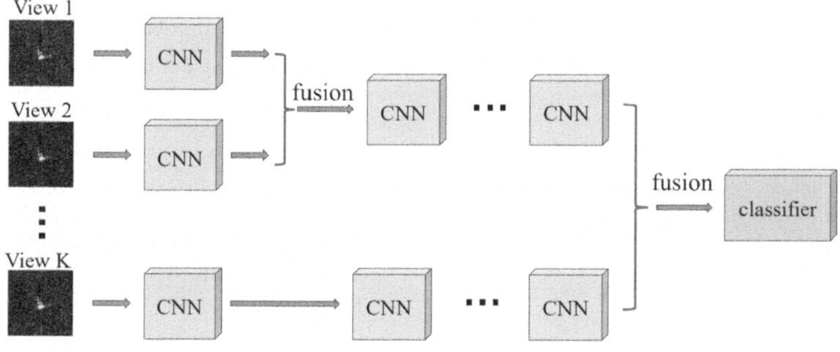

Fig. 5. The framework for multi-view SAR ATR

4 Characteristic-Driven DNNs in SAR ATR

SAR-specific DNNs can be developed by focusing on various aspects such as input data processing, loss function design, and DNN structure design. For instance, SAR images' complex-valued information can be leveraged to enhance data representation. In this context, we review related works from the perspectives of DNN design and the utilization of SAR characteristics, as summarized in Table 1. To keep the table concise, we use "Azimuth Sensitivity" to represent the azimuth sensitivity prior.

Table 1. Characteristic-driven SAR ATR

Model	Driven Characteristic	Perspective	Description
Confounding Factors knowledge enlight network [45]	Azimuth Sensitive	Loss Function	Introducing angle prediction as a constraint during target classification. It helps the network disentangle confounding factors that influence the input data and better represent classification features
SAR Target Recognition Network With Pose Angle Marginalization (SPAM-Net) [44]	Azimuth Sensitive	Structure Design & Loss Function	Consider the impact of the azimuth angle on classification. The class probability $P(Y\|X)$ can be expressed as: $P(Y\|X) = P(Y\|X, \theta_{AZ})P(\theta_{AZ}\|X)$. Two separate branches are used to estimate $P(Y\|X, \theta_{AZ})$ and $P(\theta_{AZ}\|X)$, respectively

(continued)

Table 1. (*continued*)

Model	Driven Characteristic	Perspective	Description
Data Augmentation based on azimuth [46]	Azimuth Sensitive	Data Augmentation	To generate missing azimuth angle image data $I_{\theta_{AZ*}}$, linear combinations of SAR images ($I_{\theta_{AZ1}}$ and $I_{\theta_{AZ2}}$ from adjacent azimuth angles are used, along with rotation operations to simulate SAR images
Multi-View deep learning network [41]	Azimuth Sensitive	Structure Design	First, sufficient multi-view SAR data is generated. Then, a parallel DNN topology is adopted to progressively fuse features from different views
Multi-View joint transformer network (MJT-Network) [47]	Azimuth Sensitive	Structure Design	By combining prior-based denoising, CNN feature extraction, and a transformer backbone with self-attention, the approach integrates multi-view features for joint classification at the decision level
Multi-stream complex-valued networks (MS-CVNet) [30]	Complex Data	Structure Design	A series of complex-valued operation blocks are proposed, combined with multi-scale feature extraction and fusion, to enhance target recognition
Domain knowledge-powered two-stream deep network (DKTS-N) [27]	Complex Data	Data Augmentation	Augment the SAR image to three-channel using $[B, A, R]$, which considers the imaginary part, real part, and the amplitude information with an image form
Electromagnetic scattering feature embedding network [48]	ASC	Structure Design	A deep learning network is used to learn simplified electromagnetic scattering features, enabling it to possess physical representation capabilities. During CNN training, multi-layer fusion of deep learning features and physically relevant features is incorporated
Physical Information Guided Explanation model [49]	ASC	Data Augmentation	ASC components are extracted and used to generate a mask for input perturbation. The perturbed input is then classified to produce a heatmap, which analyzes the role of components and enables the identification of key parts for physics-based target recognition.
Multilevel scattering center and deep feature fusion network (SDF-Net) [50]	ASC	Structure Design	Images are reconstructed based on ASC at both the target and component levels. Deep features and electromagnetic scattering features are complementarily fused using an end-to-end pseudo-siamese network

(*continued*)

Table 1. (*continued*)

Model	Driven Characteristic	Perspective	Description
Complex-valued network guided with sub-aperture decomposition (CGS-Net) [51]	Complex Data & Azimuth Sensitive	Structure Design & Loss Function	A complex-valued network is proposed to fully utilize the phase and amplitude information in SAR data. Additionally, a sub-aperture image reconstruction task is introduced to exploit azimuth information and assist in improving target recognition
Scattering Characteristics Analysis Network [7]	ASC & Azimuth Sensitive	Structure Design & Loss Function	An angle-adaptive classifier is proposed to mitigate image variations caused by observation angles. Additionally, a scattering extraction branch is designed to enhance SAR target feature representation

5 Future Works and Conclusion

5.1 Future Works

Integrating SAR characteristics into the development of deep learning-based ATR has become a leading research trend. In subfields such as few-shot target recognition [7,27], transfer learning [52,53], and incremental learning [54,55], these methods have demonstrated significant potential, greatly improving target recognition performance under specific conditions. In the future, characteristic-driven SAR ATR methods can be further enhanced in the following directions:

1. Explore models for domain knowledge discovery and exploitation. Leverage existing deep learning networks to discover target characteristics in SAR data [56,57] and transfer these insights to downstream tasks, thereby improving model reliability and generalization.
2. Designing physics-inspired DNNs. Incorporate physical information to design DNNs tailored to SAR imagery by considering its unique application scenes [58,59], such as complex background, enabling more meaningful feature extraction and target recognition.
3. Integrating multiple SAR characteristics. While current research often focuses on a single characteristic (e.g., ASC), future efforts should aim to integrate multiple characteristics [7,51,60] to comprehensively enhance recognition performance and model robustness.

5.2 Conclusion

This paper provides a review of characteristic-driven SAR ATR methods. It begins by introducing the factors that influence SAR image characteristics, followed by a detailed discussion of modeling approaches for target characteristics in SAR imagery. Finally, it summarizes the current state of these methods and outlines potential future development directions. We hope this review will inspire new ideas for SAR ATR and foster the development of more robust, reliable, and interpretable algorithms.

References

1. Curlander, J.C., McDonough, R.N.: Synthetic Aperture Radar, vol. 11. Wiley, New York (1991)
2. Li, J., Zhentao, Yu., Lu, Yu., Cheng, P., Chen, J., Chi, C.: A comprehensive survey on SAR ATR in deep-learning era. Remote Sens. **15**(5), 1454 (2023)
3. Cui, Y., Tang, T., Zhou, X., Ji, K.: Vehicle target detection in complex scene SAR images based on co-saliency. In: 2023 8th International Conference on Signal and Image Processing (ICSIP), pp. 99–103. IEEE (2023)
4. Tsokas, A., Rysz, M., Pardalos, P.M., Dipple, K.: SAR data applications in earth observation: an overview. Expert Syst. Appl. **205**, 117342 (2022)
5. Zhou, X., Tang, T., Sun, Z., Kuang, G., Heikkilä, J., Liu, L.: Mitigating SAR out-of-distribution overconfidence based on evidential uncertainty. IEEE Geosci. Remote Sens. Lett. 1 (2024)
6. He, Q., Zhao, L., Ji, K., Kuang, G.: Sar target recognition based on task-driven domain adaptation using simulated data. IEEE Geosci. Remote Sens. Lett. **19**, 1–5 (2022)
7. Sun, X., Lv, Y., Wang, Z., Kun, F.: SCAN: scattering characteristics analysis network for few-shot aircraft classification in high-resolution SAR images. IEEE Trans. Geosci. Remote Sens. **60**, 1–17 (2022)
8. Zhou, J., Xiao, C., Peng, B., Liu, Z., L., Liu, Y., Li, X.: DiffDet4SAR: diffusion-based aircraft target detection network for SAR images. IEEE Geosci. Remote Sens. Lett. (2024)
9. Hou, X., Ao, W., Song, Q., Lai, J., Wang, H., Feng, X.: FUSAR-ship: building a high-resolution SAR-AIS matchup dataset of Gaofen-3 for ship detection and recognition. SCI. CHINA Inf. Sci. **63**, 1–19 (2020)
10. Tan, X., Xiangguang Leng, R., Luo, Z.S., Ji, K., Kuang, G.: YOLO-RC: SAR ship detection guided by characteristics of range-compressed domain. IEEE J. Sel. Top. Appl. Earth Observ. Remote Sens. **17**, 18834–18851 (2024)
11. Zhou, X., Tang, T., Cui, Y., Kuang, G.: SAR open set recognition based on counterfactual framework. In: 2022 Photonics & Electromagnetics Research Symposium (PIERS), pp. 249–253 (2022)
12. Yanjie, X., et al.: Simulated data feature guided evolution and distillation for incremental SAR ATR. IEEE Trans. Geosci. Remote Sens. **62**, 1–17 (2024)
13. Shang, R., Wang, J., Jiao, L., Stolkin, R., Hou, B., Li, Y.: Sar targets classification based on deep memory convolution neural networks and transfer parameters. IEEE J. Sel. Top. Appl. Earth Observ. Remote Sens. **11**, 2834–2846 (2018)
14. Li, P., Feng, C., Hu, X., Tang, Z.: SAR-BagNet: an ante-hoc interpretable recognition model based on deep network for SAR image. Remote Sens. **14**(9) (2022)
15. Feng, Z., Zhu, M., Stanković, L., Ji, H.: Self-matching CAM: a novel accurate visual explanation of CNNs for SAR image interpretation. Remote Sens. **13**(9), 1772 (2021)
16. Zhang, F., Yu, Y., Ma, F., Zhou, Y.: A physically realizable adversarial attack method against SAR target recognition model. IEEE J. Sel. Top. Appl. Earth Observ. Remote Sens. **17**, 11943–11957 (2024)
17. Zhao, Y., Zhao, L., Ding, D., Hu, D., Kuang, G., Liu, L.: Few-shot class-incremental SAR target recognition via cosine prototype learning. IEEE Trans. Geosci. Remote Sens. **61**, 1–18 (2023)
18. Chen, J., Zhang, X., Wang, H., Xu, F.: A reinforcement learning framework for scattering feature extraction and SAR image interpretation. IEEE Trans. Geosci. Remote Sens. **62**, 1–14 (2024)

19. Feng, S., Ji, K., Wang, F., Zhang, L., Ma, X., Kuang, G.: PAN: part attention network integrating electromagnetic characteristics for interpretable SAR vehicle target recognition. IEEE Trans. Geosci. Remote Sens. **61**, 1–17 (2023)
20. Liao, L., Lan, D., Chen, J., Cao, Z., Zhou, K.: EMI-net: an end-to-end mechanism-driven interpretable network for SAR target recognition under EOCs. IEEE Trans. Geosci. Remote Sens. **62**, 1–18 (2024)
21. Yin, J., Duan, C., Wang, H., Yang, J.: A review on the few-shot SAR target recognition. IEEE J. Sel. Top. Appl. Earth Observ. Remote Sens. **17**, 16411–16425 (2024)
22. Su, Z., Pietikäinen, M., Liu, L.: From local binary patterns to pixel difference networks for efficient visual representation learning. In: Scandinavian Conference on Image Analysis, pp. 138–155. Springer (2023)
23. Hein, A.: Processing of SAR Data. Springer (2003)
24. Hirose, A., Yoshida, S.: Comparison of complex-and real-valued feedforward neural networks in their generalization ability. In: Neural Information Processing: 18th International Conference, ICONIP 2011, Shanghai, China, 13–17 November 2011, Proceedings, Part I 18, pp. 526–531. Springer (2011)
25. Trabelsi, C., et al.: Deep complex networks. arXiv preprint arXiv:1705.09792 (2017)
26. Bassey, J., Qian, L., Li, X.: A survey of complex-valued neural networks. arXiv preprint arXiv:2101.12249 (2021)
27. Zhang, L., et al.: Domain knowledge powered two-stream deep network for few-shot SAR vehicle recognition. IEEE Trans. Geosci. Remote Sens. **60**, 1–15 (2022)
28. Guberman, N.: On complex valued convolutional neural networks. arXiv preprint arXiv:1602.09046 (2016)
29. Zhang, Z., Wang, H., Feng, X., Jin, Y.-Q.: Complex-valued convolutional neural network and its application in polarimetric SAR image classification. IEEE Trans. Geosci. Remote Sens. **55**(12), 7177–7188 (2017)
30. Zeng, Z., Sun, J., Han, Z., Hong, W.: Sar automatic target recognition method based on multi-stream complex-valued networks. IEEE Trans. Geosci. Remote Sens. **60**, 1–18 (2022)
31. Yu, L., Hu, Y., Xie, X., Lin, Y., Hong, W.: Complex-valued full convolutional neural network for SAR target classification. IEEE Geosci. Remote Sens. Lett. **17**(10), 1752–1756 (2019)
32. Keller, J.B.: Geometrical theory of diffraction. J. Opt. Soc. Am. **52**(2), 1 (1962)
33. Xing, M., Xie, Y., Gao, Y., Zhang, J., Liu, J., Wu, Z.: Electromagnetic scattering characteristic extraction and imaging recognition algorithm: a review. J. Radars **11**(6), 921–942 (2022)
34. Gerry, M.J., Potter, L.C., Gupta, I.J., Van Der Merwe, A.: A parametric model for synthetic aperture radar measurements. IEEE Trans. Antennas Propag. **47**(7), 1179–1188 (1999)
35. Potter, L.C., Moses, R.L.: Attributed scattering centers for SAR ATR. IEEE Trans. Image Process. **6**(1), 79–91 (1997)
36. Jiang, W., Li, W.: new method for parameter estimation of attributed scattering centers based on amplitude-phase separation. J. Radars (2019)
37. Li, Z., Jin, K., Bin, X., Zhou, W., Yang, J.: An improved attributed scattering model optimized by incremental sparse Bayesian learning. IEEE Trans. Geosci. Remote Sens. **54**(5), 2973–2987 (2016)
38. Fan, J., Tomas, A.: Target reconstruction based on attributed scattering centers with application to robust SAR ATR. Remote Sens. **10**(4), 655 (2018)

39. Zhang, F., Hu, C., Yin, Q., Li, W., Li, H., Hong, W.: SAR target recognition using the multi-aspect-aware bidirectional LSTM recurrent neural networks. arXiv preprint arXiv:1707.09875 (2017)
40. Pei, J., et al.: Multiview deep feature learning network for SAR automatic target recognition. Remote Sens. **13**, 1455 (2021)
41. Pei, J., Huang, Y., Huo, W., Zhang, Y., Yang, J., Yeo, T.-S.: Sar automatic target recognition based on multiview deep learning framework. IEEE Trans. Geosci. Remote Sens. **56**(4), 2196–2210 (2017)
42. Zhou, X., Tang, T., Cui, Y., Zhang, L., Kuang, G.: Novel loss function in CNN for small sample target recognition in SAR images. IEEE Geosci. Remote Sens. Lett. **19**, 1–5 (2021)
43. Wang, Z., et al.: Multi-view SAR automatic target recognition based on deformable convolutional network. In: 2021 IEEE International Geoscience and Remote Sensing Symposium IGARSS, pp. 3585–3588. IEEE (2021)
44. Jihyong, O., Youm, G.-Y., Kim, M.: Spam-net: a CNN-based SAR target recognition network with pose angle marginalization learning. IEEE Trans. Circuits Syst. Video Technol. **31**(2), 701–714 (2020)
45. Zhong, Y., Ettinger, G.: Enlightening deep neural networks with knowledge of confounding factors. In: Proceedings of the IEEE International Conference on Computer Vision Workshops, pp. 1077–1086 (2017)
46. Ding, J., Chen, B., Liu, H., Huang, M.: Convolutional neural network with data augmentation for SAR target recognition. IEEE Geosci. Remote Sens. Lett. **13**(3), 364–368 (2016)
47. Lv, J., et al.: Recognition for SAR deformation military target from a new miniSAR dataset using multi-view joint transformer approach. ISPRS J. Photogramm. Remote. Sens. **210**, 180–197 (2024)
48. Feng, S., Ji, K., Wang, F., Zhang, L., Ma, X., Kuang, G.: Electromagnetic scattering feature (ESF) module embedded network based on ASC model for robust and interpretable sar atr. IEEE Trans. Geosci. Remote Sens. **60**, 1–15 (2022)
49. Gao, Y., Guo, W., Li, D., Yu, W.: ASC-rise: physical information guided explanation of SAR ATR models. In: IGARSS 2024-2024 IEEE International Geoscience and Remote Sensing Symposium, pp. 2159–2162. IEEE (2024)
50. Liu, Z., Wang, L., Wen, Z., Li, K., Pan, Q.: Multilevel scattering center and deep feature fusion learning framework for SAR target recognition. IEEE Trans. Geosci. Remote Sens. **60**, 1–14 (2022)
51. Wang, R., Wang, Z., Chen, Y., Kang, H., Luo, F., Liu, Y.: Target recognition in SAR images using complex-valued network guided with sub-aperture decomposition. Remote Sens. **15**(16), 4031 (2023)
52. Liu, J., Xing, M., Hanwen, Yu., Sun, G.: EFTL: complex convolutional networks with electromagnetic feature transfer learning for SAR target recognition. IEEE Trans. Geosci. Remote Sens. **60**, 1–11 (2021)
53. Shi, Yu., Lan, D., Li, C., Guo, Y., Yuang, D.: Unsupervised domain adaptation for SAR target classification based on domain- and class-level alignment: from simulated to real data. ISPRS J. Photogramm. Remote. Sens. **207**, 1–13 (2024)
54. Zhao, Y., Zhao, L., Ding, D., Hu, D., Kuang, G., Liu, L.: Few-shot class-incremental SAR target recognition via cosine prototype learning. IEEE Trans. Geosci. Remote Sens. (2023)
55. Gao, F., et al.: SAR target incremental recognition based on features with strong separability. IEEE Trans. Geosci. Remote Sens. **62**, 1–13 (2024)

56. Zhou, X., Tang, T., He, Q., Zhao, L., Kuang, G., Liu, L.: Simulated SAR prior knowledge guided evidential deep learning for reliable few-shot SAR target recognition. ISPRS J. Photogramm. Remote. Sens. **216**, 1–14 (2024)
57. Wen, Z., Liu, Z., Zhang, S., Pan, Q.: Rotation awareness based self-supervised learning for SAR target recognition with limited training samples. IEEE Trans. Image Process. **30**, 7266–7279 (2021)
58. Huang, Z., Chong, W., Yao, X., Zhao, Z., Huang, X., Han, J.: Physics inspired hybrid attention for SAR target recognition. ISPRS J. Photogramm. Remote. Sens. **207**, 164–174 (2024)
59. Zhou, J., Feng, S., Sun, H., Zhang, L., Kuang, G.: Attributed scattering center guided adversarial attack for DCNN SAR target recognition. IEEE Geosci. Remote Sens. Lett. **20**, 1–5 (2023)
60. Wen, Z., Youlan, Yu., Qian, W.: Multimodal discriminative feature learning for SAR ATR: a fusion framework of phase history, scattering topology, and image. IEEE Trans. Geosci. Remote Sens. **62**, 1–14 (2024)

Single-Image Localised Reflection Removal with k-Order Differences Term

Radim Spetlik(✉) and Jiří Matas

Faculty of Electrical Engineering, Czech Technical University in Prague, Prague, Czechia
spetlrad@fel.cvut.cz

Abstract. We introduce the problem of localized glass-reflection removal (LGRR), which targets the removal of highlights caused by light reflections on glass surfaces. Our approach uses an end-to-end convolutional neural network trained on MS-COCO image crops with synthetic reflections generated from real light source images. We propose a cost function that includes image difference terms and show that it improves reflection removal and inpainting compared to standard L_1 and L_2 losses. Experimental results demonstrate that our method effectively reduces localized reflections.

1 Introduction

Glass-reflection removal is a relevant problem in computer vision [1,11,13,34,38,40]. Given a photo taken through glass, the goal is to remove the reflection while preserving the background.

In many cases, such as indoor night photography, localized highlights from lamps occur frequently. However, most existing work focuses on globally blended reflections, leaving local bright highlights less explored.

In this paper, we address *localized glass-reflection removal* (LGRR). Unlike previous approaches, we assume that the unwanted reflection is caused by light sources with limited spatial extent. This scenario arises when taking photos indoors with lamps or flash, where the illumination of the background and the reflection often differ (see Fig. 2).

We propose a method, LGRR, that uses a convolutional neural network (see Fig. 1). Our network either removes or significantly reduces localized reflections in a single RGB image. It is trained with a loss function that includes a k-order differences term, which improves both reflection removal and inpainting performance. Our method outperforms recent approaches on the *Localised Reflections in the Wild* [37] and *Lamps* datasets.

To summarize our contributions: (i) we introduce the problem of localized glass-reflection removal (LGRR), focusing on scenarios with localized bright highlights; (ii) we propose LGRR, a convolutional neural network that removes or attenuates localized reflections from a single RGB image; (iii) we design a novel k-order differences loss term that improves performance on both reflection removal and inpainting tasks; (iv) we release the *Lamps* dataset for training and evaluating LGRR methods; and (v) we analyze limitations in existing image synthesis models and propose an improved model.

J. Petersen and V. A. Dahl (Eds.): SCIA 2025, LNCS 15726, pp. 106–119, 2025.
https://doi.org/10.1007/978-3-031-95918-9_8

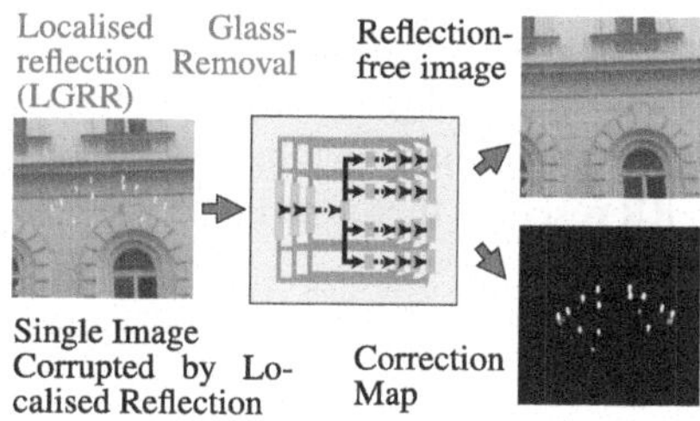

Fig. 1. Overview of the proposed LGRR method. Given a single color image, a reflection-free image and a correction map is estimated by a convolutional neural network.

Fig. 2. Images with uniform (left) and localised (right) reflections. Left: an indoor scene (taken from [38]). Right: an outdoor scene (taken from [37]).

2 Related Literature

Reflection removal has been studied extensively. Most existing methods focus on globally mixed reflections that resemble double exposure images. In contrast, we address *localized glass-reflection removal* (LGRR), where the reflection occupies only a small portion of the image. Such cases commonly occur in indoor scenes with bright lamps or other point light sources.

The LGRR problem is closely related to two established areas: inpainting and reflection removal. In inpainting, the goal is to fill in corrupted or unwanted regions with plausible content [17]. When strong localized reflections cause intensity saturation, important details behind the reflection are lost, and the task becomes similar to inpainting. However, traditional inpainting methods do not make use of any information from the corrupted region since they assume it to be unusable.

On the other hand, reflection removal methods aim to separate the reflection layer from the background layer, assuming that sufficient information is available for a successful separation. In scenarios where intensity saturation does not occur, LGRR aligns more closely with standard reflection removal problems. Although there are corner cases where the problem resembles inpainting, the connection to reflection removal is predominant in most instances.

Recent reflection removal research often assumes ideal photo acquisition conditions. Many approaches impose strict constraints to manage the ill-posed nature of the problem, which can lead to models that perform well on synthetic or controlled datasets but may struggle with real-world images. Additionally, the reflection removal community typically assumes that reflections originate from a transparent surface. In our work, we further assume that the reflective surface is a flat glass plate with multiple interfaces. Without this assumption, the task becomes a general layer separation problem, similar to handling double exposure images.

Overall, our work aims to bridge the gap between inpainting and traditional reflection removal by focusing on localized reflections in real-world scenarios.

We are aware of only three recent RR methods [11,21,25] not utilizing neural networks, which has become the dominant approach. Neural network training requires a large number of images. Data collection for the development of such methods is a tedious work. Most methods are thus trained and validated on synthesized data and tested on a small set of real-world images. Performance of RR methods on real-world images is heavily affected by particularities of the adopted image synthesis model [7]. In the following paragraphs, the discussion of the RR research therefore emphasises the image formation model.

Significant number of recent RR papers [2,3,5,6,9,30–32,40] adopted a RR problem formulation where the corrupted image is

$$\mathbf{I} = c(\alpha\mathbf{B} + \beta\mathbf{R}), \tag{1}$$

where $\mathbf{R}$ is the reflection layer, $\mathbf{B}$ is the background layer, $\alpha, \beta \in [0,1]$ are weighting coefficients, and c is a clipping function keeping brightness values in a valid range. Images $\mathbf{R}$ and $\mathbf{B}$ are of the same size, *i.e.* $\mathbf{R}, \mathbf{B} \in \mathbb{R}^{3\times H\times W}$, where H is height, and W is width. This formulation is an instance of image layer separation and since there are two layers, it is closest to the separation of two images in "double exposure photos". It is far away from a real-world setting, in which: i) the reflection surface does not commonly span the whole picture area, ii) the camera axis is rarely orthogonal to the reflection surface, and if it is, the photographer is reflected in the surface too, iii) the background and reflected scenes have different relative brightness, or contrast, iv) camera optics introduce a narrow depth of field, v) the reflective surface is a double-surface glass resulting in the "ghosting effect", and vi) the strength of a reflection depends on the angle of incidence of a light beam coming from the reflected scene with respect to a glass plane – a relation described by the Fresnel equations. The data synthesis model (1) is therefore not sufficient for the RR image synthesis.

An extension of the model (1) introduces the "blurring" kernel $\mathbf{k}$

$$\mathbf{I} = c\left(\alpha\mathbf{B} + \beta(\mathbf{k} * \mathbf{R})\right), \tag{2}$$

where $*$ denotes convolution. In this model, it is assumed that the depth of field of a lens attached to the camera is small and therefore the reflection is "blurry" [35]. By introducing the convolutional kernel $\mathbf{k}$, one can also simulate the "ghosting effect", *i.e.* multiple reflections of the scene which are spatially shifted and possibly attenuated.

Li *et al.* [19], together with [4,10,28,31], adopt the image formation model

$$\mathbf{I} = c\left(\mathbf{W} \odot \mathbf{B} + (\mathbf{1} - \mathbf{W}) \odot (\mathbf{k} * \mathbf{R})\right), \tag{3}$$

where $\odot$ denotes element-wise multiplication, and $\mathbf{W}$ is a matrix which weights the contribution of the transmission layer at each pixel. Their model does not allow for the brightness shift which would simulate scenes having different absolute brightness, a situation which is common in the real world. On the other hand, the weight matrix $\mathbf{W}$ might be used as the relative amplitude coefficient map [14], *i.e.* it might simulate the strength of the reflection when the camera axis is not orthogonal to a glass pane, a relation described by the Fresnel equations. The situation, in which the camera axis is not orthogonal to a glass pane, has not been explored yet.

In [7], an observation is made that methods training with images synthesized by models (1) and (2) "generalize poorly to real photographs". Therefore, the authors develop an image synthesis method with a new image formation model

$$\mathbf{I} = c\left(\alpha\mathbf{B} + \beta \cdot c(\mathbf{k} * \mathbf{R} + K_{\mathbf{R}})\right), \tag{4}$$

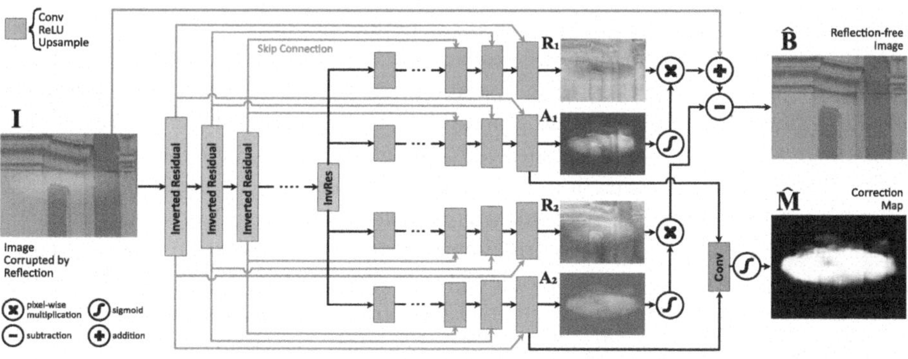

Fig. 3. Localised Reflection Removal Network. The input of the network is a single RGB image $\mathbf{I}$, the outputs are the corruption-free image $\hat{\mathbf{I}}$ and the map of correction $\hat{\mathbf{M}}$.

where $K_{\mathbf{R}} \in \mathbb{R}$ is a scalar. The method jointly simulates brightness shift of the reflection layer and the reflective surface being out-of-focus. In this work, $K_{\mathbf{R}}$ depends on brightness values of B and R. Image composition (4) approach is also used in [1, 11, 12, 15, 22, 29, 39]. Model (4) does not allow for brightness shift of the background layer, *i.e.* does not cover situation with the background being significantly darker than the reflected scene.

A purely data-driven approach is followed by Ma *et al.* [24], who propose the model

$$\mathbf{I} = G(\mathbf{B}, \mathbf{R}), \tag{5}$$

where $G(\cdot,\cdot)$ is a mapping function that generates reflection-corrupted images. A generative adversarial network [8] (GAN) setup is used. In [24], the generator is used as the mapping function G producing reflection-corrupted images. Wen *et al.* [34] utilize GAN differently in the data generation process. Instead of synthesizing corrupted images directly, a generator produces the alpha blending mask $\mathbf{W}$ and apply the model (3). Purely data-driven approaches fail to capture the statistics of the real-world data when the training set misses common classes of samples. Since one may explicitly name and simulate particular phenomena of interest, using purely-data driven image generation approaches might not be beneficial.

Different data generation choices produce methods that are inherently constrained by the image formation model they utilize. None of the discussed models simulates frequent real-world phenomena altogether. Therefore, we propose the image formation model

$$\mathbf{I} = \mathbf{W} \odot c(\alpha \cdot \mathbf{B} + K_{\mathbf{B}}) + (\mathbf{1} - \mathbf{W}) \odot c(\beta \cdot \mathbf{k} * \mathbf{R} + K_{\mathbf{R}}), \tag{6}$$

where $K_{\mathbf{R}}, K_{\mathbf{B}} \in \mathbb{R}$ are scalars simulating scene brightness shift. This model is capable of simulating all the discussed phenomena.

All digital cameras perform non-linear post-processing of captured images. This processing is another factor which makes the single-image RR very hard. To perform image synthesis correctly, one must: i) undo the non-linear image processing done by a digital camera, ii) blend the images, and iii) redo the non-linear processing. Since this processing is highly implementation-specific, only the "gamma correction" is typically undone. The gamma correction is an operation transforming brightness from a linear form, as captured by the camera, to a non-linear form natural for human vision. The gamma correction is usually parametrized by a single scalar – the *gamma*. Even if we assume that the gamma correction is the only non-linear processing performed by a camera, once the image is processed, it is virtually impossible to recover the value of gamma. Note that the gamma correction is a pixel-wise operation. The in-camera post-processing also includes global or template-based operations such as demosaicing, white balancing, and possibly many others. Lastly, training images are commonly stored in a JPEG format where another processing takes place and compression artifacts are often introduced.

Having discussed the various data generation models and their most important features, we conclude that the recent RR methods are likely to under-perform when used with real-world images taken with real cameras. Also, since the majority of recent methods is built using neural networks, and since authors commonly do not share the training codes and datasets, reproduction and validation of the published results is practically impossible.

Inpainting. In experiments, we were able to generate high-quality inpainting results with the inpainting method of Yi *et al.* [36]. The method exploits a novel *Contextual Residual Aggregation* mechanism producing high frequency residuals by weighted aggregating of residuals from contextual patches. This mechanism, when compared to common inpainting networks, reduces the network memory and computing power requirements. It is used in a two-stage neural network, which is learnt in a GAN setup. In addition, a *light weight gated convolution* is introduced reducing the number of parameters required for convolution gating. The method achieves a low ℓ_1 error of 5.439% on validation images of the Places2 dataset [41].

3 Proposed Method

This section describes the proposed LGRR method developed specifically for the LGRR task, *i.e.* for the removal of reflections from a picture taken through a glass.

Let $\mathcal{T} = \{(\mathbf{I}_j, \mathbf{B}_j, \mathbf{M}_j) \in \mathcal{I} \times \mathcal{B} \times \mathcal{M} \mid j = 1, \ldots, N\}$ be the training set that contains N tuples of RGB images $\mathbf{I} \in \mathcal{I}$ corrupted by a localised reflection, corrected RGB images $\mathbf{B} \in \mathcal{B}$ with the reflection removed, and binary "reflection maps" $\mathbf{M} \in \mathcal{M}$ containing ones in areas with reflection and zeros elsewhere. Symbol $\mathcal{I}$ denotes a set of all reflection-corrupted images, $\mathcal{B}$ is a set of all reflection-corrected images, and $\mathcal{M}$ is a set of all binary reflection maps.

3.1 Network Architecture

The architecture of LGRR neural network is inspired by the U-Net [26] and utilizes one encoder and four decoders (see Fig. 3). The decoders share skip connections from the encoder; each couple of decoders produces an RGB residual, $\hat{\mathbf{R}}$, and a single-channel attention map, $\hat{\mathbf{A}}$. Residual $\hat{\mathbf{R}}_1$ is multiplied with the corresponding attention map $\hat{\mathbf{A}}_1$ and added to input image $\mathbf{I}$. After that, the product of the second decoder couple $\hat{\mathbf{R}}_2\hat{\mathbf{A}}_2$ is subtracted from $\mathbf{I}$. Activations from the last layer of the "attention map" decoders are fed to a convolutional layer producing a binary "correction map", $\hat{\mathbf{M}}$. Ideally, ones in the map represent areas in image $\mathbf{I}$ that were corrected, zeros the areas where no change was made.

3.2 Objective Function

We exploit the specific nature of the LGRR problem and we do not follow a common approach of recent RR methods to training. In a common RR approach, a combination of three classes of losses is used in the training. The classes are usually called: i) a "pixel loss" – penalizes some norm of pixel-wise differences of the output and target images, ii) a "feature loss" – penalizes some norm of differences of selected activations in a pre-trained network, and iii) an "adversarial loss" – penalizes images having different statistics than the ground-truth images. In terms of the RR methods, we only employ the "pixel loss"

$$\mathcal{L}_{pix}(\mathbf{I}, \mathbf{B}, \Lambda; \mathbf{\Phi}) = \sum_{i,j} \lambda_1 \left\| \mathbf{B}_{i,j} - \hat{\mathbf{B}}_{i,j} \right\|_1 + \lambda_2 \left\| \mathbf{B}_{i,j} - \hat{\mathbf{B}}_{i,j} \right\|_2, \tag{7}$$

where $\hat{\mathbf{B}} = u(\mathbf{I}; \mathbf{\Phi})$ is the RGB output of the LGRR network, indices i, j denotes the i-th row and j-th column of an image, $\mathbf{\Phi}$ is a concatenation of all convolutional filter parameters, $\lambda_1, \lambda_2 \in \Lambda$ are weights of ℓ_1 and ℓ_2 terms, and $\Lambda \subset \mathbb{R}$ is a set of all weights of terms.

When trained with $\mathcal{L}_{pix}$ only, the network did not perform well in cases when the background was completely covered by a reflection. We improved the performance by introducing the k-order differences term

$$\begin{aligned} \mathcal{L}_{diff}(\mathbf{I}, \mathbf{B}, \Lambda, k; \mathbf{\Phi}) = \sum_{l=1}^{k} \gamma_l \Big(& \\ & \sum_{(i,j)\in\mathcal{V}^l} \left\| (\mathbf{B}_{i,j} - \mathbf{B}_{i+l,j}) - (\hat{\mathbf{B}}_{i,j} - \hat{\mathbf{B}}_{i+l,j}) \right\|_1 \\ & + \sum_{(i,j)\in\mathcal{H}^l} \left\| (\mathbf{B}_{i,j} - \mathbf{B}_{i,j+l}) - (\hat{\mathbf{B}}_{i,j} - \hat{\mathbf{B}}_{i,j+l}) \right\|_1 \Big), \end{aligned} \tag{8}$$

where $\mathcal{V}^l$ and $\mathcal{H}^l$ denote sets of all indices valid for vertical and horizontal shifts by l, and $\gamma_l \in \Lambda$ is a weight of the l-th difference term.

To learn the correction map $\mathbf{M}$, we apply a simple ℓ_1 of difference of the map and the prediction $\hat{\mathbf{M}}$

$$\mathcal{L}_{msk}(\mathbf{I}, \mathbf{M}, \Lambda; \mathbf{\Phi}) = \sum_{i,j} \lambda_{\mathbf{M}} \left\| \mathbf{M}_{i,j} - \hat{\mathbf{M}}_{i,j} \right\|_1, \tag{9}$$

Fig. 4. Samples from the Lamps dataset (see Sect. 4.2).

where $\hat{\mathbf{B}} = v(\mathbf{I}; \boldsymbol{\Phi})$ is the single-channel output of the LGRR network – a prediction of the correction map, and $\lambda_{\mathbf{M}} \in \Lambda$ is the weight of the map term.

The final LGRR loss is

$$\begin{aligned} \mathcal{L}_{LGRR}(\mathbf{I}, \mathbf{B}, \mathbf{M}, \Lambda, k; \boldsymbol{\Phi}) &= \mathcal{L}_{pix}(\mathbf{I}, \mathbf{B}, \Lambda; \boldsymbol{\Phi}) \\ &+ \mathcal{L}_{diff}(\mathbf{I}, \mathbf{B}, \Lambda, k; \boldsymbol{\Phi}) + \mathcal{L}_{msk}(\mathbf{I}, \mathbf{M}, \Lambda; \boldsymbol{\Phi}). \end{aligned} \tag{10}$$

Implementation Details. The encoder was built with MobileNetV2 inverted residual blocks [27]. The network was trained with the PyTorch library[1], and the AdamW optimizer [23]. The values of the following hyperparameters were found by a grid search. The weight decay was set to 10^{-3}, the learning rate to 10^{-3}. Batch size was 50. The term weights $\lambda_1, \lambda_{\mathbf{M}}, \gamma_1, \ldots, \gamma_k = 1$, $\lambda_2 = 0.3$. The set of all training tuples $\mathcal{I} \times \mathcal{B} \times \mathcal{M} = \mathbb{R}^{3 \times 128 \times 128} \times \mathbb{R}^{3 \times 128 \times 128} \times \mathbb{R}^{1 \times 128 \times 128}$. The networks were trained for 10^4 epochs.

4 Experiments

In this section, we first describe the methodology by which we assess the results of experiments. Then, we discuss the datasets used in our experiments. Afterwards, results of two quantitative and three qualitative experiments are presented.

4.1 Methodology

A common approach to RR method evaluation includes computation of: i) the peak signal-to-noise ratio (PSNR), which is the ratio between the maximum possible power

[1] Available at https://www.pytorch.org.

of a signal and the power of corrupting noise, and ii) the structural similarity index measure [33], expressing similarity between two images while taking into account perceived visual quality. We did not follow this approach as we are interested in correcting localised reflections – small areas in the image. We evaluated the difference, in the corrupted area, between output $\hat{\mathbf{B}}$ of LGRR network and $\mathbf{B}$, the ground truth background image, by ℓ_1, which is common in inpainting evaluation.

4.2 Datasets

Since we are the first to identify the problem of LGRR, there are no publicly available datasets suitable for training and evaluation of LGRR methods. In experiments, we use two datasets – a novel *Lamp* dataset (see Fig. 4), and the *Localised Reflections in the Wild* dataset.

Lamps. We created a public website and asked friends and colleagues to take photos of turned-on lamps and lights at night. We requested donors to adjust camera exposure settings to prevent brightness saturation of the lamps and to keep the borders of the image "black" so the source of light stays fully inside the photo. In this way, we collected a total of 191 *training and validation* images. The photos were uploaded exclusively from mobile phones. We stored the images in a PNG format. We then manually cropped the photos to account for different aspect ratios and different scales of light sources. The cropped images have an aspect ratio of $1:1$ and the light source fills between 20% and 50% of the image area.

Localised Reflections in the Wild. To evaluate LGRR method on real-world data, 17 $(\mathbf{I}, \mathbf{B})$ tuples out of 40 from the *Localised Reflections in the Wild* dataset [37] were randomly selected. The images were collected in one European town and capture mainly old building facades. The pictures were shot from cafes and from buses and trams. In order to create a reflection-free background, two photos of each reflection were taken from different angles of view forming a stereo pair. Between the two photos, homography was found and the images were registered and combined to avoid presence of reflections in the final image, $\mathbf{B}$. Since the dataset does not include corruption masks $\mathbf{M}$, we created them manually, and cropped the $(\mathbf{I}, \mathbf{B}, \mathbf{M})$ triplet to the $1:1$ aspect ratio. In the cropped image, $\mathbf{I}$, reflections fill 30% to 50% of the image. During the capture, the camera was set to manual exposure mode. Note that the method of obtaining the reflection-free images was not fully successful in all cases. In some ground-truth images, there are visible small alignment errors or color perturbations. Therefore, this dataset is best suited for qualitative experiments.

4.3 Synthetic Data Generation

The training and validation datasets for the LRRG network were constructed as follows. We collected 191 night-time lamp photos from the *Lamps* dataset. These images were used as reflection cutouts and were superimposed on 200K random square crops from the *train2017* subset of MS-COCO [20] to form a synthetic dataset.

Fig. 5. Qualitative comparison on the *Localised Reflections in the Wild* dataset [37]. Top to bottom: input image **I**, LGRR method, inpainting (Inp.) Yi *et al.* [36], reflection removal (RR) Li *et al.* [16].

Images were synthesized using the model in (4). We verified that this model simulates the ghosting effect, non-orthogonal camera views of a glass pane, and brightness shifts in localized reflections, capturing realistic variations such as blurring and brightness changes. Thus, the more complex model (6) is unnecessary and may be harder to use due to its additional parameters.

Since localised reflections are assumed, camera acquisition settings were set for the background, *i.e.* not for the reflections. Therefore, we set $\alpha = 1$. We sample β from a uniform distribution $\mathcal{U}(1.0, 2.0)$ to simulate different angles of incidence of the light on the reflective surface. We randomly sample $K_{\mathbf{R}}$ from $\mathcal{U}(-0.05, 0.05)$, since the brightness was normalized to $[0, 1]$. The "ghosting effect" is simulated with kernel $\mathbf{k}$ containing two Gaussian responses spatially shifted with respect to each other. One of the Gaussians is multiplied by a scalar sampled from $\mathcal{U}(0.7, 0.9)$, the shift in pixels is sampled independently for x and y axes from two uniform distributions $\mathcal{U}(-15, 15)$.

Table 1. Ablation study – mean absolute errors as a function of k, the order of the differences term in $\mathcal{L}_{LGRR}$, or its absence ($k = 0$).

	Localised Reflection Removal							Inpainting						
k	0	1	2	3	4	5	6	0	1	2	3	4	5	6
MAE	8.33	8.15	8.18	8.12	8.11	**7.98**	8.10	13.54	13.29	13.14	13.16	13.12	**13.05**	13.08

4.4 Quantitative Results

Here we present two ablations studies testing the dependence of the mean absolute error in the corrupted area on the choice of k in $\mathcal{L}_{LGRR}$ (8).

In the first experiment, the network was trained on the synthesized dataset (Sect. 4.3) to perform the localised reflection removal. As seen in the upper part of Table 1, the mean absolute difference of brightness values on the validation dataset drops from 8.33 for $k = 0$ to 7.98 for $k = 5$. The k-order differences term (8) improves the performance of the network by $\approx 5\%$.

In the second experiment, the network was trained on the synthesized dataset (Sect. 4.3) to inpaint, or to "fill in", erased parts of images. The images from the Lamps dataset (see Sect. 4.2) were thresholded to create binary masks which were then: i) utilized for erasing parts of images, and ii) concatenated with erased images. Therefore, in this experiment the input of the LGRR network had four channels – an RGB image and a mask. As in the Localised Reflection Removal experiment, the mean absolute difference on the validation datasets drops from 13.54 for $k = 0$ to 13.05 for $k = 5$, which is an improvement by $\approx 3.5\%$.

4.5 Qualitative Results

In this section, we describe four experiments inspecting: i) the performance of the LGRR network on real-world images and ii) effects of the k-order differences term on quality of correction of reflection-corrupted images.

Qualitative Comparison. We compare our LGRR method against the state-of-the-art reflection removal method of Li *et al.* [18] and the inpainting method of Yi *et al.* [36] on the Localised Reflections in the Wild dataset (Sect. 4.2). The inpainting method does not require the target areas to be zeroed out; it processes the full image along with a manually drawn mask (in the provided format).

Figure 5 shows that the inpainting method produces visually plausible content in the reflection regions. However, it fails to utilize available brightness, leading to artifacts, especially when the image has structured details.

In contrast, the reflection removal method [18] is tailored for full-image reflections and is less effective on localised ones. In the rightmost example, where the reflection covers most of the image, its performance is comparable to LGRR.

Our LGRR leverages brightness information behind localised reflections, yielding superior results. When the reflection is too strong and little brightness is available (e.g., in the leftmost image), the network attenuates the reflection and reconstructs the missing structure. However, it is less sensitive to weak, wide reflections due to its training on localised cases.

Failure Cases. Figure 7 shows five failure modes of our method:

1. **Textured Light Source:** The leftmost column shows reflections from a textured light source. Improved training diversity may mitigate this failure.
2. **Weak, Wide Reflections:** The second column illustrates weak, broad reflections. The network also shows reduced sensitivity near image borders—a result of centered training data.
3. **Dirty Glass:** In the third column, reflections are only partially removed, likely due to unmodeled effects from dirty glass in the training set.
4. **False Positives:** The fourth column reveals regions mistakenly identified as reflections. This oversensitivity stems from training on varied localized reflections and is challenging to correct without targeted data.
5. **Elongated Reflections:** The rightmost column shows failures on elongated reflections. Incorporating similarly shaped samples into the training set could improve performance.

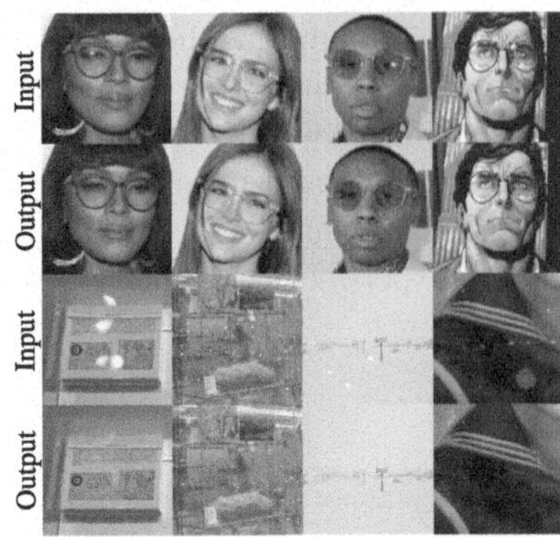

Fig. 6. Applications of LGRR method. Top two rows: glasses glint removal, two bottom rows: reflection removal on display-case cover, snow removal, dust removal.

Fig. 7. Failure cases from the *Localised Reflections in the Wild* dataset. From top to bottom: input image $\mathbf{I}$, ground truth, output of LGRR method $\hat{\mathbf{B}}$.

In summary, reflection shapes or angles not present in training data are not recognized by the model, causing false positives or incomplete removal. In principle, collecting new training examples with similarly shaped/positioned reflections alleviates this. Nonetheless, model capacity or architecture might also need adjusting if these new shapes are highly complex.

Qualitative Ablation Study. In this experiment, we were interested in how the quality of reconstruction changes when the k-order differences term is included in the objective function of the inpainting problem. Although the results for the LGRR problem were similar, they were harder to spot. For this experiment, the LGRR network was trained with k in $\mathcal{L}_{LGRR}$ ranging from 0 to 6. A pair of centered strips, see Fig. 8, was zeroed in input RGB images, and a mask marking the zeroed area was concatenated with the image forming a 4-channel input. The network was trained on the *train2017* and tested on the *test2017* subset of the MS-COCO dataset.

There are four rows in Fig. 8, each corresponding to a different input image. We magnified the areas in which the images were inpainted. When inspected closely, increasing k improves the reconstruction of edge structures in the images. In the second row, the subtle black lines in the image are beginning to appear from the bottom for $k = 5$. A gradual improvement in the reconstruction of "lines" with increasing k is visible in all four samples. Note that even though there appears to be an increase in amplitude of high frequencies in the image, LGRR network still produces inpainting results with a larger proportion of low frequency content. We believe that this is caused by a relatively low number of parameters and a relatively shallow architecture.

Fig. 8. Inpainting qualitative ablation study on the MS-COCO *test2017* dataset.

Other Applications. We demonstrate additional applications of the LGRR method. Although designed for LGRR, our model applies to tasks involving the removal of localized corruptions. Four examples are shown in Fig. 6, using the same model as in our qualitative LGRR experiments.

The top two rows show the removal of glints, including on hand-drawn images (e.g., a comic superhero). The bottom rows display removal of multiple reflections from display cases, including cases with strong, isolated reflections where brightness saturation occurs. Additional examples include snowflake and dust/spot removal.

5 Conclusion

In this work, we introduced the problem of *localized glass-reflection removal* (LGRR) from a single image. We proposed a convolutional neural network architecture with a novel k-order differences loss term. Experimental results demonstrate that incorporating the k-order differences term into the training objective improves both qualitative and quantitative performance for the LGRR and inpainting tasks. When available, the method also makes use of brightness information from behind the reflection. We observed that the addition of the k-order differences term enhances high-frequency details in the inpainted regions, leading to a reconstruction that is visually consistent with the surrounding content. Although real glass may produce complex reflections and multi-bounce effects, our experiments indicate that a simple convolution and shift blending approach is sufficient to handle strongly localized, near-saturated highlights. Moreover, since the same architecture and loss function can address various localized bright artifacts, we expect that minimal retraining or fine-tuning could extend the method to remove glints, dust specks, snowflakes, and similar corruptions.

Acknowledgements. The research reported in this paper has been funded by Czech Science Foundation grant 25-15993S.

References

1. Abiko, R., Ikehara, M.: Single image reflection removal based on GAN with gradient constraint. IEEE Access **7**, 148790–148799 (2019)
2. Ahmed, A., Kim, S., Elgharib, M., Hefeeda, M.: User-Assisted Video Reflection Removal (2020)
3. Alayrac, J.-B., Carreira, J., Arandjelović, R., Zisserman, A.: Controllable Attention for Structured Layered Video Decomposition (2019)
4. Arvanitopoulos, N., Achanta, R., Süsstrunk, S.: Single image reflection suppression. In: 2017 IEEE Conference on Computer Vision and Pattern Recognition (CVPR), pp. 1752–1760 (2017). ISSN 1063-6919
5. Chang, Y., Jung, C., Sun, J.: Joint reflection removal and depth estimation from a single image. IEEE Trans. Cybern. 1–14 (2020)
6. Chang, Y., Jung, C., Sun, J., Wang, F.: Siamese dense network for reflection removal with flash and no-flash image pairs. Int. J. Comput. Vis. **128**(6), 1673–1698 (2020)
7. Fan, Q., Yang, J., Hua, G., Chen, B., Wipf, D.: A Generic Deep Architecture for Single Image Reflection Removal and Image Smoothing, pp. 3238–3247 (2017)
8. Goodfellow, I., et al.: Generative adversarial nets. In: Advances in Neural Information Processing Systems, pp. 2672–2680 (2014)
9. Han, B., Sim, J.: Glass reflection removal using co-saliency-based image alignment and low-rank matrix completion in gradient domain. IEEE Trans. Image Process. **27**(10), 4873–4888 (2018)
10. Han, B., Sim, J.: Single image reflection removal using non-linearly synthesized glass images and semantic context. IEEE Access **7**, 170796–170806 (2019)
11. Heydecker, D., et al.: Mirror, mirror, on the wall, who's got the clearest image of them all?–a tailored approach to single image reflection removal. IEEE Trans. Image Process. **28**(12), 6185–6197 (2019)
12. Jin, M., Meishvili, G., Favaro, P.: Learning to extract a video sequence from a single motion-blurred image. In: 2018 IEEE/CVF Conference on Computer Vision and Pattern Recognition, pp. 6334–6342 (2018). ISSN 2575-7075
13. Kim, S., Huo, Y., Yoon, S.-E.: Single image reflection removal with physically-based rendering. arXiv:1904.11934, [cs] (2019)
14. Kong, N., Tai, Y., Shin, J.S.: A physically-based approach to reflection separation: from physical modeling to constrained optimization. IEEE Trans. Pattern Anal. Mach. Intell. **36**(2), 209–221 (2014)
15. Lee, D., Yang, M.-H., Oh, S.: Generative single image reflection separation. arXiv:1801.04102, [cs] (2018)
16. Li, C., Yang, Y., He, K., Lin, S., Hopcroft, J.E.: Single Image Reflection Removal Through Cascaded Refinement, pp. 3565–3574 (2020)
17. Li, J., Wang, N., Zhang, L., Du, B., Tao, D.: Recurrent Feature Reasoning for Image Inpainting, pp. 7760–7768 (2020)
18. Li, T., Lun, D.: Single-image reflection removal via a two-stage background recovery process. IEEE Signal Process. Lett. **26**(8), 1237–1241 (2019)
19. Li, Y., Brown, M.S.: Single image layer separation using relative smoothness. In: 2014 IEEE Conference on Computer Vision and Pattern Recognition, pp. 2752–2759 (2014)

20. Lin, T.-Y., et al.: Microsoft COCO: common objects in context (2015)
21. Liu, R., Jiang, Z., Fan, X., Luo, Z.: Knowledge-driven deep unrolling for robust image layer separation. IEEE Trans. Neural Netw. Learn. Syst. **31**(5), 1653–1666 (2020)
22. Liu, Y.-L., Lai, W.-S., Yang, M.-H., Chuang, Y.-Y., Huang, J.-B.: Learning to See Through Obstructions, pp. 14215–14224 (2020)
23. Loshchilov, I., Hutter, F.: Decoupled weight decay regularization. arXiv:1711.05101 (2017)
24. Ma, D., Wan, R., Shi, B., Kot, A., Duan, L: Learning to jointly generate and separate reflections. In: 2019 IEEE/CVF International Conference on Computer Vision (ICCV), pp. 2444–2452 (2019). ISSN 2380-7504
25. Punnappurath, A., Brown, M.S.: Reflection Removal Using a Dual-Pixel Sensor, pp. 1556–1565 (2019)
26. Ronneberger, O., Fischer, P., Brox, T.: U-Net: convolutional networks for biomedical image segmentation. In: Proceedings of International Conference on Medical Image Computing and Computer Assisted Intervention, pp. 234–241 (2015)
27. Sandler, M., Howard, A., Zhu, M., Zhmoginov, A., Chen, L.-C.: MobileNetV2: Inverted Residuals and Linear Bottlenecks (2019)
28. Schechner, Y.Y., Kiryati, N., Basri, R.: Separation of transparent layers using focus. Int. J. Comput. Vis. **39**(1), 25–39 (2000)
29. Sun, J., Chang, Y., Jung, C., Feng, J.: Multi-modal reflection removal using convolutional neural networks. IEEE Signal Process. Lett. **26**(7), 1011–1015 (2019)
30. Wan, R., Shi, B., Duan, L.-Y., Tan, A.-H., Gao, W., Kot, A.C.: Region-aware reflection removal with unified content and gradient priors. IEEE Trans. Image Process. **27**(6), 2927–2941 (2018)
31. Wan, R., Shi, B., Duan, L.-Y., Tan, A.-H., Kot, A.C.: CRRN: multi-scale guided concurrent reflection removal network. In: 2018 IEEE/CVF Conference on Computer Vision and Pattern Recognition, pp. 4777–4785 (2018)
32. Wan, R., Shi, B., Li, H., Duan, L.-Y., Kot, A.C.: Face image reflection removal. arXiv:1903.00865, [cs] (2019)
33. Wang, Z., Bovik, A.C., Sheikh, H.R., Simoncelli, E.P.: Image quality assessment: from error visibility to structural similarity. IEEE Trans. Image Process. **13**(4), 600–612 (2004)
34. Wen, Q., Tan, Y., Qin, J., Liu, W., Han, G., He, S.: Single Image Reflection Removal Beyond Linearity, pp. 3771–3779 (2019)
35. Yang, Y., Ma, W., Zheng, Y., Cai, J.-F., Xu, W.: Fast Single Image Reflection Suppression via Convex Optimization, pp. 8141–8149 (2019)
36. Yi, Z., Tang, Q., Azizi, S., Jang, D., Xu, Z.: Contextual residual aggregation for ultra high-resolution image inpainting. In: 2020 IEEE/CVF Conference on Computer Vision and Pattern Recognition (CVPR), pp. 7505–7514 (2020). ISSN 2575-7075
37. Zabulskyi, V.: Reflection removal with generative adversarial networks. Bachelor thesis, Ukrainian Catholic University, Lviv (2020)
38. Zhang, H., et al.: Fast user-guided single image reflection removal via edge-aware cascaded networks. IEEE Trans. Multimed. **22**, 2012–2023 (2019)
39. Zhang, X., Ng, R., Chen, Q.: Single Image Reflection Separation with Perceptual Losses (2018)
40. Zheng, Q., Shi, B., Jiang, X., Duan, L.-Y., Kot, A.C.: Denoising adversarial networks for rain removal and reflection removal. In: 2019 IEEE International Conference on Image Processing (ICIP), pp. 2766–2770 (2019). ISSN 2381-8549
41. Zhou, B., Lapedriza, A., Khosla, A., Oliva, A., Torralba, A.: Places: a 10 million image database for scene recognition. IEEE Trans. Pattern Anal. Mach. Intell. **40**(6), 1452–1464 (2018)

Detection, Recognition, Classification, and Localization in 2D and/or 3D

The Impact of Semi-supervised Learning on Line Segment Detection

Johanna Engman, Kalle Åström, and Magnus Oskarsson(✉)

CVML, Centre for Mathematical Sciences, Lund University, Lund, Sweden
{johanna.engman,karl.astrom,magnus.oskarsson}@math.lth.se

Abstract. In this paper we present a method for line segment detection in images, based on a semi-supervised framework. Leveraging the use of a consistency loss based on differently augmented and perturbed unlabeled images with a small amount of labeled data, we show comparable results to fully supervised methods. This opens up application scenarios where annotation is difficult or expensive, and for domain specific adaptation of models. We are specifically interested in real-time and online applications, and investigate small and efficient learning backbones. Our method is to our knowledge the first to target line detection using modern state-of-the-art methodologies for semi-supervised learning. We test the method on both standard benchmarks and domain specific scenarios for forestry applications, showing the tractability of the proposed method.

Keywords: Semi-supervised learning · Line detection · Domain adaptation

1 Introduction

Line segments are prevalent in images depicting scenes of both in- and outdoor human-made environments as well as in nature. Line structures are used by humans for efficient and robust interpretation of the world [4,8], and can also play a crucial role in automatic scene comprehension, encapsulating scene structure efficiently [40]. Applications and downstream tasks that are made more tractable encompass various computer vision tasks, including 3D reconstruction, Structure-from-Motion (SfM) [16,24,25], Simultaneous Localization and Mapping (SLAM) [12], visual localization [10], tracking, and vanishing point estimation [30]. Lines are also efficient intermediate representations for subsequent semantic processing. Compared to point based features, line features are more stable over time (this is true both on a small and large time scale) and more invariant to changes in scene environment such as e.g. lighting, season and weather conditions as shown in [28].

J. Engman—This work was supported by the strategic research project ELLIIT. Model training was enabled by the Berzelius resource provided by the KAW Foundation at the National Supercomputer Centre in Sweden.
Code: https://github.com/jo6815en/semi-lines.

J. Petersen and V. A. Dahl (Eds.): SCIA 2025, LNCS 15726, pp. 123–136, 2025.
https://doi.org/10.1007/978-3-031-95918-9_9

Fig. 1. From left to right: Ground truth detection, supervised model on 1/16 of data, supervised model on all data, proposed semi-supervised approach with 1/16 labeled data. The semi-supervised method improves significantly on the supervised method.

Although lines are perhaps mostly associated with man-made structures, there are many examples of linear structures in nature. In this paper we use, as a domain specific scenario, forestry images. Detecting lines in tree imagery is challenging, and many applications related to them are of particular importance. From an environmental stand-point trees can contribute with both carbon-dioxide binding and new sustainable materials in diverse industries. For many automation use cases involving forestry we would like to monitor, measure and interpret forestry scenes, which translates to both semantic computer vision tasks such as classification and scene understanding and geometric computer vision tasks such as camera pose estimation and tracking.

Fig. 2. Many trained models fail to generalize to new image domains. Top row shows output from left to right: DeepLSD [26], LETR [34], M-LSD [13], trained on the Wireframe dataset. Bottom row shows output from these models on an out of domain image.

Many newly formulated learned line segment detectors have shown both promising and impressive results [13,26,34]. However, many such large models fail to transition and generalize to new image domains without retraining, see Fig. 2. To tackle the expensive need for annotations and in contrast to large general and slower models, we propose the first method for *semi-supervised line segment detection*. Our method allows for domain specific and light-weight processing. At the same time, our use of semi-supervision enables better generalization and generates a method that better adapts to new scenes. We build on the ideas of semi-supervised learning for semantic segmentation from [29] and [37], we propose the first framework for semi-supervised line segment detection. The results is a fast, efficient, and accurate system that is trained on a smaller portion of the whole dataset and performs as well or even better than the standard supervised method.

Our main contributions are: *(i) We present the first framework for semi-supervised learning for line segment detection. (ii) We show that semi-supervised methods work well for line segment detection, improving significantly on the state-of-the-art for new environments, where there is little annotation. (iii) We show that small models, suitable for real-time applications, work well and can heavily benefit from semi-supervised methods.* Additionally we provide two new datasets for line segment detection containing forest scenes with ground truth lines, and provide code for inference and training.[1]

2 Related Work

In this paper we develop the first semi-supervised framework for line-segment detection. Here we provide background on line detection in general and on semi-supervised learning.

Line Detection. Early methods for detecting line-segments rely on detected edge features based either on first order methods, e.g. finding local maxima to smoothed image gradients [5,23] or second order methods, e.g. finding zero crossings of the Laplacian, [15]. Edge features are then typically grouped into line segments, e.g. using the Hough transform, [9,17] or line segment growing, [1,32]. Over the last years, as in most fields, learning based models have increased the performance of line segment detection. Some recent well functioning detectors are the transformer based LETR [34] and DeepLSD [26] that incorporates an optimization based refinement module for higher accuracy. A similar task to line detection is to identify wireframes, i.e. both lines and junctions. In End-to-End Wireframe Parsing [39] the authors focus on this problem. In addition to their model, they introduced a novel evaluation metric, sAP - structural average precision. This metric is more robust against overlapping line segments and multiple line detections for the same line. sAP is defined as the area under the precision recall curve computed from a scored list of the detected line segments. To evaluate our line segment detection we also use this metric in this paper. This method and other end-to-end wireframe methods, e.g. [11,18,22,35,36] have spurred new interest in more general line detection algorithms, [13]. We build our system using the line segment detector M-LSD proposed in [13], since it is fast and light-weight. Their main contribution is their segment-of-line (SoL) augmentations, where

[1] https://github.com/jo6815en/semi-lines.

they divide a line into overlapping segments that should all align. They use the Tri-point (TP) representation [19]. Another change from earlier work is that M-LSD generates line segments directly from the final feature maps in a single module process.

Semi-supervised Learning. Semi-supervised learning is a field of machine learning that tries to overcome the large need for annotated data and labels, combining the information gained in a smaller portion of labeled data together with the possibility to train on a larger dataset without labels. One of the first frameworks that succeeded with this was Mean Teacher [31], where they use two models simultaneously, one student learning fast both with labels and consistency towards the teacher, that learns slower over an average of the student. Mean Teacher uses the fact that augmenting the inputs differently should still generate the same label.

In 2018 [20] proposed an adversarial learning approach for semi-supervised learning for semantic segmentation, building on the ideas of General Adversarial Networks (GANs). Since then a number of improvements have been made, mostly abandoning the GAN-influenced ideas for contrastive learning and Mean Teacher approaches [2,7,14].

More closely related to our approach is the FixMatch model, which is a semi-supervised framework for classification using the weak-to-strong consistency regularization [29]. It uses one strongly augmented image together with one weakly augmented image to obtain information without labels. They propose to use a shared model, unlike the Mean Teacher method. This model has been further developed, creating more advanced frameworks [33].

Recently, a new semi-supervised learning framework, UniMatch, for semantic segmentation was proposed [37]. In contrast to recent development, they go back to the, fundamentally, simpler method of FixMatch [29], but add a second strongly perturbed version of the unlabeled input image. They use two strong views that simultaneously are guided by a common weak view. The role of the strong perturbed images is to minimize the distance between them, similar to contrastive learning. In [37] they perform extensive ablation studies showing that two strongly perturbed views achieve higher accuracy, compared to using one or three. In addition they add a version of the weakly augmented image and add an augmentation in the feature domain. For the unlabeled images [37] uses CutMix [38] as a regularization technique. While this is a good idea for classification and segmentation, we found that it is not an ideal technique for line detection. Using CutMix counteracts the system from finding long consistent lines, and we propose a variant more suited to line detection.

3 A Semi-supervised Line Segment Detector

In this section, we describe the details of our semi-supervised system. The main idea is to show a deep learning network a small portion of annotated data to drive the network to detect lines, and at the same time show the network more data but without any annotations. To help the network draw information from the unannotated data, a stream with slightly different versions of the same image is presented and a consistency requirement on the detection added. We visualize the system in Fig. 3, consisting of one part guided by existing ground truth, and one part utilizing information through unlabeled data. In our proposed method these two streams share the same model, F, which is guided by both the ground truth and the consistency requirement for the unlabeled data.

The idea of using multiple strongly perturbed versions of the data is often used within classification [3,6]. It has also been employed in semantic segmentation tasks [37]. There, the benefits of multiple strongly perturbed versions is explained as "regularizing two strong views with a shared weak view can be regarded as enforcing consistency between these two strong views as well" [37]. We are therefore interested in transferring this idea into our unlabeled stream and investigate if this is favorable in line segment detection as well. We validate that it is, in our ablation studies.

In the following sections we will describe the model architecture and the different loss functions in detail.

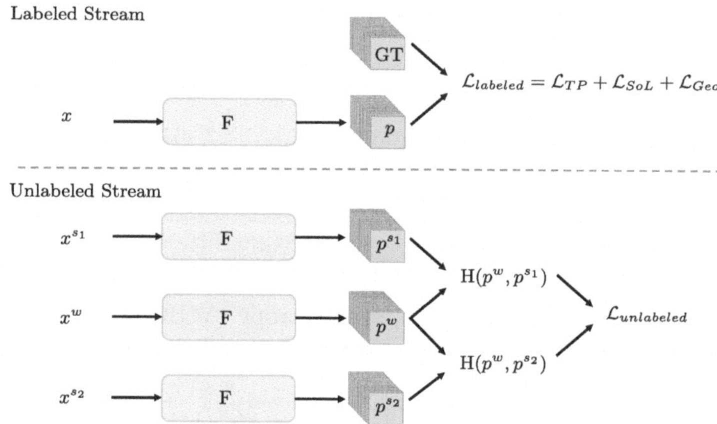

Fig. 3. System overview of our proposed method. Here F denotes the shared model, x is labeled data, x^w weakly perturbed unlabeled data, and $x^{s_{1,2}}$ strongly perturbed data. The loss function for the labeled data, $\mathcal{L}_{labeled}$, is described in Sect. 3.2, and is directly built on the loss from [13]. For the unlabeled stream, H denotes the cross-entropy loss described in Sect. 3.3 and is calculated between the weakly perturbed sample and one of the strongly perturbed samples.

3.1 Model Architecture

We want a fast, efficient and domain specific model that directly returns lines from an image. The framework M-LSD [13] meets these requirements. Utilizing the benefits of this fast line detector we build our semi-supervised framework for line segment detection with MobileNetV2 [27] to gain an easily trained, fast model, using the same tiny model as in [13]. This model has around 0.6 million model parameters, and results in a model that runs in real time even on compute constrained devices [13].

The model is described by $p = \mathrm{F}(x)$, where x is the input RGB-image rescaled to size $512 \times 512 \times 3$ and p is the output feature embedding of size $128 \times 128 \times 16$, that codes the line representation.

The network outputs 16 layers of feature maps. Seven of these are based on the Tri-point (TP) line representation [19]. The TP is an over-parameterization of the line segment, coding it by the line segment center point and two displacement vectors (to the segment end points). The seven layers consist of a length map, a degree map, four displacement maps, and one center map, all per pixel. The model uses the segment-of-line (SoL) augmentations, where each line is split into several TP representations. The next seven feature maps are the TP-maps for the SoL and follow the same configuration as the SoL feature maps. The last two feature layer maps are direct pixel-wise classification maps for line pixels and line junction pixels, respectively. All layers are used in the line loss, but only the first five feature layers are used at inference time for line detection [13].

3.2 Labeled Loss

The line segment loss for the labeled data $\mathcal{L}_{labeled} = \mathcal{L}_{TP} + \mathcal{L}_{SoL} + \mathcal{L}_{Geo}$, consists of three parts, a loss for the TP-representation $\mathcal{L}_{TP}$, a loss for the SoL augmentations, $\mathcal{L}_{SoL}$ and an additional geometric loss, $\mathcal{L}_{Geo}$ [13]. Here

$$\mathcal{L}_{TP} = \mathcal{L}_{center} + \mathcal{L}_{disp} + \mathcal{L}_{match}, \tag{1}$$

where $\mathcal{L}_{center}$ is the weighted binary cross-entropy (WBCE) loss and $\mathcal{L}_{disp}$ is the L1 loss (as proposed in [19]). [13] formulates the $\mathcal{L}_{match}$ loss as

$$\mathcal{L}_{match} = \frac{1}{|\mathbb{M}|} \sum_{(l,\hat{l}) \in \mathbb{M}} \left\| l_s - \hat{l}_s \right\|_1 + \left\| l_e - \hat{l}_e \right\|_1 + \left\| \bar{C}(\hat{l}) - (l_s + l_e)/2 \right\|_1, \tag{2}$$

where $\mathbb{M}$ is the set of matched line segments $(l, \hat{l})$ and $\bar{C}(\hat{l})$ is the center point of line $\hat{l}$ from the center map. The $\mathcal{L}_{SoL}$ loss is set up identically to $\mathcal{L}_{TP}$, but for the segmented line representation. The geometric information that [13] forms a loss for is four-fold. They formulate a segmentation loss as $\mathcal{L}_{seg} = \mathcal{L}_{junc} + \mathcal{L}_{line}$, where both the loss terms are the WBCE loss. The last part of the loss functions is a regression loss, $\mathcal{L}_{reg} = \mathcal{L}_{length} + \mathcal{L}_{degree}$, where each loss term is a L1 loss. These two parts are then combined info one geometric loss, $\mathcal{L}_{Geo} = \mathcal{L}_{seg} + \mathcal{L}_{reg}$.

3.3 Unlabeled Loss

For the unlabeled data we adopt the method used in [37] for semantic segmentation, with some changes. For every unlabeled image three versions are loaded into the model, one weakly perturbed, and two different strongly perturbed versions. A consistency loss is then calculated between each of the two strongly perturbed versions of the image and the weakly perturbed version. The loss is formulated as

$$\mathcal{L}_{unlabeled} = \frac{1}{N} \sum \mathbb{1}(max(p^w) \geq \tau)(\mathrm{H}(p^w, p^{s1}) + \mathrm{H}(p^w, p^{s2})), \tag{3}$$

where H is the cross-entropy loss and $\mathbb{1}$ is the indicator function. The threshold τ, makes sure that this loss only is applied when the prediction p^w is certain enough. Since our model outputs 16 layers we chose to look at the layer representing certainty for the center point of the lines.

The consistency loss is calculated over all of the output layers from the model, to ensure that the feature representation is kept throughout the training.

3.4 Augmentations

For the labeled data we use the same augmentations as in [13] consisting of random flip and rotation of the image, as well as the lines. A random hue, saturation, and value shift and a random brightness shift is also applied.

Following [37] for the augmentations the unlabeled data we first apply, a random flip and crop, which are recognized as the weak perturbations. Then, random blur, color jitter and gray scale are applied individually to two copies of the weakly perturbed image, the now strongly perturbed images.

Unlike the proposed method in [37] we do not have the feature augmented version of the unlabeled images, since we value a small and fast model, and chose to follow the model set up in [13] instead of having a larger bottle-neck model.

New CutMix Version. In [37] they make a strong motivation for using CutMix [38] for the unlabeled images and show that the models performance is boosted by this. One of the reasons is that this helps the model learn local image features better, without relying too much on full object or global image context. However, for line segment detection we find that using the regular CutMix is unfavorable, and leads to a bias to very short line segments. Therefore, we propose the following modification of the CutMix method. Instead of cutting out squares and patching them together for a more generalized model as in [38], we only split the images along one dimension, either in x- or in y dimension with equal probability. This still helps with generalization but prevents the model from only seeing short lines.

4 Experiments

In this section we will describe the experimental evaluation of the proposed framework. In Sects. 4 and 4 we describe the datasets and settings we used for the experiments. We carried out two kinds of experiment, *Annotation dependence* and *Domain generalization*. In Sect. 4.1 we conduct multiple test of how different portions of annotated data affect the resulting models. In Sect. 4.2 we investigate how well we can use the proposed method to generalize a trained model to a new image domain without annotations. For both the tests in Sect. 4.1 and 4.2 we compare the performances of our models to two pre-trained state-of-the-art (SOTA) models, DeepLSD [26] and LETR [34]. In Sect. 4.3 we conduct ablation studies for our proposed method. More images and extended ablation studies are provided in the supplementary material. All training was done on an NVIDIA A100 Tensor Core GPU.

Datasets and Evaluation Metrics. We have used a number of datasets to test our method. The goal was to use these datasets to test two things. First, how well semi-supervised methods work in terms of the balance between labeled and unlabeled data. Second, how well image domain adaptation works, by using primarily unlabeled data for the new domain scenario. For these reasons we use one standard benchmark dataset of man-made scenes, *the Wireframe dataset,* [18] and three domain specific datasets based on forestry imagery.

The Wireframe dataset contains 5462 images, 5000 for training and 462 for test. Since we want a validation set during training we took out 300 images from the trainingset for this.

For the non man-made scenes we use *the FinnWoodlands Dataset* [21] containing 300 RGB images, 250 for training and 50 for test, of snowy forest scenes in Finland, with corresponding segmentation masks for the trees, and surroundings. For the sake of our use we extract the silhouette lines of the trees from the segmentation masks. We split the validation set of 50 images into a validation set of 20 images and a testset of 30 images. We also provide two new datasets of forestry scenes from two different forests, with annotated lines of the tree silhouettes, Spruce A and Spruce B. Spruce A contains 975 images, where we used 775 for training and 200 images for test. Spruce B contains 104 images and we used all these as an additional testset. We will make both these datasets publicly available.

Table 1. Comparison of the performance on the Finnwoodlands testset between our method and the supervised one, for different splits. Models here are trained only on the Finnwoodlands training set. The remaining data for each split was used as unlabeled data. Values reported are mean values of three models, trained with identical settings.

Split	1/16	1/8	1/4	1/2	1	Pretrained models	
						LETR [34]	DeepLSD [26]
Supervised F^H	0.48	0.62	0.54	0.70	0.66	0.60	0.64
Ours F^H	**0.71**	**0.72**	**0.71**	**0.74**			
Supervised sAP^{10}	7.8	12.5	18.5	24.5	24.2	9.2	
Ours sAP^{10}	**24.0**	**23.7**	**22.3**	**31.1**			
Gain (Δ)	↑ 16.2	↑ 11.2	↑ 3.8	↑ 6.6			

We evaluate our models performance using two metrics, heatmap-based F^H [39] and structural average precision with a threshold of 10 (sAP^{10}). For

DeepLSD [26] no direct confidence scores are available in the line detections and therefore we omit the sAP^{10} score.

Training Settings. For both the Finnwoodlands and Wireframe dataset, we create four different splits consisting of $1/2, 1/4, 1/8$ and $1/16$ part of the whole dataset with annotations. The rest of the dataset is used as unlabeled. As a first step we train a supervised model using only the labeled data, for each split, and the labeled loss. We ran the training for 300 epochs for Finnwoodlands and 200 epochs for Wireframe, respectively. The learning rate was set to 0.001, and through-out the training we keep the best model based on the sAP^{10} score on the validation dataset. To form a baseline for comparison we made three identical runs for Finnwoodlands and two for Wireframe, reporting the mean performance. The weights from the respective supervised learning is then used for the semi-supervised training, lowering the learning rate to 0.0001 and training for another 100 epochs, again keeping the best model based on the sAP^{10} score on the labeled validation dataset. Training was performed on an NVIDIA A100 GPU.

4.1 Supervision Dependency

In Table 1 we demonstrate the performance of our proposed method in the case with not all of the data annotated. As shown, when adding unlabeled data and following our scheme, the results are better than only using the supervised model. Note also that for the 1/2 split the performance exceeds the performance of the fully supervised model, visualized in Fig. 4. We see that our model finds more lines, without increasing the false positives excessively. Interestingly, for the 1/16 split the performance of our model is similar to the fully supervised one, proving that the network is able to learn a great portion from the unlabeled stream. With sAP^{10} as metric our method for the split 1/16 is more than 200% better than the supervised model for the same split. The performance of our models also exceeds the pre-trained SOTA models, demonstrating the possibility of creating an accurate model with very little data, and even less annotated data. In addition to testing on the Finnwoodlands test dataset we also test the models performances on the Spruce B test dataset, demonstrated in Table 2, where we see that our method outperforms the fully supervised model on all the splits. This holds even for the model with only 1/16 of the labeled data, as is visualized in Fig. 1 where we note that our method allows for much greater generalization than the fully supervised. To emphasize this further, we compare the performance of our models to the SOTA models. In Table 3 we demonstrate the performance of our method on a classic line segment detection dataset, the Wireframe dataset. We see that even on a larger dataset there are additional information that the network can learn with our method, compared to the supervised one. We visualize the detected lines in one image, for the 1/16 labeled data case, in Fig. 5. Unsurprisingly LETR outperforms the other models, since this is the dataset LETR, and DeepLSD, are trained on. We got slightly lower sAP^{10} score than previously reported for LETR, which is probably due to using other thresholds than they originally did. We have kept the threshold consistent through all of our experiments.

Fig. 4. Comparison of the output results on one image from the Finnwoodlands testset. Left to right: fully supervised model, our method trained with 1/2 the dataset labeled, and the ground truth. Both models are trained on the Finnwoodlands training set.

Table 2. Comparison of the performance on the Spruce B testset between our method and the supervised one, for different splits. The models here are trained only on the Finnwoodlands training set. The remaining data for each split was used as unlabeled data. Values reported are mean values of three models, trained with identical settings.

Split	1/16	1/8	1/4	1/2	1	Pretrained models	
						LETR [34]	DeepLSD [26]
Supervised F^H	0.23	0.47	0.51	**0.71**	0.59	0.38	0.59
Ours F^H	**0.68**	**0.52**	**0.71**	0.69			
Supervised sAP^{10}	2.8	7.9	10.3	23.2	16.0	4.6	
Ours sAP^{10}	**21.0**	**17.1**	**22.9**	**32.6**			
Gain (Δ)	↑ 18.2	↑ 9.2	↑ 12.6	↑ 9.4			

4.2 Domain Adaptation

We show here how a model can be improved on a specific, previously unseen, dataset by adding data without annotation, i.e. investigating the ability to improve the model for a domain specific case. For this experiment we used the splits of Finnwoodlands with annotations and added another dataset, the completely unlabeled trainingset of Spruce A. We evaluate the results on the testset of Spruce A, in Table 4. Here we provide both the results of our method trained on only Finnwoodland data, and our method that is trained with the labeled data from Finnwoodlands and the Spruce A train dataset as unlabeled data. In most of the splits our method with the Spruce A train dataset as unlabeled data performs the best, and in the cases it doesn't, our regular model outperforms the supervised one. As demonstrated by both previous tests and this, the pretrained SOTA models fail to perform well on a novel domain.

4.3 Ablation Studies

We demonstrate in Table 5 the benefits of two strongly perturbed versions of the unlabeled images instead of one. We also illustrate the large variance that

Table 3. Comparison of the performance on the Wireframe testset between our method and the supervised one, for different splits. Remaining data was used as unlabeled data. The values reported are the mean values of two models, trained with identical settings

Split	1/16	1/8	1/4	1/2	1	Pretrained models	
						LETR [34]	DeepLSD [26]
Supervised F^H	0.75	0.76	**0.76**	0.76	0.76	0.83	0.72
Ours F^H	0.75	0.76	0.75	0.76			
Supervised sAP^{10}	23.1	32.9	41.8	49.3	53.7	50.5	
Ours sAP^{10}	**34.0**	**40.0**	**44.9**	**50.1**			
Gain (Δ)	↑ 10.9	↑ 7.1	↑ 3.1	↑ 0.8			

Fig. 5. Output results on one image from the Wireframe testset. Left to right: supervised model trained with 1/16 the dataset labeled, our method trained with 1/16 the dataset labeled and the remaining as unlabeled, and the ground truth.

comes from using the original CutMix [38], and the benefit of having our CutMix compared to without CutMix. All models are trained and tested on the Finnwoodlands dataset. To further demonstrate the benefit of our CutMix compared to the original one, and to show that ours is more suitable for lines in all kinds of scenes we make the same experiment on the Wireframe dataset, shown in Table 5. In Fig. 6 we visualize some outputs from our proposed CutMix alongside the original CutMix. Here the original CutMix detects more short and inaccurate lines than the proposed version. To ensure that the semi-supervised scheme is not simply a favorable training setting, we train on the whole Finnwoodlands dataset, both considered as labeled and unlabeled data. This lets the network see all the images, and get feedback from the ground truth, while at the same time utilizing the augmentations of the unlabeled training stream, making sure that the consistency requirement is met. In Table 6 we see that this does, however, not result in a better performing model than the one obtained with the 1/2 split, which only sees half the dataset with ground truth guidance and the other half with the consistency requirement on.

Table 4. Comparisons on the Spruce A testset between the supervised model, our method with only the remaining data from Finnwoodlands training-set as unlabeled data, and our method with the Spruce A train dataset without labels, for different splits. Models are trained only on labeled data from the Finnwoodlands training set.

Split	1/16	1/8	1/4	1/2	1	LETR	DeepLSD
Supervised F^H	0.16	0.40	0.39	0.38	0.35	0.23	0.38
Ours F^H	0.35	0.38	0.31	**0.46**			
Ours with Spruce A F^H	**0.40**	**0.42**	**0.44**	0.33	**0.40**		
Supervised sAP^{10}	2.1	4.2	5.5	4.4	6.0	2.3	
Ours sAP^{10}	**9.8**	10.0	3.9	**14.4**			
Ours with Spruce A sAP^{10}	6.5	**15.4**	**14.2**	11.4	**13.8**		

Table 5. Ablations with 1/2 the dataset labeled, for the Finnwoodlands and Wireframe datasets, trained three times, with identical setting, with reported mean of the three.

FW	A single x^s	CutMix [37]	w/o CutMix	Prop.	WF	CutMix [37]	Prop.
F^H	0.72 ± 0.03	0.72 ± 0.04	0.72 ± 0.05	0.74 ± 0.01	F^H	0.75 ± 0.01	0.76 ± 0.01
sAP^{10}	28.2 ± 0.8	33.8 ± 5.9	29.5 ± 3.4	31.1 ± 2.7	sAP^{10}	50.3 ± 0.57	50.1 ± 0.50

Table 6. Comparison between using half the training dataset with labels and half without labels, and using the whole dataset with labels, and the same whole dataset as unlabeled. The reported values are the mean of three identical runs.

	Finnwood	Spruce B		Finnwood	Spruce B
F^H Split (2/1)	0.73	**0.71**	sAP^{10} Split (2/1)	30.2	29.8
F^H Split (1/2)	**0.74**	0.69	sAP^{10} Split (1/2)	**31.1**	**32.6**

Fig. 6. Comparison between our CutMix and the original CutMix [38], using our training scheme with 1/2 the dataset as labeled and the remaining as unlabeled. Notice how the original CutMix has more short lines.

5 Conclusion

We have in this paper presented the first framework for semi-supervised learning for line segment detection. Through extensive testing on diverse datasets, we demonstrate that these methods improve significantly on the state-of-the-art for new environments or where there is little annotation. Furthermore, we show that small models, suitable for real-time applications, benefit greatly from semi-supervised methods, and indeed in some cases outperforming methods supervised on the full dataset. Future work includes testing these methods on various downstream tasks, developing augmentations, model architectures, testing new losses, as well as adapting the model to handle other low parametric features than lines.

References

1. Akinlar, C., Topal, C.: Edlines: real-time line segment detection by edge drawing. In: ICIP. IEEE (2011)
2. Alonso, I., Sabater, A., Ferstl, D., Montesano, L., Murillo, A.C.: Semi-supervised semantic segmentation with pixel-level contrastive learning from a class-wise memory bank. In: ICCV (2021)
3. Berthelot, D., et al.: Remixmatch: semi-supervised learning with distribution alignment and augmentation anchoring. arXiv preprint arXiv:1911.09785 (2019)
4. Biederman, I., Ju, G.: Surface versus edge-based determinants of visual recognition. Cogn. Psychol. **20**(1), 38–64 (1988)
5. Canny, J.: A computational approach to edge detection. IEEE PAMI **6**, 679–698 (1986)
6. Caron, M., Misra, I., Mairal, J., Goyal, P., Bojanowski, P., Joulin, A.: Unsupervised learning of visual features by contrasting cluster assignments (2020)
7. Chen, T., Kornblith, S., Norouzi, M., Hinton, G.: A simple framework for contrastive learning of visual representations. arXiv preprint arXiv:2002.05709 (2020)
8. Cole, F., et al.: Where do people draw lines? In: Seminal Graphics Papers: Pushing the Boundaries, vol. 2, pp. 409–419 (2023)
9. Elder, J.H., Almazàn, E.J., Qian, Y., Tal, R.: MCMLSD: a probabilistic algorithm and evaluation framework for line segment detection. arXiv preprint arXiv:2001.01788 (2020)
10. Gao, S., et al.: Pose refinement with joint optimization of visual points and lines. In: IROS, pp. 2888–2894. IEEE (2022)
11. Gillsjö, D., Flood, G., Åström, K.: Polygon detection for room layout estimation using heterogeneous graphs and wireframes. In: ICCV Workshops (2023)
12. Gomez-Ojeda, R., Moreno, F.A., Zuniga-Noël, D., Scaramuzza, D., Gonzalez-Jimenez, J.: PL-SLAM: a stereo slam system through the combination of points and line segments. IEEE Trans. Robot. **35**(3), 734–746 (2019)
13. Gu, G., Ko, B., Go, S., Lee, S.H., Lee, J., Shin, M.: Towards real-time and light-weight line segment detection. arXiv preprint arXiv:2106.00186 (2021)
14. Hadsell, R., Chopra, S., LeCun, Y.: Dimensionality reduction by learning an invariant mapping. In: Conference on Computer Vision and Pattern Recognition, CVPR (2006)
15. Haralick, R.M.: Zero crossing of second directional derivative edge operator. In: Robot Vision, vol. 336, pp. 91–101. SPIE (1982)
16. Hofer, M., Maurer, M., Bischof, H.: Efficient 3D scene abstraction using line segments. CVIU **157**, 167–178 (2017)
17. Hough, P.V.: Method and means for recognizing complex patterns, 18 December 1962, uS Patent 3,069,654
18. Huang, K., Wang, Y., Zhou, Z., Ding, T., Gao, S., Ma, Y.: Learning to parse wireframes in images of man-made environments. In: Computer Vision and Pattern Recognition, CVPR (2018)

19. Huang, S., Qin, F., Xiong, P., Ding, N., He, Y., Liu, X.: TP-LSD: tri-points based line segment detector. In: ECCV. Springer, Cham (2020)
20. Hung, W.C., Tsai, Y.H., Liou, Y.T., Lin, Y.Y., Yang, M.H.: Adversarial learning for semi-supervised semantic segmentation. In: BMVC (2018)
21. Lagos, J., Lempiö, U., Rahtu, E.: Finnwoodlands dataset. In: Gade, R., Felsberg, M., Kämäräinen, J.K. (eds.) Image Analysis, pp. 95–110. Springer, Cham (2023)
22. Lin, Y., Pintea, S.L., van Gemert, J.C.: Deep Hough-transform line priors. In: ECCV (2020)
23. Marr, D., Hildreth, E.: Theory of edge detection. Proc. R. Soc. Lond. Ser. B **207**(1167), 187–217 (1980)
24. Mateus, A., Tahri, O., Aguiar, A.P., Lima, P.U., Miraldo, P.: On incremental structure from motion using lines. IEEE Trans. Robot. **38**(1), 391–406 (2021)
25. Micusik, B., Wildenauer, H.: Structure from motion with line segments under relaxed endpoint constraints. Int. J. Comput. Vis. **124**, 65–79 (2017)
26. Pautrat, R., Barath, D., Larsson, V., Oswald, M.R., Pollefeys, M.: DeepLSD: line segment detection and refinement with deep image gradients. In: Computer Vision and Pattern Recognition, CVPR (2023)
27. Sandler, M., Howard, A.G., Zhu, M., Zhmoginov, A., Chen, L.: Inverted residuals and linear bottlenecks: mobile networks for classification, detection and segmentation. CoRR abs/1801.04381 (2018)
28. Sattler, T.: Benchmarking 6DoF outdoor visual localization in changing conditions. In: Computer Vision and Pattern Recognition, CVPR (2018)
29. Sohn, K., et al.: Fixmatch: simplifying semi-supervised learning with consistency and confidence. In: NeuRIPS (2020)
30. Tardif, J.P.: Non-iterative approach for fast and accurate vanishing point detection. In: ICCV (2009)
31. Tarvainen, A., Valpola, H.: Weight-averaged consistency targets improve semi-supervised deep learning results. CoRR abs/1703.01780 (2017)
32. Von Gioi, R.G., Jakubowicz, J., Morel, J.M., Randall, G.: LSD: a fast line segment detector with a false detection control. IEEE PAMI **32**(4), 722–732 (2008)
33. Wang, Y., et al.: Semi-supervised semantic segmentation using unreliable pseudo labels. In: Computer Vision and Pattern Recognition, CVPR (2022)
34. Xu, Y., Xu, W., Cheung, D., Tu, Z.: Line segment detection using transformers without edges. In: Computer Vision and Pattern Recognition, CVPR (2021)
35. Xue, N., Bai, S., Wang, F., Xia, G.S., Wu, T., Zhang, L.: Learning attraction field representation for robust line segment detection. In: Computer Vision and Pattern Recognition, CVPR (2019)
36. Xue, N., et al.: Holistically-attracted wireframe parsing. In: Computer Vision and Pattern Recognition, CVPR (2020)
37. Yang, L., Qi, L., Feng, L., Zhang, W., Shi, Y.: Revisiting weak-to-strong consistency in semi-supervised semantic segmentation. In: Computer Vision and Pattern Recognition, CVPR (2023)
38. Yun, S., Han, D., Oh, S.J., Chun, S., Choe, J., Yoo, Y.: Cutmix: regularization strategy to train strong classifiers with localizable features. In: ICCV (2019)
39. Zhou, Y., Qi, H., Ma, Y.: End-to-end wireframe parsing. In: ICCV (2019)
40. Zhou, Y., et al.: Learning to reconstruct 3D Manhattan wireframes from a single image. In: ICCV (2019)

Weak Cube R-CNN: Weakly Supervised 3D Detection Using Only 2D Bounding Boxes

Andreas Lau Hansen[1(✉)], Lukas Wanzeck[1], and Dim P. Papadopoulos[1,2]

[1] Technical University of Denmark, Kongens Lyngby, Denmark
andreaslauh@outlook.com
[2] Pioneer Center for AI, Copenhagen, Denmark
https://weakcubercnn.compute.dtu.dk/

Abstract. Monocular 3D object detection is an essential task in computer vision, and it has several applications in robotics and virtual reality. However, 3D object detectors are typically trained in a fully supervised way, relying extensively on 3D labeled data, which is labor-intensive and costly to annotate. This work focuses on weakly-supervised 3D detection to reduce data needs using a monocular method that leverages a single-camera system over expensive LiDAR sensors or multi-camera setups. We propose a general model *Weak Cube R-CNN*, which can predict objects in 3D at inference time, requiring only 2D box annotations for training by exploiting the relationship between 2D projections of 3D cubes. Our proposed method utilizes pre-trained frozen foundation 2D models to estimate depth and orientation information on a training set. We use these estimated values as pseudo-ground truths during training. We design loss functions that avoid 3D labels by incorporating information from the external models into the loss. In this way, we aim to implicitly transfer knowledge from these large foundation 2D models without having access to 3D bounding box annotations. Experimental results on the SUN RGB-D dataset show increased performance in accuracy compared to an annotation time equalized Cube R-CNN [3] baseline. While not precise for centimetre-level measurements, this method provides a strong foundation for further research.

Keywords: Weak Supervision · 3D Object Detection · Monocular Object Detection

1 Introduction

The ability to tell physical scale, distance between, and depth of objects is a very natural ability for humans and animals with binocular vision. It is, therefore, a manageable task to place objects in the three-dimensional space. However, most digital photos are taken with monocular cameras, particularly smartphones, and thus cannot benefit from the same stereoscopic effects. Still, many depth and size cues are present in an image (Fig. 1).

J. Petersen and V. A. Dahl (Eds.): SCIA 2025, LNCS 15726, pp. 137–150, 2025.
https://doi.org/10.1007/978-3-031-95918-9_10

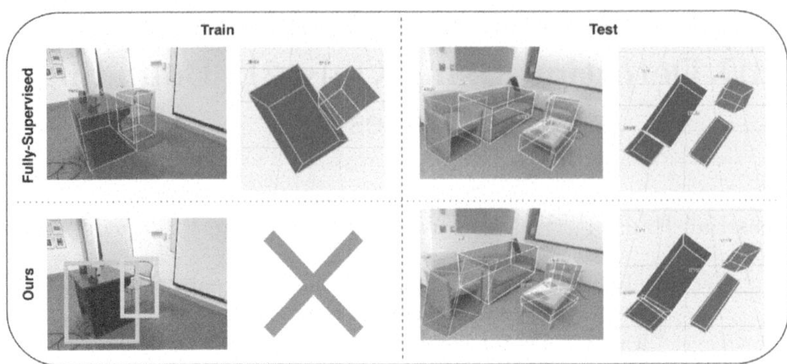

Fig. 1. Weak Cube R-CNN. In contrast to standard 3D object detectors that require 3D ground truths, our proposed method is trained using only 2D bounding boxes but can predict 3D cubes at test time. Weak Cube R-CNN significantly reduces the annotation time since 3D ground-truths require 11× more time than annotating 2D boxes. More importantly, it does not require access to LiDAR or multi-camera setups.

3D object detectors [3,15,23,26,32,44] can pick up on many depth cues nicely. The ability to locate objects in physical space with accurate dimensions provides new use cases within scene understanding, augmented reality, room mapping, and more. However, collecting 3D annotated datasets is challenging partly due to the requirement of special physical sensors; LiDAR, depth scanners, or other alternatives and partly because of the annotation time and complexity. As an example, to annotate the 3D data set SUN-RGBD [37] 2051 h were spent, which is considered to be a tiny data set with its 10,335 images. That equates to roughly 12 min per image compared to 65 s for 2D annotation (≈9%). This is not counting the extra data collection time. On the other hand, an abundance of 2D annotated datasets can be leveraged. This motivates the development of a 3D detector that relies solely on 2D annotations.

Other work has tackled this problem by using additional sensors during inference, primarily LiDAR [7,24], pseudo LiDAR [29,40], and stereo cameras [6,17,28], as they provide an accurate representation of 3D space. Most of these systems are employed in simplified driving scenarios [5,8] where visual cues are stable, for example is it easy to tell the orientation of the world as there is a large, unobstructed view of the ground and it is only required to estimate rotation about one axis. However, in a more generalized setting that we are interested in, these cues cannot be expected to be present. Thus, something more generalizable is required.

We propose a CNN-based model which only uses 2D annotated data during training to perform 3D object detection. It predicts objects' location in the image plane and then learns to place the objects in 3D by using "weak losses" using only 2D ground truths. A crucial component in this step is to use foundation models for estimating depth and the ground plane. The output of these models is used as pseudo ground truths for the weak losses.

We use a Faster R-CNN [33] type architecture to first predict 2D bounding boxes of objects. The boxes' location is then used as candidate regions for 3D cubes by pooling from backbone feature maps used in a 3D head to predict 3D cubes. Besides the 3D head, information from a metric depth estimation model is used to estimate the depth and the ground plane. The parameters of the 3D cubes are optimized, such that the 2D and 3D attributes are consistent. For image plane localization, the Generalised IoU loss [34] is employed. The pseudo ground truth depth is sampled from a depth map inferred from the image. Additionally, object size priors are incorporated through a relaxed loss, which ensures that object sizes within a particular class roughly match average-sized objects of the same class. To estimate the rotation, Pose Alignment Loss operates on the objects internally within a scene and uses the assumption that objects are typically aligned in one or more axes. A normal vector loss ensures rotational consistency with the ground, which provides a world frame of reference. Experimental results on SUN RGB-D [37] show increased performance in accuracy compared to an annotation time equalized Cube R-CNN [3] baseline.

2 Related Work

Monocular 3D Object Detection. Monocular 3D Object Detection is the task of predicting 3D bounding boxes of target objects within a single 2D RGB image. This task relies solely on RGB data without additional information such as depth, sensor data, or multiple images. The most prominent uses of 3D object detection are self-driving cars [4,22,25,32,36,39], and indoor spatial room modeling. Cube R-CNN [3] is a simple extension of the established 2D object detection methods. The method is at its core Faster R-CNN with a cube head attached to it, such that it can predict a cube for each 2D box. The idea of leveraging an existing 2D object detector and extrapolating cubes from 2D boxes is used by more methods [25] (YOLO3D). Their key assumption is the fact that a cube fits tightly into a 2D box. Their idea of proposing cubes is similar to RPNs [33], consisting of some simplifications which constrain the number of 3D proposal boxes inside each 2D box. Other methods [16,18,41,45] focus on modeling objects' depth. MonoDETR is among transformer based methods [11,44], but still uses a CNN as both the feature and depth encoder. The transformer blocks fuse the image and depth features. 3D Datasets vary in distance to objects and field of vision, from indoor [1,35,37] to outdoor scenes, where the camera is mounted on a car [5,8]. Effort has been made to homogenise datasets [3].

Weakly Supervised 3D Object Detection. The type and level of weak supervision used in other work varies substantially. Examples are: click- [24], point cloud-, direction supervision [38], and 2D box supervision [12]. Other methods [9,38] use multiple frames obtained through a video to mimic a stereo view camera. Many weakly supervised monocular 3D object detection methods rely on point cloud data obtained with LiDAR scanners [20,24,25,27,30]. Since point clouds are very accurate they can effectively be used to estimate where objects are located in 3D space by considering the density of points. Additionally, [27]

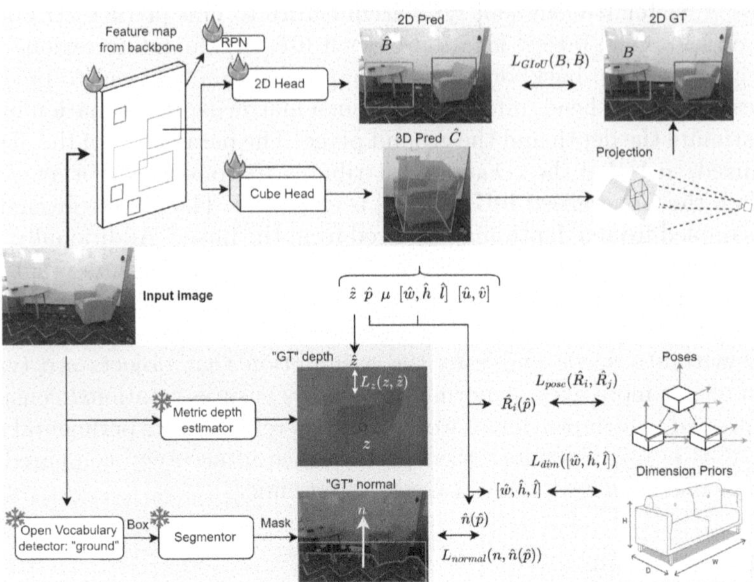

Fig. 2. Overview of *Weak Cube R-CNN*. The model extracts features from an image and predicts objects in 2D and their cubes in 3D. We split the cube into each of its attributes and optimise each attribute with regards to a pseudo ground truth information. During training, instead of the simple 3D ground truth provided in the fully supervised setting, we must use many different sources of information provided by frozen models to emulate the same ground truth annotation.

only requires the LiDAR data during training, which expands the use cases of the model. Pseudo LiDAR methods [26,40] are hybrid methods where dense data is transformed into sparse LiDAR data, where LiDAR methods work directly.

Incorporation of Foundation Models. Depth estimation models [2,42] are ideal for use in downstream tasks. Instead of incorporating depth estimation directly into the main model, only the output is used, namely as pseudo ground truths [40]. [12] does exactly this, where the pseudo ground truth depth is used as a target for a depth estimation branch of the model. Furthermore, they use both 2D and 3D sub-networks and an uncertainty distillation loss to only learn from good predictions.

3 Method

In this section, we describe the modeling approach, structure, and optimization objective of the model. Overall, the model is a derivative of Cube R-CNN [3] and borrows the structure but without any direct 3D ground truths. Thus we name our method *Weak Cube R-CNN*. Figure 2 provides an overview of the training procedure.

3.1 Overview

We use Cube R-CNN as our basis and therefore inherit all of the modeling techniques used, including virtual depth, IoUness, and allocentric rotation. Our method predicts 2D bounding boxes (bbox) with a Faster R-CNN-like architecture, based on a DLA-34 [43] FPN [19] backbone pre-trained on ImageNet. The 2D bboxes are used to pool features, with ROIAlignV2 [10]. The pooled features are used in a 3D head consisting of 2 linear layers, which predict a 3D cube expressed as 13 parameters corresponding to the image plane coordinates $[\hat{u}, \hat{v}]$, the depth $\hat{z}$, the dimensions $[\hat{w}, \hat{h}, \hat{l}]$, the 6D allocentric rotation [46] $\hat{p}$, and an uncertainty $\hat{\mu}$. A cube is predicted for each 2D box.

Due to a lack of access to the 3D ground truths, we carefully choose the appropriate sources that mimic a real 3D ground truth. When the depth is known, the physical size of an object can relatively easily be estimated based on the 2D detection and known camera intrinsics by using geometry. Finding the ground provides useful information on the rotation in a scene as it provides a frame of reference for all objects and constrains one axis of rotation. We use priors on the dimensions of objects to eliminate unrealistic cubes, though this constraint is relaxed to allow predictions to deviate from their prior sizes.

3.2 Obtaining Pseudo 3D Ground Truths

We use Depth-Anything V2 [42] fine-tuned for metric depth estimation. The model provides a depth value for each pixel in an image. A depth map offers a lot of structure in the image and we therefore use it for two downstream tasks: 1) Estimation of the ground plane in conjunction with a RANSAC algorithm and 2) The value on the depth map at the center point of the detected 2D bbox is the pseudo ground truth depth for 3D detections.

We use GroundingDINO [21] for ground detection by prompting it with the phrase "ground". This provides 2D bboxes that are passed into Segment Anything (SAM-HQ) [13,14] to get a segmentation of the ground. However, in some cases the ground is not visible or GroundingDINO fails for other reasons.

The depth map is interpreted as a point cloud by applying a simple transformation. Given the camera matrix K, we extract the focal length f and the center of the image, the principal point (c_x, c_y). Let (u, v, z) be image coordinates in pixels with depth z in meters. The conversion from the image to a set of points goes as follows: The offset of a point to the principal point is

$$\Delta u = u - c_x, \ \Delta v = v - c_y. \tag{1}$$

Let x, y, z be real-world coordinates. Each coordinate is calculated as

$$x = \frac{\Delta u \cdot z}{f}, \quad y = \frac{\Delta v \cdot z}{f}, \quad z = z. \tag{2}$$

This means that the point cloud P_{3D} is a set of all points

$$P_{3D} = (\mathbf{x}, \mathbf{y}, \mathbf{z}) \tag{3}$$

Fig. 3. Ground estimation pipeline showing the point cloud obtained through the depth map. The 2nd step selects the region in the depth map corresponding to the ground in the color image. The depth map is interpreted as a point cloud where plane-RANSAC obtains a normal vector to the ground.

The ground is detected using a simple plane RANSAC algorithm, which runs on the point cloud generated from the depth map. The algorithm finds the largest plane present in the point cloud. However, this is not always the actual floor, as it could also be a wall or any other random set of points roughly outlining a floor. This is why it is necessary to use GroundingDINO to filter the point cloud as seen in Fig. 3.

3.3 Loss Functions

Our loss functions use only 2D labels, as this is the highest level of information available to us. This entails finding a relationship between a 3D cube and its corresponding 2D box. To relate 3D cubes to 2D boxes, we project 3D cubes to the 2D image plane using the camera intrinsics. For simplicity, we project to axis-aligned 2D boxes, so some of the rotation information is lost. Occluded objects are also not well handled because the projection will overlay the object on top of the image. To convert a 3D point to 2D:

$$P_{3D} = \begin{bmatrix} \mathbf{x} \\ \mathbf{y} \\ \mathbf{z} \end{bmatrix} \rightarrow P_{2D} = \frac{f}{\mathbf{z}} \begin{bmatrix} \mathbf{x} \\ \mathbf{y} \end{bmatrix} \tag{4}$$

Training Objective. The model is designed to handle each aspect of a cube independently. We consider the following aspects: the image plane placement, the depth, the dimensions, and the rotation.

Placement Loss. For $[u, v]$ image-plane placement, we adopt the Generalised IoU loss [34] between the 2D projection of the 3D detection and the 2D bbox detection. The loss measures the 2D overlap between the projected 2D box $\hat{B}$ of the predicted 3D cube and the ground truth 2D box B.

$$L_{GIoU}(B, \hat{B}) = 1 - GIoU(B, \hat{B}) \tag{5}$$

This loss ensures that the placement of the 3D cube is correct and should align well with its 2D box.

Depth Loss. For depth, we use the depth map as a pseudo ground truth z. $\hat{z}$ is a cubes' predicted depth by the 3D head. To obtain z, we use the center point of the 2D bbox corresponding to its 3D detection to select a pixel in the depth map that has an associated depth to it. When the center point is outside the frame, the point is clamped to within 10 pixels of the image. We do this because the model severely underestimates the depth near the edges. The loss is the ℓ_1 distance between z and the predicted depth $\hat{z}$, see Fig. 2.

$$L_z(z, \hat{z}) = \|z - \hat{z}\|_1 \tag{6}$$

Prior Size Loss. To incorporate prior knowledge of the sizes of objects, we measure the z-score, *i.e.* how many standard deviations the object size is from the classes' mean size in each of the dimensions. These parameters are obtained for each class from the dataset or, if unavailable, by asking ChatGPT "average dimension of *object x*". For a given object dimension d we model the function acting as a loss:

$$\mathcal{Z} = \frac{1}{3} \sum_{d \in \{w,h,l\}} \left(\frac{|d - \mu_{prior}|}{\sigma_{prior}} \right) \tag{7}$$

$$L_{dim}(\hat{C}; \mu_{prior}, \sigma_{prior}) = \begin{cases} \mathcal{Z} & \text{if } \mathcal{Z} > 1 \\ 0 & \text{otherwise} \end{cases} \tag{8}$$

where μ_{prior} and σ_{prior} are the known mean and standard deviation dimensions of each class. This loss is a sum of z-scores for each dimension direction, a loss of 1 means the prediction is one standard deviation away from the prior of the specific class. The loss is overwritten with 0 when the loss is less than 1. This relaxation is necessary to prevent the model from learning to only predict mean sized objects. However, it makes the loss non-smooth which can make it unstable.

Normal Vector Loss. To ensure that the cubes are aligned with respect to the ground, we use the aforementioned method in Sect. 3.2 to identify the ground normal vector. We use the cosine similarity between the normal vector of the ground and the "up direction" (normal vector) of the cubes as a measure of alignment with the ground plane. The cosine similarity between two vectors $\mathbf{n}_1$ and $\mathbf{n}_2$ is given by:

$$\cos_{sim}(\mathbf{n}_1, \mathbf{n}_2) = \frac{\mathbf{n}_1 \mathbf{n}_2}{\max(\|\mathbf{n}_1\|_2 \|\mathbf{n}_2\|_2, \epsilon)} \tag{9}$$

The cosine similarity is a number in the range $[0, 1]$, 1 when the vectors are identical and 0 when they are perpendicular, to turn the measure into a loss we want to reverse the relationship. We convert the 6D allocentric rotation $\hat{p}$ to a "up" normal vector of a cube $\hat{\mathbf{n}}(\hat{p})$. For a ground normal $\mathbf{n}$ and a predicted normal $\hat{\mathbf{n}}$, we define the loss:

$$L_{normal}(\mathbf{n}, \hat{\mathbf{n}}) = (1 - \cos_{sim}(\mathbf{n}, \hat{\mathbf{n}})) \cdot \kappa_{\text{ground}} \tag{10}$$

where κ_{ground} is the confidence in whether the ground was found, inspired by [3], that is 1 when the ground is visible and 0.05 otherwise. The model still benefits from learning the placement and depth in these images, but does not get negatively affected by a poor rotation estimation.

Pose Alignment Loss. This loss ensures rotation consistency within all objects in a scene and relies on the assumption that objects typically are aligned with each other. To compute the alignment between two objects, we consider their rotation matrices. The trace of a rotation matrix R has the following relationship to the rotation angle θ:

$$Tr(R) = 1 + 2\cos(\theta) \tag{11}$$

We derive rotation matrices from the 6D representation $R(\hat{p})$, then calculate the relative angle between two rotation matrices R_1 and R_2, with the formula:

$$\cos(\theta) = \left| \frac{1}{2} \left(Tr(R_1 R_2^\top) - 1 \right) \right| . \tag{12}$$

The formula is an extension of Rodrigues' rotation formula. Since it is only possible to compare rotation matrices pairwise, we need all the unique combinations of all n instances in the image. This means we have to compute the angle for all $\frac{n(n-1)}{2}$ combinations. The number of instances in images varies greatly but is usually < 30. To not have images with many instances dominate a batch, we weight the loss inversely according to the number of objects in an image. To make a loss we flip Eq. 12:

$$L_{pose}(\theta) = 1 - \cos(\theta). \tag{13}$$

The loss is undefined for images with one instance because it doesn't make sense to align one object with itself.

Total 3D Loss. Using the superscript to denote the component the loss concerns, the final 3D loss thus becomes

$$\begin{aligned} L_{3D} = \lambda_{GIoU} L_{GIoU}^{(u,v)} + \lambda_z L_z^{(z)} + \lambda_{dim} L_{dim}^{(\hat{w},\hat{h},\hat{l})} \\ + \lambda_{normal} L_{normal}^{(\hat{p})} + \lambda_{pose} L_{pose}^{(\hat{p})} \end{aligned} \tag{14}$$

The final training objective is

$$L = L_{RPN} + L_{2D} + \sqrt{2}e^{(-\mu)} \cdot L_{3D} + \mu \tag{15}$$

All terms, except L_{3D} are the losses from [3] and L_{3D} is the collective 3D loss as described in the sections above. μ is the uncertainty predicted by the model.

4 Experimental Results

In this section, we present our experimental results, where we put special emphasis on achieving correct rotation.

Table 1. Weak Cube R-CNN ablations on SUN-RGBD. We report $AP_{3D}^{com.}$ on 10 "common categories" and on all 38 classes. Overall, not having L_z and L_{dim} significantly decreases performance.

Method	table	bed	sofa	bathtub	sink	shelves	cabinet	fridge	chair	tv	$AP_{3D}^{com.}$	AP_{3D}^{all}
w/ all (*Weak Cube R-CNN*)	12.3	26.4	20.5	20.5	10.4	1.2	6.8	5.3	18.8	5.4	**12.7**	5.4
w/o L_{GIoU}	12.3	13.1	19.8	10.9	7.7	1.4	2.6	4.2	19.6	2.9	9.5	4.4
w/o L_z	5.4	22.3	22.2	8.9	8.8	0.0	0.2	14.4	0.0	1.5	8.4	4.0
w/o L_{dim}	2.9	2.1	6.3	8.4	9.2	1.0	1.1	1.8	8.4	3.1	4.4	2.6
w/o L_{normal}	9.6	20.6	22.7	22.9	15.2	1.5	4.5	2.6	21.4	4.4	12.5	**5.7**
w/o L_{pose}	8.4	24.6	19.5	13.7	5.6	0.6	8.9	6.0	21.7	4.2	11.3	5.0

Table 2. Comparison on equal annotation time models. We report $AP_{3D}^{com.}$ on 10 "common categories". We train Cube R-CNN on SUN-RGBD based on their code. *Weak Cube R-CNN* outperforms *Cube R-CNN time eq.* overall and on all classes except table and chair, which are the 2 classes with the highest frequency in the dataset.

Method	table	bed	sofa	bathtub	sink	shelves	cabinet	fridge	chair	tv	$AP_{3D}^{com.}$	AP_{3D}^{all}
Cube R-CNN	39.8	64.4	60.0	38.3	27.4	3.1	14.1	21.6	53.5	3.8	32.6	15.1
Cube R-CNN time eq.	**13.4**	13.4	16.1	0.1	3.9	0.3	1.5	0.7	**24.2**	0.5	7.4	3.3
Weak Cube R-CNN	12.3	26.4	20.5	20.5	10.4	1.2	6.8	5.3	18.8	5.4	**12.7**	**5.4**

4.1 Setup

Datasets. We evaluate on the indoor SUN RGB-D 3D object detection dataset, which contains 10335 images, 5285 train and 5050 test images. Additionally, we use the outdoor KITTI 3D object detection dataset, with a total of 7481 images, 3712 for train and 3769 for test. We follow [3,8] and remove objects with high occlusion (>66%), truncation (>33%) and with small projections (<6.25% of image height). We do not use the full Omni3D dataset [3] due to computational constraints. We create SUN RGB-D mini and KITTI mini with 433 and 333 randomly selected images respectively, these datasets have equal annotation time to our method, *i.e.* 9% the size of the original data set.

Evaluation Metric. The most common metric for both 2D and 3D object detection is the average precision (AP). We follow the Omni3D benchmark and use *mean* AP_{3D} as our evaluation metric. It averages over all classes for different levels of IoU3D at thresholds $\tau \in [0.05, 0.10, ..., 0.5]$.

Implementation Details. Unless stated otherwise, we use the following setting for all models in this paper. We use Detectron2 and PyTorch3D [31] to implement *Weak-Cube R-CNN*. We train all models for 34 epochs with a batch size of 12 images on an A100 GPU. We use SGD with a learning rate of 0.007, which decays after 12 and 29 epochs by a factor of 10. Following [3], we use random data augmentation by horizontal flipping ($p = 0.5$) and scaling $\in [0.50, 1.25]$ during training. When indoors, we use Depth Anything V2 [42] fine-tuned for metric depth estimation on indoor scenes, with a max distance of $20m$, outdoors

Fig. 4. Qualitative examples of *Weak Cube R-CNN* predictions on SUN-RGBD test set. Images are selected to showcase behaviour in various scenarios. Only the last row is shown with ground truths in red to avoid clutter. In the last row ground truths are shown in red with predictions in green. Each image is shown side-by-side with its corresponding top-view image, where each square is 1 × 1 m. (Color figure online)

we use the outdoor model with a max of $80m$. For all our Weak Cube R-CNN models we use these loss weights: $\lambda_{GIoU} = 4$, $\lambda_z = 1$, $\lambda_{dim} = 0.1$, $\lambda_{normal} = 70$, $\lambda_{pose} = 7$, which were found by trial and error mostly to get the magnitude of each term roughly equal. Especially of note is the low weight λ_{dim}.

Training Scheme. We only require the depth and ground estimator during training. Since we only use the output of frozen models, we opt to preprocess the dataset offline. We train our model in two stages. First, the model is fine-tuned in 2D mode only. This is done because we assume the 2D head is capable of finding the objects in the image. When the 2D head nearly converges, we switch on the 3D losses and train altogether. Our reasoning for this is that the 3D head synergises with the 2D head and improves the 2D IoU, compared to freezing the 2D head.

Baselines. We compare our model to the fully supervised Cube R-CNN [3]. We also implement a version of Cube R-CNN trained on the mini datasets, which is comparative in terms of annotation time to *Weak Cube R-CNN*. Based on 12 min per 3D annotated- and 65 s per 2D annotated image.

4.2 Results

We run ablation experiments to study the impact of each loss term on the model. We show that the model performs well compared to the corresponding annotation time equalised model.

Ablations. Table 1 ablates the loss functions of *Weak Cube R-CNN*. We disclose AP_{3D} on the full dataset (all classes) and at a subset of 10 common classes. We observe an increase in precision when adding a loss, most noticeably with L_{dim}, which improves AP_{3D} by +3.1%. We believe L_{dim} is important because the model has a tendency to quickly begin predicting strange cubes that it never gets away from, but L_{dim} effectively prevents this. We observed an initial large loss on L_{dim} which decreases quickly to 0. It also has a significant effect on objectively good cubes, as $AP_{\text{3D}}^{\text{all}}$ without it is closest to 0. L_{GIoU} also has an impact on good cubes, as $AP_{\text{3D}}^{\text{com.}}$ decreases by −3.2% without it. L_{pose} has the smallest effect, which intuitively makes sense as this loss enforces matching but not necessarily correct rotation.

It stands out that removing L_z does not have the greatest impact on performance, especially for better cubes. When examining predictions, many look much better in 2D than 3D, but still have low IoU. That is mainly due to incorrect depth. As such, we must conclude that L_z does not work perfectly as a proxy for a true depth loss, yet still has a positive effect on overall predictions.

The ablations in Table 1 contradict our assumption that the ground provides a rotation frame for a scene. To validate our approach, we further test on a subset of classes to see if it really is better to include L_{normal}. Furthermore, we provide more experiments on KITTI, Table 3, which shows that the model with L_{normal} is better in terms of rotation. This is an indication that L_{normal} is indeed having the intended effect. Because we want to focus on ensuring correct rotation we thus include L_{normal} in the final model.

Comparison to Other Models. Against the fully supervised Cube R-CNN method, *Weak-Cube R-CNN* achieves about 1/3 of the performance as presented in Table 2. AP3D sees a drop from 15.1% to 5.4%, when evaluating on all classes.

Comparison with Equal Annotation Time. The goal of *Weak-Cube R-CNN* is to cut down on annotation time. We show results with our method compared to Cube R-CNN when using fully annotated 3D data but trained on SUN RGB-D mini. Furthermore, we present results on a narrow selection of classes. Table 2 demonstrates that our method exceeds baseline performance in many categories and achieves +5.3% mean AP compared to the time equalised Cube R-CNN on the reduced set of classes. It is clear that Cube R-CNN time eq. is greatly held back by very few samples of certain classes like "bathtub" where it does not learn to detect anything meaningful. Unsurprisingly, Cube R-CNN outperforms both models in all categories. However, considering the annotation time is about 11x more it does not achieve 11x the performance vs. *Weak-Cube R-CNN*.

Qualitative Results. When looking at the qualitative results in Fig. 4 we find that for indoor scenes the predictions are generally clearer for simple scenes. In the scenes with many objects and occluded objects it generally struggles. Overall, depth seems to be quite accurately predicted which we can see in the last row that is shown with ground truths. It seems that pose alignment does improve detection accuracy overall but makes the harder cases harder. Objects, like cabinets, that are not rooted in the ground are nearly all detected poorly.

Table 3. Ablation results on KITTI (with all classes), we report the mean AP_{3D}, and at thresholds 0.15, 0.25, and 0.5. We observe that using L_{normal} improves the precision considerably when the ground is clearly visible on outdoor images.

Method	AP_{3D}	AP_{3D}^{15}	AP_{3D}^{25}	AP_{3D}^{50}
w/ L_{normal}	**8.2**	**12.1**	**8.3**	**2.0**
w/o L_{normal}	6.3	10.9	6.8	0.4

Table 4. Performance with f = fully supervised methods on KITTI.

Method	f	AP_{3D}^{KITTI}
Cube R-CNN [3]	✓	36.0
SMOKE [22]	✓	25.4
ImVoxelNet [36]	✓	23.5
M3D-RPN [4]	✓	10.4
Cube R-CNN time-eq.	✓	16.4
Weak Cube R-CNN	✗	8.2

Fig. 5. Qualitative examples of *Weak Cube R-CNN* predictions on KITTI test set. KITTI predictions are shown in green and ground truth in red. Each image is shown with its corresponding top-view image, where each square is 1 × 1 m. (Color figure online)

Comparison on KITTI. The advantage that KITTI provides compared to indoor data sets is that the ground is much more consistently visible, and thus, we expect rotation to be easier to determine. For outdoor scenes, the model generally provides excellent predictions in the front view, as shown in Fig. 5. However, in the top view, it is seen that depth is often wrong. Table 4 Shows that *Weak Cube R-CNN* comes close to achieving the same precision as the older fully supervised method M3D-RPN, but even the time equalized Cube R-CNN beats it by 2x. The primary difference between KITTI and SUN RGB-D is the depth, which appears to be what *Weak Cube R-CNN* struggles with the most.

5 Conclusion

We have proposed *Weak Cube R-CNN*, a novel approach for 3D object detection that relies solely on single-view images and 2D image annotations. Our method overcomes the most prominent limitation of 3D object detection, which is the annotation availability of datasets, by leveraging weaker supervision while still achieving competitive performance. *Weak Cube R-CNN* demonstrates strong detection capabilities for objects with high visibility and simple geometric structures. Notably, on the SUN-RGBD dataset, given the same annotation time, it achieves better performance than a fully supervised Cube R-CNN model trained with 3D annotation bounding box annotations. Further work could involve exploring other weak signals, as the method is slow to converge due to the weak signals carried through the loss functions.

Acknowledgements. D. Papadopoulos was supported by the DFF Sapere Aude Starting Grant "ACHILLES".

References

1. Baruch, G., et al.: Arkitscenes - a dataset for 3D indoor scene understanding using mobile RGB-D data. In: NeurIPS (2021)
2. Bochkovskii, A., et al.: Depth pro: sharp monocular metric depth in less than a second. arXiv:2410.02073 (2024)
3. Brazil, G., Kumar, A., Straub, J., Ravi, N., Johnson, J., Gkioxari, G.: Omni3D: a large benchmark and model for 3D object detection in the wild. In: CVPR (2023)
4. Brazil, G., Liu, X.: M3D-RPN: monocular 3D region proposal network for object detection. In: CVPR (2019)
5. Caesar, H., et al.: nuscenes: a multimodal dataset for autonomous driving. In: CVPR (2020)
6. Chen, Y., Liu, S., Shen, X., Jia, J.: DSGN: deep stereo geometry network for 3D object detection. In: CVPR (2020)
7. Fan, L., Xiong, X., Wang, F., Wang, N., Zhang, Z.: Rangedet: in defense of range view for lidar-based 3D object detection. In: CVPR (2021)
8. Geiger, A., Lenz, P., Urtasun, R.: Are we ready for autonomous driving? The KITTI vision benchmark suite. In: CVPR (2012)
9. He, J., Wang, Y., Chen, Y., Zhang, Z.: Weakly supervised 3D object detection with multi-stage generalization (2024)
10. He, K., Gkioxari, G., Dollár, P., Girshick, R.: Mask R-CNN. In: ICCV (2017)
11. Huang, K.C., Wu, T.H., Su, H.T., Hsu, W.H.: Monodtr. In: CVPR (2022)
12. Jiang, X., Jin, S., Lu, L., Zhang, X., Lu, S.: Weakly supervised monocular 3D detection with a single-view image. In: CVPR (2024)
13. Ke, L., et al.: Segment anything in high quality. In: NeurIPS (2023)
14. Kirillov, A., et al.: Segment anything. In: CVPR (2023)
15. Ku, J., Pon, A.D., Waslander, S.L.: Monocular 3D object detection leveraging accurate proposals and shape reconstruction. In: CVPR (2019)
16. Kumar, A., Brazil, G., Corona, E., Parchami, A., Liu, X.: Deviant: depth equivariant network for monocular 3D object detection. In: ECCV (2022)
17. Li, P., Chen, X., Shen, S.: Stereo R-CNN based 3D object detection for autonomous driving. In: CVPR (2019)
18. Li, Z., Qu, Z., Zhou, Y., Liu, J., Wang, H., Jiang, L.: Diversity matters: fully exploiting depth clues for reliable monocular 3D object detection. In: CVPR (2022)
19. Lin, T.Y., Dollár, P., Girshick, R., He, K., Hariharan, B., Belongie, S.: Feature pyramid networks for object detection. In: CVPR (2017)
20. Liu, H., et al.: Eliminating spatial ambiguity for weakly supervised 3D object detection without spatial labels. In: ACM (2022)
21. Liu, S., et al.: Grounding dino: marrying dino with grounded pre-training for open-set object detection. In: ECCV (2024)
22. Liu, Z., Wu, Z., Tóth, R.: Smoke: single-stage monocular 3D object detection via keypoint estimation. In: CVPR Workshops (2020)

23. Liu, Z., Zhou, D., Lu, F., Fang, J., Zhang, L.: Autoshape: real-time shape-aware monocular 3D object detection. In: CVPR (2021)
24. Meng, Q., Wang, W., Zhou, T., Shen, J., Van Gool, L., Dai, D.: Weakly supervised 3D object detection from lidar point cloud. In: ECCV (2020)
25. Mousavian, A., Anguelov, D., Flynn, J., Kosecka, J.: 3D bounding box estimation using deep learning and geometry (2017)
26. Park, D., Ambrus, R., Guizilini, V., Li, J., Gaidon, A.: Is pseudo-lidar needed for monocular 3D object detection? In: CVPR (2021)
27. Peng, L., Yan, S., Wu, B., Yang, Z., He, X., Cai, D.: Weakm3D: towards weakly supervised monocular 3D object detection. arXiv preprint arXiv:2203.08332 (2022)
28. Pon, A.D., Ku, J., Li, C., Waslander, S.L.: Object-centric stereo matching for 3D object detection. In: ICRA (2020)
29. Qian, R., et al.: End-to-end pseudo-lidar for image-based 3D object detection. In: CVPR (2020)
30. Qin, Z., Wang, J., Lu, Y.: Weakly supervised 3D object detection from point clouds. In: ACM (2020)
31. Ravi, N., et al.: Pytorch3D. arXiv:2007.08501 (2020)
32. Reading, C., Harakeh, A., Chae, J., Waslander, S.L.: Categorical depth distribution network for monocular 3D object detection. In: CVPR (2021)
33. Ren, S., He, K., Girshick, R., Sun, J.: Faster R-CNN: towards real-time object detection with region proposal networks. In: NeurIPS (2015)
34. Rezatofighi, H., Tsoi, N., Gwak, J., Sadeghian, A., Reid, I., Savarese, S.: Generalized intersection over union: a metric and a loss for bounding box regression. In: WACV (2019)
35. Roberts, M., et al..: Hypersim: a photorealistic synthetic dataset for holistic indoor scene understanding. In: CVPR (2021)
36. Rukhovich, D., Vorontsova, A., Konushin, A.: Imvoxelnet: image to voxels projection for monocular and multi-view general-purpose 3dod. In: WACV (2022)
37. Song, S., Lichtenberg, S.P., Xiao, J.: Sun RGB-D: a RGB-D scene understanding benchmark suite. In: CVPR (2015)
38. Tao, R., Han, W., Qiu, Z., Xu, C., Shen, J.: Weakly supervised monocular 3dod using multi-view projection and direction consistency. In: CVPR (2023)
39. Wang, T., Zhu, X., Pang, J., Lin, D.: Fcos3D: fully convolutional one-stage monocular 3D object detection. In: CVPR (2021)
40. Wang, Y., Chao, W.L., Garg, D., Hariharan, B., Campbell, M., Weinberger, K.Q.: Pseudo-lidar from visual depth estimation: bridging the gap in 3D object detection for autonomous driving. In: CVPR (2019)
41. Xia, C., et al.: Monosaid: monocular 3D object detection based on scene-level adaptive instance depth estimation. J. Intell. Robot. Syst. (2024)
42. Yang, L., et al.: Depth anything v2 (2024)
43. Yu, F., Wang, D., Shelhamer, E., Darrell, T.: Deep layer aggregation. In: CVPR (2018)
44. Zhang, R., et al.: Monodetr. In: CVPR (2023)
45. Zhang, Y., Lu, J., Zhou, J.: Objects are different: flexible monocular 3D object detection. In: CVPR (2021)
46. Zhou, Y., Barnes, C., Lu, J., Yang, J., Li, H.: On the continuity of rotation representations in neural networks. In: CVPR (2019)

Training Binary Neural Networks with Deep Semantic Guidance

Jiehua Zhang[1], Zhuo Su[1], and Li Liu[2](✉)

[1] Center for Machine Vision and Signal Analysis, University of Oulu, Oulu, Finland
[2] National University of Defense Technology, Changsha, China
li.liu@oulu.fi

Abstract. Binary neural networks (BNNs) have gained considerable attention due to their low power consumption. However, there is still a performance gap between BNNs and their full-precision counterparts. Previous researches focus on reducing quantization errors and optimizing approximate gradients to enhance BNN performance. In this paper, we attribute the degradation of BNN performance to the destruction of deep-layer features. The limited representation capability in deep layers leads to insufficient semantic information for accurate target recognition. To address this problem, we construct a multi-branch training framework, which can guide BNN training with unimpaired deep feature supervision. Specifically, a full-precision adapter is utilized in the deep layer to refine the degraded feature representation in binarized layers. The full-precision adapter has three advantages: 1) it introduces more accurate gradient information flow to stabilize training in the shared fully binarized layers; 2) it incorporates lossless semantic information of deep layers that guides the training of binary branch; 3) it is generic and compatible to diverse BNN structures. We conduct extensive experiments on the CIFAR-100 and ImageNet datasets to evaluate the proposed method. The comprehensive experimental results demonstrate that the proposed method can significantly improve the performance of existing BNNs without incurring additional computational overhead during inference.

Keywords: Binary Neural Networks · Model Compression · Computer Vision

1 Introduction

With the rapid advancement of deep neural networks (DNNs), convolutional neural networks (CNNs) have demonstrated remarkable achievements in various visual tasks, including target recognition [9], object detection [16], semantic segmentation [31], and video understanding [35]. However, the extensive number of parameters and computational demands hinder their deployment on resource-constrained devices. To address these challenges, researchers have developed techniques such as quantization [23,25], pruning [24], distillation [6], and

J. Petersen and V. A. Dahl (Eds.): SCIA 2025, LNCS 15726, pp. 151–165, 2025.
https://doi.org/10.1007/978-3-031-95918-9_11

lightweight model design [3,30] to obtain low-cost models with minimal performance degradation. Binary neural networks (BNNs) [12,17], also referred to as 1-bit quantization, offer an extreme compression approach for model quantization. By quantizing weights and activation values to +1 or -1 and employing XNOR and PopCount operations to calculate output features, BNNs significantly reduce storage requirements and computational overhead, which is highly suitable for resource-constrained devices.

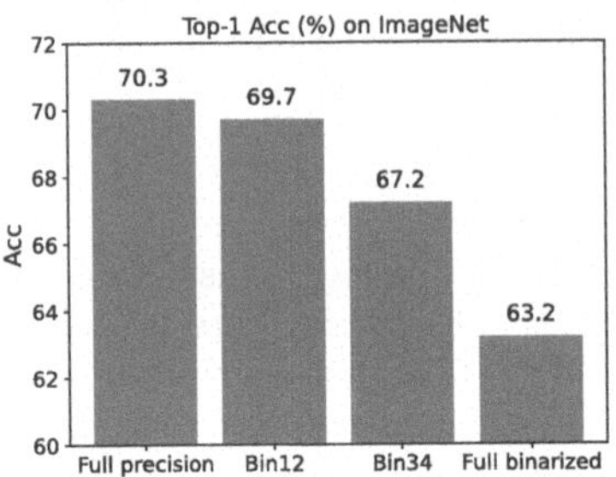

Fig. 1. The Performance comparison of ResNet18 on different settings. **Bin12** denotes that the 1st and 2nd stages of ResNet18 are binarized. **Bin34** denotes that the 3rd and 4th stages are binarized.

Despite low consumption advantages, BNNs suffer from substantial performance degradation caused by severe information loss, which hinders their application in real-world scenarios. To enhance the performance of BNNs, current researches focus on 1) mitigating quantization errors in weights and feature [13,14,36]; 2) addressing the gradient mismatch issue [20,28,33]. These methods aim to minimize the negative impact of binary quantization on model performance to narrow the gap between BNN and full-precision model training. In this paper, we attribute the performance loss of BNNs to the degradation of feature representation in the deep layers and utilize recovered deep features to guide BNNs training.

In BNNs, shallow layers demonstrate the ability to approximate feature expressions similar to those of full-precision models. However, the quantization of features in deep layers poses challenges in capturing semantic information comparable to full-precision models, thereby limiting the performance of BNNs in complex tasks. To study the impact of shallow and deep binarization on performance, we train two partially binarized ResNet18 models, namely **Bin12** and **Bin34**. Bin12 only binarizes the first two stages of ResNet18, while Bin34 only binarizes the later two stages. As depicted in Fig. 1, when only shallow layers are binarized, Bin12 achieves 69.7% top-1 accuracy with only a 0.6% accuracy drop. In contrast, Bin34 suffers a notable accuracy drop from 70.3% to 67.2%. This gap demonstrates the importance of deep layers for BNNs, which inspires us to boost deep feature representation to improve BNNs performance.

To this end, we construct a multi-branch BNN training framework. This framework simply introduces a full-precision adapter to learn rich semantic information from the middle-layer features of BNNs and then transfer the knowledge to deep layers of BNNs. The overall framework can be observed in Fig. 2. The ResNet architecture is utilized as our base model. We add a full-precision adapter in parallel with stage 4. The true label and logits from the full-precision adapter are used to guide BNN branch learning. This multi-branch training framework has three advantages: **1)** Full-precision adapter introduces more accurate gradients to stabilize training in shallow and intermediate layers; **2)** Full-precision

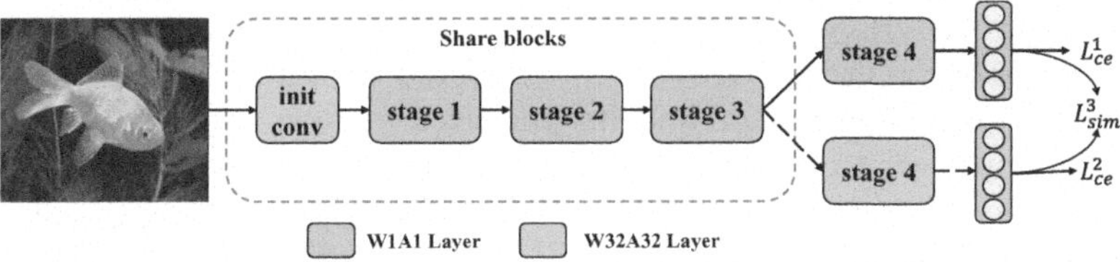

Fig. 2. The basic training framework of our proposed method. The multi-branch framework is based on ResNet architecture. The extra W32A32 stage 4 layer is regarded as the full-precision adapter in our method. **dash line** denotes this full-precision adapter branch is only used in training stage.

adapter can generate soft labels based on non-destructive deep feature representation to guide BNNs training; **3)** This framework can be incorporated into various BNNs methods. We evaluate the proposed training framework on the CIFAR-100 and ImageNet datasets. The comprehensive experiments illustrate the effectiveness of our proposed method. Our main contributions are:

- We observe that the performance degradation of BNNs comes from the homogenized deep-layer features. The experiments suggest that deep-layer features limit the representation of deep semantic information and fine-grained feature extraction of the BNNs.
- We propose a multi-branch BNN training framework and utilize unimpaired deep feature supervision as additional guidance. Specifically, a full-precision adapter is introduced to refine the degraded feature representation in binarized layers. The extracted semantic features from the adapter can be regarded as guidance to transfer the fine-grained information into BNNs.
- We conduct comprehensive experiments on the image classification task. The extensive experimental results on CIFAR-100 and ImageNet datasets suggest the superiority of our proposed method.

2 Related Work

Binary Neural Networks. Binary Neural Networks (BNNs) constrain weights and activations to two values $\{+1, -1\}$, significantly reducing memory size and improving power efficiency. Recently, BNNs have been extensively studied across various domains, such as object recognition [17,18,21], object detection [27], low-level vision [29,32]. However, BNNs are difficult to train since the gradient of a binary function is non-differentiable. To address this problem, Hubara et al. [12] proposed binarized neural network and utilized the straight-through-estimator (STE) [1] for binarization function. Although high efficiency benefits, the performance gap between full-precision neural networks is still large. To this end, previous works mainly focus on addressing two challenges: 1) large quantization error in weights and features; 2) gradient mismatch during training stage. For the first challenge, XNOR-Net [21] introduced the real-valued scale factor to minimize the binarization error. RBNN [14] considered the angle alignment between the full-precision weight vector and its binarized version. AdaBin [26] utilized an effective approach to obtain the optimal binary sets for each layer. For the second challenge, many works mainly focused on mimicking the gradients of the Sign function. IR-Net [20] proposed the EDE function to gradually approximate the sign function. FDA [33] estimated the gradient of the Sign function in the Fourier frequency domain.

Feature Representation of BNNs. The performance degradation in BNNs is caused by information loss and reduced representative capability from feature binarization. To this end, researchers have sought to preserve model capacity while minimizing information loss. Liu et al. [18] added an additional shortcut in each block to mitigate this loss. ReActNet [17] employed channel-wise learnable parameters with an activation function for a more flexible feature distribution. However, these parameters remain fixed during inference and overlook sample diversity. DyBNN and INSTA-BNN [13,36] generated learnable thresholds based on the average or skewness of the input features. In this paper, we attribute BNN performance degradation to the disruption of deep-layer features and utilize a full-precision adapter to inject unimpaired semantic information without introducing additional parameters or computational costs during inference.

3 Method

In this section, we provide detailed information on our proposed method. Firstly, we present the preliminary knowledge about BNNs. Then, we introduce our training framework and method.

3.1 Preliminary

Binary Neural Networks. BNNs employ the Sign function to binarize weights and activations to +1 or -1. We denote $W^\ell \in \mathbb{R}^{c_{out} \times c_{in} \times k \times k}$ and $X^\ell \in \mathbb{R}^{c_{in} \times h \times w}$ as the weights and input features in the ℓ-layer, where c_{in} and c_{out} are input and output channel dimension, k is the kernel size, h and w are the spatial dimensions of input feature. The input of the $\ell+1$-layer can be defined as:

$$X^{\ell+1} = \phi^\ell(\mathrm{Sign}(W^\ell) \otimes \mathrm{Sign}(X^\ell)), \tag{1}$$

$$\mathrm{Sign}(x) = \begin{cases} +1, & x > 0, \\ -1, & x \leq 0, \end{cases} \tag{2}$$

where $\otimes$ denotes convolutional operation, the $\phi^\ell(\cdot)$ denotes the nonlinear operation in the ℓ-layer (ReLU, PReLU, and BN layer, etc.). For the backpropagation, BNNs typically use the straight through estimator (STE) [1] to calculate approximate gradients in the backward process since the gradient of the Sign function is zero almost everywhere. The process [4,28] can be simply defined as:

$$\frac{\partial L}{\partial z} = \frac{\partial L}{\partial z_b}, \tag{3}$$

where L is the loss function, z is the input full-precision input, and z_b is the binarized output.The gradients related to the full-precision inputs are equal to the gradients of the binarized outputs.

3.2 Deep Semantic Supervision Framework

The severe performance degradation of BNNs is mainly attributed to the binarization of feature maps in deep layers (Fig. 1). To address this problem, we construct a multi-branch training framework, which can guide BNNs training with unimpaired deep feature supervision. Specifically, a full-precision adapter is utilized to refine the degraded feature representation in the deep layers of BNNs. The full-precision adapter is jointly trained with the original BNN structure and transfers the effective feature knowledge to deep BNN layers.

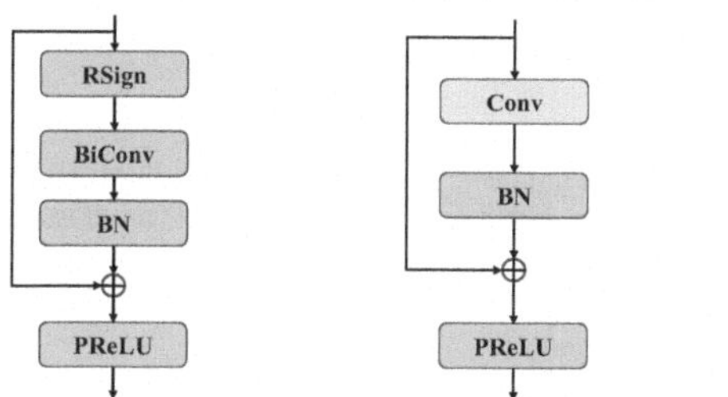

Fig. 3. The architecture of full-precision adapter

The overall framework is shown in Fig. 2. We utilize ResNet architecture as our base model. ResNet can be divided into 4 stages. We position the full-precision model after stage 3 and place it parallel to stage 4. The basic structure of a full-precision adapter is similar to BasicBlock in the BNNs. We remove the Sign function and replace binarized convolution with full-precision convolution, as shown in Fig. 3. During training, the full-precision adapter is trained jointly with BNN blocks in an end-to-end manner. In the inference stage, the full-precision adapter will be removed. Our framework features three advantages:

- **Stability.** The gradient of a full-precision adapter can help stabilize the training of the first three stages in BNNs.
- **Effective Knowledge Transfer.** The knowledge of the full-precision branch can be effectively transferred to the binary branch to guide optimization and mitigate information loss in deep features.
- **Flexibility.** Our framework is architecture-agnostic and is flexibly compatible with diverse BNN structures.

3.3 Training Methods

As shown in Fig. 2, the framework includes three loss functions: **1)** Cross entropy loss L_{ce}^1 and L_{ce}^2 from the true label to binary branch and full-precision adapter.

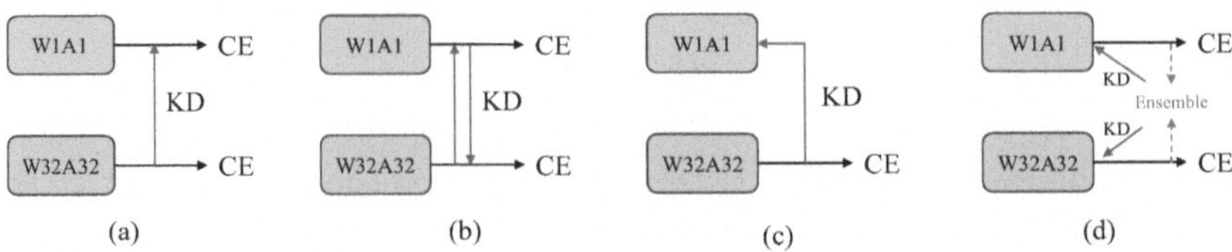

Fig. 4. Illustration of different distillation heads: (a) KD, (b) DML, (c) KD without CE in binary branch, and (d) KD with ensemble logits.

The L^1_{ce} and L^2_{ce} are calculated by the true labels from the training dataset and the output of the two branches' classifiers. The knowledge of the dataset is passed into the model through two branches, and the shared layers can be jointly optimized by binary branch and full-precision adapter; **2)** Similarity loss L^3_{sim} to measure the distance between two branches for guiding the binary branch learning, which can effectively transfer the knowledge of full-precision adapter to binary branch. We utilize L_2 Loss and KL-divergence (KL) Loss to measure the distance of two outputs. In the following, we provide the formulation of the training method.

Formulation. Given N samples $X = \{x_i\}_{i=1}^N$, the true label of X is denoted as $Y = \{y_i\}_{i=1}^M$, where M is the number of class. Our framework has two classifiers, which can be denoted as $\Theta = \{\theta_i\}_{i=1}^2$. A softmax layer is set after each classifier.

$$p_i^c = \frac{exp(z_i^c/T)}{\sum_j^c exp(z_j^c/T)}, \tag{4}$$

where z is the output of classifier, p_i^c denotes the i_{th} class probability of classifier θ_n. T is the temperature of distillation [11] ($T = 1$ for Cross entropy loss). For simple expression, we set the output of binary branch as z^b, the output of full-precision adapter as z^r, and the corresponding outputs after softmax are p^b and p^r.

We denote L^3_{sim} as the loss function to minimize the discrepancy between the outputs of two branches. L^3_{sim} can be set as L_2 loss and KL loss, which can be formulated as:

$$L^3_{sim} = \left\| z^b - z^r \right\|^2 \qquad (L_2 \text{ Loss}) \tag{5}$$

$$L^3_{sim} = -\sum_{i=1}^{C} p_i^r \log p_i^b \qquad (\text{KL Loss}) \tag{6}$$

To sum up, the loss function of the training framework consists of the loss function of each branch's classifier and the similarity loss between two branches, which can be written as:

$$loss = L^1_{ce} + \beta * L^2_{ce} + \alpha * L^3_{sim}. \tag{7}$$

where L^1_{ce} and L^2_{ce} are the Cross entropy (CE) loss for the binary and full-precision adapter, L^3_{sim} is the similarity loss between two branches, α and β are the weights used to scale the losses.

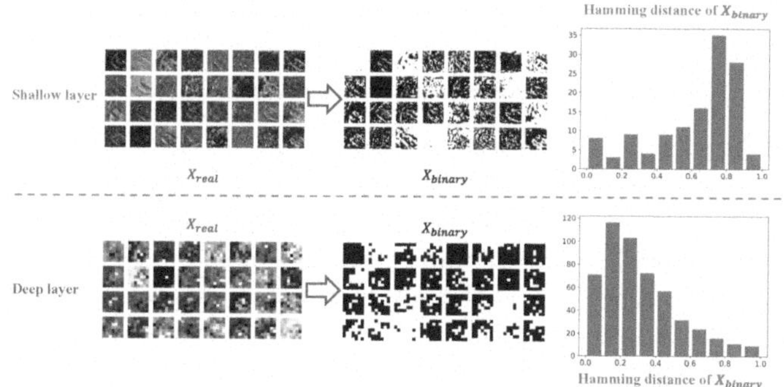

Fig. 5. Visualization of representative features in shallow layers and deep layers from ReActNet (3rd layer and 14th layer). **Left:** The visualization of feature binarization in the shallow and deep layer. **Right:** Distribution histogram of Hamming distance between binary feature channels.

3.4 Variations of Distillation Heads

To further explore our method, we tried various structures of distillation heads. Following the methods widely used in full-precision CNNs, we evaluate our method on four distillation methods: (a) Kullback-Leibler Divergence (KD), (b) Deep Mutual Learning (DML), (c) KD without CE loss in binary branch, and (d) KD with ensemble logits, as shown in Fig. 4. The detailed information is shown below.

KD. The full-precision branch is training by true label y, and the binary branch is optimized by true label y and logits from full-precision branch z^r.

DML. Following [37], the full-precision adapter is training by true label y and logits from binary branch z^b. The binary branch is optimized by true label y and logits from full-precision adapter z^r.

KD Without CE in the Binary Branch. The full-precision branch is trained by true label y, and the binary branch is only optimized by soft targets from the full-precision branch z^r.

KD with Ensemble Logits. Following previous work [8], we utilize the ensemble logits $\frac{z^b+z^r}{2}$ from two branches as guided knowledge to distill binary branch. This method utilized an ensemble of students to learn collaboratively, which can improve the generalization ability of the student network.

The detailed discussion can be found in Sect. 4.4.

4 Experiments

To evaluate the performance of proposed method, we conduct experiments on the image classification task. Two widely used datasets (CIFAR-100 and ImageNet

[5]) are included for evaluation. We first introduce the training details. Then, we analyze how deep feature binarization impacts the performance of BNNs. Finally, we provide the experimental results and analyses of our proposed method.

Training Setting. For CIFAR-100, we train binary ResNet18 for 90 epochs. The Adam optimizer is utilized for optimization. The batch size is set to 64, with an initial learning rate of 1e-3, which was reduced by a factor of 10 at the 45th and 55th epochs. The α and β are set to 0.5. The temperature T is set to 2. For ImageNet, we follow the setting in [17] and train binary ResNet18 for 128 epochs. The Adam optimizer is utilized for optimization. The batch size is 512, and the learning rate is set to 2.5e-3. The learning rate decay is set to linear rate decay. The α and β are set to 0.5. The temperature T is set to 1. The experiment is conducted on the NVIDIA V100 GPU and Pytorch 1.12 framework. **Note that we use Acc and $\mathrm{Acc}_{\mathrm{FA}}$ to represent the accuracy of the binary branch and the full-precision adapter in the subsequent sections, respectively.**

4.1 Deep Feature Binarization

Figure 1 illustrates that deep layer binarization is the main reason for degraded BNN performance. We further analyze this phenomenon from the perspective of feature binarization.

BNNs can achieve similar feature representations of full-precision models in the shallow layers. However, in the deep layers, BNNs often exhibit noticeable representation degradation [22]. We visualize both shallow and deep binary features of the BNNs in Fig. 5. Intuitively, as the scale of feature maps increases, the spatial representation ability of binary features is enhanced. In the deep layer where the feature map size is small (e.g., 7×7), binarization significantly narrows the feature space with only 2^{49} points. On the contrary, 32-bit representation allows for a significantly larger space with stronger representative ability. In the shallow layer, where the feature map size is relatively large (e.g., 56×56), the binarized feature space can provide ample expressive capability. Thus, we attribute the performance degradation of BNNs to the limited deep semantic representation in deep-layer binary features. To quantitatively demonstrate this, we calculated the spatial similarity between different channels of binary features in both shallow and deep layers. The Hamming distance is utilized as the metric to evaluate the similarity between the first channel in the binarized features and the remaining channels. As shown in Fig. 5, only a small number of features show extremely high similarity in the shallow layer, but in the deep layer, over 50% channel features are very close to the first channel features. **This phenomenon demonstrates that binary features in the deep layers of BNNs are extremely homogenized, which hinders BNNs from extracting sufficient semantic information.**

In addition, we conducted experimental analyses on the extra shortcuts in BiRealNet [18]. If deep feature representation in BNNs is limited, then removing deep shortcuts should have minimal impact on model performance. Based on

Table 1. The accuracy of removing extra shortcuts. We utilize the ReActNet based on ResNet18 and evaluate on ImageNet dataset.

Model	Top-1 Acc (%)	Top-5 Acc (%)
Baseline	63.2	84.6
Remove deep shortcuts	63.0	84.3
Remove all shortcuts	62.4	83.9

this assumption, we remove the additional shortcut in the deep layer (stage 3 and stage 4 layers). The results in Table 1 show that the accuracy only decreased by 0.2%. The slight decrease in accuracy suggests that the deep features of BNNs exhibit redundancy with minimal information loss.

Table 2. The evaluated results on CIFAR-100. The ReActNet based on ResNet18 [17] is used as base model.

Model	FA	Loss	Acc(%)	$\mathrm{Acc_{FA}}$(%)
ResNet18	×	$CE(p, y)$	66.8	×
	✓	$CE(p^b, y) + CE(p^t, y)$	67.2	69.3
	✓	$CE(p^b, y) + CE(p^t, y) + MSE(p^b, p^t)$	**69.4**	70.3
	✓	$CE(p^b, y) + CE(p^t, y) + KL(p^b, p^t, T = 1)$	68.8	69.7
	✓	$CE(p^b, y) + CE(p^t, y) + KL(p^b, p^t, T = 2)$	**69.1**	69.2
	✓	$CE(p^b, y) + CE(p^t, y) + KL(p^b, p^t, T = 3)$	67.4	69.2

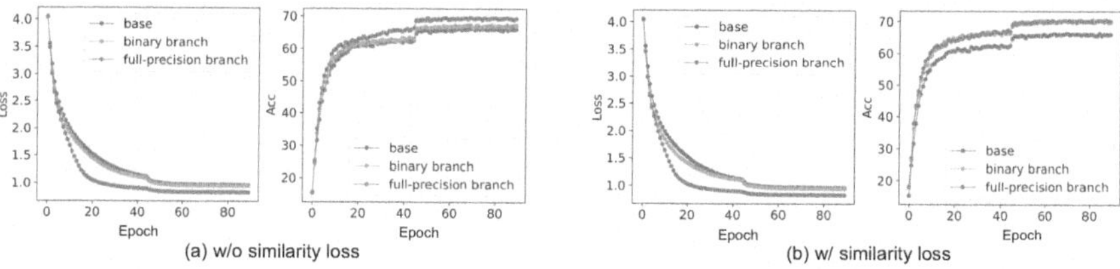

Fig. 6. Training curve without similarity loss (a) and with similarity loss (b).

4.2 Results on CIFAR-100

We first evaluate the proposed method with various similarity losses on CIFAR-100. We utilize ReActNet [17] based on ResNet18 as our baseline. The L_2 loss and KL loss with different T are used as the similarity loss.

As shown in Table 2, a simple L_2 loss significantly enhances the performance of the baseline model, resulting in a 2.6% increase in test accuracy. The training curves for loss and test accuracy are illustrated in Fig. 6. For KL Loss, we evaluated various temperature settings, finding that a temperature of 2 yields the best accuracy, improving performance by 2.3%. These results indicate that our proposed method effectively enhances BNN performance on small-scale datasets.

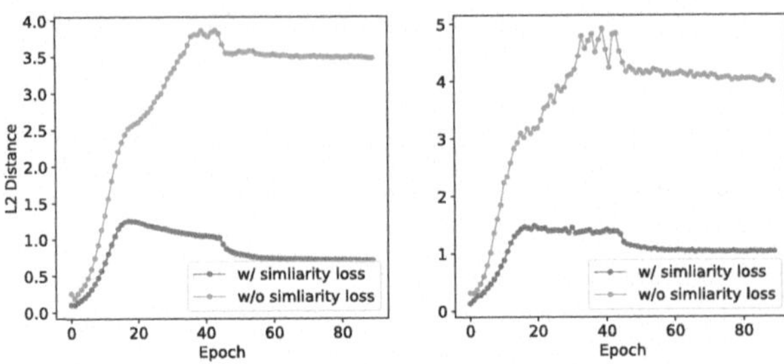

Fig. 7. L_2 distance between outputs of binary and full-precision branch on CIFAR100. **Left:** The distance is calculated on train set; **Right:** The distance is calculated on test set.

Additionally, the incorporation of a similarity loss improves the accuracy of the full-precision branch by approximately 1%, underscoring the mutual benefits between the two branches.

The primary objective of our full-precision adapter is to extract fine-grained features from the intermediate layers of BNNs, enabling the binary branch to mimic the output of the adapter and achieve performance comparable to that of the full-precision branch. To evaluate the degree of mimicry, we measured the similarity between the outputs of the two branches using L_2 distance. Figure 7 shows that the distance between the two branches' outputs first increases and then decreases due to the instability of training. When adding similarity loss, the distance between two branches is reduced, which can effectively transfer the knowledge from the full-precision adapter to the binary branch.

Table 3. Comparison of the top-1 accuracy with state-of-the-art BNN methods on ImageNet.

Binary Method	W/A	BOPs($\times 10^9$)	FLOPs($\times 10^8$)	OPs($\times 10^8$)	Acc Top-1 (%)
BNN [12]	1/1	1.70	1.20	1.47	42.2
ABC-Net [15]	1/1	–	–	–	42.7
XNOR-Net [21]	1/1	1.70	1.41	1.67	51.2
DoReFa [38]	1/2	–	–	–	53.4
Bi-RealNet-18 [18]	1/1	1.68	1.39	1.63	56.4
XNOR++ [2]	1/1	–	–	–	57.1
IR-Net [20]	1/1	–	–	–	58.1
BONN [7]	1/1	–	–	–	59.3
NoisySupervision [10]	1/1	–	–	–	59.4
ReCU [34]	1/1	1.68	1.39	1.63	61.0
ReSTE [28]	1/1	1.68	1.39	1.63	60.9
Baseline [17]	1/1	1.68	1.39	1.63	61.4
Ours	1/1	1.68	1.39	1.63	62.5($\uparrow$ 1.1)

4.3 Results on ImageNet

On the ImageNet dataset [5], we use ReActNet [17] based on ResNet18 as our baseline model. The L_2 loss can achieve superior results on CIFAR-100. However, on ImageNet, we observe that the performance was severely reduced. As shown in Fig. 8, the accuracy has decreased by 3% compared to the baseline, and the L_2 distance between two branches has also been increased without convergence. The instability of similarity loss in the early stages of training has a negative impact on models when training on large-scale datasets. Therefore, on ImageNet, we use KL loss as the similarity loss.

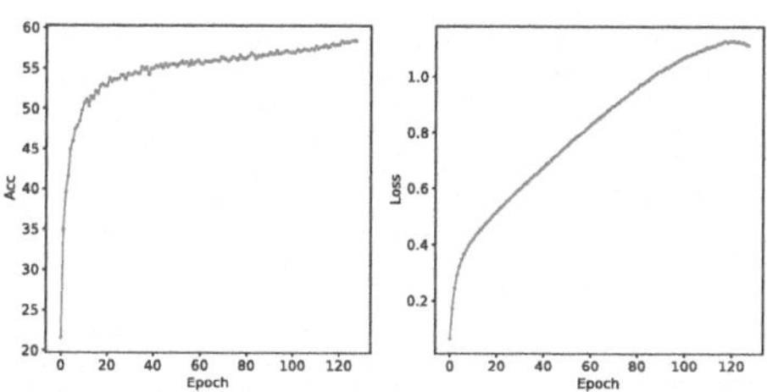

Fig. 8. The L_2 distance between the outputs of binary and full-precision adapter on ImageNet. **Left:** The distance is calculated on train set; **Right:** The distance is calculated on test set.

We compare our method with recent state-of-the-art methods without distillation. The results are presented in Table 3. With the full-precision adapter, the accuracy of the binary branch can reach 62.5% in terms of top-1 accuracy, which is 1.1% higher than the baseline. This significant performance improvement demonstrates that our full-precision adapter can effectively guide the learning of the binary branch and enhance the representation capability of binarized features.

To further illustrate the effectiveness of our method, we employ real-valued ResNet34 as the extra teacher model to binarize ResNet18. As shown in Table 4, the top-1 accuracy is further improved by 1.1%, surpassing the baseline by 0.4%. This demonstrates that our method is compatible with existing knowledge distillation methods and can achieve further performance gains.

Table 4. The accuracy when combined with full-precision ResNet34 as teacher model

	Acc_{FA}(%)	Acc(%)
Baseline	-	61.4
Ours	64.5	**62.5**(↑ 1.1)
Baseline+ResNet34	-	63.2
Ours+ResNet34	67.6	**63.6**(↑ 0.4)

Two-Step Training. Two-step training [19] is an effective method to improve the accuracy of BNNs. The accuracy of the first stage of the W32A1 model is related to the weight decay parameter, which affects the final result. We test different weight decay parameters, which can be observed in Table 5 (*69.0/66.7 denotes the accuracy of the full-precision adapter and binary branch). The result shows that large weight decay degrades the performance of two branches. When weight decay is set to 5e-6, the performance is improved by 0.5% compared with the ReActNet setting [17]. With similarity loss between the two branches, the accuracy reaches 66.7%, which is 1% higher than the baseline. Thus, we set weight decay to 5e-6 in our two-step training experiment. The results in Table 6 indicate a 0.4% improvement over the baseline.

Table 5. The ImageNet evaluation result of W32A1 training under different weight decay parameters.

	5e-6	1e-5
Baseline	66.2	65.7
Ours(Acc_{FA}/Acc)	69.0/**66.7***	68.2/65.6

Table 6. The ImageNet evaluation result of two-step training [19] compared with ReActNet [17].

	Acc_{FA}(%)	Acc(%)
Baseline	-	65.9
Ours	69.7	**66.3**(↑ 0.4)

4.4 Discussion

In this section, we discuss the impact of distillation head and depth of full-precision adapter.

Distillation Head. As shown in Sect. 3.4, we construct four different distillation heads and compare their accuracy in Fig. 4, including (a) **KD**, (b) **DML**, (c) **KD without CE in binary branch**, and (d) **KD with ensemble logits**. We compare each method on CIFAR-100 and ImageNet datasets. In Fig. 9, we can observe that the accuracy of four designs is related to the scale of the dataset. On CIFAR-100, model **(c)** achieves the best performance, reaching 69.7% top-1 accuracy. For ImageNet, model **(a)** outperforms the other three models. This disparity arises from the discriminative ability of the output logits. The limited representation capacity of BNNs hinders reliable predictions on challenging datasets. Furthermore, the excessively smooth distribution from the full-precision adapter obstructs effective knowledge transfer, further diminishing the accuracy of the binary branch. In this paper, we select model **(a)** as our baseline since it achieves superior accuracy on both small- and large-scale datasets. For

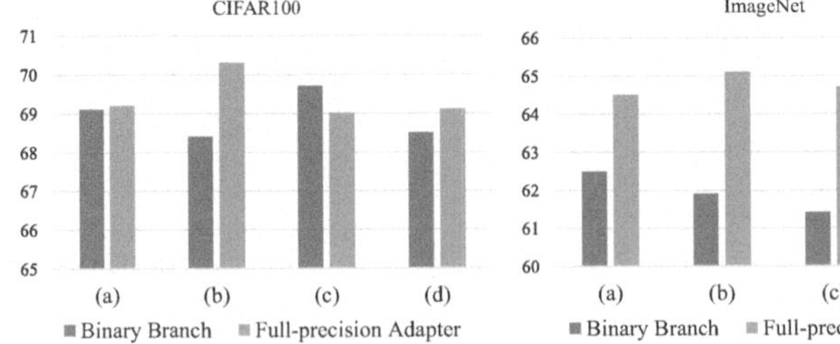

Fig. 9. The accuracy of four distillation methods on CIFAR-100 and ImageNet.

Table 7. The ablation study on the depth of full-precision adapter.

	Num Blocks	Params($\times 10^6$)	FLOPs ($\times 10^7$)	Acc (%)	Acc_{FA}(%)
CIFAR-100	2	3.54	1.42	68.4	67.5
	3	5.90	2.36	69.0	69.1
	4	8.26	3.30	69.4	70.2
	5	10.62	4.25	**70.3**	**70.5**

model **(b)** and **(d)**, DML and ensemble logits are commonly used distillation methods for full-precision CNN. Although both models have achieved certain accuracy improvements, the gap in representative ability between two branches limits the advantages of these two methods.

Depth of Full-Precision Adapter. The full-precision adapter can be regarded as a stack of basic blocks. To evaluate the impact of its depth, we evaluated the accuracy in CIFAR100 with varying numbers of basic blocks in the full-precision adapter. As shown in Table 7, the accuracy of the binary branch improves as the depth of the full-precision adapter increases. Note that our adapter is only used during training and is removed during inference without introducing any inference overhead. When the full-precision adapter has only two layers, the accuracy is only 67.5%, which is lower than the binary branch. As the number of blocks increases, the representation capability of the full-precision adapter gradually improves. In order to avoid introducing too much computation during training, we set the number of blocks to 4 in our experiments.

5 Conclusion

In this paper, we propose an effective multi-branch training framework for Binary Neural Networks (BNNs) via deep semantic supervision. Our motivation is that the deep-layer features in BNNs are extremely homogenized, resulting in insufficient semantic information for accurate target recognition. To this end, a full-precision adapter is introduced in the deep layer to guide BNN training with unimpaired deep feature knowledge. Our proposed method has three advantages: 1) introducing real gradient information flow to stabilize training of the shared fully binarized layers; 2) unimpaired deep-layer semantic information guides the training of binary branch; 3) the proposed method can be seamlessly integrated with diverse BNN structures. The extensive experiments on the CIFAR-100 and ImageNet datasets demonstrate the effectiveness of our method.

Limitation and Future Work. The introduced full-precision branch increases the training cost. In future work, we will try to design a more efficient full-precision adapter and improve the knowledge transfer from the pretrained model and full-precision adapter. Besides, we also aim to extend our framework to other tasks in future work, such as object detection and segmentation.

References

1. Bengio, Y., Léonard, N., Courville, A.: Estimating or propagating gradients through stochastic neurons for conditional computation. arXiv preprint arXiv:1308.3432 (2013)
2. Bulat, A., Tzimiropoulos, G.: Xnor-net++: improved binary neural networks. arXiv preprint arXiv:1909.13863 (2019)
3. Chen, Y., et al.: Mobile-former: bridging mobilenet and transformer. In: IEEE Conference on Computer Vision Pattern Recognition (2022)
4. Courbariaux, M., Bengio, Y., David, J.P.: Binaryconnect: training deep neural networks with binary weights during propagations. In: Advances in Neural Information Processing Systems (2015)
5. Deng, J., Dong, W., Socher, R., Li, L.J., Li, K., Fei-Fei, L.: Imagenet: a large-scale hierarchical image database. In: International Conference on Computer Vision (2009)
6. Gou, J., Yu, B., Maybank, S.J., Tao, D.: Knowledge distillation: a survey. Int. J. Comput. Vis. (2021)
7. Gu, J., et al.: Bayesian optimized 1-bit CNNs. In: International Conference on Computer Vision (2019)
8. Guo, Q., et al.: Online knowledge distillation via collaborative learning. In: IEEE Conference on Computer Vision Pattern Recognition (2020)
9. Han, K., et al.: A survey on vision transformer. IEEE Trans. Pattern Anal. Mach. Intell. (2022)
10. Han, K., Wang, Y., Xu, Y., Xu, C., Wu, E., Xu, C.: Training binary neural networks through learning with noisy supervision. In: International Conference on Machine Learning (2020)
11. Hinton, G., Vinyals, O., Dean, J.: Distilling the knowledge in a neural network. arXiv preprint arXiv:1503.02531 (2015)
12. Hubara, I., Courbariaux, M., Soudry, D., El-Yaniv, R., Bengio, Y.: Binarized neural networks. In: Advances in Neural Information Processing Systems (2016)
13. Lee, C., Kim, H., Park, E., Kim, J.J.: Insta-BNN: binary neural network with instance-aware threshold. In: International Conference on Computer Vision (2023)
14. Lin, M., et al.: Rotated binary neural network. In: Advances in Neural Information Processing Systems (2020)
15. Lin, X., Zhao, C., Pan, W.: Towards accurate binary convolutional neural network. In: Advances in Neural Information Processing Systems (2017)
16. Liu, L., et al.: Deep learning for generic object detection: a survey. Int. J. Comput. Vis. (2020)
17. Liu, Z., Shen, Z., Savvides, M., Cheng, K.T.: Reactnet: towards precise binary neural network with generalized activation functions. In: European Conference on Computer Vision (2020)
18. Liu, Z., Wu, B., Luo, W., Yang, X., Liu, W., Cheng, K.T.: Bi-real net: enhancing the performance of 1-bit CNNs with improved representational capability and advanced training algorithm. In: European Conference on Computer Vision (2018)
19. Martinez, B., Yang, J., Bulat, A., Tzimiropoulos, G.: Training binary neural networks with real-to-binary convolutions. arXiv preprint arXiv:2003.11535 (2020)
20. Qin, H., et al.: Forward and backward information retention for accurate binary neural networks. In: IEEE Conference on Computer Vision and Pattern Recognition (2020)

21. Rastegari, M., Ordonez, V., Redmon, J., Farhadi, A.: Xnor-net: imagenet classification using binary convolutional neural networks. In: European Conference on Computer Vision (2016)
22. Shen, Z., Liu, Z., Qin, J., Huang, L., Cheng, K.T., Savvides, M.: S2-BNN: bridging the gap between self-supervised real and 1-bit neural networks via guided distribution calibration. In: IEEE Conference on Computer Vision and Pattern Recognition (2021)
23. Su, Z., Welling, M., Pietikäinen, M., Liu, L.: SVNet: where so (3) equivariance meets binarization on point cloud representation. In: International Conference on 3D Vision. IEEE (2022)
24. Su, Z., et al.: Boosting convolutional neural networks with middle spectrum grouped convolution. IEEE Trans. Neural Netw. Learn. Syst. (2024)
25. Su, Z., et al.: Lightweight pixel difference networks for efficient visual representation learning. IEEE Trans. Pattern Anal. Mach. Intell. (2023)
26. Tu, Z., Chen, X., Ren, P., Wang, Y.: Adabin: improving binary neural networks with adaptive binary sets. In: European Conference on Computer Vision (2022)
27. Wang, Z., Wu, Z., Lu, J., Zhou, J.: Bidet: an efficient binarized object detector. In: IEEE Conference on Computer Vision and Pattern Recognition (2020)
28. Wu, X.M., Zheng, D., Liu, Z., Zheng, W.S.: Estimator meets equilibrium perspective: a rectified straight through estimator for binary neural networks training. In: International Conference on Computer Vision (2023)
29. Xia, B., et al.: Basic binary convolution unit for binarized image restoration network. arXiv preprint arXiv:2210.00405 (2022)
30. Xiao, C., et al.: Highly efficient and unsupervised framework for moving object detection in satellite videos. IEEE Trans. Pattern Anal. Mach. Intell. (2024)
31. Xie, E., Wang, W., Yu, Z., Anandkumar, A., Alvarez, J.M., Luo, P.: Segformer: simple and efficient design for semantic segmentation with transformers. In: Advances in Neural Information Processing Systems (2021)
32. Xin, J., Wang, N., Jiang, X., Li, J., Huang, H., Gao, X.: Binarized neural network for single image super resolution. In: European Conference on Computer Vision (2020)
33. Xu, Y., Han, K., Xu, C., Tang, Y., Xu, C., Wang, Y.: Learning frequency domain approximation for binary neural networks. In: Advances in Neural Information Processing Systems (2021)
34. Xu, Z., et al.: Recu: reviving the dead weights in binary neural networks. In: International Conference on Computer Vision (2021)
35. Yu, Z., et al.: Physformer++: facial video-based physiological measurement with slowfast temporal difference transformer. Int. J. Comput. Vis. (2023)
36. Zhang, J., Su, Z., Feng, Y., Lu, X., Pietikäinen, M., Liu, L.: Dynamic binary neural network by learning channel-wise thresholds. In: IEEE International Conference on Acoustics, Speech, and Signal Processing (2022)
37. Zhang, Y., Xiang, T., Hospedales, T.M., Lu, H.: Deep mutual learning. In: IEEE Conference on Computer Vision and Pattern Recognition (2018)
38. Zhou, S., Wu, Y., Ni, Z., Zhou, X., Wen, H., Zou, Y.: Dorefa-net: training low bitwidth convolutional neural networks with low bitwidth gradients. arXiv preprint arXiv:1606.06160 (2016)

Revisiting Likelihood-Based Out-of-Distribution Detection by Modeling Representations

Yifan Ding[1(✉)], Arturas Aleksandraus[1], Amirhossein Ahmadian[2], Jonas Unger[1], Fredrik Lindsten[2], and Gabriel Eilertsen[1]

[1] Department of Science and Technology, Linköping University, Norrköping, Sweden
{yifan.ding,arturas.aleksandraus,jonas.unger,gabriel.eilertsen}@liu.se

[2] Department of Computer and Information Science, Linköping University, Linköping, Sweden
{amirhossein.ahmadian,fredrik.lindsten}@liu.se

Abstract. Out-of-distribution (OOD) detection is critical for ensuring the reliability of deep learning systems, particularly in safety-critical applications. Likelihood-based deep generative models have historically faced criticism for their unsatisfactory performance in OOD detection, often assigning higher likelihood to OOD data than in-distribution samples when applied to image data. In this work, we demonstrate that likelihood is not inherently flawed. Rather, several properties in the images space prohibit likelihood as a valid detection score. Given a sufficiently good likelihood estimator, specifically using the probability flow formulation of a diffusion model, we show that likelihood-based methods can still perform on par with state-of-the-art methods when applied in the representation space of pre-trained encoders. The code of our work can be found at https://github.com/limchaos/Likelihood-OOD.git.

Keywords: Out-of-distribution Detection · AI Safety · Trustworthy ML

1 Introduction

Out-of-distribution (OOD) detection is the process of detecting if individual data points, e.g. images, belong to the distribution of training data or not. In machine learning, it is typically used to identify "unseen" data points that may lead to unreliable inference. The importance of OOD detection is highlighted by the fact that deep neural networks often perform poorly on input data not part of the training distribution. Detection of OOD input is thus a critical functionality to increase safety and robustness in deployed systems, particularly in real-world applications where the misclassification of OOD samples can lead to severe consequences.

OOD detection was initially developed to identify unseen data with low likelihood given training data and a statistical model [2]. Explicit likelihood-based deep generative models (DGMs), e.g., autoregressive models and normalizing

J. Petersen and V. A. Dahl (Eds.): SCIA 2025, LNCS 15726, pp. 166–179, 2025.
https://doi.org/10.1007/978-3-031-95918-9_12

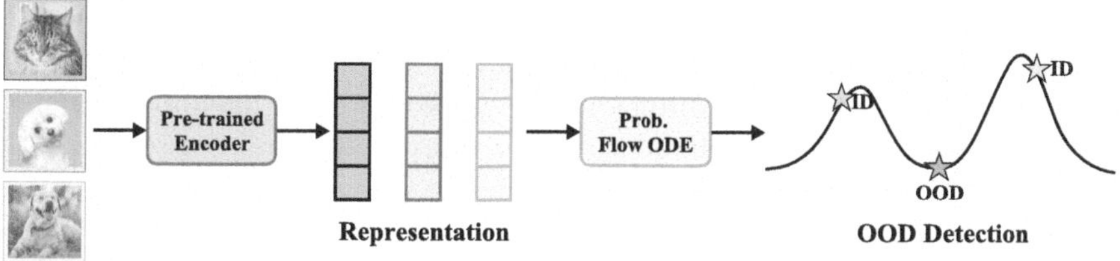

Fig. 1. Revisiting likelihood-based OOD detection, we employ a score-based diffusion model to detect OOD images in a semantically clustered representation space. The input images are first projected to the representation space. Then, a probability flow ODE [20], from the corresponding representation score-based diffusion model [28], is leveraged to calculate the precise likelihood in the representation space.

flows, appear to be well-suited for this purpose. However, it has been demonstrated that these models may assign higher likelihood to OOD images than images from the training distribution [21], e.g., models trained on FashionMNIST and CIFAR-10 incorrectly assign high likelihoods to OOD datasets like MNIST and SVHN where they should assign low likelihood. Meanwhile, it is not obvious how likelihood-based OOD detection methods could generalize to more real world settings (large-scale benchmarks). Due to the problems with likelihood-based OOD detection, the OOD community has mostly shifted the focus to post-hoc methods [9,30,32] based on classifiers. However, these methods usually require supervised training and label information of the in-distribution (ID) data. Particularly in the context of medical imaging, such as histopathology data, or other fine-grained categories, obtaining annotations is often expensive or infeasible [24] due to the requirement for expert knowledge and regulatory constraints. In such cases, many OOD detection methods that rely on logits from a supervised classifier [9,10,17–19,32] become inapplicable, necessitating alternative approaches that do not depend on labeled training data.

In this paper, we reinvestigate likelihood-based OOD detection and demonstrate that estimating likelihood in the *representation spaces of modern pre-trained image encoders* can be a very promising method for OOD detection. To this end, we make use of a score-based diffusion model [28] trained on encoded image representations as likelihood estimator. We argue that the previous failures in [14,21] can be associated with the fact that likelihood estimation carried out in image space is heavily influenced by background statistics [25] and low-level features [14]. For example, images with different semantic content may appear close in Euclidean space, while semantically similar images can be far apart due to transformations such as translation. As a result, images tend to cluster based on low-level visual features such as color, luminance, and texture, making it challenging to capture and learn meaningful semantic shifts. However, likelihood works well in semantically clustered representation spaces with accurate likelihood estimation provided by a diffusion model. Unlike [9,10,18,32], likelihood can work with self-supervised encoders, but it can also leverage ID class labels to guide the diffusion model. As illustrated in Fig. 1, we first encode images to a representation space. A score-based diffusion model is then trained only on ID representations, and likelihood is estimated for both ID and OOD data at test time. We show that the diffusion model assigns high likelihood to the ID representations and low likelihood to OOD representations. Our contributions can be summarized as follows:

- We revisit likelihood-based OOD detection by leveraging score-based diffusion models within the representation space of pre-trained image encoders.
- We conduct extensive experiments on large-scale datasets, evaluating both supervised and unsupervised encoders and benchmarking likelihood against other lines of OOD detection methods.

Compared to likelihood-based methods operating directly in image space, estimating the likelihood in a representation space with fewer dimensions results in more computationally efficient methods. The evaluations demonstrate that likelihood for representations achieves results comparable to current state-of-the-art (SOTA) methods without requiring access to labeled data. Furthermore, by employing the labels of ID data and class conditional training (guidance) for the diffusion model, we are able to surpass the performance of most SOTA methods.

2 Related Work

OOD detection was first introduced to recognize unseen data points unlikely to be part of the training distribution using statistical models [2]. Since then, it has gained significant research attention across multiple directions.

OOD Detection with Supervised Classifiers. Many post-hoc methods derive distance functions from pre-trained classifiers, including MSP [10], ODIN [17], Mahalanobis distance [15], Energy [18], ReAct [29], ViM [32], and Generalized Entropy [19]. These methods typically rely on supervised classifiers trained on ID data, using features from the penultimate layer, logits, or labels to define OOD scores. While widely adopted, their effectiveness relative to likelihood-based methods remains unclear.

DGM-Based OOD Detection. DGMs have been leveraged for OOD detection through reconstruction-based [8], likelihood-based [7,22], and synthetic OOD data generation approaches [6]. Unlike previous likelihood-based methods that model image pixels, we empirically study likelihood on image representation and benchmark with other lines of post-hoc methods. Compared to [7], our approach avoids the high computational cost associated with processing the high-dimensional image space.

OOD Detection with Self-Supervised Models. Self-supervised foundational models such as DINO [3] and DINOv2 [23] have recently been explored for OOD detection [1,33]. However, most post-hoc methods [10,19,32] depend on labels or logits, making them unsuitable for self-supervised models. Only a few methods, such as Residual [32] and KNN [30], can be adapted for these models, highlighting a gap in OOD detection research as it was originally defined [2].

3 Representation Likelihood Estimation with Diffusion Models for OOD Detection

3.1 Preliminaries

OOD Detection with ID Labels. In every image classification problem, there is a predefined set of semantic categories that the model is expected to identify, which defines the ID images. We refer to this set of labels and the associated joint distribution as $\mathcal{Y}_{\mathrm{ID}}$ and $\mathcal{D}_{\mathrm{ID}}$, respectively, where $\forall(\mathbf{x}, y) \sim \mathcal{D}_{\mathrm{ID}},\ y \in \mathcal{Y}_{\mathrm{ID}}$. In the open world, there are semantic groups that do not belong to the predefined and finite $\mathcal{Y}_{\mathrm{ID}}$, forming the OOD space $\mathcal{Y}_{\mathrm{OOD}} = \{y|y \notin \mathcal{Y}_{\mathrm{ID}}\}$ and $\mathcal{D}_{\mathrm{OOD}}$. In this setup, OOD detection has two main goals [33]. The first is to develop a discriminative model that accurately classifies ID samples drawn from $\mathcal{D}_{\mathrm{ID}}$. The second goal is to develop a detector module (typically built upon the trained classifier) that accurately identifies whether an incoming image during the inference phase is ID or OOD.

OOD Detection without ID Labels. Access to annotated ID data points is not possible in many cases. As OOD detection is initially conceptualized in [2], we can train a density model $p_\theta(\mathbf{x})$ (where θ represents the parameters) to approximate the true distribution of the training inputs $p(\mathbf{x})$, given only $\mathbf{x} \sim \mathcal{X}_{\mathrm{ID}}$ (where we use $\mathcal{X}_{\mathrm{ID}}$ for the marginal in-distribution of $\mathbf{x}$). Any $\mathbf{x}$ that has a sufficiently low density under $p_\theta(\mathbf{x})$ is assigned to the $\mathcal{D}_{\mathrm{OOD}}$ space if $p_\theta(\mathbf{x})$ is a reasonably good estimation of $p(\mathbf{x})$.

3.2 Motivating Observations

Score-based diffusion models have gained attention for likelihood estimation due to their high expressiveness compared to other likelihood estimators [28]. However, similar to earlier flow-based models, as studied by [21], likelihood estimates from diffusion models remain ineffective to classify ID from OOD data in the image space. As illustrated in Fig. 2(a), a diffusion model trained on the CIFAR-10 dataset continues to assign a lower negative log-likelihood value to the SVHN dataset, highlighting its limitations in OOD detection. In contrast, Fig. 2(b) demonstrates that extracting representations from the penultimate layer of an encoder trained on CIFAR-10 using cross-entropy loss, such as ResNet18, enhances the effectiveness of likelihood-based OOD detection. However, the performance of likelihood estimation in the representation space compared to other state-of-the-art methods remains unclear. Thus, in this study we demonstrate that likelihood estimation is effective when applied to representations from both supervised and self-supervised encoders, as validated through comprehensive experiments on large-scale benchmarks.

3.3 Pre-trained Encoder

Given the image dataset $\{\mathbf{x}_i\}_{i=0}^{N}$, representations $\{\mathbf{z}_i\}_{i=0}^{N}$ are calculated by a pre-trained encoder $\mathcal{E}$, $\mathbf{z} = \mathcal{E}(\mathbf{x})$. We extract representations for both ID and

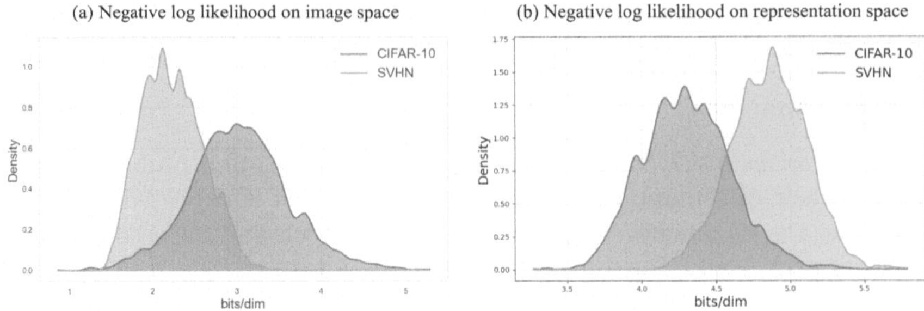

Fig. 2. Density of negative log likelihoods for CIFAR-10 (ID) vs SVHN (OOD) (a) diffusion model trained on CIFAR-10 images, (b) diffusion model trained on CIFAR-10 representations extracted from ResNet18.

OOD data using the same encoder, but only ID representations are used for training the diffusion model/likelihood estimator.

3.4 OOD Detection with Representation Likelihood Estimation

Models such as Generative Adversarial Networks (GANs) cannot calculate likelihood, while Denoising Diffusion Probabilistic Models (DDPMs) and Variational Autoencoders (VAEs) only provide a lower bound on the likelihood. However, using the (instantaneous) change of variables formula [5] with invertible generative models, such as normalizing flows, score-based diffusion models and flow matching, the likelihood can be computed exactly or estimated with high accuracy. In this work, we specifically use a score-based diffusion model [28] to estimate the likelihood, motivated by the flexibility and strong performance of this class of models for image generation. Since the model is trained on image representations, we refer to it as the Representation Diffusion Model (RDM) and its conditional version as ConRDM for brevity. Note that this is different from latent diffusion models [26] since we do not assume that we have access to a corresponding decoder. Training score-based diffusion models can be formulated as reverse-time stochastic differential equation (SDE) learning (for details, we refer to Appendix A), and the corresponding probability flow ODE [20] of such an SDE can be expressed as

$$d\mathbf{z} = \{\mathbf{f}(\mathbf{z}, t) - \frac{1}{2}g(t)g(t)^T \nabla_{\mathbf{z}} \log p_t(\mathbf{z})\}dt, \quad t \in [0, 1], \tag{1}$$

where $\mathbf{f}(\mathbf{z}, t)$ and $g(t)$ are the drift and diffusion coefficients, respectively, from the underlying SDE and $p_t(\mathbf{z})$ is the marginal distribution of $\mathbf{z}$ at time t. Note that the ODE (in this formulation, following [28]) is initialized at time 1 and runs backward in time, so dt should be seen as a *negative* infinitesimal time increment. With the instantaneous change of variables formula [5], denoting $f_\theta(\mathbf{z}, t) = \mathbf{f}(\mathbf{z}, t) - \frac{1}{2}g(t)g(t)^T s_\theta(\mathbf{z}, t)$, and assuming $\nabla_{\mathbf{z}} \log p_t(\mathbf{z}) \approx s_\theta(\mathbf{z}, t)$, we can compute the representation likelihood $p_0(\mathbf{z})$ using

$$\log p_0(\mathbf{z}(0)) = \log p_1(\mathbf{z}(1)) + \int_0^1 \nabla \cdot f_\theta(\mathbf{z}(t), t) dt. \tag{2}$$

In practice we follow [28] and make use of Hutchinson's trace estimator [12] for estimating the divergence $\nabla \cdot f_\theta(\mathbf{z}(t), t)$.

We can numerically estimate the data likelihood without prior assumption of the data distribution. After training only on ID representations, Eq. (2) is used to calculate likelihood for both ID and OOD data. We expect $\log p_0(\mathbf{z}_{ID}) > \log p_0(\mathbf{z}_{OOD})$ in general and fix a threshold λ such that we are able to make a decision for an input test image x,

$$x \in \begin{cases} \mathcal{X}_{\text{ID}} & \text{if } \log p_0(\mathbf{z}) \geq \lambda, \\ \mathcal{X}_{\text{OOD}} & \text{if } \log p_0(\mathbf{z}) < \lambda. \end{cases} \tag{3}$$

3.5 Leveraging Labeled in-Distribution Data

When ID labels are available, class-specific information can enhance OOD detection. Studies on diffusion models for density estimation and sampling indicate that conditioning the score function improves the samples quality while reducing their variance [11]. Following this principle, a class-conditioned model is expected to assign higher likelihoods to high-confidence regions of ID classes in the representation space. To incorporate labels and estimate $p(\mathbf{z}|c)$, where c denotes the class conditioning, the model learns a class-conditioned diffusion score function $s_\theta(\mathbf{z}, c, t)$. Inspired by classifier-free guidance [11], we introduce c as an additional input to the neural network. Here, we assume a pretrained encoder $\mathcal{E}$ trained via supervised classification on ID data. Given an image $\mathbf{x}$, we denote the classifier's prediction as $c = \arg\max \mathbf{W}\mathbf{z} + \mathbf{b}$, where $\mathbf{W}, \mathbf{b}$ are the last MLP layer weights. Pseudo-code for training and detection is provided in Algorithm 1 and Algorithm 2, respectively.

Algorithm 1: Training

Input Encoder $\mathcal{E}$, training images $\{\mathbf{x}_i^{id}\}_{i=0}^N$
Extract representation $\{\mathbf{z}_i^{id}\}_{i=0}^N$ by $\mathbf{z} = \mathcal{E}(\mathbf{x})$
If class condition **then**
  $c = \arg\max \mathbf{W}\mathbf{z} + \mathbf{b}$
Else $c = \emptyset$
Train RDM $\mathbf{D}(\mathbf{z}, c)$
return D

Algorithm 2: Detection

Input RDM $\mathbf{D}$, encoder $\mathcal{E}$, image $\mathbf{x}$, threshhold λ
Extract representation z by $\mathbf{z} = \mathcal{E}(\mathbf{x})$
If class condition **then**
  $c = \arg\max \mathbf{W}\mathbf{z} + \mathbf{b}$
Else $c = \emptyset$
Calculate $\log p(\mathbf{z}|c)$
If $\log p(\mathbf{z}|c) \geq \lambda$, $\mathbf{x} \in \mathcal{X}_{\text{ID}}$
Otherwise $x \in \mathcal{X}_{\text{OOD}}$

4 Experiments

In this section, we present detailed experiment settings and evaluate likelihood for both *OOD detection with ID labels* and *OOD detection without ID labels* on a large-scale benchmark and a histopathology benchmark.

Table 1. Encoders used for extracting representations and their corresponding pre-training datasets and dimension (DIM) on the representation vector **z**.

Model	Specification	Architecture	Training Method	DIM	Dataset
BiT	BiT-S-R101x1	CNN	Cross Entropy	2048	ImageNet-21K
RepVGG	RepVGG-b3	CNN	Cross Entropy	2560	ImageNet-21K
ResNet50d	ResNet-50d	CNN	Cross Entropy	2048	ImageNet-21K
Swin	Swin-B	Transformer	Cross Entropy	1024	ImageNet-21K
ViT	ViT-B/16	Transformer	Cross Entropy	768	ImageNet-21K
DeiT	ViT-B/16	Transformer	Cross Entropy	768	ImageNet-21K
MAE	ViT-B/16	Transformer	Self Supervised Learning	768	ImageNet-1K
DINO	ViT-B/16	Transformer	Self Supervised Learning	768	ImageNet-1K
DINOv2	ViT-B/14	Transformer	Self Supervised Learning	768	LVD-142M
Pathology-SSL	ViT-S/16	Transformer	Self Supervised Learning	384	TCGA&TULIP
Uni	ViT-L/16	Transformer	Self Supervised Learning	1024	Mass-100K

4.1 Experiment Settings

Data and Metrics. We use a large-scale OOD detection benchmark, utilizing ImageNet-1K as the ID dataset. As OOD data, we use four widely recognized datasets, including three far-OOD datasets: OpenImage-O, Texture and iNaturalist, and one near-OOD dataset: ImageNet-O, following the same evaluation protocol and dataset settings as in [19,32]. For the histopathology benchmark, we use the PatchCamelyon (PCam) dataset [31] and define non-tumor images as ID data and images containing tumor as OOD data, i.e. we perform unsupervised tumor detection. The histopathology benchmark represents a clinical application of OOD detection where labels of ID data are not available. We employ two conventional metrics to evaluate the OOD detection performance. The first is a threshold independent metric: Area Under the Receiver Operating Characteristic Curve (AUROC), where higher percentages reflect better performance. The second metric is the False Positive Rate at 95% True Positive Rate (FPR95), with lower percentages indicating better performance.

Encoders. In our experiments, we use both supervised and self-supervised encoders. We use the same supervised encoders as in [19,32] to allow for a fair comparison, as well as 5 self-supervised models, where 2 are trained on histopathology data. The details of the encoders are listed in Table 1.

Details of the Representation Diffusion Model. The RDM is parameterized by a time-dependent MLP with 12 residual blocks, following the architecture proposed by [16]. The model is trained using the AdamW optimizer, with a batch size of 4096 and a learning rate of 2e-3. Cosine annealing and gradient clipping are applied during training. Similar to [28], we don't use likelihood weighting. For training, we use a sub-VP SDE [28] as the default, and no significant differences were observed when comparing Variance Preserving (VP) SDE to sub-VP SDE (details in Appendix A).

Table 2. OOD detection with self-supervised encoder: AUROC and FPR95 are reported as percentages. Results for MAE, DINO and DINOv2 with ImageNet-1K as ID data and four OOD datasets: OpenImage-O, Textures, iNaturalist, and ImageNet-O. Since logits are not available, we only compare with KNN [30] and Residual [32]. The best method is marked in bold.

Method	OpenImage-O		Textures		iNaturalist		ImageNet-O		Average	
	AUROC ↑	FPR95 ↓	AUROC ↑	FPR95 ↓	AUROC ↑	FPR95 ↓	AUROC ↑	FPR95 ↓	AUROC ↑	FPR95 ↓
MAE										
KNN [30]	**60.54**	**89.03**	89.04	41.51	**48.02**	**97.69**	68.64	81.20	**66.56**	77.36
Residual w/o offset [32]	59.52	89.22	**90.33**	**38.90**	42.47	98.87	**69.60**	**79.85**	65.48	**76.71**
RDM	58.15	91.50	89.06	43.80	41.44	99.20	66.40	86.25	63.76	80.19
DINO										
KNN [30]	85.26	65.25	94.15	25.39	88.30	67.62	81.55	74.70	87.31	58.23
Residual w/o offset [32]	**87.57**	**54.77**	**97.84**	**11.10**	**92.71**	**42.76**	**81.98**	**68.40**	**90.02**	**44.25**
RDM	85.68	64.73	96.59	17.17	86.67	70.98	79.80	73.90	87.18	56.69
DINOv2										
KNN [30]	**95.05**	**25.66**	91.65	35.33	99.06	3.47	**86.67**	**57.55**	93.10	**30.50**
Residual w/o offset [32]	92.61	35.53	**93.60**	33.41	**99.32**	**1.74**	83.23	70.40	92.19	35.26
RDM	94.06	31.07	93.32	**32.50**	99.30	1.83	85.97	63.30	**93.16**	32.17

Likelihood Estimation. The likelihood is calculated on the ImageNet-1K validation set (ID dataset) and across four different OOD datasets. Regarding the PCam dataset, the likelihood is computed only on the tumor and non-tumor test splits. For likelihood estimation, we solve Eq. 2 with the RK45 ODE numerical integrators provided by *torchdiffeq*[1], where *atol = 1e-5* and *rtol = 1e-5*. For the divergence term, the Skilling-Hutchinson trace estimator [12,27] is used. Unless otherwise mentioned, we use the same settings for all encoders.

Computational Efficiency and Reproducibility. The training time of the diffusion model with 200 epochs is approximately 12 min on an RTX4090 GPU. Likelihood estimation has a throughput of 1500 image representations per second. The whole evaluation time on the ImageNet benchmark is approximately 25 s. We provide the source code of all training and evaluation implementations in the supplementary files.

4.2 Evaluation with Self-Supervised Encoders

The representation likelihood provides flexibility to detect OOD data with self-supervised encoders when label information of ID data is not available. We evaluate our approach on both the large-scale benchmark and the histopathology task with various self-supervised encoders and compare its performance against other methods that do not require labels.

ImageNet Benchmark. As illustrated in Table 2, RDM is compared to two other label-free methods, Residual [30] and KNN [32], across three different

[1] https://github.com/rtqichen/torchdiffeq.

self-supervised encoders. Likelihood on image representations shows competitive performance, particularly when paired with the best performing encoder, DINOv2. Overall, the results are mixed across the three methods; however, the differences in performance between the encoders are more significant than the differences between the OOD detection methods. We find the optimal $k = 50$ for KNN, which is selected from $k = \{1, 10, 20, 50, 100, 200, 500, 1000, 3000, 5000\}$ [30]. However, determining the optimal k requires a calibration dataset, and the optimal value may vary across different datasets or representations. The Residual score is used as part of ViM [32]. In the original implementation, an offset $\mathbf{o} = (\mathbf{W}^T)^{+}\mathbf{b}$ is subtracted to ensure the results are unbiased. However the offset is from the penultimate layer of a supervised encoder and we simply set it to 0.

Table 3. OOD detection with histopathology data: AUROC and FPR95 are reported as percentages. * The official Residual implementation is not defined for representation dimensions below 512, so the same ratio is used as in the DINO representation. The best method is marked in bold.

Method	Uni [4]		Pathology-SSL [13]		DINO [3]	
	AUROC ↑	FPR95 ↓	AUROC ↑	FPR95 ↓	AUROC ↑	FPR95 ↓
KNN [30]	88.36	50.74	**85.28**	**55.49**	60.77	98.17
Residual w/o offset [32]	89.05	48.09	75.78*	86.59*	61.14	97.72
RDM	**89.38**	**45.25**	81.49	68.54	**61.29**	**96.71**

PCam Benchmark. We also present experiments on the PCam dataset in Table 3 using three self-supervised encoders: two pre-trained on histopathology data and one pre-trained on natural image data. The results show similarly mixed performance as seen in Table 2, with the differences between encoders being larger than the differences between methods. Residual uses the smallest half to one-third of the principal subspace in the official implementation. However, the optimal number of principal components for Pathology-SSL representations spanned nearly the entire space (details in Appendix B). The diffusion model likelihood estimator, by contrast, is not sensitive to hyperparameters, as we use the same settings across all training on different representations.

4.3 Evaluation with Supervised Encoders

We now shift our attention to supervised encoders. We present results using ViT (which achieved the highest average OOD detection performance among all encoders) in Table 4, with results for other encoders provided in Table 5. We compare the likelihood-based RDM and ConRDM with a comprehensive collection of methods from the literature.

OOD Detection with ViT. In general, both RDM and ConRDM demonstrate superior performance across the benchmarks. In the label-free setup, RDM consistently outperforms the second-best method, Residual, by 4.11% in AUROC

Table 4. OOD detection with supervised encoder: AUROC and FPR95 are reported as percentages. The ID dataset is ImageNet-1K, while the OOD datasets are OpenImage-O, Texture, iNaturalist, and ImageNet-O. The supervised ViT-B/16 encoder model is used for representation extraction. The best method is marked in bold. Source refers to the information that each method requires from the encoder network: feat (feature representations), prob (softmax probabilities), or logits.

Method	Source	OpenImage-O		Textures		iNaturalist		ImageNet-O		Average	
		AUROC ↑	FPR95 ↓	AUROC ↑	FPR95 ↓	AUROC ↑	FPR95 ↓	AUROC ↑	FPR95 ↓	AUROC ↑	FPR95 ↓
With Label Information											
MSP [10]	prob	92.53	34.18	87.10	48.55	96.11	19.04	81.86	64.85	89.40	41.65
Energy [18]	logit	97.11	14.04	93.39	28.22	98.66	6.16	90.46	41.30	94.90	22.43
ODIN [17]	prob+grad	96.86	15.68	93.01	30.60	98.57	6.58	89.85	44.15	94.57	24.25
MaxLogit [9]	logit	96.87	15.68	93.01	30.60	98.57	6.58	89.85	44.15	94.57	24.25
KL Matching [9]	prob	93.80	28.49	88.76	44.09	96.88	14.79	84.12	55.70	90.89	35.77
GEN [19]	logit	96.60	17.13	92.35	34.01	98.63	5.83	89.67	47.60	94.31	23.14
ReAct [29]	feat+logit	97.38	13.50	93.34	28.49	99.00	4.31	90.71	42.60	95.11	22.22
Mahalanobis [15]	feat+label	97.48	13.54	94.24	25.17	99.54	2.12	92.81	36.95	96.02	19.45
ViM [32]	feat+logit	**97.61**	**12.61**	95.34	20.31	99.41	2.60	92.55	36.75	96.23	18.07
ConRDM	feat+logit	97.37	14.16	**95.41**	**19.07**	**99.49**	**2.21**	**93.15**	**32.80**	**96.35**	**17.06**
Without Label Information											
Residual [32]	feat	92.72	32.63	92.21	33.80	98.57	6.63	88.23	47.85	92.93	30.23
Residual w/o offset [32]	feat	91.87	36.38	92.20	33.84	98.57	6.64	88.23	47.90	92.71	31.19
KNN [30]	feat	93.54	38.92	92.95	29.40	94.68	35.65	88.86	52.80	92.51	39.20
RDM	feat	**95.98**	**21.59**	**94.25**	**25.23**	**99.12**	**4.35**	**91.41**	**40.15**	**95.18**	**22.83**

and 14.79% in FPR95. Specifically, RDM achieves notable improvements in challenging OOD datasets like ImageNet-O, where it reduces FPR95 to 40.15% compared to Residual's 47.90%. On the Textures dataset, RDM maintains competitive performance with an FPR95 of 25.23%, improving over Residual's 33.84%. With additional label information, ConRDM surpasses all methods, achieving the highest average AUROC of 96.35% and FPR95 of 17.06%. Overall, the results indicate that RDM offers robust OOD detection capabilities, with consistent performance gains across diverse datasets. Conditional likelihood further

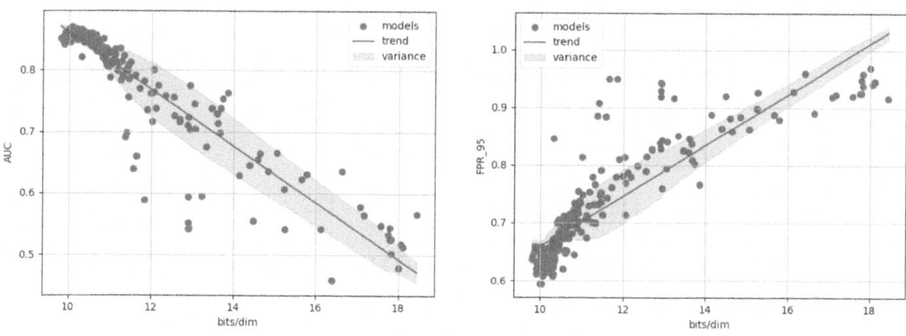

Fig. 3. More precise likelihood estimation (the lower bits/dim the better) of ID representations leads to more accurate OOD detection. The representations used in this analysis are extracted from DINOv2.

Table 5. OOD detection with other supervised encoders: AUROC and FPR95 are reported as percentages for 5 more supervised classifiers. The results are averaged on OpenImage-O, Texture, iNaturalist, and ImageNet-O for each classifier.

Method	**Swin**		**DeiT**		**RepVGG**		**ResNet50d**		**BiT**	
	AUROC ↑	FPR95 ↓	AUROC ↑	FPR95 ↓	AUROC ↑	FPR95 ↓	AUROC ↑	FPR95 ↓	AUROC ↑	FPR95 ↓
With Label Information										
MSP [10]	87.57	43.44	79.48	66.43	78.10	70.55	77.99	67.96	77.25	77.83
Energy [18]	87.77	35.08	72.80	70.14	76.38	78.99	71.08	78.39	78.48	79.68
ODIN [17]	88.00	36.58	77.13	63.92	77.72	72.68	75.27	68.56	79.24	78.63
MaxLogit [9]	88.40	35.28	76.79	64.49	77.56	73.50	75.39	69.34	79.27	78.46
KL Matching [9]	88.87	46.99	83.49	64.80	81.35	61.65	82.72	64.41	83.63	55.62
GEN [19]	91.46	32.28	84.61	**59.68**	81.33	66.00	82.75	62.08	80.00	81.00
ReAct [29]	90.17	31.36	77.37	67.00	49.14	98.96	82.93	58.63	84.53	61.38
Mahalanobis [15]	92.16	40.39	85.03	73.18	86.07	59.39	88.33	55.70	86.62	53.34
ViM [32]	**94.11**	**31.04**	**85.25**	69.95	**87.81**	**50.50**	**89.22**	**52.61**	**90.91**	**41.46**
ConRDM	91.97	42.12	83.79	77.94	83.35	64.58	84.12	66.20	82.83	57.68
Without Label Information										
Residual [32]	**92.88**	**37.38**	84.15	74.13	84.19	59.00	87.01	58.55	84.14	56.23
Residual w/o offset	92.81	37.77	84.16	74.28	83.98	59.42	86.72	59.27	84.05	56.26
KNN [30]	92.16	38.72	**85.53**	**72.16**	**87.80**	**51.80**	**89.70**	**48.34**	**85.16**	**53.93**
RDM	91.58	43.38	82.93	79.44	82.01	65.26	83.22	66.78	81.72	58.56

improves OOD detection when label information is available, making it a promising method for OOD detection in both label-free and label-augmented settings.

OOD Detection with Alternative Encoders. Table 5 presents a comprehensive comparison of OOD detection performance across various encoders, with results averaged over all OOD datasets. Representations extracted from CNNs, including BiT, RepVGG, and ResNet50d, exhibit asymmetrical representation spaces due to ReLU and are characterized by higher dimensionality compared to transformer-based encoders, with feature dimensions of 2048, 2560, and 2048, respectively. Likelihood-based methods demonstrate performance on par with state-of-the-art OOD detection approaches. ConRDM achieves a 1% improvement over RDM by leveraging label information. However, its performance is slightly lower on CNN-based encoders, such as RepVGG, ResNet50d, and BiT, potentially due to the increased dimensionality, which may necessitate larger network expressiveness for effective likelihood estimation.

4.4 Representation Space Analysis

In this part, we analyze the relationship between the accuracy of the likelihood estimation in the representation space and the OOD detection performance. Zhang et al. [34] found that sometimes DGMs with better likelihood estimation can perform worse in terms of OOD detection in certain distributions. We empirically demonstrate that this is not the case in the representation space, see Fig. 3. We find that more accurate likelihood estimation (low average bits/dim in ID validation dataset) leads to better OOD detection. Each point in Fig. 3 indicates a model trained with a randomly specified setting, selected from different selections of the number of network layers, learning rate, and training epoch.

Discussion. Likelihood-based methods have traditionally been considered ineffective for OOD detection, particularly in raw image space, where models often assign anomalously high likelihoods to OOD samples. However, our study demonstrates that when applied to well-structured representations from high-quality encoders, likelihood estimation becomes a powerful and reliable approach for OOD detection. Through extensive experiments, we show that both supervised and self-supervised encoders provide effective feature spaces where likelihood can differentiate ID from OOD samples. In particular, likelihood is highly practical for scenarios where labeled data is unavailable.

5 Conclusion

We revisited likelihood-based OOD detection using a score-based diffusion model in the representation spaces provided by pre-trained encoders. Our results show that likelihood in such spaces performs comparably to SOTA methods without requiring labels for the ID images, making it a strong contender for OOD detection with, e.g., self-supervised encoders. When a supervised encoder is available, the model can be trained with class information as conditional likelihood. This formulation outperforms most SOTAs with the ViT encoder, achieving an average AUROC of 96.35 on the large-scale OOD detection benchmark with ImageNet-1K as ID data. Our results show that likelihood-based methods can indeed be successfully used for OOD detection. While these methods often face criticism for their poor performance in image space, our approach demonstrates its effectiveness when applied in representation space. From these results, we argue that likelihood-based OOD detection, which to a large extent has been replaced by post-hoc methods, remains a powerful strategy when applied to representations from foundational models.

One of the advantages of likelihood-based OOD detection is the absence of detection related hyper-parameters, while one limitation is the sensitivity to the representation used, e.g., the RDM method does not perform well on representations from convolutional encoders. For future work, we envision that further improvements can be made both by improving the likelihood estimator and the representations. Although the score-based diffusion model excels at density estimation, it would be interesting to test other generative models that allow precise likelihood estimation. As for the representations, a promising future research direction is to fine-tune the encoder to promote characteristics that benefit OOD detection performance.

Acknowledgement. This work was partially supported by the Wallenberg AI, Autonomous Systems and Software Program (WASP) funded by the Knut and Alice Wallenberg Foundation, and the Zenith career development program at Linköping University.

References

1. Ahmadian, A., Ding, Y., Eilertsen, G., Lindsten, F.: Unsupervised novelty detection in pretrained representation space with locally adapted likelihood ratio. In: International Conference on Artificial Intelligence and Statistics, pp. 874–882. PMLR (2024)
2. Bishop, C.M.: Novelty detection and neural network validation. IEE Proc. Vis. Image Signal Process. **141**(4), 217–222 (1994)
3. Caron, M., Touvron, H., Misra, I., Jégou, H., Mairal, J., Bojanowski, P., Joulin, A.: Emerging properties in self-supervised vision transformers. In: Proceedings of the IEEE/CVF International Conference on Computer Vision, pp. 9650–9660 (2021)
4. Chen, R.J., et al.: Towards a general-purpose foundation model for computational pathology. Nat. Med. **30**(3), 850–862 (2024)
5. Chen, R.T., Rubanova, Y., Bettencourt, J., Duvenaud, D.K.: Neural ordinary differential equations. In: Advances in Neural Information Processing Systems, vol. 31 (2018)
6. Du, X., Wang, Z., Cai, M., Li, Y.: Vos: learning what you don't know by virtual outlier synthesis. arXiv preprint arXiv:2202.01197 (2022)
7. Goodier, J., Campbell, N.D.: Likelihood-based out-of-distribution detection with denoising diffusion probabilistic models. arXiv preprint arXiv:2310.17432 (2023)
8. Graham, M.S., Pinaya, W.H., Tudosiu, P.D., Nachev, P., Ourselin, S., Cardoso, J.: Denoising diffusion models for out-of-distribution detection. In: Proceedings of the IEEE/CVF Conference on Computer Vision and Pattern Recognition, pp. 2947–2956 (2023)
9. Hendrycks, D., et al.: Scaling out-of-distribution detection for real-world settings. arXiv preprint arXiv:1911.11132 (2019)
10. Hendrycks, D., Gimpel, K.: A baseline for detecting misclassified and out-of-distribution examples in neural networks. In: International Conference on Learning Representations (2017)
11. Ho, J., Salimans, T.: Classifier-free diffusion guidance. arXiv preprint arXiv:2207.12598 (2022)
12. Hutchinson, M.F.: A stochastic estimator of the trace of the influence matrix for Laplacian smoothing splines. Commun. Stat. Simul. Comput. **18**(3), 1059–1076 (1989)
13. Kang, M., Song, H., Park, S., Yoo, D., Pereira, S.: Benchmarking self-supervised learning on diverse pathology datasets. In: Proceedings of the IEEE/CVF Conference on Computer Vision and Pattern Recognition, pp. 3344–3354 (2023)
14. Kirichenko, P., Izmailov, P., Wilson, A.G.: Why normalizing flows fail to detect out-of-distribution data. Adv. Neural. Inf. Process. Syst. **33**, 20578–20589 (2020)
15. Lee, K., Lee, K., Lee, H., Shin, J.: A simple unified framework for detecting out-of-distribution samples and adversarial attacks. In: Advances in Neural Information Processing Systems, vol. 31 (2018)
16. Li, T., Katabi, D., He, K.: Self-conditioned image generation via generating representations. arXiv preprint arXiv:2312.03701 (2023)
17. Liang, S., Li, Y., Srikant, R.: Enhancing the reliability of out-of-distribution image detection in neural networks. In: International Conference on Learning Representations (2018)
18. Liu, W., Wang, X., Owens, J., Li, Y.: Energy-based out-of-distribution detection. Adv. Neural. Inf. Process. Syst. **33**, 21464–21475 (2020)

19. Liu, X., Lochman, Y., Zach, C.: Gen: pushing the limits of softmax-based out-of-distribution detection. In: Proceedings of the IEEE/CVF Conference on Computer Vision and Pattern Recognition, pp. 23946–23955 (2023)
20. Maoutsa, D., Reich, S., Opper, M.: Interacting particle solutions of fokker-planck equations through gradient-log-density estimation. Entropy **22**(8), 802 (2020)
21. Nalisnick, E., Matsukawa, A., Teh, Y.W., Gorur, D., Lakshminarayanan, B.: Do deep generative models know what they don't know? arXiv preprint arXiv:1810.09136 (2018)
22. Nalisnick, E., Matsukawa, A., Teh, Y.W., Lakshminarayanan, B.: Detecting out-of-distribution inputs to deep generative models using typicality. arXiv preprint arXiv:1906.02994 (2019)
23. Oquab, M., et al.: Dinov2: learning robust visual features without supervision. arXiv preprint arXiv:2304.07193 (2023)
24. Pocevičiūtė, M., Ding, Y., Bromée, R., Eilertsen, G.: Out-of-distribution detection in digital pathology: do foundation models bring the end to reconstruction-based approaches? Comput. Biol. Med. **184**, 109327 (2025)
25. Ren, J., et al.: Likelihood ratios for out-of-distribution detection. In: Advances in Neural Information Processing Systems, vol. 32 (2019)
26. Rombach, R., Blattmann, A., Lorenz, D., Esser, P., Ommer, B.: High-resolution image synthesis with latent diffusion models. In: Proceedings of the IEEE/CVF Conference on Computer Vision and Pattern Recognition, pp. 10684–10695 (2022)
27. Skilling, J.: The eigenvalues of mega-dimensional matrices. Maximum Entropy and Bayesian Methods: Cambridge, England **1988**, 455–466 (1989)
28. Song, Y., Sohl-Dickstein, J., Kingma, D.P., Kumar, A., Ermon, S., Poole, B.: Score-based generative modeling through stochastic differential equations. arXiv preprint arXiv:2011.13456 (2020)
29. Sun, Y., Guo, C., Li, Y.: React: Out-of-distribution detection with rectified activations. Adv. Neural. Inf. Process. Syst. **34**, 144–157 (2021)
30. Sun, Y., Ming, Y., Zhu, X., Li, Y.: Out-of-distribution detection with deep nearest neighbors. In: International Conference on Machine Learning, pp. 20827–20840. PMLR (2022)
31. Veeling, B.S., Linmans, J., Winkens, J., Cohen, T., Welling, M.: Rotation equivariant CNNs for digital pathology (2018)
32. Wang, H., Li, Z., Feng, L., Zhang, W.: Vim: out-of-distribution with virtual-logit matching. In: Proceedings of the IEEE/CVF Conference on Computer Vision and Pattern Recognition, pp. 4921–4930 (2022)
33. Zhang, J., et al.: Openood v1. 5: enhanced benchmark for out-of-distribution detection. arXiv preprint arXiv:2306.09301 (2023)
34. Zhang, L., Goldstein, M., Ranganath, R.: Understanding failures in out-of-distribution detection with deep generative models. In: International Conference on Machine Learning, pp. 12427–12436. PMLR (2021)

Quantifying Epistemic Uncertainty in Absolute Pose Regression

Fereidoon Zangeneh[1,2](✉), Amit Dekel[2], Alessandro Pieropan[2], and Patric Jensfelt[1]

[1] KTH Royal Institute of Technology, Stockholm, Sweden
{fzk,patric}@kth.se
[2] Univrses AB, Stockholm, Sweden
{fereidoon.zangeneh,amit.dekel,alessandro.pieropan}@univrses.com

Abstract. Visual relocalization is the task of estimating the camera pose given an image it views. Absolute pose regression offers a solution to this task by training a neural network, directly regressing the camera pose from image features. While an attractive solution in terms of memory and compute efficiency, absolute pose regression's predictions are inaccurate and unreliable outside the training domain. In this work, we propose a novel method for quantifying the epistemic uncertainty of an absolute pose regression model by estimating the likelihood of observations within a variational framework. Beyond providing a measure of confidence in predictions, our approach offers a unified model that also handles observation ambiguities, probabilistically localizing the camera in the presence of repetitive structures. Our method outperforms existing approaches in capturing the relation between uncertainty and prediction error.

Keywords: Camera Relocalization · Uncertainty Estimation · VAEs

1 Introduction

Visual localization is the task of estimating the pose of a camera from the image that it captures in the environment. It is a technique used for indoor navigation of robots, augmented reality devices, and autonomous driving [5]. A full visual localization pipeline consists of both frame-to-frame tracking of the camera—or relative pose estimation, as well as global relocalization—or absolute pose estimation. The latter refers to estimating the camera pose from a single image in a previously mapped environment, and is the focus of this work. Visual relocalization has been a long-standing topic of research [8,33,34], with works in search of map representations and algorithms that can most efficiently and accurately estimate the camera pose. Traditional solutions range from image databases to sparse 3D point models of the world as their map representations, used for image retrieval or keypoint matching to estimate the pose for a novel query image [1,13]. These solutions face a trade-off between efficiency and accuracy. A more recent paradigm for visual relocalization relies on using end-to-end trainable

J. Petersen and V. A. Dahl (Eds.): SCIA 2025, LNCS 15726, pp. 180–195, 2025.
https://doi.org/10.1007/978-3-031-95918-9_13

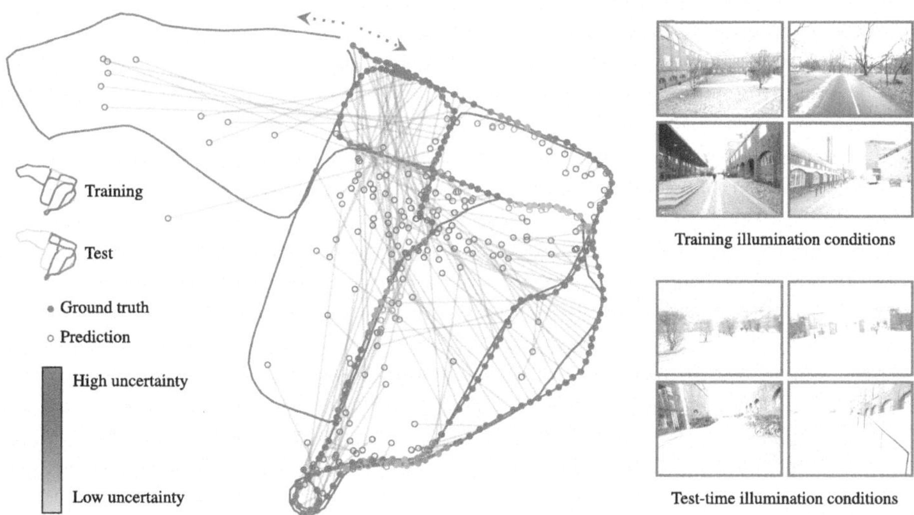

Fig. 1. Querying any absolute pose regression network on a trajectory and data domain different from training data results in high prediction errors (depicted by long lines connecting predictions and ground truth). Our proposed epistemic uncertainty quantification approach estimates the likelihood of test samples belonging to the training distribution (visualized by a color map), offering a guide for trusting the predictions. We see that the color of the predictions and their ground truths, which encodes the uncertainty, is highly correlated with the prediction error (the length of the lines).

pipelines to directly regress the camera pose from the image by a neural network [3,17,35,36]. These end-to-end methods store a map representation in the weights of a neural network, promising higher memory and compute efficiency, as well as better robustness compared to the traditional methods [39].

Although end-to-end absolute pose regression methods are appealing for their efficiency, they lag behind traditional geometric approaches in terms of accuracy, as evidenced by their performance on visual relocalization benchmarks [30,37]. While the initial motivation for using neural networks in this task was their potential to learn a spatial understanding of the scene, in practice, these models often perform similarly to pose approximations based on image retrieval [32], limiting their generalization beyond the training data. This limitation is further exacerbated by the absence of geometric constraints or verification steps in their prediction process—every input yields a prediction, which can be highly inaccurate. This contrasts with traditional methods, which can identify when there is insufficient evidence to make a reliable prediction. Despite these limitations, the fast inference speed and small memory footprint of pose regression networks has driven significant research efforts aimed at addressing its shortcomings by improving its generalization [25] and accuracy [6,35]. A remaining challenge in bridging the gap between the current state of absolute pose regression and real-world applications is to accompany network predictions with a reliable measure of confidence. Figure 1 illustrates a real-world scenario where a network is tested on a trajectory significantly different from the training/mapped region. The scenario also suffers from strong lighting variations between training and testing, further straining the visual relocalization system. In such cases, the network needs to *know when it does not know*.

Existing research on uncertainty estimation in absolute pose regression has followed two distinct tracks: one focuses on handling ambiguous observations by modeling aleatoric uncertainty [9,24,42], while the other—including this paper—aims to quantify epistemic uncertainty to provide confidence in predictions [12,14]. However, the methodologies used in these two tracks differ fundamentally leading to different frameworks. In this work, we build on a solution based on variational autoencoders (VAE) [43], originally proposed to address the challenge of handling ambiguous observations from repetitive structures. We show that VAEs can also be used for quantifying epistemic uncertainty, resulting in a unified framework that can handle both natures of uncertainty. We propose a novel method for quantifying epistemic uncertainty in absolute pose regression to estimate the reliability of the predictions, as shown in Fig. 1.

In summary: (1) We propose a unified framework for handling epistemic as well as aleatoric uncertainty in absolute pose regression. (2) We derive a formulation for quantification of epistemic uncertainty in absolute pose regression within a variational framework. (3) We show how our epistemic uncertainty quantification can estimate the reliability of network predictions. (4) We perform a thorough evaluation to show that our method outperforms the existing epistemic uncertainty quantification methods.

2 Related Work

2.1 Visual Relocalization

Traditional Methods. The well-established solutions to visual relocalization fall into two paradigms: image-based and structure-based methods. Image-based methods store a database of images and their camera poses as a representation of the map. Given a query image, they rely on global image descriptors [1] to retrieve the most similar image(s) in the database, hence approximate its pose [38]. The map representation in structure-based methods is a sparse 3D point model of the scene. For localizing a query image, 2D-3D matching is performed between the image and the sparse model, enabling accurate estimation of the camera pose [21,31]. These two paradigms have distinct strengths and properties on the accuracy-efficiency trade-off, such that they are typically combined in a hierarchical framework for large-scale relocalization [30]. Both paradigms, however, exhibit detectable signs of failure when queried with an image that deviates significantly from the map; image-based retrieval computes too little similarity between query and the map, and structure-based matching finds too small inlier sets. So each estimation can be accompanied with a measure of reliability.

Regression-Based Methods. A recent trend in visual relocalization is using end-to-end learning-based pipelines. These methods encode a map representation in the weights of a neural network that directly regresses a geometric quantity from an image. This makes them appealing alternatives to traditional methods for their space and computational efficiency in large scenes, along with data-driven handling of lighting changes, camera blur and texture-less surfaces [39]. These methods come in two main variants: scene coordinate regression and absolute pose regression. In scene coordinate regression, the neural network regresses the scene's 3D point coordinates from image patches, which are then used in a robust estimation process to compute the camera pose [36]. In absolute pose regression, the neural network directly predicts the camera pose, eliminating the need for the robust estimation step [17]. Both variants are appealing in their own right. Scene coordinate regression, due to its more local nature of predictions, shows higher generalization capabilities than absolute pose regression, resulting in more accurate predictions [3]. This is particularly beneficial in presence of occlusions, where scene coordinate regression can leverage image patches that provide consistent information to produce an accurate final pose, while the global pose predicted by an absolute pose regression network may be adversely affected. Scene coordinate regression, however, originally proposed for relocalization with RGB-D images, requires additional, more complex steps for end-to-end training with RGB images [2,3]. In contrast, absolute pose regression comes with a simpler training procedure, and despite its shortcomings [32] remains an attractive alternative. This has promoted research efforts to build upon the original work [17], including the development of more effective geometric and photometric loss functions [4,7,15], network architectures [23,39,41], extension to multiple scenes [35], and improved generalization through data synthesis during training [25–27]. Absolute pose regression in its basic form produces a point prediction for any given input through a forward pass of the network, without providing any confidence in the prediction. A complementary line of research centers on incorporating uncertainty estimation into pose regression networks, enhancing their reliability for real-world applications. Our work aligns with this direction.

2.2 Uncertainty Estimation in Absolute Pose Regression

Uncertainty in neural network predictions arises in two distinct forms: aleatoric uncertainty and epistemic uncertainty [16]. Aleatoric uncertainty arises from inherent noise and ambiguity in the data, which cannot be reduced even with an infinite amount of training data. An example of this in camera relocalization is repetitive structures that appear visually similar but correspond to different camera poses. Epistemic uncertainty, on the other hand, reflects the model's lack of knowledge about a data point—specifically, how well that point fits within the distribution learned during training.

Modeling Aleatoric Uncertainty. Aleatoric uncertainty is typically classified into two types based on its dependence on the input data: heteroscedastic uncertainty, which varies with the input; and homoscedastic uncertainty, which remains constant regardless of the input such as sensor noise. Kendall and Cipola [15] used the latter paradigm to learn the optimal weighting of translation and orientation error terms during training of each scene. On the other hand, heteroscedastic uncertainty is modeled to account for ambiguities in observations. A common method for modeling heteroscedastic aleatoric uncertainty is by predicting a unimodal parametric distribution [16]. Moreau et al. [24] applied this technique to absolute pose regression in outdoor environments, integrating it in a tracking loop. To address multimodal posterior distributions caused by repetitive structures, Deng et al. [9] proposed predicting a mixture model instead of a unimodal distribution. In contrast, Zangeneh et al. [42] adopted a non-parametric approach, representing the predicted distribution through samples rather than a mixture model. This approach was later reformulated with a conditional VAE in [43], where a VAE is trained to reconstruct camera pose samples conditioned on image observations. As a result of this, the latent space captures the space of pose ambiguities given an observation.

Quantifying Epistemic Uncertainty. A number of works have explored the quantification of epistemic uncertainty in absolute pose regression. The pioneering work of Kendall and Cipolla [14] follows Bayesian deep learning principles and incorporates dropout layers in the network to approximate variational inference for the posterior distribution of model weights [10]. At test time, Monte Carlo dropout is applied, performing stochastic forward passes over a query image, with the variance of predictions providing a measure of epistemic uncertainty. The use of dropout to generate multiple predictions was later adopted by [12], with the distinction that instead of applying dropout within a Bayesian neural network, it was applied directly to the input image pixels. This approach served a dual purpose of quantifying uncertainty along with enhancing robustness against moving objects, aim at autonomous driving scenarios. More recently, Liu et al. [20] proposed a functional method for uncertainty quantification, which directly compares the image features and the predicted pose of a query to those of the training samples. The presence of a training sample similar to the query, in both pose and appearance, is interpreted as low epistemic uncertainty for the model, with the intuition that the model must be well-informed on the query.

While conditional VAEs have been previously explored to explicitly model the aleatoric uncertainty of pose regression models [43], they can in theory also be used for likelihood estimation of test samples [18], and thus implicitly capture model's epistemic uncertainty about the prediction. Specifically, the structured latent space and the agreement between the encoder and decoder that is learned for a training set can be used to detect samples that are unlikely under the training distribution. In this work, we explore this capability of conditional VAEs for epistemic uncertainty quantification in absolute pose regression, assessing its viability as an alternative to existing methods while providing a unified framework to also capture aleatoric uncertainty.

3 Method

An ideal visual relocalization solution predicts the true camera pose $y \in \mathrm{SE}(3)$ for any given image $\boldsymbol{x} \in \mathbb{R}^{H \times W \times 3}$ sampled in the scene, where $(\boldsymbol{x}, y) \sim p_{\text{true}}$.

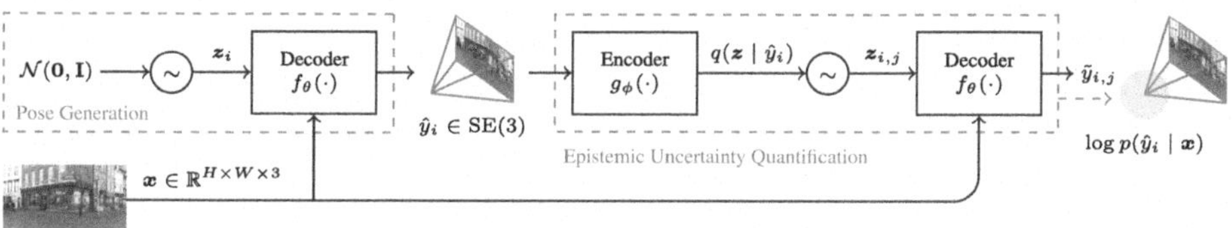

Fig. 2. Our pipeline models a scene in a conditional VAE setup. Given a test-time image observation $\boldsymbol{x}$, the decoder is used to sample poses $\hat{y}$ from the posterior distribution of camera poses $p(y \mid \boldsymbol{x})$. Reconstruction of $\hat{y}$ through the VAE pipeline then gives an estimate on $\log p(\hat{y} \mid \boldsymbol{x})$, that is the likelihood of the test sample under the training distribution. This reflects model's epistemic uncertainty about the observation $\boldsymbol{x}$.

In practice, the real-world distribution p_{true} can only be modeled by a finite set of training samples that form an empirical distribution p_{train}. Successful modeling of p_{train} implies generalization to any other unseen empirical distribution p_{test} from the underlying p_{true}. However, there is in practice always a model performance gap between queries from p_{train} and p_{test}. We are interested in a visual relocalization model that once trained, can measure how well a test sample $(\boldsymbol{x}, y) \sim p_{\text{test}}$ adheres to the modeled distribution p_{train}, hence estimate how reliable its predictions are. To this end, we frame the visual relocalization task as learning the distribution $p_{\text{train}}(y \mid \boldsymbol{x})$. Sampling from this conditional distribution performs the task itself [43], while estimating its likelihood for a given sample $(\boldsymbol{x}, y)$ quantifies the degree of its conformity to the modeled distribution p_{train}. We discuss the training procedure to model and sample from $p_{\text{train}}(y \mid \boldsymbol{x})$ in Sect. 3.1. We then lay down our proposed approach to estimate the likelihood of samples and quantify model's epistemic uncertainty in Sect. 3.2. We hereafter refer to $p_{\text{train}}(y \mid \boldsymbol{x})$ as $p(y \mid \boldsymbol{x})$.

3.1 Learning a Generative Model

We intend to train a neural network $f_\theta(\cdot)$ that conditioned on a given image $\boldsymbol{x} \in \mathbb{R}^{H \times W \times 3}$ transforms samples $\boldsymbol{z} \in \mathbb{R}^d$ from a noise distribution $p(\boldsymbol{z}) = \mathcal{N}(\mathbf{0}, \mathbf{I})$ to the posterior distribution over camera poses $y \in \mathrm{SE}(3) \sim p(y \mid \boldsymbol{x})$. A shown in [43], such a generative network can be trained as the decoder in a conditional VAE pipeline that reconstructs camera poses given images in the scene. We briefly outline this setup and training procedure below, while referring the reader to [43] for its design principles.

Setup and Optimization. The conditional VAE consists of an encoder $g_\phi(\cdot)$ and a conditional decoder network $f_\theta(\cdot)$ that are optimized to, together, reconstruct the camera pose y_i for a given image $\boldsymbol{x}_i$ from the training set $(\boldsymbol{x}_i, y_i) \in \mathcal{D}_{\text{train}}$. The encoder maps each training pose y_i to the mean and covariance of a Gaussian posterior $q(\boldsymbol{z} \mid y_i)$ that is intended to resemble the true latent posterior distribution $p(\boldsymbol{z} \mid y_i)$. The decoder, conditioned on the corresponding image $\boldsymbol{x}_i$, then maps samples drawn from this inferred posterior $\boldsymbol{z}_j \sim q(\boldsymbol{z} \mid y_i)$ to reconstructions $\hat{y}_{i,j} \in \mathrm{SE}(3)$ of the original pose y_i.

The optimization objective of the pipeline is the evidence lower bound (ELBO) [19]. From the likelihood of training samples it is possible to derive that

$$\begin{aligned} &\log p(y|\boldsymbol{x}) - \overbrace{D_{\mathrm{KL}}\big(q(\boldsymbol{z} \mid y) \parallel p(\boldsymbol{z} \mid y)\big)}^{\geq 0} \\ &= \underbrace{\mathbb{E}_{q_\phi(\boldsymbol{z}|y)} \log p_\theta(y \mid \boldsymbol{z}, \boldsymbol{x}) - D_{\mathrm{KL}}\big(q_\phi(\boldsymbol{z} \mid y) \parallel p(\boldsymbol{z})\big)}_{\text{ELBO}}, \end{aligned} \tag{1}$$

which indicates that maximizing ELBO with optimization variables ϕ and θ over $\mathcal{D}_{\text{train}}$ optimizes the networks $g_\phi(\cdot)$ and $f_\theta(\cdot)$ that model $p(y \mid \boldsymbol{x})$. In the above expression the subscripts ϕ, θ denote the relation between distributions and weights of the networks they are materialized in. We compute the expected value of the reconstruction likelihood with Monte Carlo samples from $q_\phi(\boldsymbol{z} \mid y)$, following the Gaussian model

$$\log p_\theta\big(y \mid \boldsymbol{z}, \boldsymbol{x}\big) = -1/2\big(6 \log 2\pi + \log\det(\boldsymbol{\Sigma}) + \boldsymbol{\xi}^T \boldsymbol{\Sigma}^{-1} \boldsymbol{\xi}\big), \tag{2}$$

where a homoscedastic 6×6 covariance matrix $\boldsymbol{\Sigma}$ is shared for training all samples and optimized alongside network weights θ and ϕ. We take the reconstruction error to be a local correction on the prediction such that $y = \hat{y} \exp(\boldsymbol{\xi}^\wedge)$, hence define the error vector as $\boldsymbol{\xi} = \log(\hat{y}^{-1} y)^\vee \in \mathbb{R}^6$ in the Lie algebra $\mathfrak{se}(3)$[1]. Defining the error in the tangent space and learning a shared $\boldsymbol{\Sigma}$ enables automatic adjustment of loss weights across the different degrees of freedom of the camera pose throughout training. This automatic adjustment is inspired by [15] and eliminates the need for the suboptimal manual tuning performed per dataset [9,42,43]. However, while [15] explored learning only two loss weight parameters between translation and rotation error components, we extend this approach to all six degrees of freedom by modeling the full covariance matrix.

Sample Generation. We can easily sample from $p(y \mid \boldsymbol{x})$ by conditioning the decoder on the image $\boldsymbol{x}$ and passing samples from the prior through the decoder to get $\mathcal{Y} = \{\hat{y}_i = f_\theta(\boldsymbol{z}_i, \boldsymbol{x}) \mid \boldsymbol{z}_i \sim \mathcal{N}(\boldsymbol{0}, \mathbf{I})\}$. This is illustrated in the left dotted box in Fig. 2. As discussed in [43], the decoder learns the space of aleatoric pose ambiguities associated with each image, and for ambiguous images, it appropriately splits the latent space such that different latent regions are mapped to distinct camera poses for an image.

3.2 Likelihood Estimation

We propose to quantify the epistemic uncertainty of a trained model about a test sample by computing its marginal likelihood, which we can estimate by importance sampling:

[1] $\exp : \mathfrak{se}(3) \mapsto \mathrm{SE}(3)$ is the exponential map from Lie algebra to Lie group and log is its inverse. The operator $\wedge$ turns $\boldsymbol{\xi}$ into a member of the Lie algebra $\mathfrak{se}(3)$ and $\vee$ is its inverse.

$$\begin{aligned}\log p(y \mid \boldsymbol{x}) &= \log \int p_\theta(y \mid \boldsymbol{z}, \boldsymbol{x}) \overbrace{p(\boldsymbol{z} \mid \boldsymbol{x})}^{=p(\boldsymbol{z})} d\boldsymbol{z} \\ &= \log \int q_\phi(\boldsymbol{z} \mid y) \frac{p_\theta(y \mid \boldsymbol{z}, \boldsymbol{x}) p(\boldsymbol{z})}{q_\phi(\boldsymbol{z} \mid y)} d\boldsymbol{z} \\ &= \log \mathbb{E}_{q_\phi(\boldsymbol{z}\mid y)} \frac{p_\theta(y \mid \boldsymbol{z}, \boldsymbol{x}) p(\boldsymbol{z})}{q_\phi(\boldsymbol{z} \mid y)} \\ &\approx \log \frac{1}{M} \sum_{\boldsymbol{z}_j} \frac{p_\theta(y \mid \boldsymbol{z}_j, \boldsymbol{x}) p(\boldsymbol{z}_j)}{q_\phi(\boldsymbol{z}_j \mid y)} \qquad {\scriptstyle \boldsymbol{z}_j \sim q_\phi(\boldsymbol{z}\mid y) \atop j=1:M}\end{aligned} \tag{3}$$

The equality $p(\boldsymbol{z} \mid \boldsymbol{x}) = p(\boldsymbol{z})$ stems from the assumption that all observations $\boldsymbol{x}$ share the same prior distribution $p(\boldsymbol{z})$ [40]. At test time, only the image $\boldsymbol{x}$ is available while the true pose y is unknown. Therefore, we propose estimating the expected $\log p(\hat{y}_i|\boldsymbol{x})$ with Monte Carlo generations $\hat{y}_i = f_\theta(\boldsymbol{z}_i, \boldsymbol{x}), \boldsymbol{z}_i \sim \mathcal{N}(\mathbf{0}, \mathbf{I})$. This likelihood measure quantifies the agreement between encoder and decoder when conditioned on $\boldsymbol{x}$, which is how a VAE implicitly captures epistemic uncertainty. The two networks are expected to be in agreement only for in-distribution samples, as the latent space is structured based on the training distribution. In other words, we interpret the likelihood $\log p(\hat{y}|\boldsymbol{x})$ as the network's confidence in its prediction for observation $\boldsymbol{x}$. Our proposed pipeline for pose estimation and epistemic uncertainty quantification is outlined in Fig. 2.

4 Implementation Details

We implement the encoder and decoder networks both as multilayer perceptrons with 5 layers of 512 neurons each, and residual connections between their input and third layers. All neurons (except at output) go through LeakyReLU activations. A pretrained ResNet-18 [11] backbone is used to extract 512-dimensional conditioning image feature vectors. The input to the decoder is a concatenation of image feature and latent sample vectors. We opt for a 2-dimensional latent space in all experiments. As the encoder (if expressive enough) is able to learn the surjection of poses to appearances, it is not strictly necessary to condition the encoder on the image features and we obtained similar results for both encoder configurations. We parameterize an SE(3) element at encoder input and decoder output with a 3-vector for translation, and a 6-vector for rotation that can be continuously mapped to a valid rotation matrix by a Gram-Schmidt process [44]. The 2×2 latent posterior covariance matrix is parameterized by its log-variances and an SO(2) rotation. The 6×6 homoscedastic noise covariance matrix is parameterized by its lower triangular matrix from Cholesky decomposition.

We train the networks for 30k iterations with a learning rate of 1e$-$4 and Adam optimizer with a decoupled weight decay of 1e$-$2 [22]. We use a batch size of 128 and draw 100 Monte Carlo samples to estimate ELBO's expected reconstruction likelihood term in (1). We found a $0 \rightarrow 1$ warm-up of KL divergence term after 10k iterations to help with better convergence. Following previous works [9,17,43] the images are first resized such that the smaller edge is 256 pixels, then randomly cropped to 244×244 squares during training, and center-cropped to the same size at test time. To improve the conditioning of the decoder output, a linear transformation is computed for each scene, allowing the network to predict values within the $[0, 1]$ range for each translation component, which are then mapped to the true metric scale. To enhance robustness to motion blur and lighting variations, we apply random Gaussian blurring with $\sigma \in [0.05, 5]$ during training, along with brightness and color jittering using PyTorch's [29] ColorJitter function, with brightness, contrast, and saturation set to 0.2, and hue to 0.1.

5 Experiments

We aim to assess the efficacy of our proposed likelihood-based quantification of epistemic uncertainty. Defining a measurable ground truth for epistemic uncertainty is inherently challenging. However, we identify and design an evaluation protocol around a key statistical property that the uncertainties should exhibit. We evaluate our method across a variety of operation conditions, comparing it against existing techniques.

5.1 Evaluation Protocol

Given a trained absolute pose regression model, predictions tend to diverge from the ground truth when test samples deviate from the training distribution. In other words, we expect to observe higher prediction errors for test samples that the model is less informed about and assigns lower likelihood values for. Therefore, the quantified uncertainty should ideally increase with the prediction error. While the exact nature of this relationship remains unknown, we can except one to be a monotonically increasing function of the other. We evaluate this by computing the Spearman's rank correlation coefficient between the estimated negative log-likelihood and prediction error of test samples. This coefficient ranges from -1 to $+1$, where $+1$ is the ideal scenario where negative log-likelihood (epistemic uncertainty) is a perfect monotonically increasing function of prediction error. Through this score, we offer a quantitative counterpart to the qualitative uncertainty-error tables and plots used in [12, 20]. As model predictions are made in SE(3) that does not have a natural metric to measure prediction errors, we report the correlation coefficients for translation and rotation error separately.

5.2 Comparison Baselines

We compare our method with existing approaches for epistemic uncertainty estimation in absolute pose regression: Bayesian PoseNet (BPN) [14], RVL [12], and HR-APR [21]. For a fair comparison and to rule out the predictive power of different architectures as a factor, we train all methods with the same network architecture and training setup as our own. Given that RVL's dropout statistics are primarily suited for driving scenarios, which we do not evaluate, we use a

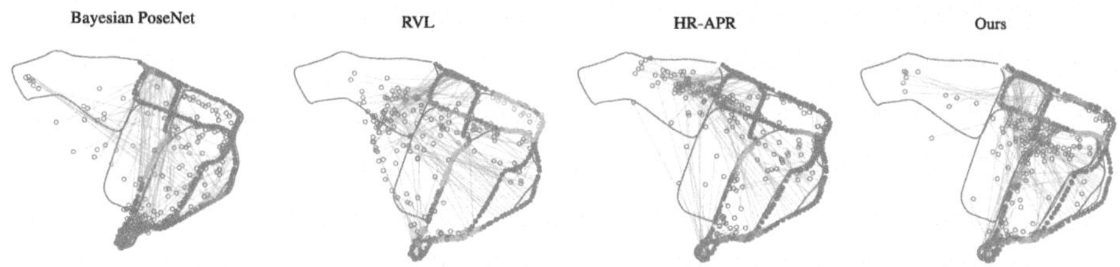

Fig. 3. Testing sequence `kth_day_09` on models trained on `kth_day_06` from [28]. The lines depict the translation error between the predictions (○) and ground truth (•), which are colored from cyan to magenta by the epistemic uncertainty. The length of the lines is expected to correlate with epistemic uncertainty, as is the case in our predictions, that is cyan for short lines (small error) to magenta for long lines (large error). (Color figure online)

uniform distribution for their dropout instead. We obtain the distribution of predicted poses for BPN and RVL by 100 Monte Carlo forward passes. While they propose variance of predictions as the measure of epistemic uncertainty, we found that using negative log-likelihood of the mean prediction benefits both methods in our evaluation metric. So we opt to use that instead to quantify their uncertainties. HR-APR quantifies epistemic uncertainty by a visual dissimilarity measure, which saturates for positionally unlikely samples.

5.3 Datasets

We evaluate our method on the test splits of the real-world datasets Cambridge Landmarks [17], 7-Scenes [36], and ambiguous relocalization sequences of [9,42], covering a variety of outdoor and indoor environments. Additionally, we use two sequences from the Multi-Campus Dataset [28] to evaluate the performance of absolute pose regression and uncertainty quantification on test samples far outside the training trajectory.

Table 1. Median prediction error—translation and rotation separately (lower is better)

Dataset	BPN [14]		RVL [12]		HR-APR [20]		Ours	
	Tra / m	Rot / $\circ$	Tra / m	Rot / $\circ$	Tra / m	Rot / $\circ$	Tra / m	Rot / $\circ$
Ambiguous [9] + Ceiling [42]	0.94	13.13	0.87	28.15	1.36	15.47	**0.21**	**9.06**
7-Scenes [36]	**0.19**	**9.48**	0.23	10.98	0.23	11.05	0.23	11.46
Cambridge Landmarks [17]	**0.67**	11.61	1.23	7.69	1.29	7.38	1.21	**7.17**
Multi-Campus: KTH Day [28]	**22.73**	61.20	29.84	36.02	23.34	43.73	24.28	**32.24**

Table 2. Spearman's rank correlation coefficient for quantified epistemic uncertainties and prediction errors—translation and rotation separately (−1 to +1, higher is better)

Scene	BPN [14]		RVL [12]		HR-APR [20]		Ours	
	Tra	Rot	Tra	Rot	Tra	Rot	Tra	Rot
Ceiling	0.16	**0.68**	0.18	0.59	0.25	0.01	**0.38**	0.21
Blue Chairs	0.02	−0.07	−0.46	0.49	0.28	0.23	**0.64**	**0.65**
Meeting Table	0.08	0.01	−0.12	−0.11	**0.21**	**0.20**	0.16	−0.01
Seminar	0.24	0.14	**0.82**	**0.82**	0.56	0.32	0.76	0.56
Staircase	0.24	0.39	0.47	0.53	0.55	0.55	**0.67**	**0.64**
Staircase Extra	0.10	0.20	0.19	**0.28**	0.24	0.19	**0.26**	0.09
Chess	0.25	−0.17	0.43	0.31	0.16	0.32	**0.43**	**0.34**
Fire	−0.06	0.13	**0.57**	0.60	0.45	**0.62**	0.38	0.53
Heads	0.03	−0.27	0.43	0.31	0.52	0.58	**0.70**	**0.61**
Office	0.25	0.19	**0.50**	0.45	0.48	**0.49**	0.41	0.44
Pumpkin	0.12	−0.07	**0.53**	0.46	0.44	**0.47**	0.50	0.43
Redkitchen	0.01	−0.02	**0.55**	**0.52**	0.50	0.39	0.48	0.51
Stairs	−0.16	0.25	0.05	0.27	**0.44**	0.30	0.39	**0.39**
King's College	**0.64**	**0.76**	0.12	0.41	0.12	0.09	0.06	−0.17
Old Hospital	0.27	**0.15**	0.16	0.00	0.35	**0.15**	**0.52**	0.12
Shop Façade	−0.28	−0.30	0.55	0.10	**0.58**	0.08	0.37	**0.12**
St Mary's Church	0.15	0.15	0.21	0.24	**0.42**	**0.32**	0.39	**0.32**
KTH Day 6 → 9	0.10	0.03	0.45	0.41	0.51	0.43	**0.62**	**0.52**
Average	0.12	0.12	0.31	**0.37**	0.39	0.32	**0.45**	0.35

6 Results and Discussion

We start our discussion by examining the behavior of an absolute pose regression network when its test distribution only partially overlaps with the training distribution. This is illustrated for our method in Fig. 1 and alongside other baselines in Fig. 3, where the network trained on sequence `kth_day_06` is tested on `kth_day_09` from Multi-Campus Dataset [28]. The test camera trajectory only partially overlaps with the training trajectory considering both position and orientation of the cameras. Additionally, as shown in Fig. 1, test images are captured under significantly different illumination conditions. We can see that this distribution shift results in generally large prediction errors across all methods, with median translation errors exceeding $20m$, as reported in Table 1. However, Fig. 1 and 3 also highlight that in overlapping stretches of the two trajectories the predictions are better aligned with the ground truth. It is critical for an uncertainty quantification method to distinguish the in- from out-of-distribution regions. Qualitatively, our method shows the highest degree of success at this, with its estimated uncertainty showing the strongest correlation with the prediction error. Next, we quantitatively compare the correlation scores across different methods.

Figure 4 presents the scatter plot of quantified uncertainty against translation and rotation errors for all methods on `kth_day_09` test samples. The scattered points should ideally form a monotonically increasing function, corresponding to a Spearman rank correlation of +1. Starting from our method, we observe a generally increasing trend in uncertainty for small prediction errors. However, this pattern breaks for larger errors—uncertainty becomes *numb* to increases in error. We hypothesize that this occurs because, once critically far from the training distribution, the VAE model cannot differentiate between different test distributions. In the third and fourth columns of the same figure we filter out data points with translation errors larger than an empirical value of $50m$, which increases correlation coefficients from 0.62 and 0.52 to 0.75 and 0.67, showing the good performance of our method for test samples are not too far from the training distribution. Turning our attention to the results of other methods in Fig. 4, we see that other methods show generally lower correlations compared to ours. We also observe the same *numbing* phenomenon in their epistemic uncertainties, though even more severe, as it is harder to empirically filter out their insensitive data points. In the case of HR-APR we can see uncertainty values that saturate at a maximum dissimilarity for predicted poses that deviate significantly from the training samples. We note that while this strategy penalizes the monotonic relationship we evaluate, it can be a practical and cost-effective technique for outlier rejection.

Table 2 summarizes the correlation coefficients across different methods and environment. We see that the best performing method can differ between different scenes; BPN performs its best in outdoor scenes, while RVL and HR-APR generally perform better indoors. Our method, however, shows the most consistent performance across all datasets and has the highest average correlation between the quantified epistemic uncertainty and the prediction error. Importantly, our method achieves this while significantly outperforming the baselines methods in ambiguous scenarios with repetitive structures in terms of prediction accuracy, as seen in Table 1. In computing the median prediction errors, except for HR-APR that is deterministic, we draw 100 pose prediction samples per image observation and compute the error for the closest 10% of predicted samples to the ground truth. This is to handle ambiguous scenarios that result in multi-modal posterior distribution of camera poses given an image [43]. As evidenced by Table 1, on unambiguous datasets our method's accuracy is on par with the baselines.

As an outlook on future work, we reflect on the large variance of correlations across different scenes. Table 2 shows that, despite the improvements of our method, some scenes still exhibit correlations close to zero. This highlights the need to enhance the robustness of quantified epistemic uncertainty, potentially by incorporating temporal constraints. Another avenue for future research is addressing the *numbing* effect observed in Fig. 4. One possible approach is to

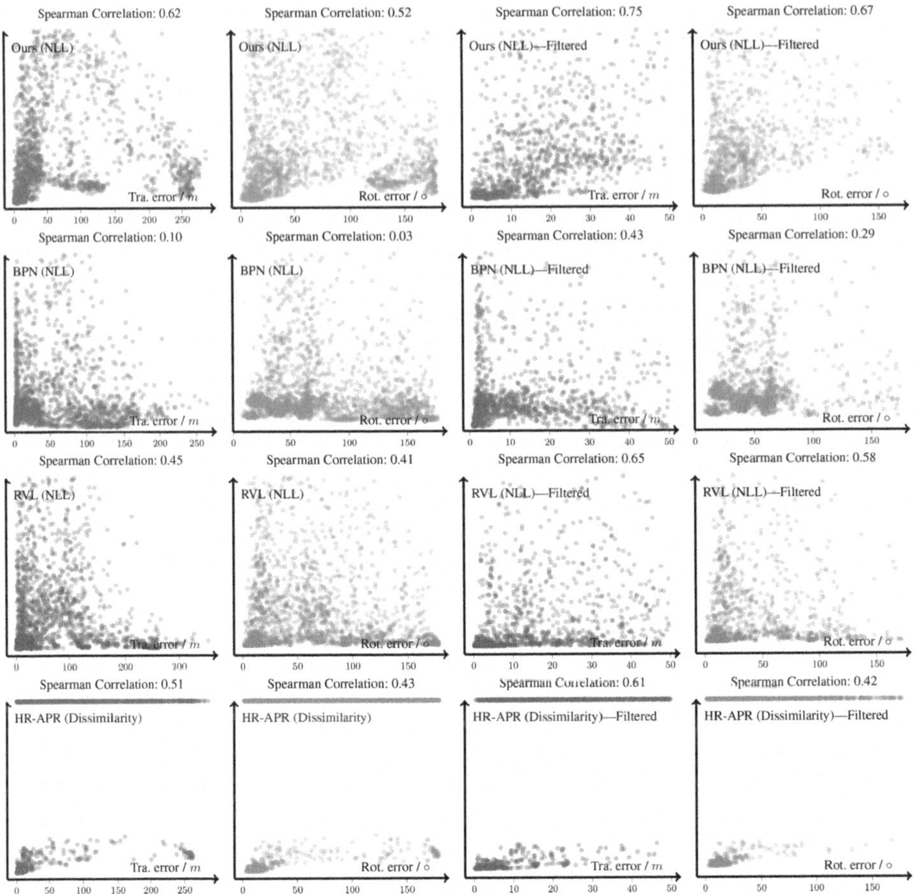

Fig. 4. The relationship between quantified epistemic uncertainty and prediction error across methods for our test sequence from Multi-Campus Dataset. The first two columns present scatter plots of all test samples, while the last two columns focus on samples with a translation error of less than $50m$. Since Spearman's rank correlation is invariant to affine changes, we scale and shift the uncertainty distribution in each plot so that their 10th and 90th percentiles align with those of the error distributions for better visualization, hence removing y-axis ticks. Our proposed approach demonstrates higher correlations between uncertainty and error, both with and without sample filtering.

frame this as an out-of-distribution binary classification problem, leveraging a method conceptually similar to HR-APR [21].

7 Conclusion

In this work, we address a key challenge in using absolute pose regression for visual relocalization: estimating the reliability of predictions for queries outside the training distribution. We propose a method for quantifying epistemic uncertainty by estimating the likelihood of observations belonging to the training distribution. We leverage variational autoencoders for modeling this distribution and performing likelihood estimation, which acts as a unified framework for also modeling aleatoric uncertainty to handle observation ambiguities. Through extensive evaluation, we show the effectiveness of our approach in comparison to existing methods, and point out directions for future work.

Acknowledgements. This work was partially supported by the Wallenberg AI, Autonomous Systems and Software Program (WASP) funded by the Knut and Alice Wallenberg Foundation.

References

1. Arandjelovic, R., Gronat, P., Torii, A., Pajdla, T., Sivic, J.: NetVLAD: CNN architecture for weakly supervised place recognition. In: CVPR, pp. 5297–5307 (2016)
2. Brachmann, E., Rother, C.: Learning less is more-6d camera localization via 3d surface regression. In: CVPR, pp. 4654–4662 (2018)
3. Brachmann, E., Rother, C.: Visual camera re-localization from RGB and RGB-D images using DSAC. TPAMI **44**(9), 5847–5865 (2021)
4. Brahmbhatt, S., Gu, J., Kim, K., Hays, J., Kautz, J.: Geometry-aware learning of maps for camera localization. In: CVPR, pp. 2616–2625 (2018)
5. Bürki, M., et al.: Vizard: reliable visual localization for autonomous vehicles in urban outdoor environments. In: 2019 IEEE Intelligent Vehicles Symposium (IV), pp. 1124–1130. IEEE (2019)
6. Chen, S., Li, X., Wang, Z., Prisacariu, V.A.: DFNet: enhance absolute pose regression with direct feature matching. In: ECCV, pp. 1–17. Springer (2022)
7. Chen, S., Wang, Z., Prisacariu, V.: Direct-PoseNet: absolute pose regression with photometric consistency. In: 3DV, pp. 1175–1185. IEEE (2021)
8. Cummins, M., Newman, P.: FAB-MAP: probabilistic localization and mapping in the space of appearance. IJRR **27**(6), 647–665 (2008)
9. Deng, H., Bui, M., Navab, N., Guibas, L., Ilic, S., Birdal, T.: Deep bingham networks: dealing with uncertainty and ambiguity in pose estimation. IJCV, 1–28 (2022)
10. Gal, Y., Ghahramani, Z.: Dropout as a Bayesian approximation: representing model uncertainty in deep learning. In: ICML, pp. 1050–1059. PMLR (2016)
11. He, K., Zhang, X., Ren, S., Sun, J.: Deep residual learning for image recognition. In: CVPR, pp. 770–778 (2016)
12. Huang, Z., Xu, Y., Shi, J., Zhou, X., Bao, H., Zhang, G.: Prior guided dropout for robust visual localization in dynamic environments. In: ICCV, pp. 2791–2800 (2019)
13. Jin, Y., Mishkin, D., Mishchuk, A., Matas, J., Fua, P., Yi, K.M., Trulls, E.: Image matching across wide baselines: from paper to practice. IJCV **129**(2), 517–547 (2021)
14. Kendall, A., Cipolla, R.: Modelling uncertainty in deep learning for camera relocalization. In: ICRA, pp. 4762–4769 (2016)
15. Kendall, A., Cipolla, R.: Geometric loss functions for camera pose regression with deep learning. In: CVPR, pp. 5974–5983 (2017)

16. Kendall, A., Gal, Y.: What uncertainties do we need in Bayesian deep learning for computer vision? NeurIPS **30** (2017)
17. Kendall, A., Grimes, M., Cipolla, R.: PoseNet: a convolutional network for real-time 6-DOF camera relocalization. In: CVPR, pp. 2938–2946 (2015)
18. Kingma, D.P., Welling, M.: Auto-encoding variational Bayes. In: ICLR (2014)
19. Kingma, D.P., Welling, M., et al.: An introduction to variational autoencoders. Found. Trends® Mach. Learn. **12**(4), 307–392 (2019)
20. Liu, C., Chen, S., Zhao, Y., Huang, H., Prisacariu, V., Braud, T.: HR-APR: apr-agnostic framework with uncertainty estimation and hierarchical refinement for camera relocalisation. In: ICRA, pp. 8544–8550 (2024)
21. Liu, L., Li, H., Dai, Y.: Efficient global 2D-3D matching for camera localization in a large-scale 3D map. In: ICCV, pp. 2372–2381 (2017)
22. Loshchilov, I., Hutter, F.: Decoupled weight decay regularization. arXiv preprint arXiv:1711.05101 (2017)
23. Melekhov, I., Ylioinas, J., Kannala, J., Rahtu, E.: Image-based localization using hourglass networks. In: ICCV Workshops, pp. 879–886 (2017)
24. Moreau, A., Piasco, N., Tsishkou, D., Stanciulescu, B., de La Fortelle, A.: CoordiNet: uncertainty-aware pose regressor for reliable vehicle localization. In: CVPR, pp. 2229–2238 (2022)
25. Moreau, A., Piasco, N., Tsishkou, D., Stanciulescu, B., de La Fortelle, A.: LENS: localization enhanced by NeRF synthesis. In: CoRL, pp. 1347–1356. PMLR (2022)
26. Naseer, T., Burgard, W.: Deep regression for monocular camera-based 6-DoF global localization in outdoor environments. In: IROS, pp. 1525–1530 (2017)
27. Ng, T., Lopez-Rodriguez, A., Balntas, V., Mikolajczyk, K.: Reassessing the limitations of CNN methods for camera pose regression. arXiv preprint arXiv:2108.07260 (2021)
28. Nguyen, T.M., et al.: MCD: diverse large-scale multi-campus dataset for robot perception. In: CVPR, pp. 22304–22313 (2024)
29. Paszke, A., et al.: PyTorch: an imperative style, high-performance deep learning library. In: Wallach, H., Larochelle, H., Beygelzimer, A., d'Alché-Buc, F., Fox, E., Garnett, R. (eds.) NeurIPS, pp. 8024–8035. Curran Associates, Inc. (2019)
30. Sarlin, P.E., Cadena, C., Siegwart, R., Dymczyk, M.: From coarse to fine: robust hierarchical localization at large scale. In: CVPR, pp. 12716–12725 (2019)
31. Sattler, T., Leibe, B., Kobbelt, L.: Efficient & effective prioritized matching for large-scale image-based localization. TPAMI **39**(9), 1744–1756 (2016)
32. Sattler, T., Zhou, Q., Pollefeys, M., Leal-Taixe, L.: Understanding the limitations of CNN-based absolute camera pose regression. In: CVPR, pp. 3302–3312 (2019)
33. Schindler, G., Brown, M., Szeliski, R.: City-scale location recognition. In: CVPR, pp. 1–7. IEEE (2007)
34. Se, S., Lowe, D.G., Little, J.J.: Vision-based global localization and mapping for mobile robots. IEEE Trans. Rob. **21**(3), 364–375 (2005)
35. Shavit, Y., Ferens, R., Keller, Y.: Learning multi-scene absolute pose regression with transformers. In: ICCV, pp. 2733–2742 (2021)
36. Shotton, J., Glocker, B., Zach, C., Izadi, S., Criminisi, A., Fitzgibbon, A.: Scene coordinate regression forests for camera relocalization in RGB-D images. In: CVPR, pp. 2930–2937 (2013)
37. Toft, C., Maddern, W., Torii, A., Hammarstrand, L., Stenborg, E., Safari, D., Okutomi, M., Pollefeys, M., Sivic, J., Pajdla, T., et al.: Long-term visual localization revisited. TPAMI **44**(4), 2074–2088 (2020)
38. Torii, A., Arandjelovic, R., Sivic, J., Okutomi, M., Pajdla, T.: 24/7 place recognition by view synthesis. In: CVPR, pp. 1808–1817 (2015)

39. Walch, F., Hazirbas, C., Leal-Taixe, L., Sattler, T., Hilsenbeck, S., Cremers, D.: Image-based localization using LSTMs for structured feature correlation. In: ICCV, pp. 627–637 (2017)
40. Walker, J., Doersch, C., Gupta, A., Hebert, M.: An uncertain future: forecasting from static images using variational autoencoders. In: ECCV, pp. 835–851. Springer (2016)
41. Wang, B., Chen, C., Lu, C.X., Zhao, P., Trigoni, N., Markham, A.: AtLoc: attention guided camera localization. In: AAAI, vol. 34, pp. 10393–10401 (2020)
42. Zangeneh, F., Bruns, L., Dekel, A., Pieropan, A., Jensfelt, P.: A probabilistic framework for visual localization in ambiguous scenes. In: ICRA, pp. 3969–3975 (2023)
43. Zangeneh, F., Bruns, L., Dekel, A., Pieropan, A., Jensfelt, P.: Conditional variational autoencoders for probabilistic pose regression. In: IROS (2024)
44. Zhou, Y., Barnes, C., Lu, J., Yang, J., Li, H.: On the continuity of rotation representations in neural networks. In: CVPR, pp. 5745–5753 (2019)

EfficientPose 6D: Scalable and Efficient 6D Object Pose Estimation

Zixuan Fang[1,2], Thomas Pöllabauer[1,2], Tristan Wirth[1], Sarah Berkei[1], Volker Knauthe[1(✉)], and Arjan Kuijper[1,2]

[1] Technical University of Darmstadt, Darmstadt, Germany
volker.knauthe@tu-darmstadt.de
[2] Fraunhofer IGD, Darmstadt, Germany

Abstract. In industrial applications requiring real-time feedback, such as quality control and robotic manipulation, the demand for high-speed and accurate pose estimation remains critical. Despite advances improving speed and accuracy in pose estimation, finding a balance between computational efficiency and accuracy poses significant challenges in dynamic environments. Most current algorithms lack scalability in estimation time, especially for diverse datasets, and the state-of-the-art methods are often too slow. This study focuses on developing a fast and scalable set of pose estimators based on GDRNPP to meet or exceed current benchmarks in accuracy and robustness, particularly addressing the efficiency-accuracy trade-off essential in real-time scenarios. We propose the AMIS algorithm to tailor the utilized model according to an application-specific trade-off between inference time and accuracy. We further show the effectiveness of the AMIS-based model choice on four prominent benchmark datasets (LM-O, YCB-V, T-LESS, and ITODD).

Keywords: Monocular 6D pose estimation · efficient · fast

1 Introduction

In the field of computer vision, object pose estimation can be considered one of the most important tasks. It aims to determine the rotation and translation of an object in three dimensions respectively. Due to its wide range of applications such as robotics, autonomous driving, and augmented reality, researchers have attempted to tackle this problem. In robotics, for example, high-quality 6D pose estimation plays a crucial role in autonomous robotic object manipulation in real-world scenarios such as picking up industrial bins [3].

In applications requiring real-time feedback, low inference times play an important role in pose estimation. During robot navigation or autonomous driving, pose estimation systems must be capable of analyzing and responding rapidly and accurately to changes in the environment in order to prevent collisions and ensure the safety of the robot or passenger. Fast pose estimation can be used for sports analysis and video surveillance to capture critical motions

J. Petersen and V. A. Dahl (Eds.): SCIA 2025, LNCS 15726, pp. 196–209, 2025.
https://doi.org/10.1007/978-3-031-95918-9_14

and behaviors, which can be utilized for real-time decision support, as well. As a result, enhancing the processing speed of a pose estimation algorithm is essential to broaden its practical applications.

Often the inference time of a model is closely linked to its accuracy. Based on this observation, there are a lot of applications that have strong constraints regarding their inference time budget. For these applications, finding the most accurate model, that fulfils its inference time constraints is key. Tan et al. [21] introduce a family of algorithms for Object Detection, that perform well under different time budgets, giving the user the possibility to chose the variation, that achieves the highest accuracy given their time budget. Pöllabauer et al. [18] propose such a family of architectures for 6D pose estimation.

In this work, we propose multiple candidate architectures based on GDRNPP [15], which constitutes an enhanced version of GDR-Net [26], that optimize inference time while maintaining or surpassing its accuracy on multiple BOP challenge [1] datasets. Furthermore, we introduce the Adaptive Margin-Dependent Iterative Selection (AMIS) algorithm that selects a subset out of candidate architectures, that constitute a beneficial trade-off between inference time and accuracy over multiple datasets. The proposed AMIS algorithm can be applied to a diverse range of task-specific datasets, allowing the choice of a model that reflects the domain-specific requirements with regard to a trade-off between inference time and accuracy.

In summary,

- we propose 40 candidate architectures based on modifications of GDRNPP [15] by adapting backbone and Geo Head architecture with the primary goal to enhance the resulting inference time while maintaining high accuracy,
- we present the AMIS algorithm, which identifies a suitable set of candidate models that constitute an optimal trade-off between inference time and their 6D pose estimation quality over multiple datasets, and
- we present quantitative results of the candidate models identified by the proposed AMIS algorithm for the LM-O, YCB-V, T-LESS, and ITODD datasets [4,8,28].

2 Related Work

2.1 6D Object Pose Estimation

The estimation of 6D positions from monocular RGB-D pictures has a multitude of applications in the industry, e.g., robot grasping, and therefore constitutes an important computer vision task overall. Especially in the industrial area the need for real-time 6D object pose estimation is prevalent [6], therefore this work focuses on high accuracy applications under constraint inference time.

Hodan et al. [8] give a comprehensive overview over the recent developments in 6D object pose estimation as benchmarked by the BOP challenge, that uses a wide variety of datasets and relevant metrics. Li et al. [12] increase accuracy and robustness by introducing Coordinates-based Disentangled Pose

Network (CDPN) that uniquely separates the prediction of rotation from the prediction of translation, demonstrating a high level of flexibility and efficiency even with texture-less and occluded objects. Labbe et al. [10] propose Cosypose, which is able to estimate the 6D pose of multiple objects in scenes captured based on unknown camera viewpoints. From individual images, hypotheses of object poses are generated, which are matched across different views to estimate both the camera viewpoints and the poses in a unified scene framework. By employing an object-level bundle adjustment, the method innovatively manages object symmetries without requiring depth information, improving the accuracy and robustness by minimizing reprojection errors in complex multi-object, multi-view scenarios. Wang et al. [26] propose GDR-Net, introducing a dense correspondence-based intermediate geometric representations, which in contrast to earlier strategies that set up 2D-3D correspondences and subsequently apply PnP/RANSAC strategies, allows for end-to-end training. As a result of this approach, the 6D pose can be directly regressed, combining the advantages of both direct and indirect methods. GDRNPP [15] upgrades GDR-Net by introducing stronger domain randomization operations, such as background replacement and color enhancements, and most prominently replacing the ResNet-34 backbone with ConvNeXt. We discuss the architecture of GDR-Net and GDRNPP in more detail in Sect. 3. Haugaard and Buch [7] propose an unsupervised approach, that is employed to learn dense, continuous 2D-3D correspondence distributions on object surfaces without prior knowledge of visual ambiguities such as symmetry. They utilize a compact, fully connected key model and an encoder-decoder query model, both operating within object-specific latent spaces. Zebrapose [20] introduces a discrete descriptor that provides a more detailed and accurate mapping of the object surface as compared to previous methods, which allows encoding the surface of an object more efficiently by incorporating a hierarchical binary grouping. Furthermore, they propose a novel coarse-to-fine training strategy that enhances the accuracy by enabling fine-grained correspondence prediction.

2.2 Speed and Efficiency in Deep Learning

Modern deep learning architectures have been specifically tailored to address the computational bottlenecks in 6D pose estimation. For instance, lightweight neural networks that incorporate depthwise separable convolutions, such as MobileNet and EfficientNet, have demonstrated significant reductions in computational complexity and latency without compromising accuracy by reducing the number of parameters and operations, thereby speeding up the inference time and allowing for scaling and using different configurations [13,25].

Beyond architectural changes, algorithmic adjustments also play a crucial role. For example, employing more sophisticated loss functions that focus on critical parts of the pose estimation task or the application of better regularization techniques can lead to more efficient learning dynamics and prevent overfitting. [5,14].

Pöllabauer et al. [18] propose a set of scalable 6d pose estimation architectures over a wide scale of inference time budgets based on a fixed set of datasets from the BOP challenge [1]. In contrast to that, we propose a strategy that is able to automatically identify architectures with a beneficial trade-off between inference time and accuracy for arbitrary 6D pose estimation datasets.

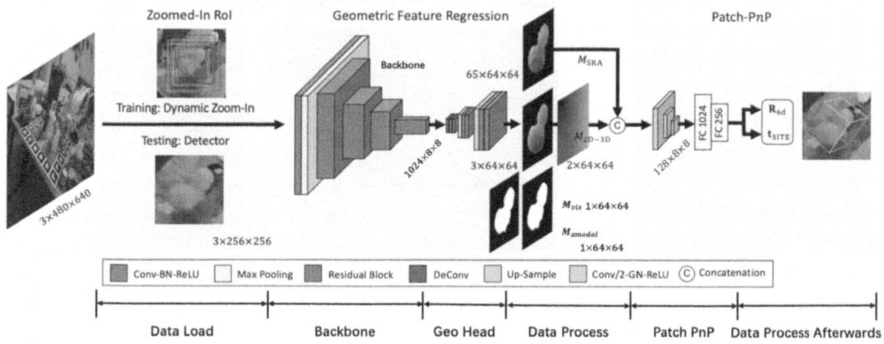

Fig. 1. Architecture of GDRNPP/GDR-Net [26]

3 Preliminaries

GDRNPP [15] is a 6D object pose estimation architecture, that constitutes an enhanced version of GDR-Net [26]. It estimates the pose of an object given an RGB image, by firstly detecting relevant image regions, containing the object region, then predicting relevant features in these image regions using Convolutional Neural Networks (CNN) in the form of a backbone and a subsequent Geo Net, based on which a PnP-Module directly regresses the rotation and translation from the learned features. Subsequently a depth-based pose refinement can be performed as an optional step.

We choose to adopt the GDRNPP architecture as a base for the proposed scalable 6D object pose estmation models, due to its high performance in the BOP Challenge [1] and for its highly adoptable architecture, as demonstrated in a range of diverse extensions [16,17]. To identify relevant parts of the architecture for inference time optimization, we subdivide the process of GDRNPP into six conceptual stages, i.e., *Data Load*, *Backbone*, *Geo Head*, *Data Process*, *Patch PnP* and *Data Process Afterwards*, which are illustrated with the architecture of GDR-Net (respectively GDRNPP) in Fig. 1. Preliminary experiments show that the major part of inference time are caused by the Data Load, Backbone and Geo Head stages (Fig. 2). Therefore, in the following we propose multiple changes to the architecture in these conceptual stages to reduce inference time, while aiming to preserve accuracy.

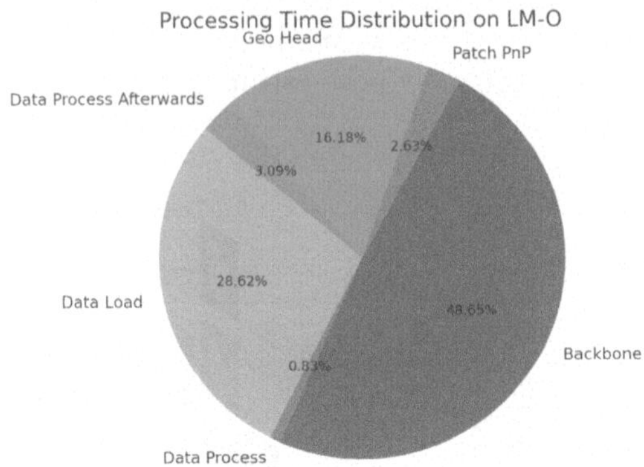

Fig. 2. Relative required runtime of the six conceptual stages during inference of GDR-Net/GDRNPP.

4 Methodology

In this section, we show potential optimizations to GDRNPP [15] especially by choosing suitable backbone architectures (see Sect. 4.1) and proposing alterations to the applied Geo head architecture (see Sect. 4.2) akin to the proposed optimizations of Pöllabauer et al. [18]. Therefore, we select the five backbones from a pool of 22 candidate backbones which exhibit outstanding performance under a specified GMAC budget (see Sect. 4.1). Furthermore, we propose 11 potential alterations of the Geo Head architecture, from which we chose 4 according to their performance on the LM-O dataset (see Sect. 4.2). The combination of selected backbones and Geo Head architecture results in a pool of 40 candidate architectures. We propose the AMIS algorithm, that selects a subset of these architectures, which allows efficient inference time and estimate quality trade-off by choosing the appropriate architecture from that pool (see Sect. 4.3). In the end, we propose some implementation optimizations regarding the GDRNPP net architecture, that we applied to further increase inference time (see Sect. 4.4).

4.1 Backbone

The impact of choosing different backbones for feature extraction on inference time and accuracy is crucial, prompting experiments with various models available in the Timm [22] package, which offers a broad array of pre-trained CNN and ViT [29] models, enabling access to advanced architectures and pre-trained weights which facilitate transfer learning and accelerate model development.

To select the most effective models, we set our criteria based on balancing performance speed and accuracy. Faster models with a *Giga Multiply-Accumulate Operations per Second* ratio up to the standard set by GDRNPP were prioritized.

Among models with similar sizes, those offering superior accuracy were chosen. Considering Transformer models' effectiveness in computer vision, variants like Nextvit [11] and Maxvit [23] were selected for their compact sizes suitable for the 256×256 input image size used in our model. This approach ensures a reasonable selection of backbones that optimize both computational efficiency and task performance.

We identified 22 candidates for the backbone architecture by analyzing their inference time distribution, which appeared to cluster into three distinct groups, as shown in Fig. 3a [2,23,24,27]. Accordingly, we employed the k-means clustering algorithm to categorize these candidates into three groups. For each group, we utilized the successive halving algorithm to select the best-performing backbone candidate in the group (Fig. 3b), selecting the five best performing backbones after five training epochs, the best three backbones after ten epochs, until the best performing backbone remained.

Notably, we also included Maxvit [23] for its competitive accuracy and FastVit [24] for ts superior speed. This strategic approach enables a focused evaluation of backbones that potentially optimize both speed and accuracy in our model.

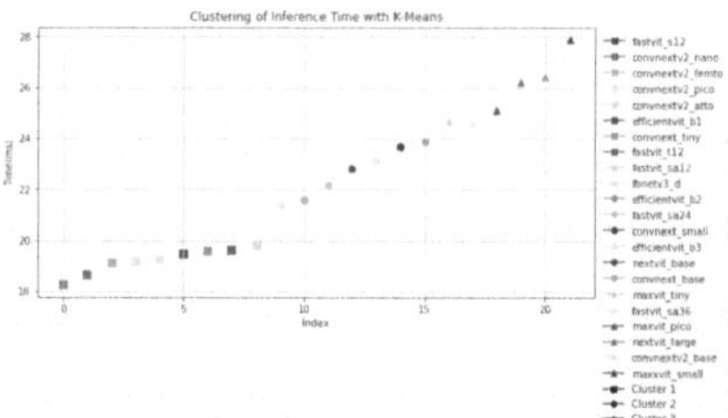

(a) K-means Clustering of Backbone according to inference time

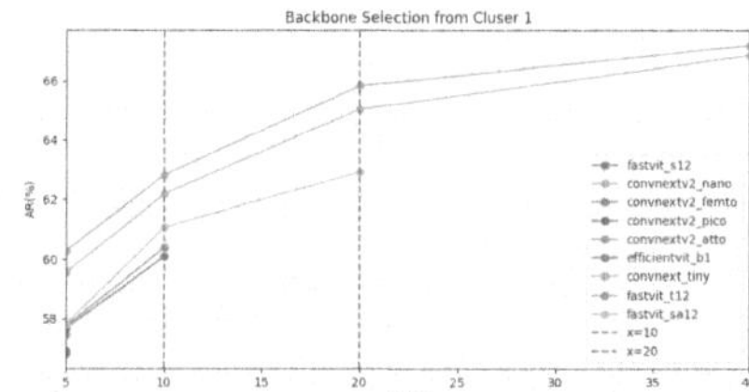

(b) Successive halving algorithm on Cluster 1

Fig. 3. Intermediate results during backbone selection using k-means clustering and the Successive Halving Algorithm with the mean of MSPD, MSSD and VSD as selection criterion.

4.2 Geo Head

The Geo Head part of GDRNPP is one of the most time consuming parts during inference. The standard structure of the Geo Head comprises three convolution blocks processing the $1024 \times 8 \times 8$ output from the Backbone including upsampling and convolution layers. To reduce the inference time requirements, we propose two variants of the vanilla Geo Head architecture: *Geo Head Variation 1*, reducing the convolution blocks to two and modifying the layer setup to handle changes in the feature map size without upsampling, and *Geo Head Variation 2*, which further streamlines this process by eliminating an additional upsampling

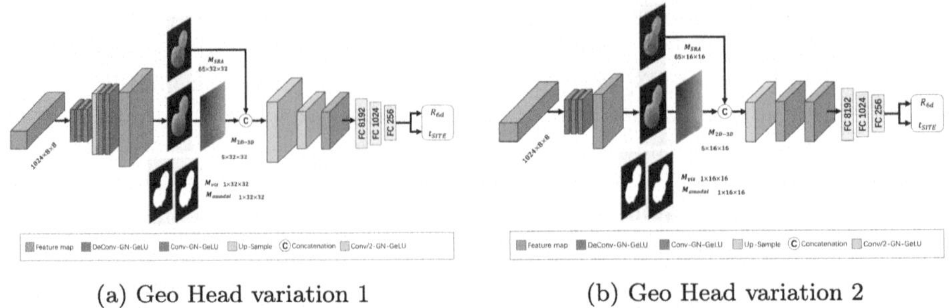

(a) Geo Head variation 1

(b) Geo Head variation 2

Fig. 4. Geo Head Variations GDR-Net architecture with the adaptations Geo Head variation 1 and 2, which reduce the number of up-sample layer of the vanilla Geo Head part.

layer and adjusting the convolution sequence, resulting in smaller feature map sizes and faster data processing. These modifications are aimed at decreasing the computational load and accelerating inference time by reducing the number of operations the GPU processes and by lowering the input size to the Data Process part, enhancing overall performance. The proposed architecture variations are illustrated in Fig. 4.

Inspired by U-Net [19], we implemented skip connections in our neural network architectures to address the gradient vanishing problem and enhance learning capabilities by directly connecting layers across the network. These connections facilitate detail recovery and image segmentation by leveraging rich contextual information during up-sampling. However, introducing skip connections adds complexity to the model, increases computational demands, and may impact the network's ability to generalize to unseen data. To balance these factors, we simplified the skip connection structure to involve minimal additional computations, ensuring that connected feature maps match in size to avoid unnecessary computations.

For each of the three Geo Head Variants (*Vanilla*, *Variation 1* and *Variation 2*), there are three different candidate locations, where skip connections could be added regarding the aforementioned criteria. These locations are illustrated in Fig. 5.

4.3 Adaptive Margin-Dependent Iterative Selection (AMIS) Algorithm

We propose the Adaptive Margin-Dependent Iterative Selection (AMIS) algorithm to identify a subset of models that excel regarding their estimate quality in comparison to their time budget, i.e., inference time, over arbitrary datasets. Existing strategies often average results over datasets, which essentially overweights the results of datasets that require high inference time, due to their absolute inference time differences being higher. In contrast to that, we opt for a strategy that reduces the influence of this effect. The proposed strategy pinpoints

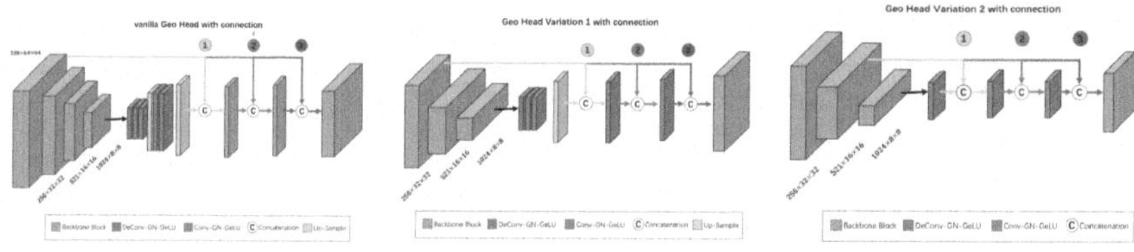

(a) vanilla Geo Head with connection (b) Geo Head Variation 1 with connection (c) Geo Head Variation 2 with connection

Fig. 5. Candidates for adding connection in vanilla Geo Head, Geo Head variation 1, Geo Head variation 2. The connection can be added in one of positions labeled ①, ② and ③.

models that are part of an optimal wrap-line (Fig. 6a) in the space of inference time and estimation quality, which show substantial accuracy improvement over its quicker counterparts with only minimal gains compared to slower successors. Recognizing these models aids in selecting models that effectively balance speed and accuracy, enhancing model selection strategies.

In an initial phase, we measure inference time and accuracy for each candidate model and each dataset. For each dataset we fit a straight line in the 2D space spanned by inference time and accuracy metrics using linear regression, which we call the *default slope*. For each model and dataset, we calculate the distance of its result from this default slope, normalizing them for each dataset on a scale of 0 to 100. These scores are weighted depending on the desired dataset weight, resulting in a final score for each model. We rank these scores assigning every model a fixed amount of scoring points depending on their rank. We then iteratively repeat that ranking process for 100 *adjustment factors*, between 0.001 and 3, which are multiplied with the *default slope* (Fig. 6b). We accumulate the ranking points per model for different adjustment factors except from the case, where the 10 best performing models do not change in comparison to the previous adjustment factor. This procedure ensures, that the resulting model selection is robust against diverse trade-off preferences. After completing this step, we select the best ranking model repeating the process with the repeating models, until the number of selected models meets the task-specific requirements.

4.4 Other Optimization

To reduce extended inference times linked with dynamic tensor creation, which is resource-intensive due to repeated memory allocation and initialization, we have adopted alternative measures. Our strategy minimizes new tensor creation by pre-allocating tensor memory before inference, using a reusable tensor pool throughout the inference cycle to avoid frequent new allocations. This approach enhances efficiency, which is particularly critical in real-time applications.

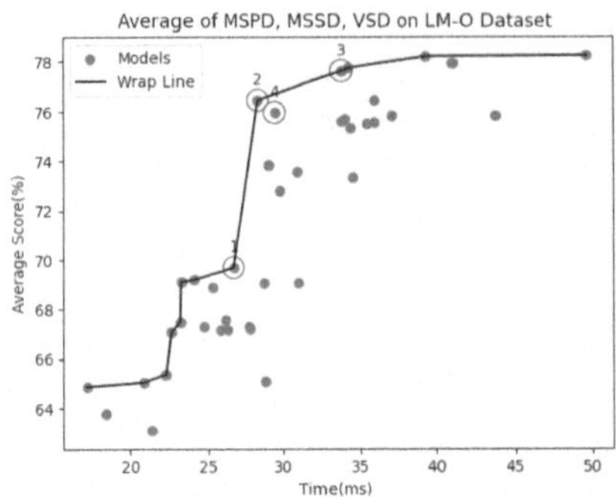

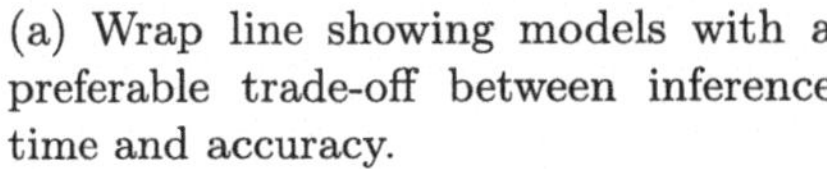
(a) Wrap line showing models with a preferable trade-off between inference time and accuracy.

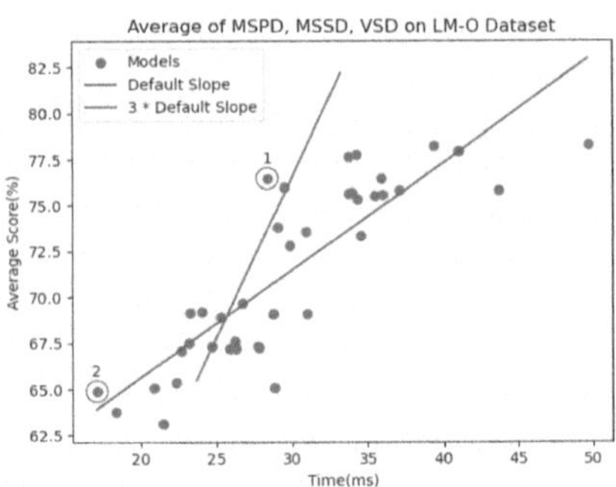

(b) *Default slope* and slope with *adjustment factor* in the AMIS algorithm.

Fig. 6. Visualization of substages of the AMIS algorithm.

5 Results

In this section, we provide a quantitative analysis of the proposed Geo Head architecture improvements (see Sect. 5.1), proposing 40 candidate architectures. Furthermore, we show the results of the models selected by the proposed AMIS for multiple datasets (see Sect. 5.3), showing that the selected models outperform GDRNPP and provide a reasonable model selection regarding a desired trade-off between inference time and accuracy on multiple datasets.

5.1 Preliminary Experiments

In Sects. 4.2 and 4.4 we proposed alterations to the Geo Head architecture of GDRNPP with the goal of reducing inference time while maintaining a high level of accuracy. We evaluate the performance of GDRNPP using these Geo Head architectures on the LM-O dataset. The results of this evaluation are illustrated in Table 1. These results indicate, that the proposed general optimizations *B0* and *C0* positively influence the inference time, while maintaining the desired level of precision, measured by MSPD, MSSD, VSD and AR, which give insight into a wide variety of aspects of 6D object pose estimation provided by the BOP Toolkit [9]. Furthermore, the results indicate, that the introduced variations of the Geo Head architecture reduce the inference time even further. We ablate the influence of architectural changes in more detail in Sect. 5.2.

Based on the results, we chose the Geo Head architectures *C0*, *E0*, *F0*, and *F2* due to their high performance in comparison to the required inference time. We combine those architectures with the previously identified candidate backbones to obtain 20 combinations. Those candidates are assigned with a number for their identification according to Table 2. We evaluate each of these architecture combinations with and without the optional refinement step of GDR-NET in the subsequent experiments, essentially resulting in 40 candidate architectures.

Table 1. Quantitative results of the proposed Geo Head candidates ond the LM-O dataset regarding MSPD, MSSD, VSD, AR and Inference time. The Geo Head architectures that we chose to construct candidate architecture are indicated in **bold**.

Row	Method	MSPD %	MSSD %	VSD %	AR %	Time ms
A0	GDRNPP	87.14	67.03	52.38	68.85	28.8
B0	A0: Optimizations	87.14	67.07	52.38	68.86	26.72
C0	**B0: Data Process Optimization**	87.14	67.03	52.38	68.86	**25.29**
D0	C0: Add connection in location 1	**87.66**	**67.29**	**52.53**	**69.16**	28.79
D1	C0: Add connection in location 2	86.89	66.75	51.91	68.82	29.17
D2	C0: Add connection in location 3	87.39	66.64	52.01	68.68	25.68
E0	**C0: Vanilla Geo Head→ variation 1**	**87.11**	**67.78**	**52.58**	**69.15**	**24.05**
E1	E0: Add connection in location 1	86.71	67.2	52.11	68.7	24.47
E2	E0: Add connection in location 2	87.16	67.13	52.42	68.9	24.32
E3	E0: Add connection in location 3	87.30	66.83	51.90	68.68	24.13
F0	**C0: Vanilla Geo Head→ variation 2**	85.23	65.40	50.61	67.08	22.66
F1	F0: Add connection in location 1	85.29	65.72	50.88	67.3	22.22
F2	**F0: Add connection in location 2**	**84.96**	**66.26**	**51.27**	**67.49**	23.20
F3	F0: Add connection in location 3	81.46	58.30	44.82	61.53	23.11
G0	F2: convnext_base→convnextv2_nano	83.47	63.00	48.15	64.87	**17.12**
H0	E0: convnext_base→convnextv2_base	87.31	67.62	52.63	**69.18**	24.05

Table 2. Candidate Models consisting of one of the previously identified backbones and adapted Geo Head candidates (F0, F1, E0, E1).

Backbone	configuration			
	F0	F2	E0	C0
fastvit_s12	1	2	3	4
convnextv2_nano	5	6	7	8
convnext_base	9	10	11	12
convnextv2_base	13	14	15	16
maxxvit_small	17	18	19	20

5.2 Geo Head

In this section, we summarize the findings from the proposed alterations to the Geo Head part of GDR-Net/GDRNPP as evaluated on the LM-O dataset (see Table 1). Our results indicate that the proposed optimizations (see Sect. 4.4) improve the inference time without a relevant impact on the measured accuracy. The proposed Geo Head variation 1 leads to a speed-up in inference time and an improvement in accuracy, constituting it a successful optimization to the Geo Head architecture. In contrast to that, the proposed Geo Head variation 2, further increased inference time, while reducing accuracy significantly. The availability of such an adaption, however, is highly beneficial when looking for highly accurate 6D pose estimation models under varying inference time budgets. Adding connections within the Geo Head was generally found to negatively impact inference speed, particularly in the vanilla Geo Head setup, where it was deemed not worthwhile due to significant slowdowns without notable accuracy benefits. However, in the second variation of Geo Head, adding a connection at the first location improved both speed and accuracy, making it a promising adjustment for balancing performance metrics.

5.3 AMIS

We employ the aforementioned AMIS algorithm (see Sect. 4.3) to identify a suitable subset of models within the previously identified candidate architectures on the IMO, LM-O, YCB-V, T-LESS, and ITODD datasets [4,8,28]. The results of the five identified candidates are illustrated in Fig. 7 on scatterplot showing inference time and 6D object pose estimation accuracy. Furthermore, the average results of the models are illustrated in Table 3.

The experimental results show, that even with minimal time budget, i.e., when we expect our model to perform the fastest inference, the required inference time is reduced by 35% in comparison to GDRNPP, while the achieved performance only drops by 3% measured by the average of MSPD, MSSD, and VSD. As the time budget increases, the performance of our candidates also gradually improves. This variation can adapt to the complex scenarios of different time and accuracy requirements in industrial environments. It is noteworthy that compared to GDRNPP, using about 31% additional time can lead to approximately a 25% improvement in performance. Figure 7 furthermore shows that the selected models show an increase in accuracy with increased inference time for all 4 datasets.

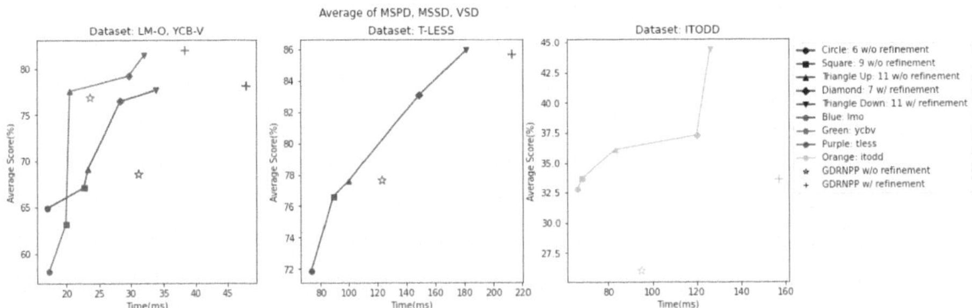

Fig. 7. Average over MSPD, MSSD, VSD of the 5 candidate architectures with desireable trade-off between inference time and accuracy using the AMIS algorithm on LM-O (left), YCB-V(left), T-LESS (center) and ITODD (right) dataset.

Table 3. Quantitative results of the candidate architectures identified by the AMIS algorithm, showing that they constitute a desirable trade-off between inference time and accuracy.

5 selected Candidates vs GDRNPP				
Candidates Number	Average of MSPD, MSSD VSD %	ADD %	Average of MSPD, MSSD, VSD, ADD %	Time %
6 without ref.	−2.82	−16.62	−3.42	−35.71
9 without ref.	+2.06	−9.82	+1.66	−24.86
11 without ref.	+10.12	+3.0	+10.44	−17.61
7 with ref.	+16.24	+41.28	+20.56	+15.81
11 with ref.	+25.14	+50.83	+30.35	+30.74

6 Conclusion

We presented a fast and scalable pose estimator, dynamically adjusting to custom needs of estimation quality versus inference times trade-offs. To that end, we proposed 40 candidate architectures which aim to otimize the trade-off between inference time and 6D object pose estimation accuracy based on GDRNPP, which were curated by choosing promising backbones and identifying beneficial architectural changes to the Geo Head part of the GDRNPP architecture. Additionally, we proposed the AMIS algorithm, a tool designed to quantitatively identify model architectures that represent sweet spots which effectively address the challenges posed by the diverse scales of precision and time across various datasets. In our comparison with GDRNPP, we demonstrate the candidates selected by the AMIS algorithm exhibit an outstanding inference time and progressively enhancing model accuracy for a wide variety of inference time budgets.

Future directions might include temporal trajectory composition of different frequencies for long-horizon tasks, policy composition methods on large-scale datasets, and policy distillation from composed policies. For further research, there are some potential areas of improvement: adopting end-to-end methods that integrate detection or segmentation and exploring other lightweight methods like teacher-student models.

References

1. BOP: Benchmark for 6D Object Pose Estimation. https://bop.felk.cvut.cz/leaderboards/
2. Cai, H., Gan, C., Han, S.: Efficientvit: enhanced linear attention for high-resolution low-computation visual recognition. arXiv preprint arXiv:2205.14756 (2022)
3. Deng, X., Xiang, Y., Mousavian, A., Eppner, C., Bretl, T., Fox, D.: Self-supervised 6d object pose estimation for robot manipulation. In: 2020 IEEE International Conference on Robotics and Automation (ICRA), pp. 3665–3671. IEEE (2020)

4. Drost, B., Ulrich, M., Bergmann, P., Hartinger, P., Steger, C.:Introducing mvtec itodd-a dataset for 3d object recognition in industry. In: Proceedings of the IEEE International Conference on Computer Vision Workshops, pp.2200–2208 (017)
5. Gonzalez, S., Miikkulainen, R.: Effective regularization through loss-function metalearning. arXiv:2010.00788 (2020)
6. Gorschlüter, F., Rojtberg, P., Pöllabauer, T.: A survey of 6d object detection based on 3d models for industrial applications. J. Imaging **8**(3), 53 (2022)
7. Haugaard, R.L., Buch, A.G.: Surfemb: dense and continuous correspondence distributions for object pose estimation with learnt surface embeddings. In: Proceedings of the IEEE/CVF Conference on Computer Vision and Pattern Recognition, pp. 6749–6758 (2022)
8. Hodan, T., Haluza, P., Obdržálek, Š., Matas, J., Lourakis, M., Zabulis, X.: T-less: an RGB-D dataset for 6d pose estimation of texture-less objects. In: 2017 IEEE Winter Conference on Applications of Computer Vision (WACV), pp. 880–888. IEEE (2017)
9. Hodan, T., et al.: Bop challenge 2023 on detection segmentation and pose estimation of seen and unseen rigid objects. In: Proceedings of the IEEE/CVF Conference on Computer Vision and Pattern Recognition, pp. 5610–5619 (2024)
10. Labbé, Y., Carpentier, J., Aubry, M., Sivic, J.: CosyPose: consistent multi-view multi-object 6D Pose estimation. In: Vedaldi, A., Bischof, H., Brox, T., Frahm, J.-M. (eds.) ECCV 2020. LNCS, vol. 12362, pp. 574–591. Springer, Cham (2020). https://doi.org/10.1007/978-3-030-58520-4_34
11. Li, J., et al.: Next-vit: next generation vision transformer for efficient deployment in realistic industrial scenarios. arXiv:2207.05501 (2022)
12. Li, Z., Wang, G., Ji. X.: CDPN: coordinates-based disentangled pose network for real-time RGB-based 6-DOF object pose estimation. In: Proceedings of the IEEE/CVF International Conference on Computer Vision, pp. 7678–7687 (2019)
13. Liang, Y., Chen, F., Liang, G., Wu, X., Feng, W.: An efficient lightweight deep neural network for real-time object 6d pose estimation with RGB-D inputs. In: 2021 International Joint Conference on Neural Networks (IJCNN), pp. 1–8. IEEE (2021)
14. Liu, F., Hu, Y., Salzmann, M.: Linear-covariance loss for end-to-end learning of 6D pose estimation. In: Proceedings of the IEEE/CVF International Conference on Computer Vision, pp. 14107–14117 (2023)
15. Liu, X.: Gdrnpp for bp2022 (2024). https://github.com/shanice-l/gdrnpp_bop2022
16. Pöllabauer, T., Li, J., Knauthe, V., Berkei, S., Kuijper, A.: End-to-end probabilistic geometry-guided regression for 6dof object pose estimation. arXiv:2409.11819 (2024)
17. Pöllabauer, T., Emrich, J., Knauthe, V., Kuijper, A.: Extending 6d object pose estimators for stereo vision. In: Pattern Recognition and Artificial Intelligence. LNCS, vol. 14893 (2024)
18. Pöllabauer, T., Pramod, A., Knauthe, V., Wahl, M.: Fast gdrnpp: improving the speed of state-of-the-art 6d object pose estimation. arXiv:2409.12720 (2024)
19. Ronneberger, O., Fischer, P., Brox, T.: U-Net: convolutional networks for biomedical image segmentation. In: Navab, N., Hornegger, J., Wells, W.M., Frangi, A.F. (eds.) MICCAI 2015. LNCS, vol. 9351, pp. 234–241. Springer, Cham (2015). https://doi.org/10.1007/978-3-319-24574-4_28
20. Su, Y., et al.: Zebrapose: coarse to fine surface encoding for 6DOF object pose estimation. In: Proceedings of the IEEE/CVF Conference on Computer Vision and Pattern Recognition, pp. 6738–6748 (2022)

21. Tan, M., Pang, R., Le, Q.V.: Efficientdet: scalable and efficient object detection. In: Proceedings of the IEEE/CVF Conference on Computer Vision and Pattern Recognition, pp. 10781–10790 (2020)
22. Timm (PyTorch Image Models) (2024). https://huggingface.co/timm
23. Tu, Z., et al.: Maxvit: multi-axis vision transformer. In: European Conference on Computer Vision, pp. 459–479. Springer (2022)
24. Vasu, P.K.A., Gabriel, J., Zhu, J., Tuzel, O., Ranjan, A.: Fastvit: a fast hybrid vision transformer using structural reparameterization. In: Proceedings of the IEEE/CVF International Conference on Computer Vision, pp. 5785–5795 (2023)
25. Wang, F.: A lightweight 6d pose estimation network based on improved atrous spatial pyramid pooling. Electronics **13**(7), 1321 (2024)
26. Wang, G., Manhardt, F., Tombari, F., Ji, X.: Gdr-net: geometry-guided direct regression network for monocular 6d object pose estimation. In: Proceedings of the IEEE/CVF Conference on Computer Vision and Pattern Recognition, pp. 16611–16621 (2021)
27. Woo, S., et al.: Convnext v2: co-designing and scaling convnets with masked autoencoders. In: Proceedings of the IEEE/CVF Conference on Computer Vision and Pattern Recognition, pp. 16133–16142 (2023)
28. Xiang, Y., Schmidt, T., Narayanan, V., Fox, D.: Posecnn: a convolutional neural network for 6d object pose estimation in cluttered scenes. arXiv:1711.00199 (2017)
29. Yuan, L., et al.: Tokens-to-token vit: training vision transformers from scratch on imagenet. In: Proceedings of the IEEE/CVF International Conference on Computer Vision, pp. 558–567 (2021)

Subspace-Based Embedded Feature Reduction for Fast Anomaly Detection

Naoki Murakami(✉), Naoto Hiramatsu, Hiroki Kobayashi, Shuichi Akizuki, and Manabu Hashimoto(✉)

Chukyo University, 101-2 Yagoto Honmachi, Nagoya, Aichi 466-8666, Japan
murakami@asmi.sist.chukyo-u.ac.jp, mana@isl.sist.chukyo-u.ac.jp

Abstract. Recently, anomaly detection methods that use pre-trained CNNs to extract image embeddings have attracted much attention. The objective of this research is to increase processing speed while maintaining accuracy in embedding-based anomaly detection methods. Previous methods extract features by concatenating the intermediate outputs of pre-trained CNNs. However, these features utilize the output of the deeper layers of the CNN, and the dimensionality of the features grows in proportion to the depth of the layers. This is a reason for the increased processing time. The proposed method speeds up the anomaly inspection process by efficiently reducing the dimensionality of image embeddings using the subspace generated by PCA. In evaluation experiments using the MVTec AD dataset, the proposed method achieved approximately 3.5 times faster processing speed (61.4 fps) than the previous methods PaDiM (6.9 fps) and PatchCore (17.4 fps), while maintaining a high anomaly detection accuracy of 99.0% on average AUROC. This result is more than twice faster than the typical video rate of 30 fps, achieving a near practical level of both inspection accuracy and inspection speed.

Keywords: Anomaly detection · Visual inspection · Subspace · PCA · Dimensionality reduction

1 Introduction

Image anomaly detection is an important task in the manufacturing. In this task, unsupervised methods that use only normal images for training are commonly used because anomalous samples cannot be gotten in this field. In the field of image anomaly detection, the approaches of state-of-the-art methods are changing: methods [1–7] using image reconstruction models such as autoencoders, and methods [8–11] that generate artificial anomaly images and train them to neural networks. Recently, embedding-based methods [12–17] have shown state-of-the-art results in comparisons using the MVTec AD dataset [18]. Embedding in the image field refers to the transformation of images into features using convolutional neural networks (CNNs) that have been pre-trained in ImageNet [19] or other datasets.

J. Petersen and V. A. Dahl (Eds.): SCIA 2025, LNCS 15726, pp. 210–223, 2025.
https://doi.org/10.1007/978-3-031-95918-9_15

In this embedding-based method, a normal image is first input to the pre-trained CNN, and image embeddings are extracted. Next, distributions of these normal features is modeled. There are two main methods of modeling. One is to use a statistical model such as a Gaussian distribution. The other method is to store the set of normal features as a memory bank. During inference, anomalies are determined based on the distance between the test features and these models. These methods have achieved an average anomaly detection accuracy of AUROC 99%, which approaches the practical level depending on field needs. However, these methods have a practical problem in that they require processing time when calculating anomalies. For example, the average processing speed of PatchCore, a state-of-the-art embedding-based method, is 17.4 fps. This is slower than the typical video rate of 30 fps. For practical use, it is important to be able to inspect at a processing speed of 30 fps or higher.

To solve this problem, we propose a faster embedding-based anomaly detection method while maintaining anomaly detection accuracy. The proposed method speeds up the computation of anomaly scores by efficient dimensionality reduction of image embeddings using subspaces. The proposed method showed an average AUROC of 99.0% in the performance evaluation on the MVTec AD dataset. This result is comparable to that of PatchCore. In terms of processing time, the proposed method computed anomaly scores at an average of 61.4 fps. This is about 3.5 times faster than PatchCore's average of 17.4 fps. Furthermore, the speed is faster than 30 fps, achieving results close to the practical level in terms of both inspection accuracy and inspection speed.

2 Related Work

In recent years, embedding-based anomaly detection has become the mainstream. Embedding-based methods model normal image embeddings. During inference, an anomaly is determined based on the distance between the test features and these normal models. There are two main methods of modeling. One is to use a statistical model and the other is to use a memory bank.

First, the aproach of using statistical models [12,13] to model normal feature distributions are explained. PaDiM [13] is the most prominent method in this approach. PaDiM extracts feature maps by concatenating the intermediate outputs from layers 1 to 3 of a pre-trained CNN. The feature map consists of multiple patch embeddings. Next, for each position in the feature map, features of normal images are modeled by a multivariate Gaussian distribution. During inference, the Mahalanobis distances between the test features and these distributions are computed for each location in the feature map. PaDiM's modeling has the advantage of requiring fewer parameters to be stored and PaDiM's calculation of anomaly scores has the advantage of a simple algorithm. It is also versatile enough to be applied to a variety of objects. However, PaDiM has the practical problem of taking a long time to calculate anomaly scores. This is due to the calculation of the Mahalanobis distance for each feature map location and the large dimensionality of the features. Therefore, PaDiM has a problem in that it is unable to achieve both anomaly detection accuracy and processing speed.

Next, an aproach of using memory bank [15–17] to model normal feature distribution are explained. PatchCore [17] is the most prominent method in this approach. PatchCore extracts feature maps by concatenating the intermediate outputs of the second and third layers of a pre-trained CNN. Next, these normal features are stored as a memory bank. During inference, the Euclidean distance between the memory bank and the test feature is calculated for each feature map location. PatchCore uses the stored normal features directly during inference. Therefore, compared to statistical modeling methods, the error with the real distribution is smaller, resulting in higher anomaly detection accuracy. However, this results in a time-consuming distance calculation between the test features and the memory bank, at the expense of processing time.

As described above, embedding-based methods have not been able to achieve both anomaly detection accuracy and processing speed. In the next section, a proposed method is described to solve this problem.

3 Proposed Method

3.1 Idea of Proposed Method

In order to increase processing speed while maintaining anomaly detection accuracy, this study adopts the idea of dimensionality compression of image embeddings. Previous methods concatenate intermediate outputs from different layers of pre-trained CNNs based on the idea that features at different semantic levels are important, but they do not consider which feature dimensions are important in the representation of normality. Therefore, this study reduces redundant dimensions by selecting only the important information in the feature map after concatenation. As a result, the dimensionality of image embeddings is reduced, which speeds up the anomaly computation process and the accuracy of anomaly detection is maintained because this idea reduces only unnecessary information. An anomaly detection method that reflects the above idea is described in the next and subsequent sections.

3.2 Feature Extraction and Feature Dimensionality Reduction

First, the feature extraction method is described. The proposed method uses a CNN pre-trained on ImageNet as a feature extractor, similar to the previous methods such as PaDiM and PatchCore. The specific feature extraction method is shown in Fig. 1. The proposed method extracts features by concatenating the outputs of two intermediate layers obtained when an image $\boldsymbol{x}$ is input to a pre-trained CNN. In this study, the intermediate outputs of the second and third layers of the CNN are concatenated, similar to PatchCore. The output of the second layer is a C dimensional vector for $H \times W$. The output of the third layer yields C' dimensional vectors of $H' \times W'$. When combining feature maps of different resolutions, the proposed method performs averaging filter processing on the second and third layer feature maps. This is intended to refer

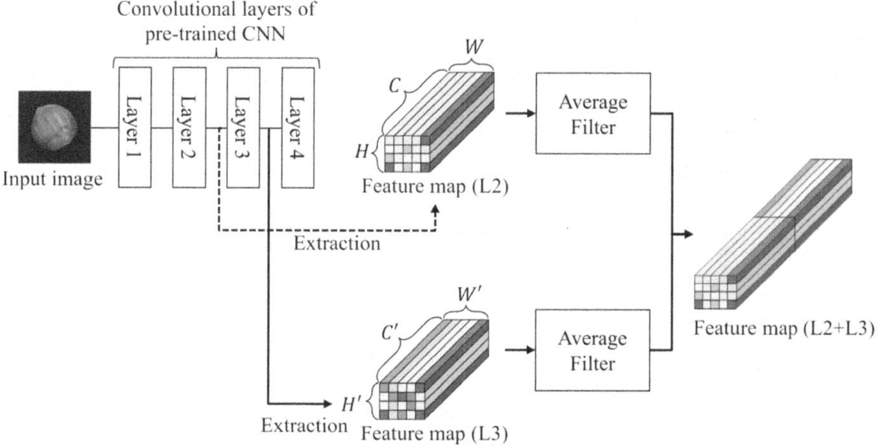

Fig. 1. Overview of the feature extraction approach. Proposed method uses a CNN pre-trained on ImageNet as a feature extractor and extracts features by concatenating the outputs of two intermediate layers.

to nearby features and is expected to be robust against misalignment. Then, by concatenating the feature maps, a $C + C'$ dimensional vector is obtained.

Next, a method of dimensionality reduction of image embeddings is described. In this study, the normal subspace generated from only normal features is used for dimensionality reduction. The normal subspace is generated by Principal Component Analysis (PCA). The reason for using PCA for dimensionality reduction is its high computational efficiency, which enables dimensionality reduction of features using linear operations. Although there is a method of dimensionality reduction using an autoencoder, we did not use it because it requires training of a neural network and has low compatibility with embedding-based methods. The feature maps $\boldsymbol{M}_n$ is obtained by inputting N normal images $\boldsymbol{x}_n$ into the pre-trained CNN and extracting the intermediate output. A normal subspace is generated for each (i, j) local position of this feature maps. First, the average vector $\bar{\boldsymbol{f}}(i, j)$ of N feature vectors $\boldsymbol{f}_n(i, j)$ at position (i, j) is calculated. Next, using this mean vector $\bar{\boldsymbol{f}}(i, j)$, the features are centered. Let $\hat{\boldsymbol{f}}_n(i, j)$ denote the centered features. Using this centered feature data matrix $\hat{\boldsymbol{F}}(i, j)$, the covariance matrix $\boldsymbol{S}(i, j)$ at position (i, j) is obtained. By solving the eigenvalue problem for the obtained covariance matrix $\boldsymbol{S}(i, j)$, eigenvalues $\lambda_k(i, j)$ and a set of eigenvectors $\boldsymbol{a}_k(i, j)$ corresponding to $\lambda_k(i, j)$. The subspace $\boldsymbol{A}(i, j)$ is obtained by choosing K eigenvectors $\boldsymbol{a}_k(i, j)$ from the larger eigenvalues $\lambda_k(i, j)$. It is common to determine this K based on the cumulative contribution rate. This operation is performed at each position of the feature map to obtain $H \times W$ normal subspaces. The generation of the normal subspace at position (i, j) is shown in Fig. 2. Since this normal subspace has the ability to approximate fluctuations in normal features with low dimension, projecting $C + C'$ dimensional features onto this space effectively reduces the dimension of the features to K dimensions. These are the methods of feature extraction and dimensionality reduction.

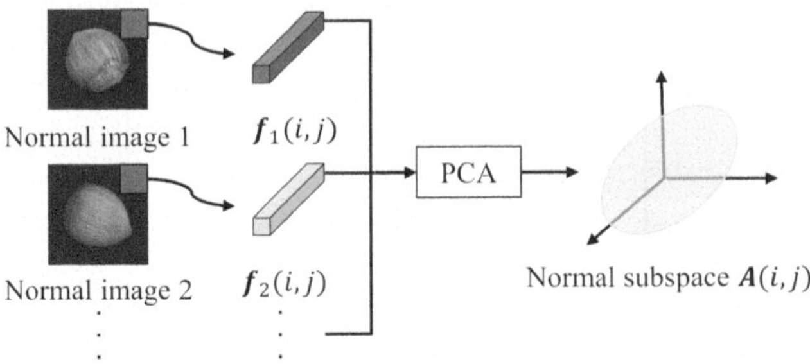

Fig. 2. Generation of the normal subspace at position (i, j) of feature map. This normal subspace has the ability to approximate fluctuations in normal features.

3.3 Definition of Anomaly Score

In this study, the anomaly score is defined not by the conventional Mahalanobis or Euclidean distances, but by the reconstruct error of the test feature itself. This was set by focusing on the feature information that is lost due to dimensionality reduction. First, the test image $\boldsymbol{x}_{test}$ is transformed into a feature map $\boldsymbol{M}_{test}$ by the method of Fig. 1. Anomaly score is calculated for each (i, j) position in the feature map. Next, a test feature $\boldsymbol{f}_{test}(i, j)$ is centered using the mean vector $\bar{\boldsymbol{f}}(i, j)$ to obtain $\hat{\boldsymbol{f}}_{test}(i, j)$. Then, by projecting the test feature $\hat{\boldsymbol{f}}_{test}(i, j)$ onto the normal subspace $\boldsymbol{A}(i, j)$, the test feature is transformed into K-dimensional feature $\boldsymbol{z}_{test}(i, j)$. This normal subspace has the ability to approximate the original normal features in low dimensions, but it cannot approximate abnormal features because anomaly features are not used to generate the normal subspace. Thus, if the test feature $\hat{\boldsymbol{f}}_{test}(i, j)$ is abnormal, the lower-order test feature $\boldsymbol{z}_{test}(i, j)$ cannot not represent the anomaly information. Then, based on the fact that $\tilde{\boldsymbol{f}}_{test}(i, j)$, which is the back-projection of this low-order test feature $\boldsymbol{z}_{test}(i, j)$ onto the original feature space, becomes a normal feature, we define an anomaly score $D(i, j)$ in the following formula $D(i, j)$.The process of computing anomaly scores for the test features is shown in Fig. 3.

$$D(i,j) = \| \hat{\boldsymbol{f}}_{test}(i,j) - \tilde{\boldsymbol{f}}_{test}(i,j) \| = \| \hat{\boldsymbol{f}}_{test}(i,j) - \boldsymbol{A}(i,j)\boldsymbol{A}(i,j)^T \hat{\boldsymbol{f}}_{test}(i,j) \| \quad (1)$$

The proposed anomaly score is simpler to compute than conventional anomaly scores because it is defined from a linear transformation of feature vectors and a difference calculation. The anomaly scores are obtained at all locations in the feature map, and the maximum anomaly score is output as the representative value of the test image.

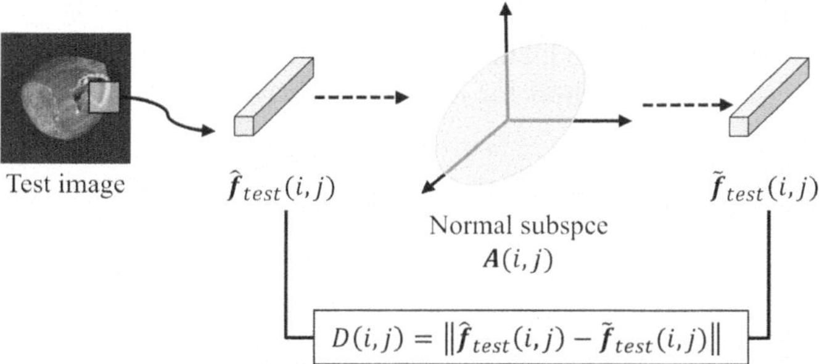

Fig. 3. Calculation of anomaly score based on the inverse projection error of test feature vectors. This anomaly score is computed at every location in the feature map.

4 Experiment

4.1 Verification of Speedup by Dimensionality Reduction

The purpose of this experiment is to demonstrate that, in dimensional reduction using subspaces, the higher the dimensional reduction ratio, the faster the processing speed of the anomaly computation.

Experimental Setting. MVTec AD [18], a public dataset for unsupervised anomaly detection, was used for this experiment. MVTec AD is an image dataset consisting of 15 categories. MVTec AD's 15 categories consist of 5 Texture categories and 10 Object categories. For each category, there is a training set of images consisting of normal images only, and a test set of images consisting of normal images, anomaly images, and Ground Truth (GT) images of anomaly regions. In this experiment, the MVTec AD images were resized to 256 pixels and center-cropped at 224 pixels.

Wide-ResNet50 [20], which was pre-trained by ImageNet [19], was used as the feature extractor. The intermediate outputs of Layer2 and Layer3 of Wide-ResNet50 were used for the features. The size of the feature map was 28 × 28 × 1536. In this experiment, we compared the anomaly detection accuracy and processing time of the proposed method when the cumulative contribution rate in generating the normal subspace was set at five different levels: 70%, 80%, 90%, 99%, and 100%. The area under the receiver operating characteristic curve (AUROC) was used for anomaly detection accuracy. Image-AUROC, which is calculated from representative values of anomaly, was used for AUROC. As for the processing time, we have measured only pure processing time which required to obtain the anomaly score defined in Sect. 3.3. So, the input and feature extraction portions of the image are not measured. The indicator is the number of computed images per second (fps). The PC specifications used in this experiment were OS: Ubuntu 24.04.1 LTS, CPU: Intel Core i9-14900KF, GPU: NVIDIA GeForce RTX 4070 Ti SUPER.

Results and Discussion of the Experiment. The results of the experiment are shown in Table 1. Table 1 shows a comparison of AUROC and processing speed for the proposed method for each cumulative contribution ratio.

Table 1. Comparison of Image-AUROC [%] and fps based on the magnitude of cumulative contribution rate (CCR).

CCR	CCR-70%		CCR-80%		CCR-90%		CCR-99%		CCR-100%	
Category name	AUROC	fps	AUROC	fps	AUROC	fps	AUROC	fps	AUROC	fps
Carpet	99.1	71.0	99.1	63.2	99.2	59.0	99.2	48.5	99.1	1.0
Grid	98.5	78.9	97.8	68.4	98.3	57.1	98.2	53.0	97.7	0.54
Leather	100	73.1	100	65.2	100	59.2	100	52.4	100	1.03
Tile	99.0	72.5	98.9	62.2	99.0	57.7	99.0	51.5	99.1	1.04
Wood	99.2	73.8	99.2	60.1	99.3	56.1	99.2	50.4	99.4	0.85
Bottle	100	75.1	100	53.8	100	59.0	100	54.8	100	0.65
Cable	98.7	73.7	99.1	69.2	99.4	64.3	99.5	53.9	99.4	1.4
Capsule	95.1	76.7	97.5	68.3	97.3	60.5	97.0	55.7	96.5	1.2
Hazelnut	100	74.2	100	63.1	100	58.3	100	41.1	100	1.0
Metal nut	97.3	85.2	99.8	68.5	99.9	64.1	100	53.3	100	0.6
Pill	97.4	73.0	97.3	68.8	97.3	64.0	97.1	50.3	97.0	0.8
Screw	41.9	88.1	75.8	75.9	96.5	68.1	97.4	52.9	96.9	1.4
Toothbrush	100	78.0	100	71.3	100	65.9	100	56.7	100	0.3
Transistor	100	76.5	100	65.0	100	61.8	100	53.9	100	0.5
Zipper	98.8	74.3	98.7	69.1	98.3	65.2	97.7	55.3	97.3	1.0
Average	95.0	76.3	97.5	66.1	99.0	61.4	99.0	52.2	98.8	0.89

From this table, it can be seen that the speed of anomaly computation increased when the cumulative contribution ratio, which determines the dimension of the feature, was reduced and the compression ratio of the dimension was increased. For example, comparing the results when all dimensions were used and when features with cumulative contributions up to 90% were used, the processing speed of anomaly computation was accelerated from an average of 0.89 fps to an average of 61.4 fps. Comparison of anomaly detection accuracy showed that accuracy was maintained even when the dimension of the features was reduced, since the average AUROC was 98.8% when all dimensions were used and the average AUROC was 99.0% when features with cumulative contribution rates up to 90% were used. This is considered to be due to the efficient compression of features by conversion to subspace. However, in the Screw category, a significant decrease in accuracy was observed with an AUROC of 41.9% at a cumulative contribution rate of 70%. The causes of this were analyzed by comparing anomaly maps for the Screw category. The anomaly maps for the Screw category at the cumulative contribution rates of 70%, 80%, 90%, and 99% are shown in Fig. 4. It can be confirmed from the Fig. 4 that Screw's own anomaly level is higher when the cumulative contribution rate is 70%. Conversely, when the cumulative contribution is 99%, it can be confirmed that only the anomaly areas are heated. The results suggest that when the object rotates, dimensional reduction may cause a loss of information about the object's orientation, which in turn reduces the detection accuracy. Therefore, if the object rotates, setting the cumulative contribution in the range of 90% to 99% can speed up the processing time which required to obtain the anomaly score while avoiding a significant loss of accuracy. These results confirm that the proposed method's idea of dimensional reduction of embedded features is effective in speeding up processing time while maintaining accuracy.

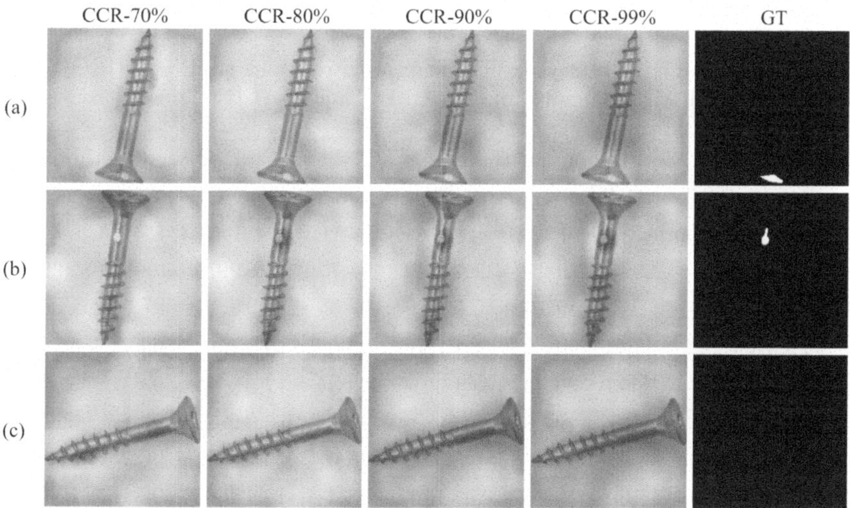

Fig. 4. Comparison of anomaly map based on the magnitude of cumulative contribution rate (CCR). (a) and (b) show the anomaly images of Screw. (c) is a normal image.

4.2 Comparative Experiment with Previous Methods

The purpose of this experiment is to demonstrate that the proposed method is faster than previous methods in terms of processing speed, while maintaining the same anomaly detection accuracy as previous methods.

Experimental Setting. The dataset and feature extractor used in the experiment were the same as in the previous section. For the proposed method, feature dimensions with cumulative contributions up to 90% were used. The effectiveness of the proposed method was evaluated by comparing its anomaly detection accuracy and processing speed with those of two typical embedding-based anomaly detection methods, PaDiM [13] and PatchCore [17]. The feature extractors for PaDiM and PatchCore were Wide-ResNet50 pre-trained on ImageNet, as in the proposed method. The layers used were Layer2 and Layer3. In PaDiM, the inverse of the covariance matrices were obtained a priori and used during inference. The selection percentage of the PatchCore's core set sample was selected to 1%.

We evaluated anomaly detection using two evaluation indices: Image-AUROC, which is calculated from the anomaly score for each image, and Pixel-AUROC, which is calculated from the anomaly score for each pixel. The method of measuring processing time is the same as in the previous section. The specifications of the PC used in this study were the same as in the previous section.

Table 2. Results of comparative experiments with previous studies (AUROC [%] and fps). In the table, I-AUC refers to Image-AUROC and P-AUC refers to Pixel-AUROC.

	Previous method						Proposed method		
Method	PaDiM			PatchCore-1%			CCR-90%		
Category name	I-AUC	P-AUC	fps	I-AUC	P-AUC	fps	I-AUC	P-AUC	fps
Carpet	99.2	98.4	7.2	98.9	98.7	14.6	99.2	98.4	59.0
Grid	98.2	97.3	5.5	98.0	97.6	15.0	98.3	97.3	57.1
Leather	100	98.6	7.1	100	99.0	16.4	100	98.7	59.2
Tile	99.0	94.3	7.1	99.0	95.3	17.0	99.0	94.4	57.7
Wood	99.2	92.8	5.4	98.8	93.7	16.3	99.3	93.1	56.1
Avg. texture	99.1	96.3	6.5	98.9	96.9	15.9	99.2	96.4	57.8
Bottle	100	98.0	5.7	100	98.4	18.1	100	98.1	59.0
Cable	99.5	98.3	8.3	99.7	98.3	17.5	99.4	98.4	64.3
Capsule	97.0	98.5	7.7	97.4	98.7	17.8	97.3	98.6	60.5
Hazelnut	100	98.3	6.8	100	98.6	11.3	100	98.3	58.3
Metal nut	100	97.8	7.0	100	98.5	17.7	99.9	97.0	64.1
Pill	97.1	98.1	8.8	95.6	98.1	15.0	97.3	98.1	64.0
Screw	97.2	98.7	8.6	98.1	99.0	13.2	96.5	98.8	68.1
Toothbrush	100	98.5	3.3	100	98.6	35.8	100	98.5	65.9
Transistor	100	98.4	6.5	100	97.5	18.0	100	98.5	61.8
Zipper	97.8	97.5	8.3	98.9	98.1	16.7	98.3	97.7	65.2
Avg. object	98.9	98.2	7.1	99.0	98.4	18.1	98.9	98.2	63.1
Avg. all	98.9	97.6	6.9	99.0	97.9	17.4	99.0	97.6	61.4

Results and Discussion of the Experiment. The results of the comparison experiment are shown in Table 2. Table 2 shows the processing speed, Image-AUROC and Pixel-AUROC in each category.

First, we compare the accuracy of anomaly detection. For both the Texture and Object categories, the proposed method was not found to be significantly inferior to the previous method. This may indicate that the proposed method is versatile enough to be applied to various objects.

Next, average processing speeds of PaDiM and PatchCore were 6.9 fps and 17.4 fps, respectively, while the average speed of the proposed method was 61.4 fps. This indicates that the proposed method is approximately 3.5 times faster than previous methods for calculating anomalies. In the comparison of processing speeds, the idea of effective dimension reduction was valid for all objects because proposed method won in all 15 categories of MVTec AD. These results show that the proposed acceleration idea can accelerate processing time which required to obtain the anomaly score to more than 3.5 times faster than previous methods while maintaining state-of-the-art anomaly detection accuracy. In addition, the proposed method is more than twice faster than the typical video rate of 30 fps, showing results closer to the level of practical application.

4.3 Experimental Comparison with Previous Methods When the Number of Dimensions are Matched

PaDiM [13] compares dimensionality reduction using PCA with using random selection and shows that random reduction is more effective. So, in this experiment, we compare the proposed method of dimensionality reduction by PCA with previous studies of dimensionality reduction by random selection.

Experimental Setting. The dataset and feature extractor used in the experiment were the same as in the previous section. The methods compared to the proposed method are PaDiM and PatchCore. The number of dimensions of the proposed method employs feature dimensions with cumulative contributions up to 90%. Since the number of dimensions differs for each location in the feature map using this method, the number of dimensions for the other locations is unified to the maximum value of dimensions. The number of dimensions of PaDiM and PatchCore were randomly selected to match the proposed method. To account for the effect of random seeding, PaDiM and PatchCore show the average of the results of changing the seed value three times.We evaluated anomaly detection using two evaluation indices: Image-AUROC and Pixel-AUROC.

Results and Discussion of the Experiment. The results of the comparison experiment are shown in Table 3 and Fig. 5. Table 3 shows Image-AUROC and Pixel-AUROC in each category. Figure 5 shows the anomaly score maps for each of the various methods for the test images.

First, a quantitative comparison is made based on the table. Comparing the average Image-AUROCs, the proposed method had 99.0%, PaDiM had 95.8%,

Table 3. Results of comparative experiments with previous studies (AUROC [%]). In the table, I-AUC refers to Image-AUROC and P-AUC refers to Pixel-AUROC. The column dim indicates the number of dimensions after reduction.

		Previous method				Proposed method	
Method		PaDiM-Random		PatchCore-Random		CCR-90%	
Category name	dim	I-AUC	P-AUC	I-AUC	P-AUC	I-AUC	P-AUC
Carpet	102	99.0	98.2	97.7	98.4	99.2	98.4
Grid	65	90.2	92.3	89.5	93.4	98.2	97.4
Leather	96	100	98.4	100	98.9	100	98.7
Tile	106	98.7	94.4	98.9	94.9	99.0	94.4
Wood	98	97.5	92.1	97.1	91.7	99.1	93.1
Avg. texture	-	97.1	95.1	96.6	95.5	99.1	96.4
Bottle	73	100	97.3	100	97.7	100	98.0
Cable	90	95.9	96.4	94.7	96.1	99.5	98.3
Capsule	79	93.8	97.9	93.2	98.0	97.2	98.5
Hazelnut	117	97.3	97.6	99.9	98.0	100	98.3
Metal nut	85	99.7	95.8	97.8	95.9	100	97.8
Pill	102	94.2	97.2	90.6	97.0	97.1	97.9
Screw	76	81.1	95.3	88.6	97.5	97.2	98.8
Toothbrush	34	93.8	97.2	94.0	97.4	100	98.5
Transistor	83	99.1	98.1	97.0	93.9	100	98.4
Zipper	82	96.1	96.5	96.6	96.3	97.9	97.5
Avg. object	-	95.1	96.9	95.2	96.8	98.9	98.2
Avg. all	-	95.8	96.3	95.7	96.3	99.0	97.6

and PatchCore had 95.7%. In other words, previous methods have shown that a significant reduction in dimensionality reduces accuracy by more than 3.0%. In particular, the accuracy loss in the Grid, Screw, and Toohbrush categories is significant. Compared to previous methods using all dimensions and after dimensionality reduction, Grid, Screw, and Toothbrush decreased in accuracy by more than 4%, 9.5%, and 6.2%, respectively. On the other hand, the proposed method is highly accurate, averaging AUROC 99%, even when the dimension is reduced to 90% of the cumulative contribution. This indicates the efficiency of dimensionality reduction by PCA.

The qualitative results in Fig. 5 indicate that previous studies using random reduction missed anomalies or calculated low anomalies. In contrast, the proposed method correctly detects anomalies. This difference is created by the definition of abnormality. Since previous studies define anomaly score by the distance between features, accuracy is reduced when anomaly information is missing due to dimensionality reduction. The proposed method calculates the anomaly score

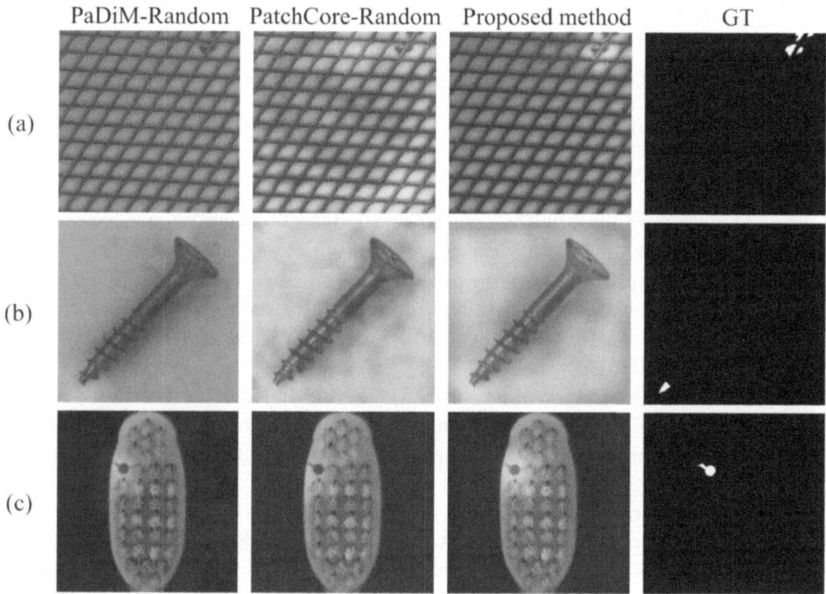

Fig. 5. Comparison of anomaly maps. This figure shows the anomaly scoremaps and the GTs for (a) Grid, (b) Screw, and (c) Toothbrush test images.

by focusing on the amount of information lost due to dimensionality reduction, thus enabling significant dimensionality reduction.

Finally, we present the limitations of this study. This study uses PCA for dimensionality reduction, and the number of dimensions after compression is determined by setting the cumulative contribution ratio. Therefore, the optimal number of dimensions for the balance between processing time and accuracy is determined experimentally, rather than automatically. In the future, it is necessary to study an algorithm that automatically calculates the optimal number of dimensions.

5 Conclusion

In this study, we proposed an anomaly detection method based on dimensionality reduction of embedded features using subspaces, aiming to speed up the process of embedding-based anomaly detection methods while maintaining accuracy. In an experiment to verify the effect of dimensionality reduction, the processing speed of anomaly computation was accelerated from 0.89 fps to 61.4 fps when all dimensions were used and when features with cumulative contributions of up to 90% were used, confirming the effectiveness of the idea. The proposed method (AUROC99.0%, 61.4 fps) was approximately 3.5 times faster than the previous methods, PaDiM (AUROC98.9%, 6.9 fps) and PatchCore(AUROC99.0%, 17.4 fps), while maintaining high anomaly detection accuracy in experiments using the MVTec AD dataset. The proposed method is more than twice faster than the typical video rate of 30 fps, achieving results close to the practical level in terms of both inspection accuracy and inspection speed.

References

1. Gong, D., et al.: Memorizing normality to detect anomaly: memory-augmented deep autoencoder for unsupervised anomaly detection. In: Proceedings of the IEEE/CVF International Conference on Computer Vision, pp. 1705–1714 (2019)
2. Liu, W., et al.: Towards visually explaining variational autoencoders. In: Proceedings of the IEEE/CVF Conference on Computer Vision and Pattern Recognition, pp. 8642–8651 (2020)
3. Schlegl, T., Seeböck, P., Waldstein, S.M., Schmidt-Erfurth, U., Langs, G.: Unsupervised anomaly detection with generative adversarial networks to guide marker discovery. In: Information Processing in Medical Imaging, vol. 10265, pp. 146–157. Springer, Cham (2017)
4. Akcay, S., Atapour-Abarghouei, A., Breckon, T.P.: GANomaly: semi-supervised anomaly detection via adversarial training. In: Jawahar, C.V., Li, H., Mori, G., Schindler, K. (eds.) ACCV 2018. LNCS, vol. 11363, pp. 622–637. Springer, Cham (2019). https://doi.org/10.1007/978-3-030-20893-6_39
5. Akçay, S., Atapour-Abarghouei, A., Breckon, T.P.: Skip-ganomaly: skip connected and adversarially trained encoder-decoder anomaly detection. In: 2019 International Joint Conference on Neural Networks (IJCNN), pp. 1–8. IEEE (2019)
6. Perera, P., Nallapati, R., Xiang, B.: OCGAN: one-class novelty detection using gans with constrained latent representations. In: Proceedings of the IEEE/CVF Conference on Computer Vision and Pattern Recognition, pp. 2898–2906 (2019)
7. Tang, T.W., Kuo, W.H., Lan, J.H., Ding, C.F., Hsu, H., Young, H.T.: Anomaly detection neural network with dual auto-encoders GAN and its industrial inspection applications. Sensors **20**(12), 3336 (2020)
8. Li, C.-L., Sohn, K., Yoon, J., Pfister, T.: CutPaste: self-supervised learning for anomaly detection and localization. In: Proceedings of the IEEE/CVF Conference on Computer Vision and Pattern Recognition, pp. 9664–9674 (2021)
9. Schlüter, H.M., Tan, J., Hou, B., Kainz, B.: Natural synthetic anomalies for self-supervised anomaly detection and localization. In: European Conference on Computer Vision, pp. 474–489 (2022)
10. Collin, A.-S., Vleeschouwer, C.D.: Improved anomaly detection by training an autoencoder with skip connections on images corrupted with Stain-shaped noise. In: 25th International Conference on Pattern Recognition, pp. 7915–7922. IEEE (2021)
11. Zavrtanik, V., Kristan, M., Skočaj, D.: DRAEM-a discriminatively trained reconstruction embedding for surface anomaly detection. In: Proceedings of the IEEE/CVF International Conference on Computer Vision, pp. 8330–8339 (2021)
12. Rippel, O., Mertens, P., Merhof, D.: Modeling the distribution of normal data in pre-trained deep features for anomaly detection. In: International Conference on Pattern Recognition, pp. 6726–6733 (2021)
13. Defard, T., Setkov, A., Loesch, A., Audigier, R.: PaDiM: a patch distribution modeling framework for anomaly detection and localization. In: International Conference on Pattern Recognition, pp. 475–489 (2021)

14. Rudolph, M., Wandt, B., Rosenhahn, B.: Same same but differnet: semi-supervised defect detection with normalizing flows. In: Proceedings of the IEEE/CVF Winter Conference on Applications of Computer Vision, pp. 1907–1916 (2021)
15. Bergman, L., Cohen, N., Hoshen, Y.: Deep nearest neighbor anomaly detection, arXiv preprint arXiv:2002.10445 (2020)
16. Cohen, N., Hoshen, Y.: Sub-image anomaly detection with deep pyramid correspondences, arXiv preprint arXiv:2005.02357 (2020)
17. Roth, K., Pemula, L., Zepeda, J., Schölkopf, B., Brox, T., Gehler, P.: Towards total recall in industrial anomaly detection. In: Proceedings of the IEEE/CVF Conference on Computer Vision and Pattern Recognition, pp. 14318–14328 (2022)
18. Bergmann, P., Batzner, K., Fauser, M., Sattlegger, D., Steger, C.: The MVTec anomaly detection dataset: a comprehensive real-world dataset for unsupervised anomaly detection. Int. J. Comput. Vision **129**(4), 1038–1059 (2021)
19. Deng, J., Dong, W., Socher, R., Li, L.J., Li, K., Fei-Fei, L.: Imagenet: a large-scale hierarchical image database. In: 2009 IEEE Conference on Computer Vision and Pattern Recognition, pp. 248–255. IEEE (2009)
20. Zagoruyko, S.: Wide residual networks, arXiv preprint arXiv:1605.07146 (2016)

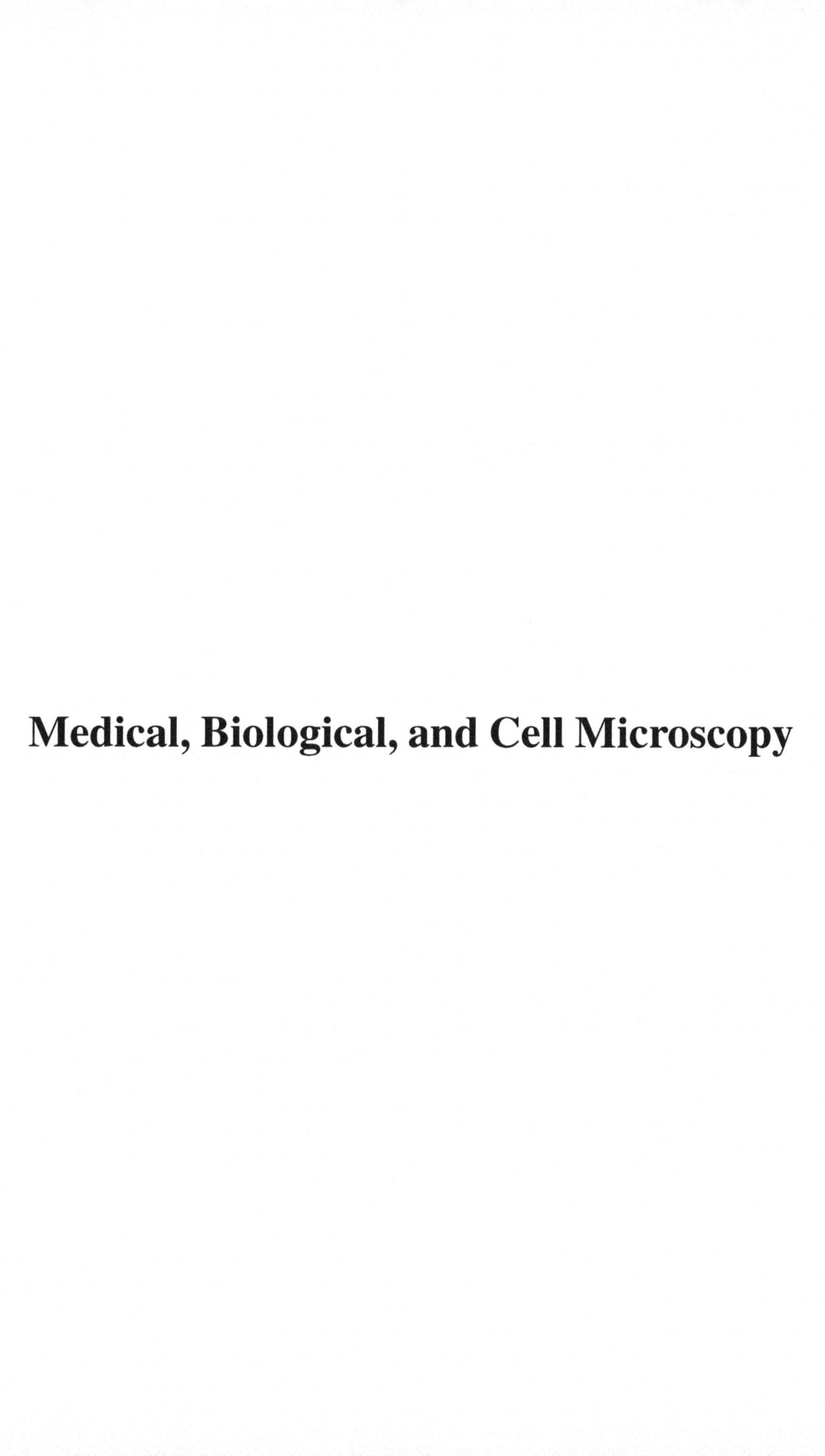

Medical, Biological, and Cell Microscopy

Active Learning with nnUNet for Coronary Artery Lumen Segmentation Using a Centerline Prior

Anna Bøgevang Ekner[1], Mathias Micheelsen Lowes[1], Rasmus R. Paulsen[1], Klaus Fuglsang Kofoed[2], Andreas Ohrt Johansen[2], Kristine Aavild Sørensen[1,3], and Josefine Vilsbøll Sundgaard[1,3](✉)

[1] Technical University of Denmark, Lyngby, Denmark
josh@dtu.dk

[2] The Heart Center, Rigshospitalet, University of Copenhagen, Copenhagen, Denmark

[3] Novo Nordisk A/S, Bagsværd, Denmark

Abstract. Annotating medical images for segmentation is both costly and time-consuming, making it crucial to identify the most informative images for annotation. Active learning aims to address this challenge by selecting samples that maximize model performance while minimizing labeling effort. This paper presents an active learning framework that incorporates an anatomical prior for coronary artery segmentation, using nnUNet as the segmentation model. We introduce two novel centerline-based sampling strategies, Lowest Weighted Overlap (LWOV) and Highest Weighted Overlap (HWOV), designed to enhance structural consistency in model predictions. The method is evaluated on Left Anterior Descending (LAD) artery segmentation from Computed Tomography (CT) images. Our results show that although all the active learning strategies evaluated performed well with marginal differences, random sampling achieved the highest performance, highlighting the challenges of designing optimal selection strategies. Furthermore, we demonstrate that with only 16.6% of the available data, we achieve segmentation accuracy comparable to training on the full dataset.

Keywords: Active learning · Coronary artery segmentation · nnUnet

1 Introduction

Cardiovascular diseases are the leading cause of death worldwide and represent 32% of global deaths annually [18]. Coronary artery disease (CAD) is a common type of cardiovascular disease caused by a build-up of fat, cholesterol, and other substances in the artery walls. The coronary arteries are vessels that supply blood and provide essential oxygen and nutrients to the myocardium (the heart muscle). In stable CAD, coronary vessel narrowing causes reversible myocardial ischemia due to a mismatch of inadequate myocardial perfusion compared to increased demands in situations such as physical exertion or emotion. Chest

J. Petersen and V. A. Dahl (Eds.): SCIA 2025, LNCS 15726, pp. 227–239, 2025.
https://doi.org/10.1007/978-3-031-95918-9_16

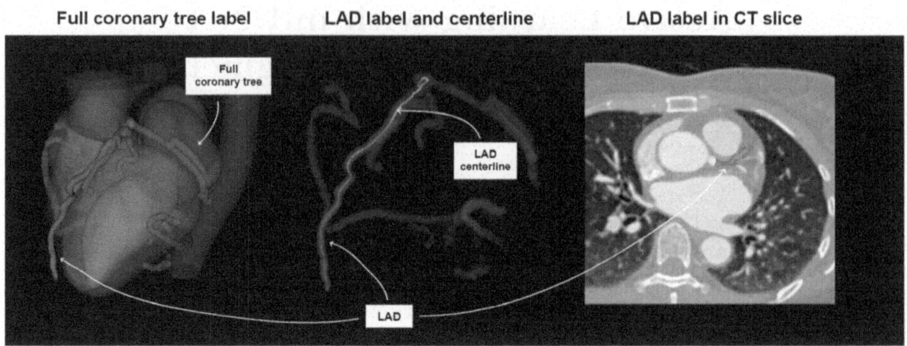

Fig. 1. Visualization of the coronary artery segmentation. From left to right: **(Left)** Volume rendering of the entire coronary tree overlaid on the whole heart, **(Middle)** LAD artery segmentation with its associated centerline highlighted within the coronary tree, **(Right)** Cross-sectional view of the LAD segmentation within a CT slice.

discomfort, also known as angina pectoris, is one of the primary symptoms of CAD. Typical symptoms include pain in the chest, arm, or jaw, characterized as dull or heavy, but symptoms can also present as sharp or burning pain in the upper abdomen or back, along with fatigue, shortness of breath, and nausea, making the diagnostic process and medical evaluation difficult [1]. Silent myocardial ischemia is defined as asymptomatic CAD which further complicates the diagnostic process and is particularly observed in elderly patients, patients with diabetes mellitus, and perioperative patients [12].

Coronary Computed Tomography Angiography (CCTA) is recommended as the preferred diagnostic modality in patients with a low or moderate pre-test likelihood of stable CAD [13]. Using CCTA allows for a manual assessment of the site and degree of stenosis in the coronary arteries. However, such manual assessment is time consuming and prone to misdiagnosis, which has led to an interest in automatic coronary artery segmentation [4]. Figure 1 shows the coronary artery tree in relation to the heart.

The coronary arteries originate from the aortic root and consist of the right and left main coronary arteries (RCA and LMCA). The LMCA divides into the left anterior descending artery (LAD) and the left circumflex artery (LCX). The main coronary arteries are generally between 3 and 5 mm in diameter and branch into vessels with decreasing diameter [2]. The LAD is the largest coronary artery responsible for supplying nearly 50% of myocardial perfusion and provides two sets of branches: diagonal branches that supply the anterior-lateral left ventricle and septal branches that supply the inter-ventricular septum [16]. The thin vessel and the large variation in where and when the arteries branch make it difficult to accurately track the arteries and assess their volumes. Current guidelines define patients at high risk for adverse events if the CCTA indicates $\geq$50% LMCA stenosis, $\geq$70% proximal three-vessel CAD, or single- or two-vessel CAD with

$\geq$70% proximal LAD stenosis [13], thus emphasizing the importance of accurate LAD segmentation in CAD assessment.

Automatic coronary artery tracing and segmentation has been an active research field for more than two decades, and several commercial solutions exists. Very broadly, the methods can be divided into three categories: region growing, graph-based, and segmentation-based. Region growing is typically based on a level set that starts at a seed point and grows the segmentation until a local criterion stops the evolution [22]. The graph based methods typically recursively trace centerlines from a starting point and keep track of branch points [10,17]. The segmentation-based methods are typically variants of convolutional neural networks that assign voxel labels based on probabilities. The field is large and there exist a variety of hybrid approaches. A brief overview can be found in [23]. Although the choice of method is abundant, the availability of high-quality data is sparse due to data privacy concerns. In this paper, we are interested in the case where an institution has access to a large, but uncurated, dataset and would like to find the optimal way to trace and segment the coronary arteries in this set. For simplification, we focus solely on LAD, however, the approach generalizes to the whole coronary artery tree.

Active learning attempts to accelerate model training by iteratively selecting the most informative samples in an unlabeled data set for labeling, with the purpose of reducing the cost of labeling while maintaining performance [9]. The selected sampling strategy thus heavily influences the effectiveness and performance of an active learning training process. Sampling strategies based on uncertainty, diversity, and representativeness have primarily been developed for image classification. For example, uncertainty-based approaches utilize the idea that a lower model certainty for a specific example indicates that using the example for training will add significant information to the classifier [19]. A measurement of uncertainty sampling can, for example, be entropy [11,20]. Cost-effective active learning [14] enriches the data set by assigning pseudo-labels to the samples with high confidence in addition to the labeled examples. However, these methods are not directly applicable in image segmentation. Yang et al. [21] suggests uncertainty-based active learning to identify the most effective annotation areas in medical images. Contrary, Zhao et al. [24] defines a sampling criteria based on the segmentation performance of the hidden layers in the segmentation model compared to the output of the final layer under the assumption that the higher performing samples will be more informative samples. A challenging aspect of active learning is that the effectiveness of the training process depends on data preprocessing, model architecture, training procedure, and requires robust models [7]. To ensure robust models and limit the effect of hyperparameter tuning etc. in the training process, Föllmer et al. [3] proposed to employ nnUNet [5] in an active learning setting to increase reproducibility. The nnUNet is a self-configuring pipeline for segmentation tasks in both 2D and 3D. It has become a standard in segmentation tasks, due to its ability to make robust design choices in both preprocessing, model architecture, and hyperparameters, while achieving state-of-the-art performance [8]. Föllmer et al. [3] show that using nnUNet in an active learning setting is an effective strategy to reduce labeling costs and avoid the cumbersome configuration of the training process. They furthermore propose the Uncertainty-aware Submodular mutual Information Measure (USIM) criteria for active learning sampling, showing superior results for 2D segmentation.

In this paper, we extend the work by Föllmer et al. [3] to the task of coronary artery segmentation. Our main contributions are

- An active learning framework that incorporates an anatomical prior for coronary artery segmentation.
- Evaluation of existing sample selection strategies for active learning.
- Proposal of two novel centerline-overlap-based sample selection strategies: Lowest Weighted Overlap (LWOV) and Highest Weighted Overlap (HWOV).
- Validation of the proposed method on a publicly available coronary CT angiography dataset of varying data and annotation quality.

2 Data

The active learning pipeline is evaluated using coronary CT angiography (CTA) images from the publicly available ImageCAS dataset [23]. The dataset consists of 1000 3D CTA scans with annotations of the coronary artery tree, including key branches such as the LAD, LCX, and RCA arteries. In this study, we focus on the LAD and its centerline. The centerlines of the entire left coronary tree are computed using the Vascular Modeling Toolkit (VMTK) [6]. Secondly, the LAD is identified as the centerline that has the smallest average distance to the right ventricle. The segmentation of the right ventricle is found using the TotalSegmentator segmentation framework [15]. In the following, the centerline computed using VMTK is defined as the *reference centerline*. From the full coronary tree, ground truth labels for the LAD artery were derived by dilating the estimated LAD reference centerline and identifying the overlap with the reference coronary tree segmentation. Some scans were found to have incomplete or noisy coronary tree labels, often due to under- or oversegmentation, and the extraction of LAD centerlines occasionally lead to inaccuracies. In general, the ImageCAS dataset contains scans and segmentations of varying quality. The scans are acquired with a scanner that does not cover the entire heart and a heart scan is therefore stitched together based on several sub-scans acquired over multiple heart beats. This can lead to step artifacts, where there is physical movement between the sub-scans or that the contrast agent is partly washed out in the later stage sub-scans leading to contrast jumps. See Fig. 2 for an example. Therefore, the dataset was filtered based on the quality of the reference centerline, and a refined subset of 884 scans was used for our work. The data was split into 100 images and segmentations used to evaluate the segmentation accuracy of the models. The active learning pipeline is trained on the remaining 784 images. In the following, we assume that a state-of-the-art centerline extraction tool as for example [17] can provide a good centerline estimate on previously unseen scans.

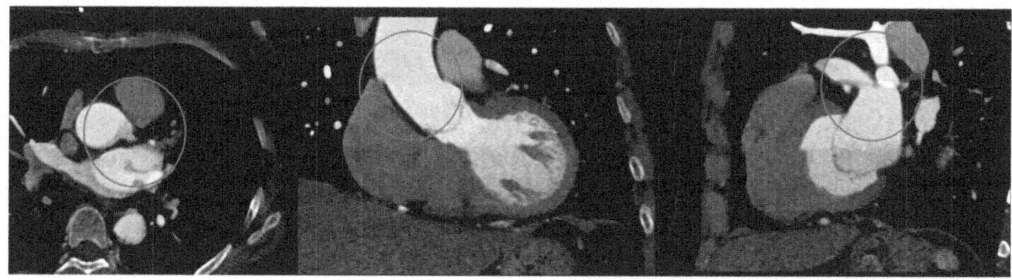

Fig. 2. An example scan from the ImageCAS dataset with severe step artefacts. It can be seen that the scan is divided close to the aortic annulus where the coronary arteries originates (red circles). The LAD is cut by the step artefact and a continuous centerline can not be extracted. (Color figure online)

3 Method

3.1 Active Learning with the NnUNet Segmentation Model

Initially, we consider the full 784 images in the training set as an unlabeled data pool $\mathcal{U} = \{\mathbf{x}_j\}_{j=1}^{N_U}$, where $\mathbf{x} \in \mathbb{R}^{H \times W \times D}$ represents a 3D CT scan. To initiate the active learning pipeline, we include the binary segmentation mask $\mathbf{y} \in \mathbb{Z}^{H \times W \times D}$ from a small, randomly selected subset of the training data and denote this our initial labeled dataset $\mathcal{L}_0 = \{(\mathbf{x}_j, \mathbf{y}_j)\}_{j=1}^{N_{L_0}}$. In the first active learning iteration, a segmentation model is trained with the labeled data in $\mathcal{L}_0$ and the model is applied to the remaining unlabeled data $\mathcal{U} = \{\mathbf{x}_j\}_{j=1}^{N_U - N_{L_0}}$. As a part of the active learning pipeline the labeled data $\mathcal{L} = \{(\mathbf{x}_j, \mathbf{y}_j)\}_{j=1}^{N_L}$ is iteratively expanded with images from the unlabeled data pool $\mathcal{U}$ and their corresponding segmentations until an acceptable performance is reached.

Figure 3 shows an overview of the active learning pipeline with the nnUNet segmentation model. At each iteration of the active learning pipeline, the nnUNet 3D full resolution model [5] is used as the segmentation model. The labeled dataset is processed using the nnUNet "plan and preprocessing" function to automatically configure the model architecture and training parameters. The configured model is trained on the labeled dataset $\mathcal{L}$ and used to generate predictions $\mathbf{y}_j^*$ for the unlabeled pool $\mathcal{U}$. These predictions are evaluated using the metrics outlined in Sect. 3.2 to select a batch $\mathcal{B} \subseteq U$ of the $N_{\mathcal{B}}$ most informative samples. The selected samples are annotated (simulated using pre-existing annotations), and the corresponding labeled data $\{(\mathbf{x}_j, \mathbf{y}_j)\}_{j \in \mathcal{B}}$ is added to the labeled dataset $\mathcal{L}$, expanding it for the next iteration. To prevent bias towards samples selected early in the process, the segmentation model is retrained from scratch at each active learning iteration. The performance of the model is evaluated using the test set after each iteration. This iterative pipeline of preprocessing, training, sample selection, and annotation continues for a specified number of iterations K.

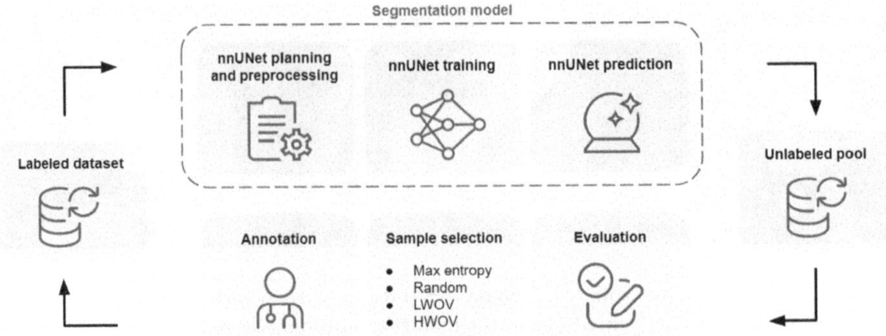

Fig. 3. Active learning pipeline with nnUNet as the segmentation model. The nnUNet model is iteratively self-configured and trained on the labeled dataset, then used to identify and select the most informative samples from the unlabeled pool based on a selection strategy, which are annotated and incorporated into the labeled dataset.

3.2 Sampling Strategy Based on Centerline Overlap Measure

We propose a sampling strategy based on a centerline overlap measure. Each data point $\mathbf{x}_j$ is associated with an estimated reference centerline computed using VMTK, $\mathbf{c}_{\text{ref},j}$. This reference centerline is compared to the segmentation-derived centerline, $\mathbf{c}_{\text{pred},j}$, which is obtained by skeletonizing the predicted segmentation $\mathbf{y}_j^*$. The overlap between $\mathbf{c}_{\text{ref},j}$ and $\mathbf{c}_{\text{pred},j}$ is measured through a point-to-point correspondence between the centerlines, as described by Schaap et al. [10]. For each point on $\mathbf{c}_{\text{ref},j}$ the closest point on $\mathbf{c}_{\text{pred},j}$ is identified. If the distance between these points is less than a predefined radius r_{OV}, the reference point is classified as a true positive for the reference (TPR_j). Otherwise, it is considered a false negative (FN_j). Similarly, for each point on $\mathbf{c}_{\text{pred},j}$ the closest point on $\mathbf{c}_{\text{ref},j}$ is identified. If the distance between these points is less than r_{OV} the point on $\mathbf{c}_{\text{pred},j}$ is a true positive for the prediction (TPP_j) else it is a false positive (FP_j). With these definitions we can define the overlap measure (OV) as

$$\text{OV}_j = \frac{\|\text{TPP}_j\| + \|\text{TPR}_j\|}{\|\text{TPP}_j\| + \|\text{TPR}_j\| + \|\text{FN}_j\| + \|\text{FP}_j\|}, \tag{1}$$

where $\|\cdot\|$ denotes the number of points in the category.

To account for the structural consistency of the predicted segmentation, we extend the OV measure by introducing a weighted overlap (WOV) measure. This weighting penalizes predictions with multiple connected components, which typically indicate fragmentation. The WOV is defined as

$$\text{WOV}_j = \text{OV}_j \cdot (1 - (N_{cc,j} - 1) \cdot 0.1), \tag{2}$$

where $N_{cc,j}$ represents the number of connected components in $\mathbf{y}_j^*$ that exceed a specified size threshold. For accurate segmentations, $N_{cc,j} = 1$, making the WOV

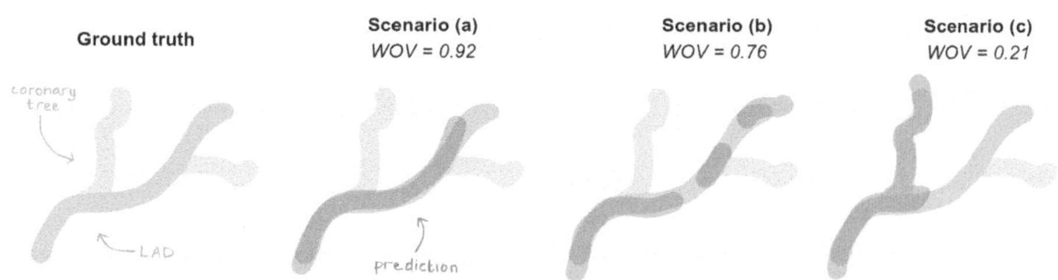

Fig. 4. Examples of predicted segmentations with their corresponding WOV scores. The scenarios illustrate: **(a)** an accurate, continuous segmentation of the LAD artery, **(b)** a fragmented segmentation of the LAD artery, and **(c)** an inaccurate segmentation that follows a different artery in the coronary tree.

equivalent to the OV. Predictions with $N_{cc,j} > 1$ are thus penalized proportionally to the number of connected components. This connected-component-based weighting ensures that segmentations with fragmented arteries receive lower scores. Examples of different scenarios and their corresponding WOV scores are sketched in Fig. 4.

We propose two novel sampling strategies based on the WOV measure for selecting a batch $\mathcal{B} \subseteq \mathcal{U}$ of the N_B most informative samples for annotation. The first strategy, lowest-WOV (LWOV) sampling, selects the N_B samples from the unlabeled pool with the lowest WOV scores, targeting samples where the model performs poorly. The selection process is defined as

$$\mathcal{B}_{\text{lowest}} = \underset{\mathcal{B} \subseteq \mathcal{U}, |\mathcal{B}| = N_B}{\arg\min} \ \text{WOV}(\mathbf{x}_j), \quad \forall \mathbf{x}_j \in \mathcal{B}. \tag{3}$$

The second strategy, highest-WOV (HWOV) sampling, selects the N_B samples from the unlabeled pool with the highest WOV scores. Hence, prioritizing samples where the model performs well. It is defined as

$$\mathcal{B}_{\text{highest}} = \underset{\mathcal{B} \subseteq \mathcal{U}, |\mathcal{B}| = N_B}{\arg\max} \ \text{WOV}(\mathbf{x}_j), \quad \forall \mathbf{x}_j \in \mathcal{B}. \tag{4}$$

These strategies represent two contrasting approaches to improving model performance. HWOV sampling focuses on reinforcing the model's strength by including its most confident predictions, while LWOV sampling aims to address the model's weaknesses by exposing it to its most challenging cases.

4 Results

The HWOV and LWOV sampling strategies outlined in Sect. 3.2 are evaluated alongside random sampling, maximum entropy sampling [11], and a baseline model trained on the full dataset. All experiments start with the same small initial labeled dataset, $\mathcal{L}_0$, containing $N_{L_0} = 5$ randomly selected samples, as well as a fixed test set of 100 samples. The WOV scores used for our sampling

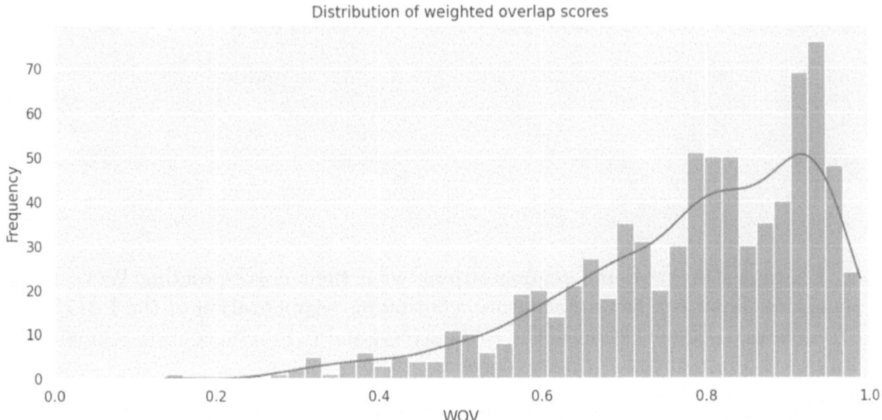

Fig. 5. Histogram of WOV scores for the full unlabeled pool after the first active learning iteration.

strategies of the full unlabeled pool after the first active learning iteration are found in Fig. 5. In each iteration we select $N_B = 25$ additional samples from the unlabeled data pool based on the respective sampling strategy, and add these samples together with their segmentation mask to the labeled training data. The active learning pipeline runs for $K = 6$ iterations, where the segmentation accuracy is on par with what was achieved with the full dataset. The code is available at https://github.com/annaekner/Active-Learning-Coronary-Arteries.

Figure 6 shows the model performance for all selection strategies based on the WOV measure across the six active learning iterations. Table 1 shows multiple evaluation metrics for each selection strategy in the final active learning iteration. The results show that all active learning sampling strategies achieve a mean WOV value in the final iteration within the range of [0.870, 0.885], closely matching the baseline model trained on the full dataset, which achieves a mean WOV

Table 1. Evaluation results for the metrics: DICE score, weighted overlap (WOV), Hausdorff distance, and number of connected components (N_{cc}). The table reports the mean values (± standard deviation) from the final iteration of the active learning pipeline. Arrows indicate whether a higher (↑) or lower (↓) score is better.

	Dice ↑	**WOV ↑**	**Hausdorff dist. ↓**	N_{cc} ↓
LWOV	$0.874 \pm 5.1 \cdot 10^{-2}$	$0.872 \pm 9.9 \cdot 10^{-2}$	12.78 ± 11.66	1.260 ± 0.482
HWOV	$0.872 \pm 5.6 \cdot 10^{-2}$	$0.870 \pm 1.0 \cdot 10^{-1}$	14.63 ± 13.71	1.270 ± 0.526
Random	$0.879 \pm 5.3 \cdot 10^{-2}$	$0.885 \pm 9.2 \cdot 10^{-2}$	11.36 ± 11.29	1.230 ± 0.466
Max entropy	$0.874 \pm 5.2 \cdot 10^{-2}$	$0.875 \pm 1.0 \cdot 10^{-1}$	12.79 ± 12.00	1.280 ± 0.531
Full dataset	$0.882 \pm 4.9 \cdot 10^{-2}$	$0.884 \pm 9.5 \cdot 10^{-2}$	11.29 ± 11.08	1.230 ± 0.526

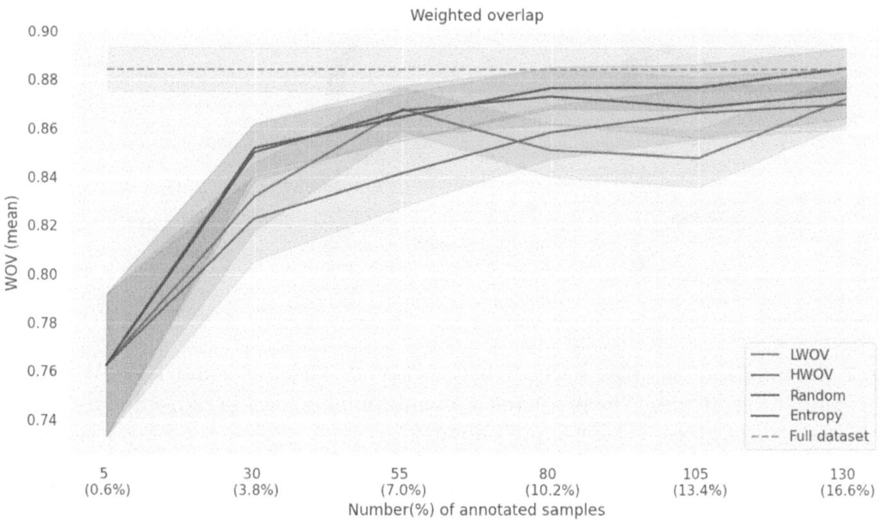

Fig. 6. Comparison of model performance across the active learning sampling strategies: lowest weighted overlap (LWOV), highest weighted overlap (HWOV), random, max entropy, and the fully labeled dataset. Performance is shown with respect to the number of annotated samples in each iteration, and the corresponding fraction of the full dataset that these samples represent.

of 0.884. Notably, the random sampling strategy slightly exceeds the baseline performance. This suggests that even when trained on just 130 samples (corresponding to 16.6% of the entire dataset), the active learning strategies achieve good performance.

Among active learning sampling strategies, the random sampling strategy outperformed all others in every metric listed in Table 1. Although the WOV-based sampling strategies performed worse than the max entropy sampling in terms of the WOV measure, they both outperform max entropy sampling in the number of connected components.

Figure 7 displays examples of samples selected by the LWOV and HWOV strategies, showing the predicted segmentation (red) overlaid on the ground truth (green). It is evident that samples with the lowest WOV scores typically consist of multiple disconnected components, suggesting that the segmentation model struggles to correctly identify and connect the LAD artery across CT slices. In contrast, samples with the highest WOV scores show near-perfect predicted segmentations.

5 Discussion

Among the active learning strategies, the random sampling strategy achieved the highest performance, although the differences between the strategies were

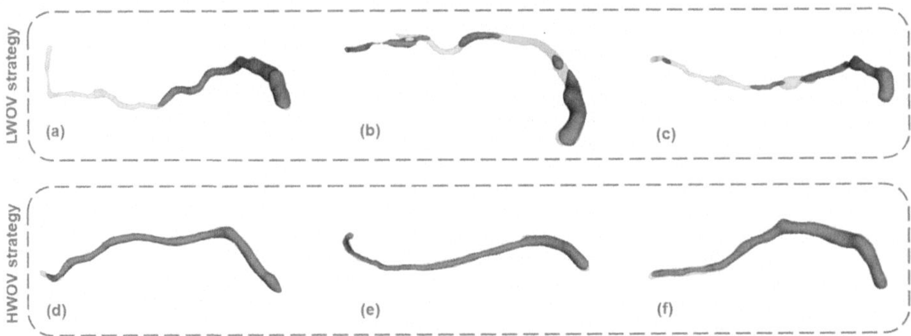

Fig. 7. Examples of samples selected by the LWOV (*top row*) and HWOW (*bottom row*) sampling strategies. The predicted segmentation is shown in red, while the ground truth is shown in green. (Color figure online)

marginal. Notably, all strategies achieved strong results and are all comparable to the performance of a model trained on the entire dataset, even when using only 130 samples (16.6% of the dataset). This shows that active learning can significantly reduce the amount of labeled data required while still achieving good performance for the task of coronary artery segmentation.

In active learning, the optimal sampling strategy depends on the specific task. Structural consistency is crucial for coronary artery segmentation, as we know that the LAD artery should form a single connected component. By targeting structural issues, LWOV effectively prioritizes learning from challenging and fragmented samples and ultimately achieves the lowest number of connected components in model predictions compared to other strategies, aside from random sampling. LWOV achieves comparable Dice and Hausdorff distance performance to max entropy sampling, which is a widely used sampling strategy in active learning, while outperforming on number of connected components, which is a key performance metric for our task. However, because LWOV iteratively selects the worst predictions for retraining, it is particularly sensitive to annotation errors or inaccuracies in the dataset. The ImageCAS dataset contains varying annotation quality, which led to filtering the dataset from 1000 to 884 samples. While this may have improved label reliability, it also reduced variability, potentially affecting the generalizability of the findings. Additionally, the accuracy of the reference centerline estimates used in LWOV and HWOV is crucial, as errors in these estimates could misguide the selection process and lead to suboptimal model improvements. Our centerline-based strategies assume the availability of a reliable reference centerline estimate for both training and testing. In practice, this assumption may not always hold, particularly if the tools used to extract centerlines are not robust. These factors highlight the importance of prioritizing label quality over quantity when allocating resources, while also ensuring reliable reference data. Given that impressive performance can be achieved with a small fraction of labeled data, ensuring accurate annotations is essential to maximize the benefits of active learning strategies.

The effectiveness of these active learning strategies also depends on the anatomical variation of the structure of interest. For structures with more variation than the LAD artery, LWOV might struggle by focusing too much on outliers, while HWOV might fail to recognize uncommon cases. Balancing these trade-offs remains a key challenge for active learning. Since LWOV and HWOV represent opposite extremes of sampling based on the centerline WOV metric, their limitations suggest that a hybrid selection strategy combining elements of both could provide a more balanced approach. Although active learning shows potential for medical imaging applications, it can be challenging to consistently outperform random sampling, especially when there is high anatomical variability and limited annotated samples.

6 Conclusion

In this paper, we proposed an active learning framework that incorporates an anatomical prior while leveraging the self-configuring nnUNet as segmentation model. We evaluated the framework on the large-scale, publicly available ImageCAS dataset. In addition to evaluating existing sampling strategies, we proposed two novel centerline-overlap-based strategies, LWOV and HWOV, which prioritize structural consistency in the segmentation process. Among the strategies evaluated, random sampling achieved the best performance; however, we show that our centerline-overlap-based strategies ensure structural consistency in coronary artery segmentation compared to max entropy sampling. Our findings demonstrate that active learning can effectively reduce the amount of required labeled data, as only 16.6% of the available data was needed to achieve the same performance as a network trained on the full dataset.

References

1. DeVon, H.A., Mirzaei, S., Zègre-Hemsey, J.: Typical and atypical symptoms of acute coronary syndrome: time to retire the terms? J. Am. Heart Assoc. **9**(7), e015539 (2020). https://doi.org/10.1161/JAHA.119.015539
2. Dodge Jr., J.T., Brown, B.G., Bolson, E.L., Dodge, H.T.: Lumen diameter of normal human coronary arteries. Influence of age, sex, anatomic variation, and left ventricular hypertrophy or dilation. Circulation **86**(1), 232–246 (1992)
3. Föllmer, B., Schulze, K., Wald, C., Stober, S., Samek, W., Dewey, M.: Active learning with the nnUNet and sample selection with uncertainty-aware submodular mutual information measure. In: Medical Imaging with Deep Learning (2024)
4. Ghekiere, O., et al.: Image quality in coronary CT angiography: challenges and technical solutions. Br. J. Radiol. **90**(1072) (2017). https://doi.org/10.1259/bjr.20160567
5. Isensee, F., Jaeger, P.F., Kohl, S.A., Petersen, J., Maier-Hein, K.H.: nnU-net: a self-configuring method for deep learning-based biomedical image segmentation. Nat. Methods **18**(2), 203–211 (2021)

6. Izzo, R., Steinman, D., Manini, S., Antiga, L.: The vascular modeling toolkit: a python library for the analysis of tubular structures in medical images. J. Open Sour. Softw. **3**(25), 745 (2018)
7. Ji, Y., Kaestner, D., Wirth, O., Wressnegger, C.: Randomness is the root of all evil: more reliable evaluation of deep active learning. In: Proceedings of the IEEE/CVF Winter Conference on Applications of Computer Vision, pp. 3943–3952 (2023)
8. Munjal, P., Hayat, N., Hayat, M., Sourati, J., Khan, S.: Towards robust and reproducible active learning using neural networks. In: Proceedings of the IEEE/CVF Conference on Computer Vision and Pattern Recognition, pp. 223–232 (2022)
9. Ren, P., et al.: A survey of deep active learning. arXiv preprint arXiv:2009.00236 (2020)
10. Schaap, M., et al.: Standardized evaluation methodology and reference database for evaluating coronary artery centerline extraction algorithms. Med. Image Anal. **13**(5), 701–714 (2009)
11. Shannon, C.E.: A mathematical theory of communication. ACM SIGMOBILE Mob. Comput. Commun. Rev. **5**(1), 3–55 (2001)
12. Theofilis, P., et al.: Silent myocardial ischemia: from pathophysiology to diagnosis and treatment. Biomedicines **12**(2) (2024). https://doi.org/10.3390/biomedicines12020259
13. Vrints, C., et al.: 2024 esc guidelines for the management of chronic coronary syndromes: developed by the task force for the management of chronic coronary syndromes of the European society of cardiology (ESC) endorsed by the European association for cardio-thoracic surgery (EACTS). Eur. Heart J. **45**(36), 3415–3537 (2024)
14. Wang, K., Zhang, D., Li, Y., Zhang, R., Lin, L.: Cost-effective active learning for deep image classification. IEEE Trans. Circuits Syst. Video Technol. **27**(12), 2591–2600 (2016)
15. Wasserthal, J., et al.: Totalsegmentator: robust segmentation of 104 anatomic structures in CT images. Radiol.: Artif. Intell. **5**(5) (2023)
16. Weber, C., Brown, K.N., Borger, J.: Anatomy, Thorax, Heart Anomalous Left Anterior Descending (LAD) Artery. StatPearls Publishing (2019)
17. Wolterink, J.M., van Hamersvelt, R.W., Viergever, M.A., Leiner, T., Išgum, I.: Coronary artery centerline extraction in cardiac CT angiography using a CNN-based orientation classifier. Med. Image Anal. **51**, 46–60 (2019)
18. World Health Organization: Cardiovascular diseases (CVDs). https://www.who.int/news-room/fact-sheets/detail/cardiovascular-diseases-(cvds). Accessed 09 Dec 2024
19. Wu, J., et al.: Multi-label active learning algorithms for image classification: overview and future promise. ACM Comput. Surv. (CSUR) **53**(2), 1–35 (2020)
20. Wu, J., Chen, J., Huang, D.: Entropy-based active learning for object detection with progressive diversity constraint. In: Proceedings of the IEEE/CVF Conference on Computer Vision and Pattern Recognition, pp. 9397–9406 (2022)
21. Yang, L., Zhang, Y., Chen, J., Zhang, S., Chen, D.Z.: Suggestive annotation: a deep active learning framework for biomedical image segmentation. In: Medical Image Computing and Computer Assisted Intervention- MICCAI 2017: 20th International Conference, Quebec City, QC, Canada, 11–13 September 2017, Proceedings, Part III 20, pp. 399–407. Springer (2017)
22. Yang, Y., Tannenbaum, A., Giddens, D., Stillman, A.: Automatic segmentation of coronary arteries using Bayesian driven implicit surfaces. In: 2007 4th IEEE International Symposium on Biomedical Imaging: From Nano to Macro, pp. 189–192. IEEE (2007)

23. Zeng, A., et al.: ImageCAS: a large-scale dataset and benchmark for coronary artery segmentation based on computed tomography angiography images. Comput. Med. Imaging Graph. **109**, 102287 (2023)
24. Zhao, Z., Yang, X., Veeravalli, B., Zeng, Z.: Deeply supervised active learning for finger bones segmentation. In: 2020 42nd Annual International Conference of the IEEE Engineering in Medicine & Biology Society (EMBC), pp. 1620–1623. IEEE (2020)

Automated Cardiac Adipose Tissue Segmentation in Computed Tomography: A Literature Review

Andreas W. Aspe[1(✉)], Jonas Jalili Pedersen[2], Andreas Ohrt Johansen[2], Klaus Fuglsang Kofoed[2], Kristine Aavild Sørensen[1,3], Rasmus Reinhold Paulsen[1], and Josefine Vilsbøll Sundgaard[1,3]

[1] DTU Compute, Technical University of Denmark, Kongens Lyngby, Denmark
awias@dtu.dk
[2] Heart Centre, Rigshospitalet, Copenhagen, Denmark
[3] Novo Nordisk A/S, Søborg, Denmark

Abstract. This review provides an overview of recent advancements in automated segmentation methods on Computed Tomography (CT) for two types of cardiac fat: Epicardial adipose Tissue (EAT) and Pericardial Adipose Tissue (PAT). These fat deposits, separated by the pericardium, have been linked to various cardiovascular diseases, with EAT receiving the most research attention. Their complex anatomical context makes manual quantification highly time-consuming and prone to considerable inter-observer variability. Automated methods effectively address these complications, offering a more efficient and consistent solution. This study encompasses a broad range of methods, spanning AI as well as non-AI approaches. Additionally, it presents the remaining challenges, including the need for larger annotated public datasets and optimized attenuation thresholds for contrast-enhanced CT. It is demonstrated that automated methods are able to achieve segmentation results comparable to the quality of human annotation, proving their potential as a clinical tool for discovering new biomarkers and enhancing patient outcomes.

Keywords: Epicardial Adipose Tissue · Pericardial Adipose Tissue · Review · AI · Medical Imaging · Deep Learning

1 Introduction

Cardiovascular Diseases (CVDs) represent the leading cause of mortality and disability, accounting for more than 42.5% of all annual deaths in the European Region [51]. Major CVDs include ischemic heart disease, stroke, and cardiac arrhythmogenic disorders [28]. Adiposity and obesity play a central role in cardiovascular health and disease by influencing systemic metabolism, vascular function, and inflammatory processes [33]. Adiposity is commonly assessed using Body Mass Index (BMI), a widely used but simplistic measure that does

The original version of the chapter has been revised. A correction to this chapter can be found at https://doi.org/10.1007/978-3-031-95918-9_29

J. Petersen and V. A. Dahl (Eds.): SCIA 2025, LNCS 15726, pp. 240–253, 2026.
https://doi.org/10.1007/978-3-031-95918-9_17

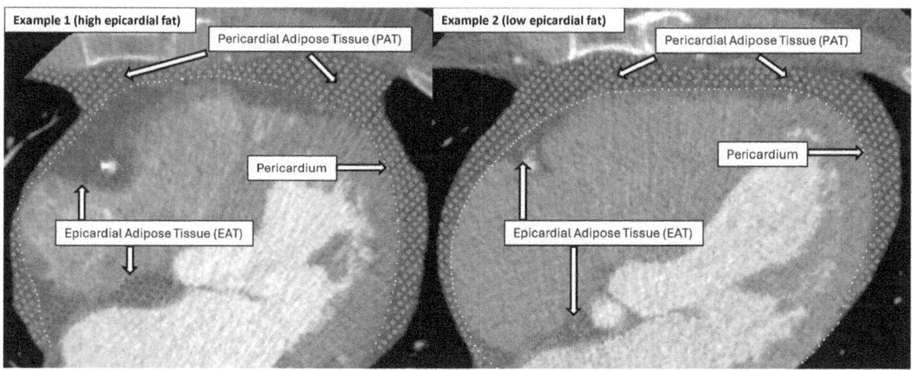

Fig. 1. The anatomy of EAT, PAT and pericardium on CCTA [12].

not account for fat distribution in the body. Adipose tissue may be further categorized according to anatomical location, typically separated into subcutaneous adipose tissue underneath the skin and visceral adipose tissue surrounding the inner organs. Recent advances in medical imaging allow for quantification of visceral adipose tissue volumes. Among these, cardiac adipose tissue has been a matter of growing interest since pioneering research described the anatomy and clinical relationship with the heart early in this century [20]. Adipose tissue located underneath the pericardium, Epicardial Adipose Tissue (EAT), has been associated with cardiac diseases including heart failure, ischemic heart disease, and atrial arrhythmogenic disorder [19,22]. Although potentially less clinically prominent, Pericardial Adipose Tissue (PAT), situated above the pericardium, has also been associated with atrial fibrillation and inflammation [14].

Computed tomography (CT) has been established as an important clinical tool in the assessment of cardiac disease [48]. Modern CT is standardized in clinical practice as a time-efficient image acquisition method, delivering low radiation doses. Contrast-enhanced Cardiac CT Angiography (CCTA) is an effective imaging modality for cardiac segmentation, offering superior visualization of cardiac structures such as the pericardium, also compared to other image modalities such as MRI due to higher resolution [25,29]. Figure 1 shows CCTA scans from two patients in the Copenhagen General Population Study (CGPS) [12] with EAT and PAT highlighted in red and blue, respectively. Today, segmenting EAT and PAT from CT largely relies on labor-intensive visual qualitative and semi-quantitative assessments with notable limitations, including observer variability [3,13].

Recently, Artificial Intelligence (AI) has enabled automated analysis of EAT and PAT, decreasing manual labor while showing competitive performance compared to manual annotations [6,13]. These methods facilitate large-scale studies and clinical correlation analyses. For example, Kroll et al. [23] utilized a 3D U-Net to segment EAT and PAT in just 5–10 s per patient, investigating its association with the Coronary Artery Calcium Score in a large cohort.

The purpose of this paper is to give a comprehensive technical overview of recent image analysis methods for cardiac adipose tissue segmentation using CT.

This is a highly active research field with many recent publications. The latest reviews are from 2022 [6,13] and many contributions have been made since, which appeals for an updated review. Additionally, there is a need to establish a clear and consistent terminology for the different types of cardiac adipose tissues. Methods within the last 10 years are covered and were found through citation chasing and search databases such as PubMed and Google Scholar, applying numerous relevant keywords. Only methods developed for CT and methods distinguishing between EAT and PAT are considered.

2 Cardiac Adipose Tissue and Anatomy

Cardiac adipose tissue can be divided into segments, anatomically separated by the pericardial sac (the pericardium), a thin protective layer around the heart. The pericardium has two main parts: the fibrous pericardium and the serous pericardium. Additionally, the serous pericardium has two layers: the outer parietal layer which is attached to the fibrous pericardium, and the inner visceral layer, also known as the epicardium, which lies directly on the myocardium (the heart muscle) [41]. There is a small fluid-filled space between the parietal and the visceral layers called the pericardial cavity. Fat located underneath the epicardium is termed Epicardial Adipose Tissue (EAT), whereas fat located around the heart, outside the epicardium, is called Pericardial Adipose Tissue (PAT). The anatomy of the pericardium and cardiac adipose tissues is illustrated in Fig. 2. The distinction between EAT and PAT is important, as they have different anatomic relationships and different biomolecular properties [20].

Adipose tissue on CT is characterized by a distinct unique Hounsfield Unit (HU) range, usually defined as $(-190, -30)$ HU [54]. This attenuation range is substantially lower than other tissues, such as muscles, myocardium and blood vessels, making adipose tissue easily identifiable through thresholding.

There is terminological disagreement in the medical literature regarding the definition of the adipose tissues surrounding the outside of the heart. Some lit-

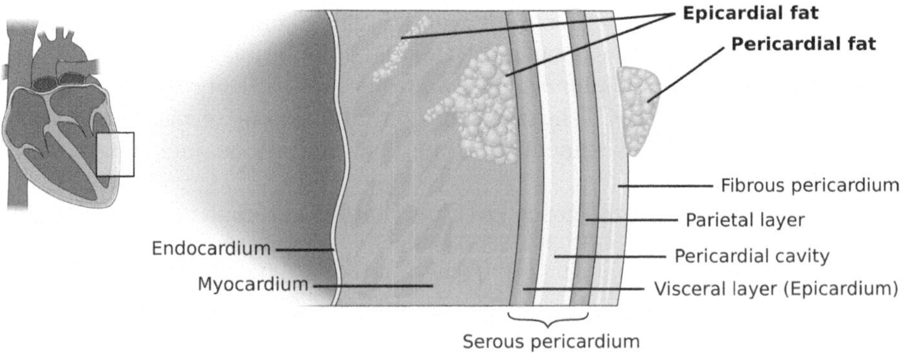

Fig. 2. The pericardium and location of cardiac adipose tissues.

erature defines pericardial adipose tissue to be the union of fat located inside and outside the pericardium [6]. Other studies refer to the fat outside the pericardium as paracardial adipose tissue and define pericardial adipose tissue as fat specifically located inside the pericardium itself and within the pericardial cavity [13,22]. However, since the pericardium is a very thin structure, the layers cannot be distinguished on CT and the pericardium is typically only slightly visible due to a small increase in attenuation of the fibrous layer. In terms of reviewing previous papers, we will acknowledge any wording or misconceptions and adjust the descriptions to fit the terminology defined in this paper, which adheres to the clinical consensus statement from the ESC Working Group [1].

3 Methods in Literature

An overview of the most important methods and results for EAT segmentation can be found in Table 1. The 'Dataset' column notes the total number of CT volumes used to train and test each segmentation model. The label '(CE)' indicates exclusive use of contrast-enhanced scans, no label implies non-contrast scans, while '(Both)' specifies the use of both non-contrast and contrast-enhanced scans. The star (*) indicates if a given method is based on the same publicly available dataset by VisualLab [47]. While the vast majority of methods focus on EAT, some studies are also concerned with PAT. These are listed in Table 2.

There are generally two approaches for doing EAT and PAT segmentation on CT: 1) direct segmentation of adipose tissue or 2) pericardium delineation followed by thresholding. Direct segmentation is the conventional approach, where a model generates a mask or probability map directly for the target. However, due to the irregular distribution of cardiac adipose tissue, it may be beneficial to instead segment the pericardium and apply thresholding to the inner and outer regions, in which EAT and PAT are defined, using the HU range of fat.

The earliest approaches for automated EAT segmentation emerged in the 2000s, using traditional image analysis methods based on intensities, textures, and region growing algorithms [2,10]. Subsequently, atlas-based methods gained prominence, while also beginning to differentiate between EAT and PAT [11,43]. This paradigm shifted as machine learning gained popularity [38] and soon after, deep learning took over. Nowadays, most of the proposed work uses some kind of deep learning, while a few methods still rely on traditional non-AI techniques.

3.1 Non-AI Methods

Zlokolica et al. [56] propose a slice-based semi-automatic method for EAT segmentation. The method is based on morphological operations to extract the heart, followed by a fuzzy c-means clustering algorithm and geometric ellipse fitting. The method requires the user to place a seed manually once for each patient. The authors only present the Dice Similarity Coefficient (DSC) of 8 patients, showing considerable variation. The mean is presented in Table 1.

Table 1. Overview of methods for EAT segmentation. CE = contrast-enhanced scans, Both = non-contrast and contrast-enhanced scans, ρ_c = Lin concordance correlation, ρ_s = Spearman correlation, DSC = Dice Coefficient, * = Visual Lab dataset [47].

Authors	Year	Dim.	Method	Dataset	Results
			Deep Learning		
Commandeur et al. [8]	2018	2D	Two CNNs	250	DSC = 82.3%
Commandeur et al. [9]	2019	2.5D	CNN	850	DSC = 87.3%
Zhang et al. [55]	2020	2D	Two U-Nets	20*	DSC = 91.2%
He et al. [16]	2020	3D	U-Net	200 (CE)	DSC = 88.7%
Li et al. [25]	2021	2.5D	U-Net	103 (CE)	DSC = 93.3%
Siriapisith et al. [45]	2021	3D	U-Net	260 (Both)	DSC = 90.1%
Molnar et al. [31]	2021	3D	CNN	1811	DSC = 90%
Benčević et al. [5]	2021	2D	U-Nets, Polar transform	20*	DSC = 78.4%
Benčević et al. [7]	2021	2.5D	U-Net	20*	DSC = 85.7%
Li et al. [26]	2022	3D	U-Net	70 (Both)	DSC = 92.6%
Qu et al. [36]	2022	2D	CNN, U-Net	103	DSC = 88.3%
Hoori et al. [17]	2022	2.5D	DeepLab-v3-plus + bisect	89	DSC = 88.5%
Bartoli et al. [4]	2022	2D	U-Net + xHRNet	115	DSC = 85%
West et al. [50]	2023	3D	U-Net	3720 (CE)	$\rho_c = 0.972$
Qu et al. [35]	2023	3D	CNN, Swin UN-ETR	724	DSC = 83.9%
Silva et al. [42]	2024	2D	pix2pix	20*	DSC = 98.7%
Miller et al. [30]	2024	2.5D	Conv. LSTM	1138	$\rho_s = 0.91$
Tang et al. [46]	2024	2D	U-Net	108 (CE)	DSC = 89.6%
Kuo et al. [24]	2024	3D	U-Net	157 (CE)	DSC = 87.0%
Wang et al. [49]	2025	2.5D	U-Net	50 (CE)	DSC = 93.9%
			Machine Learning		
Rodrigues et al. [39]	2016	2D	Atlas, RF	20*	DSC = 98.1%
Norlén et al. [32]	2016	3D	Atlas, RF, graph cuts	30 (CE)	DSC = 91%
Priya and Sudha [34]	2019	2D	GLCM, region growing	20*	DSC = 98.7%
Kazemi et al. [21]	2020	2D	GLCM, RF	20*	DSC = 97.8%
			Non-AI methods		
Zlokolica et al. [56]	2017	2D	Fuzzy c-means, ellipse fitting	10	DSC = 69%
Militello et al. [29]	2019	2D	Manual ROIs, interpolation	145 (Both)	DSC = 93.1%
Rebelo et al. [37]	2022	2D	Template matching, Convex Hull	20*	DSC = 77.3%

Table 2. Overview of methods for PAT segmentation. CE = contrast-enhanced scans, DSC = Dice Coefficient, * = Visual Lab dataset [47].

Authors	Year	Dim.	Method	Dataset	Results
			Deep Learning		
Commandeur et al. [8]	2018	2D	Two CNNs	250	-
He et al. [15]	2020	3D	U-Net	422 (CE)	DSC $\geq$ 0.85 %
Hoori et al. [18]	2023	2.5D	DeepLab-v3-plus + bisect	89	DSC = 82.5%
Silva et al. [42]	2024	2D	pix2pix	20*	DSC = 98.4%
			Machine Learning		
Rodrigues et al. [39]	2016	2D	Atlas, RF	20*	-
Priya and Sudha [34]	2019	2D	GLCM, Region growing	20*	DSC = 98.3%
Kazemi et al. [21]	2020	2D	GLCM, RF	20*	DSC = 98.4%

Militello et al. [29] propose a technically simple semi-automatic method to assist manual annotation, making it faster and more precise. The authors developed a GUI, requesting the user to manually draw pericardium delineations on a self-chosen number of slices. Linear interpolation is performed to fill out the delineation gaps, with the option for users to refine the curves and, when satisfied, define the adipose tissue threshold range for extracting EAT. The process is 4–8 times faster than slice-based manual segmentation.

Rebelo et al. [37] isolate the heart region by using Otsu's Method, connected component analysis, and template matching. The pericardium contour is refined by computing the convex hull and using morphological operations, allowing the authors to access EAT by thresholding. The method is implemented in an open-source software called HARTA.

3.2 Machine Learning Methods

Both the method by Rodrigues et al. [39] and Norlén et al. [32] utilize a multi-atlas approach to do registration, followed by training a Random Forest (RF) classifier for tissue classification. Whereas Rodrigues et al. classify EAT and PAT directly, Norlén et al. use Random Forest to estimate if a voxel is just inside, just at the boundary, or just outside of the pericardium or neither of the three. Segmentation is completed using graph cuts from a Markov random field.

Kazemi et al. [21] enhance CT slices using Contrast Limited Adaptive Histogram Equalization (CLAHE) and extract features such as contrast, correlation, energy, etc. by using the gray-level co-occurrence matrix (GLCM) and Gabor filters. A Random Forest classifier is trained on these features to directly segment four classes, among them being EAT and PAT. Priya and Sudha [34] also utilize GLCM for feature extraction, which serves as input to a shallow Grey Wolf-Optimized (GWO) neural network for classifying fat or non-fat slices. A region-growing algorithm combined with Fruitfly optimization is then applied to segment EAT, PAT, and the pericardium.

3.3 Deep Learning Methods

Network Types. Most networks in this review are based on U-Nets. A few methods [5,7,24,55] use the conventional U-Nets architecture as originally presented in Ref. [40], while others add slight modifications such as batch normalization or residual connections [26,46,50]. Some methods present more complex U-Net variants incorporating concepts like attention and transformers [16,35,45,49]. Molnar et al. [31] utilize a Convolutional Neural Network (CNN) with an architecture very similar to a U-Net, however, instead of downsampling, a raw CT volume patch is provided as input across multiple layers at different resolutions.

Other networks than the U-Net are also proposed. Miller et al. [30] employ a convolutional Long-Short Term Memory model (LSTM). Hoori et al. [17,18] apply DeepLab-v3-plus which has a ResNet-18 backbone. Bartoli et al. [4] propose the network called xHR-NET, which maintains high-resolution representations throughout the network. Lastly, Silva et al. [42] apply pix2pix, a generative conditional adversarial network designed for image-to-image translation tasks.

Multi-step Approaches. A common approach for solving the segmentation task is to first determine whether a slice or patch contains the heart or not [4,8,9,35,36]. Commandeur et al. [8] use two sequential CNN's for EAT and PAT segmentation, where the first network is responsible for slice selection. Upsampling of hidden layers is then used as input to the second network, responsible for the delineation of the pericardium. Combining the outputs, the segmentation masks are obtained by thresholding. A year later, the same authors propose a simplified method in which both tasks are handled by a single condensed CNN. Similarly, Qu et al. [36] use a CNN to do slice classification and a variant of the U-Net for the actual segmentation of EAT; in this case direct. Bartoli et al. [4] implement a classification head in the center of a U-Net to identify slices containing the heart, before performing the actual segmentation with the network xHR-Net.

Attention. Two methods utilize 3D U-Nets with incorporated attention gates to focus on EAT boundary structures. [16,45]. He et al. adopt the same attention-based 3D U-Net in Ref. [15] as in Ref. [16] to do EAT and PAT segmentation, respectively. Qu et al. [36] present a multi-scale residual attention U-Net, and Miller et al. [30] incorporate attention head into the LSTM for doing pericardium segmentation. Although Miller et al. do not report DSC, they note a promising Spearman volume correlation of 0.91 compared to expert manual segmentations.

Transformer. The first transformer-based architecture is presented by Qu et al. [35] in 2023. As in their earlier work [36], a 2D CNN selects candidate slices for EAT to reduce the consumption of GPU memory. For the segmentation, the authors employ a 3D transformer model Swin UNETR for its effectiveness in modeling global dependencies. Recently, Wang et al. [49] introduce Convolutional Transformer modules into the classical U-Net, yielding state-of-art results; however, the method was tested on only 10 patients.

Transfer Learning. Transfer learning is used by two methods to accommodate the shortcoming of few data samples. Siriapisith et al. [45] train on 220 CT scans and fine-tune on a smaller dataset of 40 CCTA scans, achieving a DSC of 90.1%, but a slightly lower number, 88.2%, is reported for CCTA. This last result closely matches the performance reported by He et al. [16], who use a similar network trained exclusively on CCTA data, with both studies utilizing datasets of comparable total size. Hoori et al. [17] perform segmentation with the network DeepLab-v3-plus, which is pre-trained on the ImageNet dataset.

2.5D Methods. Some methods [7,9,17,25,30,49] balance spatial precision with computational efficiency by employing multi-slice approaches, i.e. multiple slices are inputted to segment a single target slice. Benčević et al. [7] input the CT slice, alongside an additional image with every pixel representing the slice depth, to a U-Net, arguing that the pericardial region strongly depends on slice depth. Hoori et al. [17] give three consecutive CT slices as input, segmenting the middle one. They also apply 'bisection', which entails splitting the heart into two halves at its widest point. This ensures the network processes slices with increasing pericardial curvature, simplifying learning through a consistent image progression. The same method is applied by Hoori et al. to segment PAT in Ref. [18]. The model by Li et al. [25] is implemented in an open-source software named TIMESlice.

4 Discussion

4.1 Data

Many previous methods rely on very small datasets, making it difficult to assess the model's ability to generalize. While training set size is important, the size of the test set is equally critical. For example, Siriapisith et al. [45] use 200 non-contrast scans for training but only 20 scans for testing. Although this dataset is larger than in many other studies, the test set most likely does not capture a representative variation in anatomy. Contrary, Miller et al. [30] utilize a much larger test set comprising 638 patients from four different sites, increasing the likelihood that the test set better reflects a real-life distribution.

To avoid overfitting, it is crucial to create a test set without data leakage. Data leakage might happen if slices from the same patient are present in both the training and test set as the method adapts to a certain anatomy. Silva et al. [42] conduct random slice-based testing rather than patient-based testing, which likely explains why it achieves a significantly higher DSC compared to other methods. A similar approach may have been used in methods presenting similar high results [21,34,39], as their data splitting is not thoroughly explained.

A major limitation in the field of cardiac fat segmentation is the lack of publicly available datasets with manual annotations. The only open source dataset with labeled EAT and PAT volumes is provided by Visual Lab [47], created by Rodrigues et al. [38]. This scarcity has led numerous methods to rely on and train their models using this dataset [5,7,21,34,37,39,42,55]. However, it is very small, comprising only 20 non-contrast CT scans, which is insufficient for developing robust, generalizable models and for proper validation. High performance on this dataset may be a result of overfitting. Additionally, the annotations lack clarity: true labels are represented with an RGB (Red, Green and Blue) color, but 31% of voxels exhibit overlapping color codes. This ambiguity affects EAT and PAT quantification across models, as no standardized method for differentiation in such cases has been defined.

It is evident that more open-source data is needed. It is of great interest to have large common datasets of high quality that can be used for benchmarking. Without this, it will be difficult to compare the quality of different models. Also, there is currently no publicly available dataset on contrast-enhanced scans. This would be of high interest, as European Guidelines have implemented CCTA as a recommended imaging tool for detecting heart disease in symptomatic individuals [48], and also the pericardium is more visible on CCTA [24], potentially easing the job for data-driven segmentation models.

4.2 Cardiac Fat Attenuation Range

Many of the reviewed methods rely on the characteristic attenuation values of adipose tissue, especially when segmenting adipose tissue by pericardium delineation and thresholding, making the choice of an accurate interval crucial. While this interval is typically defined as (−190, −30) HU [54], the exact threshold values are subject to debate, as they vary with e.g. acquisition parameters and the use of contrast medium [27,52,53]. This variability is also reflected in methods using non-contrast scans, where proposed HU intervals include (−200, −30) [5,7,21,37,39,42], (−210, −10) [36], and (−140, −30) [4], with the most methods adhering to the standard interval (−190, −30).

While variations in the lower threshold appear to have minimal impact, the upper threshold is particularly critical for contrast-enhanced CT. Studies report that using a standard fixed upper threshold of −30 significantly underestimates EAT in contrast-enhanced scans compared to non-contrast CT [27,52], whereas thresholds closer to 0 HU are more accurate [27,52]. Figure 3 illustrates this difference using a contrast-enhanced CT-slice from the CGPS dataset [12], annotated locally by a medical doctor. EAT regions are marked for two upper thresholds: −30 HU (green) and 0 HU (orange), with −190 HU as the lower limit in both cases. In this case, increasing the upper threshold from −30 HU to 0 HU result in an approximate 20% increase in estimated EAT volume.

For methods using contrast-enhanced CT scans, the proposed intervals are (−190, −30) HU [16,24,26,46,49], (−192, −30) HU [32] and (−175, −15) HU

[25]. These intervals fail to account for the appropriate upper threshold range, resulting in systematic under-segmentation of EAT.

4.3 Model Performance and Challenges

Multiple studies report better segmentation performance at slices in the middle of the heart, with poorer performance near the upper and lower end [8,37,46]. This is to be expected, as these regions are also more difficult to segment manually. At the lower end, the pericardium is typically thinner and may therefore be harder to identify. The anatomy becomes more intricate at the upper end, where the pericardium connects to the aorta, pulmonary veins, and arteries, introducing ambiguity about its exact boundary. To truly compare the performance of the methods, a clear definition of the heart limits should be defined. For instance, West et al. [50] define the limits to be the bifurcation of the pulmonary trunk superiorly and the apex of the pericardium inferiorly, whereas Bartoli et al. [4] set the limits to be the top of the left atrium and the last slice in which the left ventricle was identified. A consensus is needed to compare models and give a robust measure of model performance in the extremities.

Variation in anatomy is a major issue for automated methods in this field, as EAT and PAT exhibit significant individual differences in distribution and composition. Moreover, several methods report reduced performance in cases with high EAT and PAT [8,30], likely due to anatomical variations in patients with higher BMI. Similarly, Kazemi et al. [21] report poor performance when attempting to segment the pericardial border due to anatomical variability.

Cardiac adipose tissue lacks a neat, uniform structure. Sometimes it appears as clusters, other times as fragments of smaller chunks. For this reason, most methods in this review segment EAT by delineating the pericardium rather than performing direct segmentation. The pericardium provides a relatively smooth and consistent outline, making it easier to identify. However, with this approach, segmentation inaccuracies tend to occur near the pericardial borders, and since

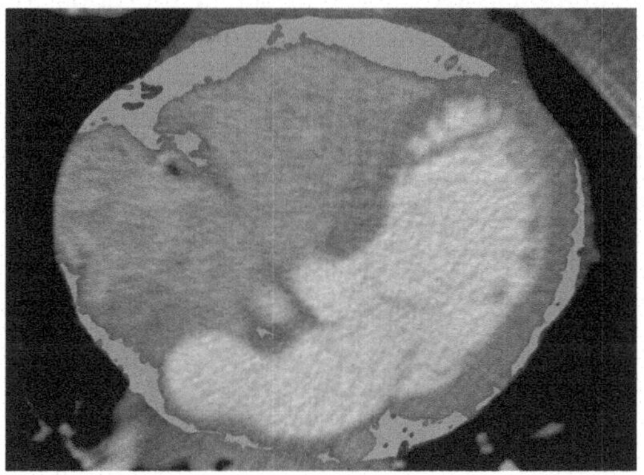

Green
HU attenuation range: $[-190, -30]$
Volume: 4.37 cm^3

Orange
HU attenuation range: $[-190, 0]$
Volume: 5.33 cm^3 ($\sim$ 20% increase)

Fig. 3. The difference between EAT attenuation thresholds on a CCTA slice [12].

the majority of both EAT and PAT reside close to the pericardial sac, even small misalignments can significantly impact segmentation performance.

In general, many methods demonstrate a performance competitive with manual segmentation. Strong correlations with manual estimations of EAT and PAT volume and attenuation are reported. Molnar et al. [31] concluded 99.4% of the automated segmentations to be successful. However, the last few unsuccessful scans did not have any easily identifiable characteristics. While automated approaches show some variability, manual segmentation does as well, to a comparable extent. Bartoli et al. [4] accounts for an inter-observer DSC of ~3%, whereas others report an inter-observer variation in volume of 10.4% [3].

In the past, processing time was a major limiting factor for segmentation methods, but advancements in computational power and model efficiency have largely resolved this issue. While Rodrigues et al. [39] reported processing times of several hours, recent methods perform interference in a few seconds [30,36,46], enabling real-time clinical use.

4.4 Future Directions and Clinical Perspectives

With the growing number of precise and efficient deep-learning algorithms for the segmentation of cardiac adipose tissue, more research should be put into identifying novel biomarkers and uncovering new clinical outcomes. Currently, most work is focused on total EAT volume and/or attenuation, however, not many are focusing on the effects of possible regional accumulations, making this an intriguing area for future exploration. Clinically, there is a great interest in adipose tissue located around the coronary arteries [44], a subset of EAT. However, automated methods directly addressing this have yet to emerge.

5 Conclusion

Developing accurate methods to measure EAT and PAT is vital for enhancing CVD risk assessment and understanding their role in cardiovascular health and disease progression. This review demonstrates a broad set of methods for automated segmentation, including deep learning, machine learning and non-AI methods. Several of these are highly accurate with low processing times, highlighting the potential of automated models based on medical image analysis and AI to support clinicians and contribute to medical diagnostics.

References

1. Antoniades, C., et al.: Perivascular adipose tissue as a source of therapeutic targets and clinical biomarkers. Eur. Heart J. **44**(38), 3827–3844 (2023)
2. Bandekar, A.N., Naghavi, M., Kakadiaris, I.A.: Automated pericardial fat quantification in CT data. In: Proceedings of EMBC. IEEE (2006)
3. Barbosa, J.G., Figueiredo, B., Bettencourt, N., Tavares, J.: Towards automatic quantification of the epicardial fat in non-contrasted CT images. Comput. Methods Biomech. Biomed. Eng. **14**(10), 905–914 (2011)

4. Bartoli, A., et al.: Automatic deep-learning segmentation of epicardial adipose tissue from low-dose chest CT and prognosis impact on Covid-19. Cells **11**(6), 1034 (2022)
5. Benčević, M., Galić, I., Habijan, M., Babin, D.: Training on polar image transformations improves biomedical image segmentation. IEEE Access **9** (2021)
6. Benčević, M., Galić, I., Habijan, M., Pižurica, A.: Recent progress in epicardial and pericardial adipose tissue segmentation and quantification based on deep learning: a systematic review. Appl. Sci. **12**(10) (2022)
7. Benčević, M., Habijan, M., Galić, I.: Epicardial adipose tissue segmentation from CT images with a semi-3D neural network. In: 2021 International Symposium ELMAR, pp. 87–90. IEEE (2021)
8. Commandeur, F., et al.: Deep learning for quantification of epicardial and thoracic adipose tissue from Non-Contrast CT. IEEE Trans. Med. Imaging **37**(8), 1835–1846 (2018)
9. Commandeur, F., et al.: Fully automated CT quantification of epicardial adipose tissue by deep learning: a multicenter study. Radiol. Artif. Intell. **1**(6), e190045 (2019)
10. Dey, D., et al.: Automated quantitation of pericardiac fat from noncontrast CT. I. Rad. **43**(2), 145–153 (2008)
11. Ding, X., Terzopoulos, D., Diaz-Zamudio, M., Berman, D.S., Slomka, P.J., Dey, D.: Automated epicardial fat volume quantification fromnon-contrast CT. In: Progress in Biomedical Optics and Imaging - Proceedings of SPIE, vol. 9034 (2014)
12. Fuchs, A., et al.: Subclinical coronary atherosclerosis and risk for myocardial infarction in a Danish cohort. Ann. Int. Med. **176**(4), 433–442 (2023)
13. Greco, F., et al.: Epicardial and pericardial fat analysis on CT images and artificial intelligence: a literature review. Quant. Imaging Med. Surg. **12**(3), 2075–2089 (2022)
14. Greif, M., et al.: Increased pericardial adipose tissue is correlated with atrial fibrillation and left atrial dilatation. Clin. Res. Cardiol. **102**, 555–562 (2013)
15. He, X., et al.: Automatic quantification of myocardium and pericardial fat from coronary computed tomography angiography: a multicenter study. Eur. Radiol. **31**, 3826–3836 (2021)
16. He, X., et al.: Automatic segmentation and quantification of epicardial adipose tissue from coronary computed tomography angiography. Phys. Med. Biol. **65**(9), 095012 (2020)
17. Hoori, A., et al.: Deep learning segmentation and quantification method for assessing epicardial adipose tissue in CT calcium score scans. Sci. Rep. **12**(1), 2276 (2022)
18. Hoori, A., Hu, T., Lee, J., Al-Kindi, S., Rajagopalan, S., Wilson, D.L.: Analysis of paracardial adipose tissues using deep learning segmentation in CT calcium score images. In: Medical Imaging 2023, vol. 12468, pp. 19–24. SPIE (2023)
19. Iacobellis, G.: Epicardial adipose tissue in contemporary cardiology. Nat. Rev. Cardiol. **19**(9), 593–606 (2022)
20. Iacobellis, G., Corradi, D., Sharma, A.M.: Epicardial adipose tissue: anatomic, biomolecular and clinical relationships with the heart. Nat. Clin. Pract. Cardiovasc. Med. **2**(10), 536–543 (2005)
21. Kazemi, A., Keshtkar, A., Rashidi, S., Aslanabadi, N., Khodadad, B., Esmaeili, M.: Automated segmentation of cardiac fats based on extraction of textural features from non-contrast CT images. In: Proceedings of CSICC, pp. 1–7 (2020)
22. Konwerski, M., Gąsecka, A., Opolski, G., Grabowski, M., Mazurek, T.: Role of epicardial adipose tissue in cardiovascular diseases. Biology (Basel) **11**(3) (2022)

23. Kroll, L., Nassenstein, K., Jochims, M., Koitka, S., Nensa, F.: Assessing the role of pericardial fat as a biomarker connected to coronary calcification. J. Clin. Med. **10**(2) (2021)
24. Kuo, L., et al.: Deep learning-based workflow for automatic extraction of atria and epicardial adipose tissue on cardiac computed tomography in atrial fibrillation. J. Chin. Med. Assoc. **87**(5), 471–479 (2024)
25. Li, X., et al.: Automatic quantification of epicardial adipose tissue volume. Med. Phys. **48**(8), 4279–4290 (2021)
26. Li, Y., et al.: Segmentation and volume quantification of epicardial adipose tissue in computed tomography images. Med. Phys. **49**(10), 6477–6490 (2022)
27. Marwan, M., et al.: Quantification of epicardial adipose tissue by cardiac CT: influence of acquisition parameters and contrast enhancement. Eur. J. Radiol. **121**, 108732 (2019)
28. Mensah, G.A., Fuster, V., Murray, C., Roth, G.A.: Global Burden of Cardiovascular Diseases and Risks Collaborators: global burden of cardiovascular diseases and risks, 1990–2022. J. Am. Coll. Cardiol. **82**(25), 2350–2473 (2023)
29. Militello, C., et al.: A semi-automatic approach for epicardial adipose tissue segmentation and quantification on cardiac CT scans. Comput. Biol. Med. **114**, 103424 (2019)
30. Miller, R.J., et al.: Ai-derived epicardial fat measurements improve cardiovascular risk prediction from myocardial perfusion imaging. NPJ Dig. Med. **7**(1), 24 (2024)
31. Molnar, D., et al.: Artificial intelligence based automatic quantification of epicardial adipose tissue suitable for large scale population studies. Sci. Rep. **11**(1), 23905 (2021)
32. Norlén, A., et al.: Automatic pericardium segmentation and quantification of epicardial fat from computed tomography angiography. J. Med. Img. **3**(3), 034003 (2016)
33. Oikonomou, E.K., Antoniades, C.: The role of adipose tissue in cardiovascular health and disease. Nat. Rev. Cardiol. **16**(2), 83–99 (2019)
34. Priya, C., Sudha, S.: Adaptive fruitfly based modified region growing algorithm for cardiac fat segmentation using optimal neural network. J. Med. Syst. **43**(5), 104 (2019)
35. Qu, J., et al.: Epicardial adipose tissue segmentation and quantification based on transformer model. In: Proceedings of BIC 2023, pp. 325–330 (2023)
36. Qu, J., et al.: Deep learning-based approach for the automatic quantification of epicardial adipose tissue from non-contrast CT. Cogn. Comput. **14**(4), 1392–1404 (2022)
37. Rebelo, A.F., Ferreira, A.M., Fonseca, J.M.: Automatic epicardial fat segmentation and volume quantification on non-contrast cardiac computed tomography. Comput. Methods Programs Biomed. Update **2**, 100079 (2022)
38. Rodrigues, É.O., Conci, A., Morais, F.F.C., PÉrez, M.G.: Towards the automated segmentation of epicardial and mediastinal fats. In: 2015 IEEE International Conference on Industrial Technology (ICIT), pp. 1779–1785 (2015)
39. Rodrigues, É.O., Morais, F., Morais, N., Conci, L.S., Neto, L.V., Conci, A.: A novel approach for the automated segmentation and volume quantification of cardiac fats on computed tomography. Comput. Methods Programs Biomed. **123**, 109–128 (2016)
40. Ronneberger, O., Fischer, P., Brox, T.: U-net: convolutional networks for biomedical image segmentation. In: Proceedings of MICCAI, pp. 234–241. Springer, Cham (2015)

41. Sacks, H.S., Fain, J.N.: Human epicardial adipose tissue: a review. Am. Heart J. **153**(6), 907–917 (2007)
42. Santos da Silva, G., Casanova, D., Oliva, J.T., Rodrigues, E.O.: Cardiac fat segmentation using computed tomography and an image-to-image conditional generative adversarial neural network. Med. Eng. Phys. **124**, 104104 (2024)
43. Shahzad, R., et al.: Automatic quantification of epicardial fat volume on non-enhanced cardiac CT scans using a multi-atlas segmentation approach. Med. Phys. **40**(9), 091910 (2013)
44. Simantiris, S., et al.: Perivascular fat: a novel risk factor for coronary artery disease. Diagnostics (Basel) **14**(16) (2024)
45. Siriapisith, T., Kusakunniran, W., Haddawy, P.: A 3D deep learning approach to epicardial fat segmentation in non-contrast and post-contrast cardiac CT images. PeerJ Comput. Sci. **7**(e806), e806 (2021)
46. Tang, K.X., et al.: An enhanced deep learning method for the quantification of epicardial adipose tissue. Sci. Rep. **14**(1), 24947 (2024)
47. Visual Lab, Computing Institute IC/UFF: Cardiac fat database—computed tomography (2015). https://visual.ic.uff.br/en/cardio/ctfat/. Accessed Nov 2024
48. Vrints, C., et al.: 2024 ESC guidelines for the management of chronic coronary syndromes. Eur. Heart J. **45**(36), 3415–3537 (2024)
49. Wang, Y., et al.: Automated pericardium segmentation and epicardial adipose tissue quantification from computed tomography images. Biomed. Signal Process. Control **100**, 107167 (2025)
50. West, H.W., et al.: Deep-learning for epicardial adipose tissue assessment with computed tomography: implications for cardiovascular risk prediction. JACC Cardiovasc. Imaging **16**(6), 800–816 (2023)
51. WHO: Cardiovascular diseases kill 10 000 people in the WHO European region every day (2024). https://www.who.int/europe/news-room/15-05-2024-cardiovascular-diseases-kill-10-000-people-in-the-who-european-region-every-day--with-men-dying-more-frequently-than-women. Accessed 14 Jan 2025
52. Xu, L., et al.: Comparison of epicardial adipose tissue radiodensity threshold between contrast and non-contrast enhanced computed tomography scans: a cohort study of derivation and validation. Atherosclerosis **275**, 74–79 (2018)
53. Yin, L., et al.: Measurement of epicardial adipose tissue using non-contrast routine chest-CT: a consideration of threshold adjustment for fatty attenuation. BMC Med. Imaging **22**(1), 114 (2022)
54. Yoshizumi, T., et al.: Abdominal fat: standardized technique for measurement at CT. Radiology **211**(1), 283–286 (1999)
55. Zhang, Q., Zhou, J., Zhang, B., Jia, W., Wu, E.: Automatic epicardial fat segmentation and quantification of CT scans using dual u-nets with a morphological processing layer. IEEE Access **8**, 128032–128041 (2020)
56. Zlokolica, V., et al.: Semiautomatic epicardial fat segmentation based on fuzzy c-means clustering and geometric ellipse fitting. J. Healthcare Eng. **2017**(1), 5817970 (2017)

Determining Fetal Orientations From Blind Sweep Ultrasound Video

Jakub Maciej Wiśniewski[1], Anders Nymark Christensen[1(✉)], Mary Le Ngo[2,3,4], Martin Grønnebæk Tolsgaard[2,4], and Chun Kit Wong[1,5(✉)]

[1] Technical University of Denmark, Kongens Lyngby, Denmark
{anym,ckwo}@dtu.dk
[2] University of Copenhagen, Copenhagen, Denmark
[3] Slagelse Hospital, Slagelse, Denmark
[4] CAMES Rigshospitalet, Copenhagen, Denmark
[5] Pioneer Centre for AI, Copenhagen, Denmark

Abstract. Cognitive demands of fetal ultrasound examinations pose unique challenges among clinicians. With the goal of providing an assistive tool, we developed an automated pipeline for predicting fetal orientation from ultrasound videos acquired following a simple blind sweep protocol. Leveraging on a pre-trained head detection and segmentation model, this is achieved by first determining the fetal presentation (cephalic or breech) with a template matching approach, followed by the fetal lie (facing left or right) by analyzing the spatial distribution of segmented brain anatomies. Evaluation on a dataset of third-trimester ultrasound scans demonstrated the promising accuracy of our pipeline. This work distinguishes itself by introducing automated fetal lie prediction and by proposing an assistive paradigm that augments sonographer expertise rather than replacing it. Future research will focus on enhancing acquisition efficiency, and exploring real-time clinical integration to improve workflow and support for obstetric clinicians.

Keywords: Fetal Orientation · Obstetric Ultrasound

1 Introduction

Ultrasound imaging stands as a ubiquitous tool in modern maternal healthcare, playing a crucial role in prenatal monitoring and fetal well-being assessment. During a typical ultrasound examination, sonographers skillfully navigate the probe across the maternal abdomen, mentally constructing a dynamic three-dimensional representation of the fetus [20]. This mental model allows them to track fetal position, orientation, and the location of specific anatomical structures, guiding their probe movements to capture the necessary diagnostic information. However, this process demands significant cognitive effort and sustained concentration from the sonographer. High patient volumes, lengthy examination

J. Petersen and V. A. Dahl (Eds.): SCIA 2025, LNCS 15726, pp. 254–263, 2025.
https://doi.org/10.1007/978-3-031-95918-9_18

times, and the need for meticulous image interpretation contribute to substantial mental workload [4,10,17,21] and potential for cognitive fatigue. This is particularly pronounced for sonographers in training or those with less experience [16], who may find it challenging to rapidly build and maintain an accurate mental 3D model of the fetus while simultaneously acquiring and interpreting ultrasound images. This cognitive burden can impact workflow efficiency and potentially increase the risk of diagnostic errors.

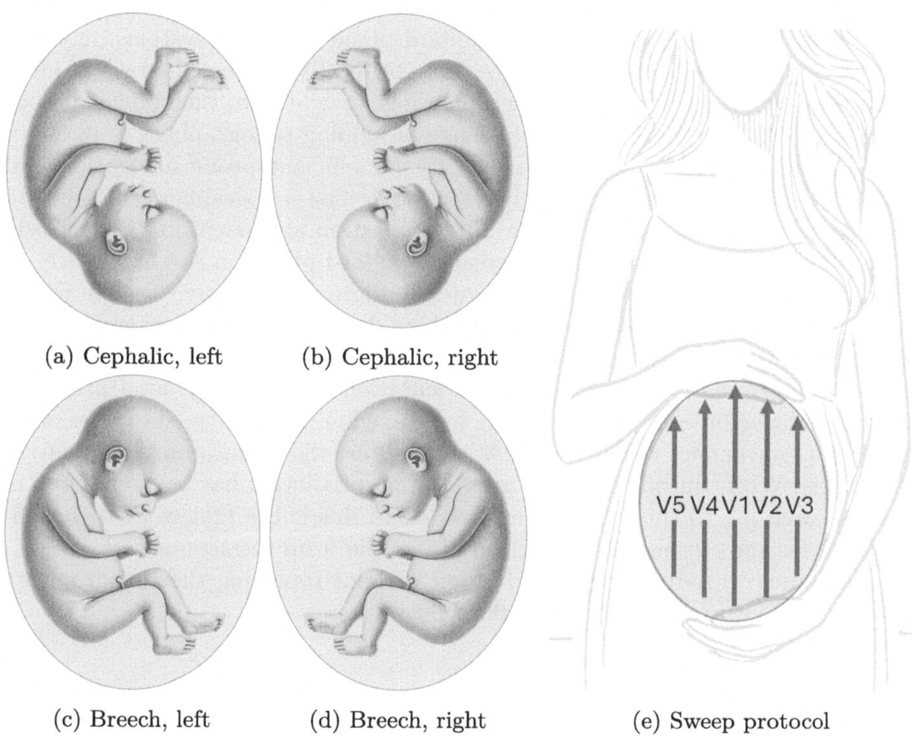

(a) Cephalic, left (b) Cephalic, right (c) Breech, left (d) Breech, right (e) Sweep protocol

Fig. 1. (a-d) The four fetal orientations. (e) Blind sweep protocol across the pregnant abdomen. The sweeps start from the center marked with V1, followed by two on the mother's left (V2, V3) and finally two on the right (V4, V5).

To address these challenges, numerous prior works have explored automated solutions for analysis of ultrasound examinations [3,12,18]. A significant portion of this research has focused on automating fetal biometry measurements, such as head [9,23] or abdominal [11] circumference and femur length [8], which is crucial for downstream tasks such as fetal weight estimation and delivery planning. More recent works also explored gestational age estimation from ultrasound videos [7,13,15], often acquired via some blind sweep protocols. While these automated systems offer the promise of increased efficiency and reduced operator dependency, they often lack the flexibility required in dynamic clinical settings. Blind sweep protocols, while standardized, may not always capture all necessary information, and sonographer intervention to adjust scanning technique is sometimes necessary. In scenarios where the sonographer deviates from the prescribed protocol to obtain optimal images, these rigid automated systems may fail.

Meanwhile, another significant line of research focuses on fetal standard plane detection [1,2,19], which aim to identify and classify key diagnostic planes. Such standard plane detection systems often form an integral part of the fully automated ultrasound analysis systems described previously, or function as retrospective analysis tools. While these standard plane detection systems may also be used to provide real-time assistance, they primarily address the task of identifying specific planes after the sonographer has navigated to them. They do not directly address the fundamental challenge of helping sonographers in developing the crucial mental 3D model necessary for efficient probe navigation and anatomical localization during the live scan [22]. The sonographer still needs to mentally keep track of the fetus's overall orientation and position to effectively guide the probe and acquire the necessary standard planes in the first place.

In this work, we propose a different paradigm: an assistive tool designed to alleviate cognitive loading on sonographers during ultrasound examinations. Instead of automated estimation of fetal biometry from blind sweep videos, we focus on determination of coarse fetal orientation - specifically, fetal presentation (cephalic or breech) and lie (facing left or right), with the goal of presenting them as a continuous visual reminder throughout the ultrasound examination (see Fig. 1). While automated fetal presentation prediction has been explored in prior research [6], to the best of our knowledge, this is the first work to specifically address the automated prediction of fetal lie from ultrasound videos. This approach aims to reduce cognitive burden without replacing the sonographer's expertise or limiting their scanning flexibility.

2 Methods

2.1 Blind Sweep Protocol

For data acquisition, we followed a simple blind sweep protocol inspired by [15], which involved performing five vertical ultrasound sweeps across the pregnant abdomen in the caudal-cranio direction, each traversing from the pubis to the uterine fundus in 10 s (see Fig. 1e). This ensures comprehensive coverage of the maternal abdomen, and thereby maximizing the probability of capturing the fetal head within the acquired ultrasound videos.

2.2 Head Detection and Segmentation

The core of our data processing pipeline relies on a pre-trained progressive concept bottleneck model (PCBM) [14], a deep learning model designed for explainable standard plane detection in fetal ultrasound. While PCBM offers a wide array of functionalities, we specifically leverage two of its key capabilities in this work. First, we utilize PCBM's plane classification to identify video frames containing the fetal head within our blind sweep acquisitions. Second, we employ the model's integrated head segmentation module, which provides segmentation masks for two anatomical structures within the head: the thalamus and the cavum septum pellucidum (CSP).

2.3 Fetal Presentation Classification

Inspired by [6], we use a simple template matching algorithm to classify the fetal presentation (cephalic versus breech) by making use of temporal information from PCBM's prediction on our blind sweep videos. This is based on the expectation that in cephalic presentations (i.e. head down), the fetal head is encountered earlier in the caudal-cranio sweep, while in breech presentations, the head appears later in the sweep. For each sweep video i, we first define a function $f_{v_i}(t)$ as the softmax probability output by PCBM for the video frame at temporal point t being classified as containing a fetal head. We then define two exponential template functions, $f_c(t)$ and $f_b(t)$, to represent the expected temporal distribution of head detections for cephalic and breech presentations:

$$f_c(t) = \frac{\exp(N_f - t)}{\exp(N_f)}, \qquad f_b(t) = \frac{\exp(t)}{\exp(N_f)} \tag{1}$$

where N_f is the total number of frames in the sweep video. These reflect the decreasing likelihood of head detection as the sweep progresses in cephalic cases, and vice versa in breech cases. Finally, we calculate the cosine similarity between the function $f_{v_i}(t)$ and each of the template functions, $f_c(t)$ and $f_b(t)$. The fetal presentation is then classified as cephalic or breech based on which template function yields a higher cosine similarity with $f_{v_i}(t)$, followed by majority voting across i as an aggregation step.

2.4 Fetal Lie Classification

For the classification of fetal lying (i.e. facing left versus right), we utilized the segmentation masks of the thalamus and CSP obtained from PCBM, and leveraged the anatomical knowledge that the thalamus is situated posteriorly to the CSP within the fetal head. Consequently, the vector pointing from the thalamus towards the CSP provides an indication of the fetal facing direction. We developed an image processing algorithm to automate this process (see Fig. 2).

For the thalamus mask, we first determined its geodesic center, which is opted over the centroid. This is because the thalamus exhibits a crescent-shaped morphology, which can lead to the centroid falling outside the actual thalamus structure. On the other hand, the geodesic center, by definition, is guaranteed to lie within the structure. To compute the geodesic center, we first skeletonized the thalamus mask, followed by constructing an adjacency graph of the skeleton pixels. The geodesic center can then be identified as the pixel with the minimum sum of shortest path distances to all other pixels within this graph. For the

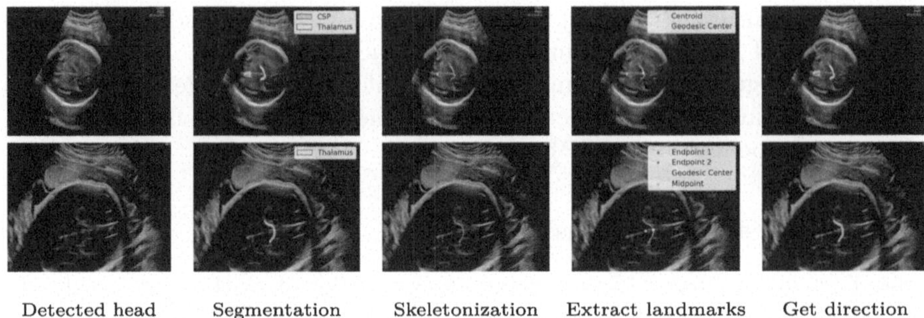

Fig. 2. The process of estimating fetal lie depicted from left to right. The top row illustrates the main approach of utilizing both thalamus and CSP, while the bottom row illustrates the thalamus-only fallback approach.

CSP mask, given its typically blob-like shape, we calculated the centroid. The vector originating from the thalamus geodesic center and terminating at the CSP centroid was then taken as our primary prediction for the fetal facing direction.

To enhance the robustness of our lie determination pipeline, we also incorporated a fallback mechanism to handle scenarios where only the thalamus is reliably segmented, but the CSP is not. In such cases, we again began by skeletonizing the thalamus mask. Alongside the geodesic center, we also identified the two endpoints of the thalamus skeleton. These endpoints were detected by convolving a kernel function over the skeletonized image. From the straight line segment connecting these two endpoints, we calculated the orthogonal direction. This orthogonal direction then served as our predicted fetal facing direction when the CSP was not available. However, to ensure the reliability of this thalamus-only approach, we empirically established from early experiments a set of criteria that the thalamus mask must satisfy:

- the thalamus mask must consist of a single connected component
- it must contain a minimum of 55 pixels
- it must exhibit a minimum solidity of 0.82
- the distance between the midpoint of the straight line joining the two endpoints and the geodesic center of the thalamus must be at least 1.09 pixels

If these criteria were not met, the pipeline would abstain from predicting the lie.

3 Experiments and Results

3.1 Dataset

To evaluate the performance of our proposed pipeline, we utilized a dataset acquired with full consent from 13 pregnant subjects in their third trimester. Ultrasound data were collected using our blind sweep protocol (see Sect. 2.1), with recordings performed using a research software deployment setup integrated within the clinical ultrasound environment [22]. Following each blind sweep acquisition, the sonographer who performed the scan also provided the annotation labels immediately following the acquisition. These labels indicated the fetal presentation, categorized as "down" (cephalic) or "up" (breech), and fetal lie, categorized as "left" or "right". In total, 12 cases were annotated as "down" presentation and 1 as "up" presentation, while for fetal lie 5 cases were annotated as "left" and 8 as "right". We acknowledge that this dataset exhibits an imbalance in presentation classes, with a notably smaller number of breech cases. However, this class distribution is reflective of the real-world prevalence of fetal presentations in the third trimester, where cephalic presentation is considerably more common than breech presentation [5].

3.2 Fetal Presentation Classification

We evaluated the performance of our fetal presentation classification method (see Sect. 2.3) on the dataset of 13 subjects. In 2 out of the 13 cases, our pipeline failed to detect the presence of a fetal head within the blind sweep videos, and consequently, presentation classification could not be performed for these cases. For the remaining 11 subjects, where fetal head detection was successful, our template matching algorithm (see Fig. 3) was able to classify all 10 "down" and the single "up" presentations correctly within this subset of 11 cases.

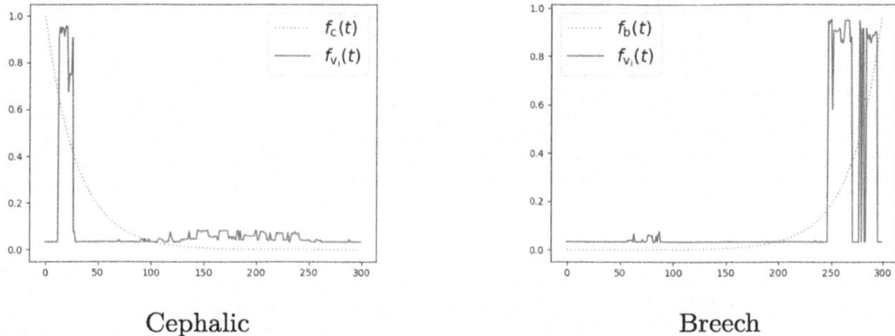

Fig. 3. Presentation prediction visualized with sweeps from 2 subjects. The temporal softmax probability for head from PCBM is indicated in blue, while the cephalic and breech template functions (see 1) are indicated in gray. (Color figure online)

3.3 Fetal Lie Classification

Next, we evaluated the performance of our fetal lie determination method. Our method, as described in Sect. 2.4, predicts a continuous direction vector representing the fetal facing direction. To align with the binary "left" versus "right" lie annotations provided in our dataset, we first discretized our directional predictions by binning them into "left-facing" and "right-facing" categories based on the vector's orientation relative to the image axes. In addition to the two cases where fetal head detection failed, there were two further cases among the 13 where, despite successful head detection, the PCBM model did not reliably segment the thalamus and CSP, precluding lie determination for those cases. For the remaining 9 subjects, where our pipeline successfully segmented the necessary anatomical structures and generated a lie prediction, we were able to classify all of them correctly when comparing our binned predictions to the "left" or "right" lie annotations. Among these lie predictions, as visualized in Fig. 4, some were derived using both thalamus and CSP segmentations, while others relied on the thalamus-only fallback method in cases where CSP segmentation was not reliable.

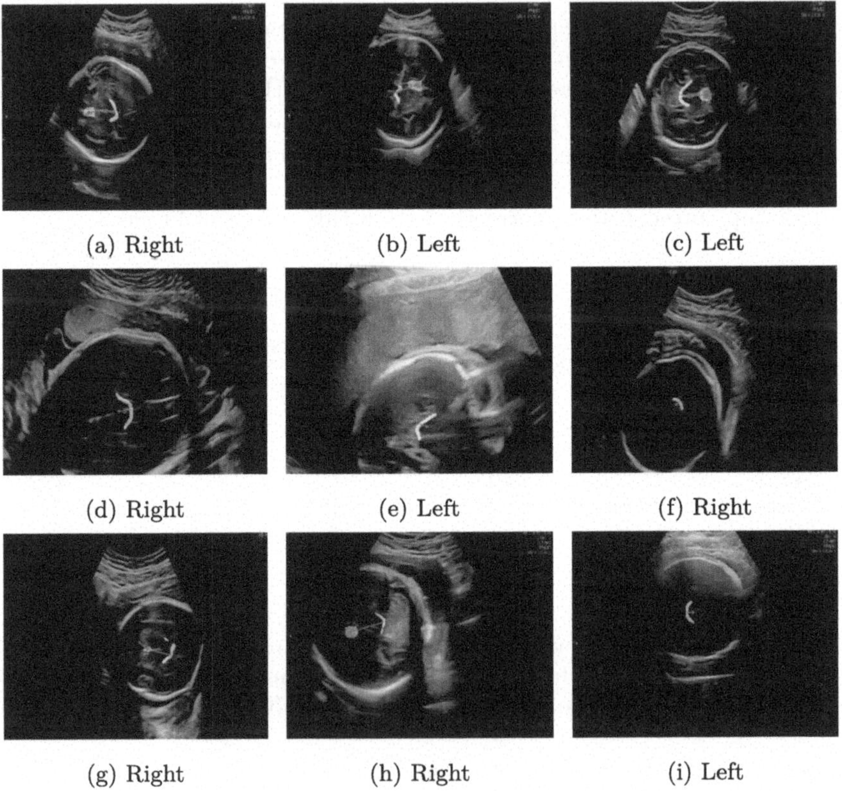

Fig. 4. Lie prediction visualized for 9 subjects, with ground-truth in the captions.

4 Discussions

4.1 Limitations

As shown in Sect. 3, our proposed pipeline relies on the successful detection and segmentation of the fetal head and its internal anatomical structures (thalamus and CSP). While PCBM demonstrates robust performance in our experience, fetal head visualization in ultrasound is not always guaranteed. In later trimesters, the fetal head may descend into the pelvis, becoming partially or fully obscured and thus hindering detection during the blind sweep. Furthermore, even when the fetal head is detected, the visibility of internal structures like the thalamus and CSP can be compromised by shadowing artifacts or suboptimal image quality, potentially impacting the accuracy of segmentation and subsequent lie determination. It is important to acknowledge that while our pipeline achieves high accuracy when these steps are successful, its applicability is contingent upon the quality of the acquired ultrasound data.

Another inherent limitation stems from our assumption of limited fetal movement during the blind sweep acquisition. While fetal movement is a biological reality, it is often negligible for the purposes of coarse presentation and lie determination within the short duration of a sweep. Fetuses are unlikely to transition from cephalic to breech or left to right lie within a few seconds of scanning. However, we acknowledge that significant or rapid fetal movements during the sweep could potentially introduce inaccuracies in our method.

4.2 Future Works

One promising avenue for future development is to further simplify our blind sweep protocol to enhance its efficiency and practicality in busy clinical settings. Our current blind sweep protocol, involving five vertical sweeps, was initially designed to maximize the chances of capturing the fetal head regardless of its position within the maternal abdomen. This comprehensive approach aimed to ensure robust head detection and subsequent analysis. However, upon detailed review of our acquired dataset, we observed a recurring pattern: in a majority of successful cases, the fetal head and the anatomical structures necessary for lie determination were adequately visualized within a single vertical sweep, particularly when the sweep was performed along the midline of the abdomen. This suggests that the redundancy provided by five sweeps may not always be necessary for successful presentation and lie assessment, as long as the fetal head can be detected and segmented effectively within that single sweep.

Therefore, the protocol can potentially be simplified to a single vertical sweep together with an adaptive acquisition strategy. In this refined workflow, the system would initially prompt the sonographer to perform only the single midline vertical sweep. The automated pipeline would then analyze the acquired data. If the system encounters challenges - such as failure to detect the fetal head or reliably segment anatomical structures - the system would guide the sonographer to acquire additional vertical sweep lines to improve data quality; otherwise, it would terminate with presentation of the predictions following a successful automated analysis. This adaptive approach aims to strike a balance between minimizing scanning time in typical cases and ensuring reliable results even in more challenging scenarios, ultimately enhancing the clinical utility of our automated pipeline.

Acknowledgment. This work was supported by the Danish Pioneer Centre for AI (DNRF grant number P1) and SONAI - a Danish Regions' AI Signature Project.

References

1. Baumgartner, C.F., et al.: Sononet: real-time detection and localisation of fetal standard scan planes in freehand ultrasound. IEEE Trans. Med. Imaging **36**(11), 2204–2215 (2017)
2. Chen, L., Bentley, P., Mori, K., Misawa, K., Fujiwara, M., Rueckert, D.: Self-supervised learning for medical image analysis using image context restoration. Med. Image Anal. **58**, 101539 (2019)
3. Chen, Z., Liu, Z., Du, M., Wang, Z.: Artificial intelligence in obstetric ultrasound: an update and future applications. Front. Med. **8**, 733468 (2021)
4. Daugherty, J.M.: Burnout: how sonographers and vascular technologists react to chronic stress. J. Diagnostic Med. Sonography **18**(5), 305–312 (2002)
5. Fox, A., Chapman, M.G.: Longitudinal ultrasound assessment of fetal presentation: a review of 1010 consecutive cases. Aust. N. Z. J. Obstet. Gynaecol. **46**(4), 341–344 (2006)
6. Gleed, A.D., et al.: Statistical characterisation of fetal anatomy in simple obstetric ultrasound video sweeps. Ultrasound Medi. Biol. **50**(7), 985–993 (2024)
7. Gomes, R.G., et al.: A mobile-optimized artificial intelligence system for gestational age and fetal malpresentation assessment. Commun. Med. **2**(1), 128 (2022)
8. Hermawati, F.A., Tjandrasa, H., Sugiono, Sari, G.P., Azis, A.: Automatic femur length measurement for fetal ultrasound image using localizing region-based active contour method. In: Journal of Physics: Conference Series, vol. 1230, p. 012002. IOP Publishing (2019)
9. van den Heuvel, T.L., Petros, H., Santini, S., de Korte, C.L., van Ginneken, B.: Automated fetal head detection and circumference estimation from free-hand ultrasound sweeps using deep learning in resource-limited countries. Ultrasound Med. Biol. **45**(3), 773–785 (2019)
10. Johnson, J., Arezina, J., McGuinness, A., Culpan, A.M., Hall, L.: Breaking bad and difficult news in obstetric ultrasound and sonographer burnout: is training helpful? Ultrasound **27**(1), 55–63 (2019)
11. Kim, B., Kim, K.C., Park, Y., Kwon, J.Y., Jang, J., Seo, J.K.: Machine-learning-based automatic identification of fetal abdominal circumference from ultrasound images. Physiol. Meas. **39**(10), 105007 (2018)
12. Komatsu, M., et al.: Towards clinical application of artificial intelligence in ultrasound imaging. Biomedicines **9**(7), 720 (2021)
13. Lee, C., et al.: Development of a machine learning model for sonographic assessment of gestational age. JAMA Netw. Open **6**(1), e2248685–e2248685 (2023)
14. Lin, M., Feragen, A., Bashir, Z., Tolsgaard, M.G., Christensen, A.N.: I saw, I conceived, I concluded: progressive concepts as bottlenecks. arXiv preprint arXiv:2211.10630 (2022)

15. Pokaprakarn, T., et al.: AI estimation of gestational age from blind ultrasound sweeps in low-resource settings. NEJM Evidence **1**(5), EVIDoa2100058 (2022)
16. Sharma, H., Drukker, L., Papageorghiou, A.T., Noble, J.A.: Machine learning-based analysis of operator pupillary response to assess cognitive workload in clinical ultrasound imaging. Comput. Biol. Med. **135**, 104589 (2021)
17. Singh, N., et al.: Occupational burnout among radiographers, sonographers and radiologists in Australia and New Zealand: findings from a national survey. J. Med. Imaging Radiat. Oncol. **61**(3), 304–310 (2017)
18. Song, K., Feng, J., Chen, D.: A survey on deep learning in medical ultrasound imaging. Front. Phys. **12**, 1398393 (2024)
19. Tan, J., Au, A., Meng, Q., Kainz, B.: Semi-supervised learning of fetal anatomy from ultrasound. In: Wang, Q., et al. (eds.) DART/MIL3ID -2019. LNCS, vol. 11795, pp. 157–164. Springer, Cham (2019). https://doi.org/10.1007/978-3-030-33391-1_18
20. Tolsgaard, M.G.: Assessment and learning of ultrasound skills in Obstetrics & Gynecology. University of Copenhagen, Faculty of Health and Medical Sciences (2017)
21. Tran, M.: Incidence and cause of occupational burnout syndrome among sonographers. J. Diagnostic Med. Sonography **40**(3), 233–239 (2024)
22. Wong, C.K., et al.: Deployment of deep learning model in real world clinical setting: a case study in obstetric ultrasound. arXiv preprint arXiv:2404.00032 (2024)
23. Zeng, Y., Tsui, P.H., Wu, W., Zhou, Z., Wu, S.: Fetal ultrasound image segmentation for automatic head circumference biometry using deeply supervised attention-gated v-net. J. Digit. Imaging **34**, 134–148 (2021)

A Comparison of Deep Learning Methods for Cell Detection in Digital Cytology

Marco Acerbis(✉), Nataša Sladoje, and Joakim Lindblad

Center for Image Analysis, Department of Information Technology, Uppsala University, Uppsala, Sweden
marco.acerbis@it.uu.se

Abstract. Accurate and efficient cell detection is crucial in many biomedical image analysis tasks. We evaluate the performance of several Deep Learning (DL) methods for cell detection in Papanicolaou-stained cytological Whole Slide Images (WSIs), focusing on accuracy of predictions and computational efficiency. We examine recent *off-the-shelf* algorithms as well as custom-designed detectors, applying them to two datasets: the CNSeg Dataset and the Oral Cancer (OC) Dataset. Our comparison includes well-established segmentation methods such as StarDist, Cellpose, and the Segment Anything Model 2 (SAM2), alongside centroid-based Fully Convolutional Regression Network (FCRN) approaches. We introduce a suitable evaluation metric to assess the accuracy of predictions based on the distance from ground truth positions. We also explore the impact of dataset size and data augmentation techniques on model performance. Results show that centroid-based methods, particularly the Improved Fully Convolutional Regression Network (IFCRN) method, outperform segmentation-based methods in terms of both detection accuracy and computational efficiency. This study highlights the potential of centroid-based detectors as a preferred option for cell detection in resource-limited environments, offering faster processing times and lower GPU memory usage without compromising accuracy.

Keywords: Cell Detection · Digital Cytology · Deep Learning · Whole Slide Imaging

1 Introduction

Early stage cancer detection is crucial in order to limit and effectively treat tumor formation. Cytopathological analysis can be a powerful and minimally invasive tool to identify anomalies and suspicious cells. However, to accurately analyze hundreds of thousands of cells in Whole Slide Images (WSIs) is a tedious and difficult task even for the most capable cytologist. DL-based methods allow to identify, extract and classify cells in WSIs, and to support the pathologist in detecting malignancy. The first step in such a pipeline is to detect cells and distinguish them from other formations or external materials. In this work we compare different methods for cell detection, ranging from *off-the-shelf* algorithms

J. Petersen and V. A. Dahl (Eds.): SCIA 2025, LNCS 15726, pp. 264–277, 2025.
https://doi.org/10.1007/978-3-031-95918-9_19

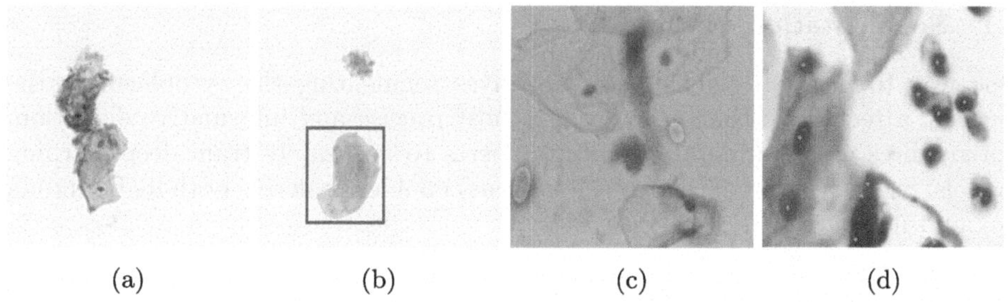

(a) (b) (c) (d)

Fig. 1. (a) Part of cytology image extracted from the OC Dataset; (b) MedSam [19] used to segment the object inside a given bounding box; (c) Example of nuclei segmentation using StarDist [26]; (d) Cell detection via a centroid-based method (FCRN [18,29]).

to custom designed and trained detectors. An important aspect for cell detection algorithms is the output format that determines how a detected cell is localized. Segmentation based techniques [12,19,22,26,27], as exemplified in Fig. 1(b)–(c), are commonly used, and provide pixel-level masks of detected objects of interest. However, such methods require precise and detailed annotations to be effectively trained. Generating such annotations is time consuming and typically requires the supervision of experts. In many situations, a detailed and precise delineation is not necessary and localizing each cell is enough to analyze it. An often used approach to localize an object in an image is to draw its bounding box [11,16,17,28]. An even less laborious alternative is provided by centroid-based methods. An example is shown in Fig. 1(d). These methods rely only on center-point annotations which are very fast and easy to collect. An example of such a method is proposed in Lu et al. [18]. We evaluate centroid-based approaches in comparison with alternatives relying on segmentation.

We conduct a series of experiments on two cytology datasets. To evaluate the performances, we develop a custom metric, *Localization Error*, that incorporates the distance between prediction and ground truth in the performance score. In addition, we study the impact of limited training data on performance of non-pretrained methods. To facilitate reproducibility, we share our complete implementation and evaluation framework as open source: https://github.com/MIDA-group/Cell-Detection.

2 Background and Related Work

Cell detection is an important step in medical image analysis. Development of efficient and fast ways to locate cell nuclei has been stimulated by competitions like the *2018 Data Science Bowl* [1] or the *CoNIC Challenge* [6]. The existing approaches include methods that result in (i) cell segmentation, (ii) a bounding box around the cell, and (iii) a centroid of a cell. We briefly describe methods for segmentation and centroid-based detection. We do not include any of the bounding-box detectors, due to their observed weaker performance compared to the other approaches, as presented in Gräbel et al. [4].

2.1 Segmentation Methods

One way to perform cell detection involves segmenting the cytoplasm or the nucleus, a technique that provides the most precise and informative detection, but requires costly annotations from experts to accurately train deep learning models. Segmentation of cell (or cell nucleus) enables to derive both its bounding box and its centroid.

StarDist. StarDist [26] makes use of star-convex polygons and leverages the capabilities of the U-Net architecture [23] to rely on the nucleus position rather than the cell boundaries. The proposed method particularly focuses on the challenge of overlapping cells.

Cellpose. Proposed in 2020, Cellpose [27] replaces classical methods based on the watershed algorithm for cell segmentation by implementing a U-Net architecture. It leverages a set-up for two-channel images: the first common channel is used for the cytoplasmic label, and the second optional channel shares information about the nucleus structure and position. From these inputs, Cellpose generates maps to recognize which pixels should be grouped together in the final masks.

HoVerNet. HoVerNet [5] implements a one-network architecture that simultaneously segments and classifies cells. HoVerNet does not rely on the U-Net architecture, but utilizes its own custom architecture inspired by the Preact-ResNet50 [7] model, to generate horizontal and vertical gradient maps used to reconstruct each cell mask.

MedSAM. Segment Anything Model (SAM) [12] affirmed itself among the *state-of-the-art* methods for general image segmentation. Ma et al. [19] proposes a modified version, MedSAM, that is fine-tuned on images from different medical fields. MedSAM requires user-drawn bounding box prompts, making it unsuitable for extracting nuclei from batches of images. An example of a segmented cell within a user-defined bounding-box is shown in Fig. 1(b).

Segment Anything Model 2 (SAM 2). SAM 2 [22] is the second generation of the foundation model Segment Anything and introduces an enhanced architecture to handle both image and video segmentation. Notably, it implements an MAE [8] pre-trained Hiera [24] to capture high-resolution details, and a Memory Mechanism to handle information from previous and promped frames.

2.2 Centroids-Based Methods

The main drawback of segmentation models is their need for precisely annotated data to be properly trained. However, detailed segmentation is often not required; detected centroids of cell nuclei are often sufficient to locate and cut-out the cells for further processing. Collecting simpler annotations could significantly reduce the required workload and even the need for experts.

Fully Convolutional Regression Networks (FCRNs). Lu et al. [18] proposed to extract the centroids of cell nuclei by means of a Fully Convolutional Regression Network (FCRN) [29] approach. In their pipeline, a modified U-Net [23] was trained to generate density maps that highlights the position of each detected nucleus. The centroid positions are then extracted from the predicted blobs. Lian et al. [15] further develop the FCRN model, reducing the U-Net size and replacing thresholding by local maxima detection, however without presenting any explicit performance evaluation of the performed modifications.

ACFormer. In Huang et al. [9] the authors underline that cell segmentation is (in many cases) not a necessary step in a cell analysis pipeline. They propose a centroid-based detector that leverages the capabilities of the transformer architecture. The introduced *Affine Consistent Transformer* (ACFormer) aims to locate and classify cell nuclei by leveraging the capabilities of two sub-networks, a local and a global network. The first learns how to handle objects at smaller scales, while the second one handles the large-scale predictions.

Cell-DETR. Segmentation algorithms usually struggle to efficiently process large WSIs, especially in resource-limited settings. Cell-DETR [21] proposes to adapt detection transformers (DETR) [2] as a cost-effective and fast solution to locate and classify cells without the need of segmentation. It uses a hierarchical backbone to generate a four level feature pyramid of the input, that is further processed by a multi-scale DETR [31] composed of 6 encoder and 6 decoder layers.

3 Data

We evaluate the considered methods on two datasets: the *CNSeg Dataset* and the *Oral Cancer (OC) Dataset*. We create four different training/validation/test splits of each dataset to perform a 4-folded cross-validation. The validation set is used to tune hyperparameters during the training phase and to calibrate specific algorithms, for example by finding the value of the *radius* parameter for Cellpose. The training set is used to train the centroid-based detectors.

Table 1. (a) Number of images in train/validation/test sets for each of the four splits of the CNSeg Dataset. (b) Number of ground truth nuclei in each of the four non-overlapping test sets (each containing 477 images) of the CNSeg Dataset. (c) Number of images in train/validation/test sets of the OC Dataset. (d) Number of ground truth nuclei in each test set of the OC Dataset (each containing 78 patches from one out of four WSIs).

Subset	Images
Training	2462
Validation	500
Test	477
Total	3439

(a)CNSeg Dataset

Fold	Nuclei
1	8221
2	9075
3	7600
4	9419

(b) CNSeg Dataset

Subset	Images
Training	156
Validation	78
Test	78
Total	312

(c)OC Dataset

Fold	Nuclei
1	381
2	364
3	571
4	236

(d)OC Dataset

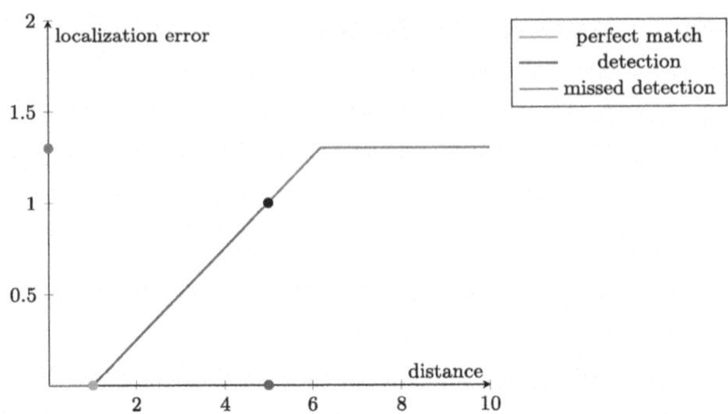

Fig. 2. Localization Error ε_l as a function of distance between detection and ground truth, for $s = 1$, $t = 5$ and $\alpha = 0.3$.

CNSeg Dataset. The CNSeg Dataset [30] is a collections of cervical cell images. All images in this dataset were prepared in a standardized manner: Papanicolaou staining was used, and the images have a resolution of 0.25 μm/px. Figure 1(c)–(d) shows segmentation and centroid detection on two images from the dataset. We use the subset *PatchSeg*, comprising a total of 3439 annotated images of size 512 × 512 px. Tables 1(a)–(b) summarize how we divide the samples in each subset and the total number of ground truth nuclei in each test set. Each nucleus annotation consists of a polygon; we calculated the ground truth centroid position (C_x, C_y) as the geometric centroid of the polygon.

Oral Cancer Dataset. The Oral Cancer (OC) Dataset consists of 312 image patches of size 256 × 256 px extracted from 4 WSIs obtained from LBC-prepared Papanicolaou stained slides of brush-sampled cells from the oral cavity. The slides have been imaged under white light using a NanoZoomer S60 digital slide scanner, providing a resolution of 0.23 μm/px. Ground truth centroids in each sample have been annotated by non-specialists using the CytoBrowser [25] tool. Figure 1(a)–(b) shows examples of tiles of the OC dataset. Tables 1(c)–(d) summarize how the dataset is used to create different subsets for each split and the number of ground truth nuclei in the test sets. Cells from one WSI do not appear in more than one set per fold.

4 Evaluation Metrics

Localization Error (ε_l). A standard follow-up to nucleus detection involves cutting out the cell image using a fixed square window centered on the centroid position. Inspired by previous works (Huang et al. [3,9]) on centroid-based detection, we define an evaluation metric that accounts for the distance between the predicted centroid and the corresponding ground truth, in addition to measuring missed detections, False Negatives (FN), and extra predictions, False Positives (FP). After calculating the Euclidean distance between each predicted centroid

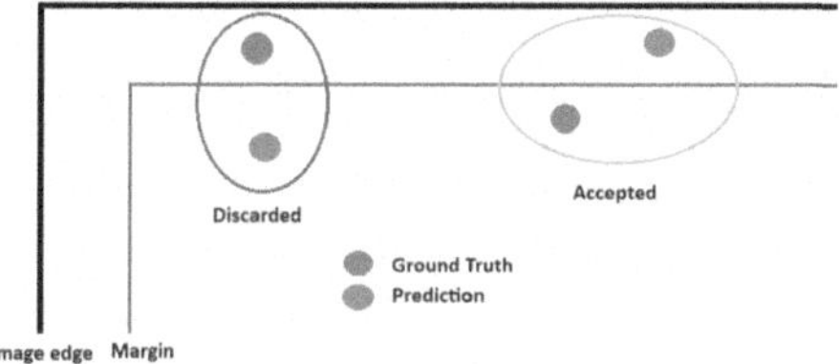

Fig. 3. Examples of discarded (red, left) and accepted (green, right) detections. If the ground truth lies in the margin, the centroid and any associated prediction are discarded and not included in the analysis. (Color figure online)

and ground truth centroid, matching is performed using the *Hungarian Algorithm* [13]. Equation (1) defines how the error is assigned based on the distance d between the ground truth and the matched centroid:

$$\varepsilon_l(d) = max(0, min(1 + \alpha, \frac{d - s}{t - s})) . \tag{1}$$

- If the distance d is less than or equal to the *slack* s, defined as 0.25 of the average nucleus diameter, the error is 0, indicating a perfect match;
- If d is greater than s but less than or equal to the *threshold* t, defined as the average nucleus diameter, the error increases linearly with d up to 1;
- If the distance d exceeds t, the detection is considered a miss and both FP and FN counts are incremented by 1. To avoid a discontinuity in the error measure, $\varepsilon_l(d)$ continues to increase linearly until it saturates at $1 + \alpha$.

Figure 2 shows an example of the localization error as a function of the distance from the matched ground truth.

FNs (missed cells) are assigned a Localization Error of 1, whereas unmatched FPs (spurious detections) are assigned an error of α. By varying the value of α, we can modulate the cost imposed by FP detections. While FPs are typically handled well by a following classification step, an excessive number can become challenging and computationally too expensive to process. The overall Localization Error $\mathcal{E}_l$ is the sum of Localization Errors for all images divided by the total number N of ground truth nuclei, according to Eq. (2):

$$\mathcal{E}_l = \frac{1}{N} \sum_{N} \varepsilon_l . \tag{2}$$

Near-Edge Detections. Cells close to the edge of the image may be captured only partially, thus being unsuitable for further analysis. We therefore discard each ground truth nucleus whose distance from the image edge is less than average nucleus diameter. Figure 3 shows that independently from the predicted centroid position, the pair is accepted and included in the results only if the ground truth lies inside the margin.

Precision, Recall, and F-Score. Common metrics to evaluate object detection methods are Precision, Recall and F-score. To make our findings comparable with past and future similar works, we also include these metrics in our analysis.

Inference Rate and GPU Memory. In cytology studies, the usual approach is to process thousands of images in order to extract tens to hundreds of cells from each. To process these large collections of images, machines with dedicated GPUs are typically used. Since the access to such resources is generally limited, we also include the inference rate (IR) and GPU memory usage at inference time in our analysis.

5 Methods

In this section, we describe in more detail how we implement the methods introduced in Sect. 2 that we include in our study.

Fully Convolutional Regression Network (FCRN). The FCRN approach described in [18] presents a way to perform nuclei detection relying only on centroid annotations to train a modified U-Net [23] for which the final softmax is replaced by a linear activation function. From the annotations, ground truth binary masks are constructed and then dilated by a disk of radius r, followed by convolution with a 2D Gaussian filter to generate a fuzzy ground truth D. The FCRN learns the mapping between the original image and the corresponding fuzzy ground truth. At inference, the model predicts a probability map for each image which is binarized at a threshold T and the centroid of each connected foreground component is extracted as a detected nucleus location.

In our tests, we replicate the U-Net architecture of [18][1], converting the original implementation to the most recent *Keras* release to overcome compatibility issues. We also implement the centroid extraction using Python instead of ImageJ Macro. We train the model following the authors instructions by generating ground truth binary masks around the centroid position, followed by Gaussian blurring. Each training run lasts 100 epochs and has a batch size of 32. A relevant hyperparameter for this model is the *threshold* value T used to binarize the network output. Based on empirical evaluation on the validation set, we set it to 0.58 for images from the CNSeg dataset, and to 0.65 for the OC dataset.

Improved FCRN (IFCRN). It is observed that the binarization of the network output with a global threshold T does not provide reliable results for heterogeneous dataset. In Lian et al. [15], nuclei locations are instead detected at local maxima of height $h > 0.5$ in the prediction output, leading to improved performance. Further, the original 23 convolution layers of the U-Net are reduced to 8 convolution layers, providing a much leaner model. Images are downsampled by a factor 4×4, further reducing the computational burden. The two steps for

[1] https://github.com/MIDA-group/OralScreen.

generating the fuzzy ground truth of Lu et al. [18], dilation followed by Gaussian blur, are replaced by only Gaussian blur ($\sigma = 3$).

In our tests, we replicate the architecture of [15]. Different from [15], we work with only one focus level, to reproduce the results on images from different datasets, like the CNSeg Dataset [30]. Further, we tune the detection sensitivity by adjusting the required minimal height h for the detected local maxima. Based on empirical evaluation on the validation set, we set it to 0.4 for images from both the CNSeg and the OC datasets.

Cellpose. Cellpose [27] is made available through a Python library with the same name that relies on PyTorch. There are a two different models: *"cyto"* and *"nuclei"*. The more complex *"cyto"* model, which uses both cell boundaries and nuclei information to segment the cell, performed significantly better on our validation sets, and we therefore use that model for our evaluation. The value of the parameter *diameter* requested by the algorithm is set to 30 on the CNSeg dataset and to *Auto* on the OC dataset. In this second case, the helper functions of Cellpose automatically calculate the value of the parameter. We select these settings since they obtained the best performance on the validation set.

StarDist. StarDist [26] is released in two different versions, accessible via its own Python library, and it is implemented with Tensorflow. One is trained on brightfield data of H&E stained cells extracted from the *MoNuSeg* 2018 [14] dataset and the *TNBC* dataset presented in Naylor et al. [20]. The second model is trained on fluorescence images using nuclear markers extracted from the *2018 Data Science Bowl* [1] dataset. Since both datasets in our study collect brightfield images, we select the former model for our experiments.

Segment Anything Model 2. Segment Anything 2 [22] model and weights can be directly downloaded from the official FAIR GitHub repository[2]. It leverages a transformer architecture trained on the SA-V dataset [22]. We study SAM2 L and SAM2 T. We include both versions because Huang et al. [10] highlights the superior overall performance of the non-fine-tuned large model compared to the tiny counterpart, whereas Ma et al. [19] focuses on a fine-tuned SAM Tiny [12].

6 Experiments and Results

In this section we present the results of the evaluation experiments. All experiments are performed on an Intel(R) Core(TM) i9-9940X running Gentoo Linux 6.6.67 with Python 3.12.8, PyTorch 2.5.1+cu124, and Tensorflow 2.18.0. The machine is equipped with a Nvidia GeForce RTX 4090 GPU card with 24 GB of memory.

[2] https://github.com/facebookresearch/sam2.

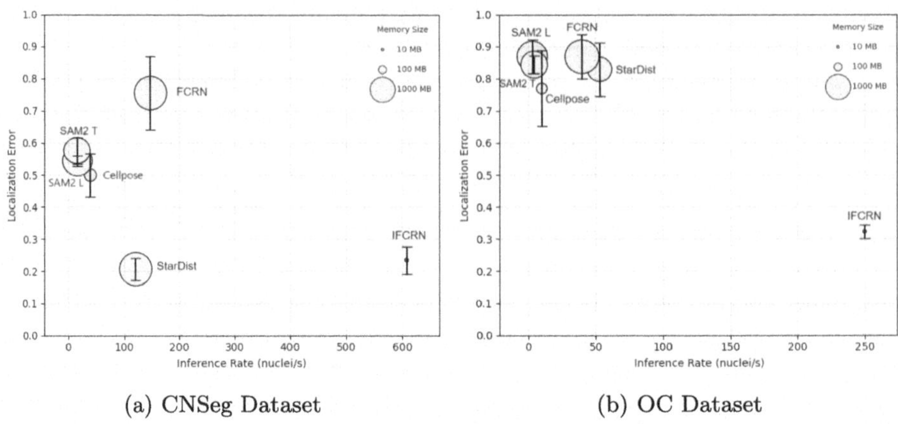

(a) CNSeg Dataset (b) OC Dataset

Fig. 4. Localization Error $\mathcal{E}_l$ ($\alpha = 0.3$) for the evaluated methods on (a) the CNSeg Dataset and on (b) the OC Dataset. Best performance is located in the right bottom part of the graph. The size of the circle for each method is proportional to the memory requirements of that model.

6.1 Comparison of Different Detection Techniques

The main experiments conducted in this study aim to compare the performance of different methods on the task of cell detection in cytology images. For both datasets, CNSeg and OC, we implemented the following pipeline:

- **Parameter tuning.** When needed, we select the values for a given method parameters by running different configurations on the validation set. The best performing configuration are then selected to run the experiments;
- **Zero-shot evaluation.** Segmentation methods are evaluated in zero-shot experiments without any re-training or fine-tuning;
- **Training from scratch.** For each split, FCRN and IFCRN are *trained from-scratch* and then evaluated on the corresponding test set.

The results, reported in Tables 2 and 3, are the weighted average of the results for each split, where the number of ground truth nuclei in the test set, reported in Table 1(b) and (d), is used as weight. Each result is followed by $\pm$ the standard deviation computed over the four folds.

CNSeg Results. From Table 2 and Fig. 4(a) we observe that the overall best performing methods on the CNSegn Dataset are StarDist and IFCRN. IFCRN achieves comparable results to StarDist and the overall best F-Score. It is also much faster with an inference rate of 608 nuclei$/s$, and requires only a fraction of the GPU memory at inference time of the other methods.

OC Results. Table 3 and Fig. 4(b) present the results for the OC Dataset. For this dataset IFCRN turns out to be the best performing method. The $\mathcal{E}_l$ for

Table 2. Results on the CNSeg dataset. Inference rate (IR) is measured in nuclei/s. Memory refers to the GPU memory required at inference time.

Model	$\mathcal{E}_l$ ($\alpha = 0.3$)	$\mathcal{E}_l$ ($\alpha = 1$)	Precision	Recall	F-Score	IR (n/s)	Memory (MB)
Cellpose	0.50 ± 0.07	0.67 ± 0.09	79.3 ± 3.0	76.5 ± 2.9	77.8 ± 2.8	41	526
StarDist	**0.21** ± 0.03	**0.31** ± 0.06	86.2 ± 3.4	**91.2** ± 1.6	88.6 ± 2.5	121	4317
SAM2 L	0.54 ± 0.02	0.74 ± 0.03	77.2 ± 2.4	75.8 ± 0.85	76.7 ± 1.0	17	3418
SAM2 T	0.57 ± 0.04	0.75 ± 0.05	81.1 ± 2.5	73.0 ± 2.4	76.8 ± 1.0	16	2687
FCRN	0.75 ± 0.11	0.79 ± 0.12	**94.7** ± 2.4	60.6 ± 4.3	73.8 ± 3.7	149	4297
IFCRN	0.23 ± 0.04	0.32 ± 0.05	88.8 ± 1.8	89.1 ± 2.6	**88.9** ± 2.0	**608**	**3**

Table 3. Results on the OC dataset. Inference rate (IR) is measured in nuclei/s. Memory refers to the GPU memory required at inference time.

Model	$\mathcal{E}_l$ ($\alpha = 0.3$)	$\mathcal{E}_l$ ($\alpha = 1$)	Precision	Recall	F-Score	IR (n/s)	Memory (MB)
Cellpose	0.77 ± 0.12	1.2 ± 0.29	64.6 ± 11	72.1 ± 4.5	67.8 ± 6.7	10	403
StarDist	0.83 ± 0.08	0.86 ± 0.09	**93.7** ± 3.3	60.5 ± 5.2	73.4 ± 4.0	53	2203
SAM2 L	0.87 ± 0.05	1.7 ± 0.1	47.4 ± 1.6	71.1 ± 2.7	56.9 ± 1.7	3	3506
SAM2 T	0.84 ± 0.03	1.3 ± 0.1	61.4 ± 5.6	65.7 ± 3.2	63.3 ± 2.0	5	2726
FCRN	0.87 ± 0.07	0.99 ± 0.13	86.6± 8.5	60.8 ± 2.3	71.3 ± 3.4	40	4293
IFCRN	**0.32** ± 0.02	**0.50** ± 0.14	80.5 ± 1.1	**86.6** ± 2.3	**83.0** ± 6.4	**250**	**1**

$\alpha = 0.3$ is 42% better than the results for the second best performing model, Cellpose. At the same time, IFCRN inference rate is 5 times the second fastest method, StarDist, while only using a small fraction of GPU memory at inference time.

6.2 Impact of the Amount of Training Data

Off-the-shelf methods, and foundation models (such as SAM 2) in particular, are trained on large datasets, making them a powerful tool even when fine-tuning is not possible. Smaller customizable models, like FCRN, can be trained on relatively small annotated data. In this experiment we use training sets, extracted from the CNSeg Dataset [30], of varying size to train IFCRN. Each training run has 450 epochs and a batch size of 32.

Data Augmentation. A key aspect to achieve good results is the amount of data available. It is often beneficial to include data augmentation to train a more robust model. We explore different augmentations; from simple *Random Rotations* and *Flips* to *Color Jitter*, *GaussianBlur*, and *GaussianNoise*.

Results. Figure 5 shows the positive impact of increasing the number of samples in the training set. IFCRN results are comparable to those of Cellpose and SAM2 with only 100 images in the training set, while with 1000 images IFCRN reaches

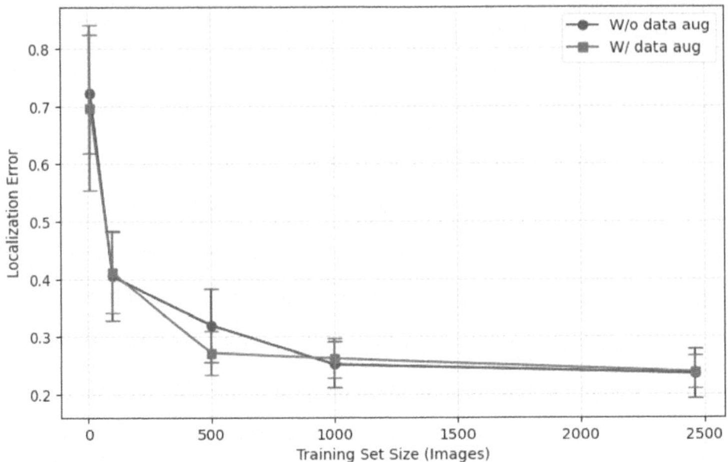

Fig. 5. Impact of different size of the training set for IFCRN on the CNSeg Dataset, without and with data augmentation.

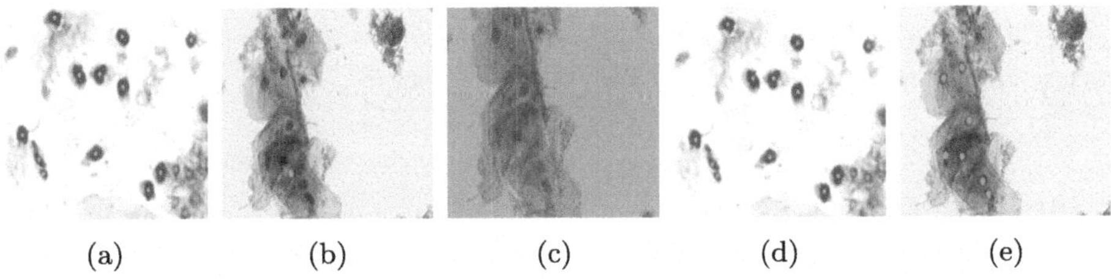

(a) (b) (c) (d) (e)

Fig. 6. Example results of StarDist and IFCRN on images from CNSeg and OC Datasets: (a) StarDist successfully detects the nuclei in an image from the CNSeg Dataset. (b) StarDist fails to locate the cell nuclei in an image from the OC Dataset. (c) Failed segmentation attempt of StarDist on the same OC image. (d) IFCRN successfully detects the nuclei in an image from the CNSeg Dataset. (e) IFCRN successfully detects the nuclei in an image from the OC Dataset.

results in the range of the top performing models, like StarDist. As expected, data augmentations can contribute to improve the results, especially when the number of samples is very limited.

7 Discussion

From the results presented in Sect. 6.1, we observe a much reduced performance of all the segmentation methods on OC data, compared to CNSeg Dataset. Figure 6 shows two examples. Images extracted from the CNSeg Dataset generally present well defined and isolated cells whose nuclei can also be easily identified; possibly explaining the good results of the segmentation algorithms such as StarDist, as shown in Fig. 6(a). On the contrary, images from the OC Dataset may present groups of overlapping cells, whose boundaries and nuclei are not always clear and easy to recognize. This may cause segmentation algorithms to fail in properly detecting the cell nucleus as shown in Fig. 6(b)–(c). Both FCRN and IFCRN maintain similar results on both datasets. This result can be explained by how these methods approach the task: instead of outlining the cell boundary, that is not always well defined, they search for the usually more highlighted nucleus. This way they can still perform well even when the cells are clustered or overlapping. Figure 6(d)–(e) shows two example of IFCRN results on images from both datasets.

A WSI typically contains 10,000–150,000 cells, making inference speed important when processing multiple WSIs. In terms of inference rate and GPU memory usage, IFCRN is by far the best performing method. While it is expected that a foundation model such as SAM 2 would be more computationally demanding, even other U-Net based architectures under-perform. For Cellpose or StarDist this may be due to the extra resources and time needed to generate segmentation masks, from which sub-regions the centroids are extracted. FCRN fails to efficiently implement nucleus detection by using a deeper network than IFCRN and by implementing a threshold-based extraction of the predicted centroids.

From the second experiment, presented in Sect. 6.2, we observe that not only, as expected, more samples in the training set leads to improved results, but also that already rather small datasets are enough to obtain results comparable to, or better than those of *off-the-shelf* methods. This important result shows how custom solutions trained on small dataset provide a valid and strong alternative to more generally (pre-)trained solutions. The impact of data augmentations can also be relevant, especially with very limited datasets, even if it requires more testing and parameter tuning to find the optimal configuration.

8 Conclusion

We present a comparison between contemporary deep learning-based segmentation and centroid-based cell detectors. The aim of our study is to find efficient solutions to extract cells from WSIs for further analysis. We observe that centroid-based methods, and in particular the IFCRN method, perform on par or better than segmentation-based approaches, especially when cells are clustered or overlapping. The IFCRN method delivers the overall best performance in our tests performed. The IFCRN method also requires much less computational power and can process nuclei up to 50× faster than the other methods.

We observe that, under the constraint of limited data, a simpler and more tailored solution may be a better choice for the task of cell detection, reaching results comparable with those of more complex pre-trained method.

Acknowledgements. This work is supported by: Sweden's Innovation Agency (VINNOVA) through the Analytic Imaging Diagnostics Arena (AIDA), grant 2021-01420, the Swedish Research Council, grant 2022-03580, and Cancerfonden projects 22 2353 Pj and 22 2357 Pj. We are grateful to M. Matić for accurate cell location annotations.

Disclosure of Interests. The authors have no competing interests to declare that are relevant to the content of this article.

References

1. Caicedo, J.C., Goodman, A., Karhohs, K.W., et al.: Nucleus segmentation across imaging experiments: the 2018 data science bowl. Nat. Methods **16**(12), 1247–1253 (2019). https://doi.org/10.1038/s41592-019-0612-7
2. Carion, N., Massa, F., Synnaeve, G., Usunier, N., Kirillov, A., Zagoruyko, S.: End-to-end object detection with transformers. In: Vedaldi, A., Bischof, H., Brox, T., Frahm, J.-M. (eds.) ECCV 2020. LNCS, vol. 12346, pp. 213–229. Springer, Cham (2020). https://doi.org/10.1007/978-3-030-58452-8_13
3. Dolezel, P., Skrabanek, P., Stursa, D., et al.: Centroid based person detection using pixelwise prediction of the position. J. Comput. Sci. **63**, 101760 (2022). https://doi.org/10.1016/j.jocs.2022.101760
4. Gräbel, P., Özkan, Ö., Crysandt, M., et al.: State of the art cell detection in bone marrow whole slide images. J. Pathol. Inform. **12**, 36 (2021). https://doi.org/10.4103/jpi.jpi_71_20
5. Graham, S., Vu, Q.D., Raza, S., et al.: Hover-Net: simultaneous segmentation and classification of nuclei in multi-tissue histology images. Med. Image Anal. **58**, 101563 (2019). https://doi.org/10.1016/j.media.2019.101563
6. Graham, S., Vu, Q.D., Jahanifar, M., et al.: CoNIC challenge: pushing the frontiers of nuclear detection, segmentation, classification and counting. Med. Image Anal. **92**, 103047 (2024). https://doi.org/10.1016/j.media.2023.103047
7. He, K., Zhang, X., Ren, S., Sun, J.: Identity mappings in deep residual networks. In: Leibe, B., Matas, J., Sebe, N., Welling, M. (eds.) ECCV 2016. LNCS, vol. 9908, pp. 630–645. Springer, Cham (2016). https://doi.org/10.1007/978-3-319-46493-0_38
8. He, K., Chen, X., Xie, S., et al.: Masked autoencoders are scalable vision learners. In: 2022 IEEE/CVF Conference on Computer Vision and Pattern Recognition (CVPR), pp. 15979–15988 (2022). https://doi.org/10.1109/CVPR52688.2022.01553
9. Huang, J., Li, H., Wan, X., et al.: Affine-consistent transformer for multi-class cell nuclei detection. In: 2021 ICCV, pp. 21327–21336 (2023). https://doi.org/10.1109/iccv51070.2023.01955
10. Huang, Y., Yang, X., Liu, L., et al.: Segment anything model for medical images? Med. Image Anal. **92**, 103061 (2024). https://doi.org/10.1016/j.media.2023.103061
11. Huang, Z., Patel, B., Lu, W., et al.: Yeast cell detection using fuzzy automatic contrast enhancement (FACE) and you only look once (YOLO). Sci. Rep. **13**(1), 16222 (2023). https://doi.org/10.1038/s41598-023-43452-9
12. Kirillov, A., Mintun, E., Ravi, N., et al.: Segment anything. arXiv preprint arXiv:2304.02643 (2023)
13. Kuhn, H.W.: The Hungarian method for the assignment problem. Nav. Res. Logist. Q. **2**(1–2), 83–97 (1955). https://doi.org/10.1002/nav.3800020109
14. Kumar, N., Verma, R., Sharma, S., et al.: A dataset and a technique for generalized nuclear segmentation for computational pathology. IEEE Trans. Med. Imaging **36**(7), 1550–1560 (2017)
15. Lian, W., Lindblad, J., Runow Stark, C., et al.: Let it shine: autofluorescence of papanicolaou-stain improves AI-based cytological oral cancer detection. Comput. Biol. Med. **185**, 109498 (2025). https://doi.org/10.1016/j.compbiomed.2024.109498

16. Liang, H., Cheng, Z., Zhong, H., et al.: A region-based convolutional network for nuclei detection and segmentation in microscopy images. Biomed. Signal Process. Control **71**, 103276 (2022). https://doi.org/10.1016/j.bspc.2021.103276
17. Liu, C., Li, D., Huang, P.: ISE-YOLO: improved squeeze-and-excitation attention module based YOLO for blood cells detection. In: 2021 IEEE International Conference on Big Data, pp. 3911–3916 (2021). https://doi.org/10.1109/BigData52589.2021.9672069
18. Lu, J., Sladoje, N., Runow Stark, C., Darai Ramqvist, E., Hirsch, J.-M., Lindblad, J.: A deep learning based pipeline for efficient oral cancer screening on whole slide images. In: Campilho, A., Karray, F., Wang, Z. (eds.) ICIAR 2020. LNCS, vol. 12132, pp. 249–261. Springer, Cham (2020). https://doi.org/10.1007/978-3-030-50516-5_22
19. Ma, J., He, Y., Li, F., et al.: Segment anything in medical images. Nat. Commun. **15**(1), 654 (2024)
20. Naylor, P., Lae, M., Reyal, F., et al.: Segmentation of nuclei in histopathology images by deep regression of the distance map. IEEE Trans. Med. Imaging **38**(2), 448–459 (2019)
21. Pina, O., Dorca, E., Vilaplana, V.: Cell-DETR: efficient cell detection and classification in WSIs with transformers. In: Medical Imaging with Deep Learning (2024)
22. Ravi, N., Gabeur, V., Hu, Y., et al.: SAM 2: segment anything in images and videos. arXiv preprint arXiv:2408.00714 (2024)
23. Ronneberger, O., Fischer, P., Brox, T.: U-Net: convolutional networks for biomedical image segmentation. In: Navab, N., Hornegger, J., Wells, W.M., Frangi, A.F. (eds.) MICCAI 2015. LNCS, vol. 9351, pp. 234–241. Springer, Cham (2015). https://doi.org/10.1007/978-3-319-24574-4_28
24. Ryali, C., Hu, Y., Bolya, D., et al.: Hiera: a hierarchical vision transformer without the bells-and-whistles. In: International Conference on Machine Learning (2023). https://doi.org/10.5555/3618408.3619632
25. Rydell, C., Lindblad, J.: CytoBrowser: a browser-based collaborative annotation platform for whole slide images. F1000Research **10**, 226 (2021). https://doi.org/10.12688/f1000research.51916.1
26. Schmidt, U., Weigert, M., Broaddus, C., Myers, G.: Cell detection with star-convex polygons. In: Frangi, A.F., Schnabel, J.A., Davatzikos, C., Alberola-López, C., Fichtinger, G. (eds.) MICCAI 2018. LNCS, vol. 11071, pp. 265–273. Springer, Cham (2018). https://doi.org/10.1007/978-3-030-00934-2_30
27. Stringer, C., Wang, T., Michaelos, M., et al.: Cellpose: a generalist algorithm for cellular segmentation. Nat. Methods **18**, 100–106 (2021). https://doi.org/10.1038/s41592-020-01018-x
28. Sun, Y., Huang, X., Zhou, H., et al.: SRPN: similarity-based region proposal networks for nuclei and cells detection in histology images. Med. Image Anal. **72**, 102142 (2021). https://doi.org/10.1016/j.media.2021.102142
29. Weidi Xie, J., Noble, A., Zisserman, A.: Microscopy cell counting and detection with fully convolutional regression networks. Comput. Methods Biomech. Biomed. Eng. Imaging Visualization **6**(3), 283–292 (2018). https://doi.org/10.1080/21681163.2016.1149104
30. Zhao, J., He, Y., Zhou, S., et al.: CNSeg: a dataset for cervical nuclear segmentation. Comput. Methods Programs Biomed. **241**, 107732 (2023). https://doi.org/10.1016/j.cmpb.2023.107732
31. Zhu, W., Su, W., Lu, L., et al.: Deformable DETR: deformable transformers for end-to-end object detection (2020)

Assessing the Efficacy of Multi-task Learning in Mammographic Density Classification: A Study on Class Imbalance and Model Performance

Suaiba A. Salahuddin[1(✉)], Elisabeth Wetzer[1], Kristoffer Wickstrøm[1], Solveig Thrun[1], Michael Kampffmeyer[1,2], and Robert Jenssen[1,2,3]

[1] Department of Physics and Technology, UiT The Arctic University of Norway, Tromsø, Norway
suaiba.a.salahuddin@uit.no
[2] Norwegian Computing Center, Oslo, Norway
[3] Pioneer Centre for AI, University of Copenhagen, Copenhagen, Denmark

Abstract. Multi-task learning in mammography has gained attention in recent years as a strategy to improve multi-label breast cancer and density classifications by leveraging shared representations. While previous studies have suggested that combining cancer and density labels in a multi-task learning framework can be beneficial, empirical evidence supporting these claims over single-task baselines remains limited. As breast density is a crucial factor for breast cancer risk but is less explored compared to cancer classification, this study focuses on density classification within the multi-task learning framework. Our findings indicate that single-task models can be beneficial over multi-task models, challenging the assumption that multi-task learning inherently provides performance benefits. Furthermore, we explore various sampling strategies in the multi-task setting to address class imbalance, a prevalent issue in mammography datasets, but find that none surpass the single-task baseline. Our findings suggest that the additional complexity of multi-task learning does not yield substantial benefits for density classification, highlighting the need for further research into alternative strategies for improving mammography classification performance.

Keywords: Mammography · Multi-Task Learning · Breast Density Estimation · Deep Learning · Artificial Intelligence

1 Introduction

Breast cancer (BC) is the most common cancer diagnosed worldwide, with its burden steadily increasing [34]. In the U.S., this trend is reflected in a notable rise

Supplementary Information The online version contains supplementary material available at https://doi.org/10.1007/978-3-031-95918-9_20.

J. Petersen and V. A. Dahl (Eds.): SCIA 2025, LNCS 15726, pp. 278–292, 2025.
https://doi.org/10.1007/978-3-031-95918-9_20

in BC diagnoses among women under 50 over the past two decades [34]. X-ray mammography, a non-invasive and low-dose screening method, plays a critical role in early BC detection [14]. Standard screening involves imaging both breasts from two projections, resulting in four images per subject.

Breast density is a well-known, independent risk factor for BC influencing the sensitivity of mammograms. It is defined as the amount of fibroglandular tissue relative to fatty tissue. Individuals with higher breast density are more than 4–6 times more likely to develop BC compared to those with fatty breast tissue [20]. Despite its critical role in BC risk assessment, breast density classification has received less attention than cancer detection. This is partly due to the complexity and subjectivity involved in determining density from mammograms, which can vary significantly among radiologists [29].

Both breast density and BC findings can be assessed using a standardised numerical system known as Breast Imaging-Reporting and Data System, BI-RADS [1]. The currently used 5th edition of BI-RADS distinguishes density into four classes; A: almost entirely fatty to D: extremely dense, while cancer findings are classified into seven categories: BI-RADS 0 (incomplete), BI-RADS 1 (negative), BI-RADS 2 (benign findings), BI-RADS 3 (likely benign), BI-RADS 4 (suspicious abnormality), BI-RADS 5 (strongly indicative of malignancy), and BI-RADS 6 (confirmed malignancy through biopsy). Figure 1 shows example images from each BI-RADS density category.

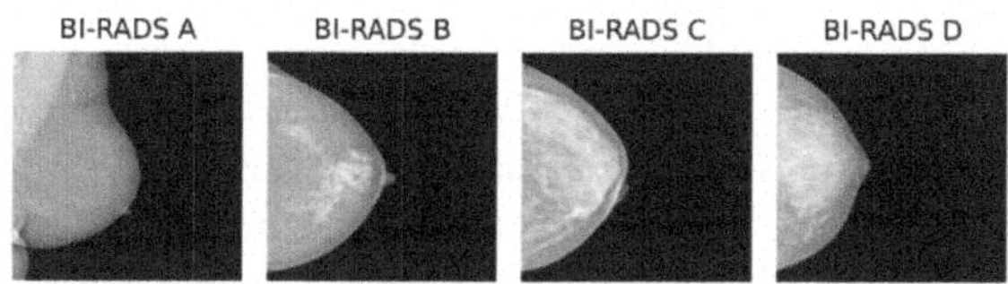

Fig. 1. Example VinDr-Mammo images from the four different BI-RADS density categories A-D showcasing increasing levels of fibroglandular tissue and potential masking effects with higher density levels.

Despite the success of mammographic screening, challenges remain including the reliance on radiologists for assessment which can lead to inconsistent interpretations and varying diagnosis and treatment plans [5,29].

Computer-aided diagnosis using deep learning (DL), has shown great promise in combating these issues and assisting radiologists in analysing mammograms across different tasks, including lesion detection, classification, and segmentation [19]. DL approaches have been proposed to improve the accuracy and consistency of breast density evaluations, offering valuable support to radiologists in their diagnostic processes [4,17,22,24].

Classifying BC BI-RADS findings remains more widely explored in literature than mammographic density. Notably, several studies have demonstrated the capability of DL systems as independent readers which reduce radiologist workloads and can perform on par if not better than radiologists to detect BC from mammograms and determine BC risk [2,15,18,26].

Typically, breast cancer and density are classified in a single-task fashion. However, recent literature increasingly supports the notion that generalization improves when multiple tasks are addressed simultaneously, a method known as Multi-Task Learning (MTL) [13,32,33,35,37]. Combining breast density and cancer classification into a MTL framework could provide a more comprehensive assessment of a patient, as breast density affects both cancer risk and mammographic sensitivity. However, most studies on MTL in mammography have not explored this combination. One notable challenge attributed to this is the limited availability of public, large-scale datasets that contain relevant information for both tasks. Instead of breast density and cancer prediction, most MTL mammography research focuses on the combination of other tasks such as incorporating text reports and mammograms or mass segmentation and classification [8,33,35]. Alternatively, studies that leverage breast density information tend to do so as one of several auxiliary factors to enhance the primary task of BC classification [13,32]. Further, most of these tasks do not involve multi-label classification. More recently, there has been a shift towards employing MTL with multi-label approaches specifically for density and cancer classification [22,24]. However, these works lack comparisons to single-task baselines, leaving a gap in understanding the true benefits of MTL in this context.

In this study, we critically evaluate the performance of MTL frameworks for BC and breast density classification. Our primary aim is to assess the effectiveness of MTL, particularly for breast density classification, compared to traditional single-task approaches. We conduct extensive investigations into model architectures and explore data balancing strategies to address significant class imbalances inherent in the datasets. Our findings challenge the prevailing assumption that MTL inherently enhances performance, offering new insights into the complexities of multi-task versus single-task learning in mammographic density classification. Key contributions of this work include an in-depth analysis of single-task learning versus MTL across multiple network architectures and datasets, a thorough evaluation of balancing strategies to mitigate severe class imbalance, and new insights into the challenges of MTL in contexts of high-class imbalance and potentially conflicting task interplay.

2 Related Work

Since this paper primarily focuses on density classification, we review work most closely related to that topic.

2.1 Breast Density Classification

Within this section, we highlight some recent directions towards improving breast density classification.

In a two-stage approach, the authors in [27] first segmented breast tissue and then classified breast density into BI-RADS using a convolutional neural network (CNN). The model was trained and evaluated on the INBreast [21] database. The work demonstrated that the balanced datasets yielded the best classification performance. In [17], the authors developed a deep residual CNN tailored for BI-RADS density classification and showed that using the BI-RADS Atlas distribution (A: $10\%, B : 40\%, C : 40\%, D : 10\%$) yielded more accurate test results compared to both a uniform distribution and the dataset's native distribution. Their findings underscore the importance of addressing data imbalance to enhance breast density prediction.

Li et al. [16] incorporated dilated convolution and channel-wise attention mechanisms, improving density classification on a private dataset – an effect that did not generalize to the smaller INBreast dataset. Similarly, building on dilated convolutions and attention, [4] proposed Fusion Left-Right (FLeft-FRight) and Fusion Cranial Caudal-Mediolateral Oblique (FCC-FMLO), which are based on DenseNet201+LGBM [25], achieving a macro F1-score of 0.69 on VinDr-Mammo.

Another recent approach, [6], introduced an evidential deep learning method known as MV-DEFEAT, leveraging Dempster-Shafer evidential theory and subjective logic to provide calibrated uncertainty estimations for density classification. The study utilized four open-source datasets, including VinDr-Mammo [23]. MV-DEFEAT demonstrated superior performance in terms of weighted macro-average area under the ROC curve (AUC).

2.2 Multi-task Learning for Mammography

MTL is a deep learning strategy designed to simultaneously learn multiple tasks by leveraging shared knowledge among them [37] and has been beneficial across a range of different DL-based medical imaging applications including mammography classification [9].

In the context of mammography, very few studies have explored the synergy of jointly learning density and cancer classification. Current approaches typically focus on other mammographic aspects [33,35] or prioritize BC or mass classification while treating breast density as one of several auxiliary tasks [13,32].

For instance, a multitask DL model was developed to classify mammography findings and predict extensive intraductal component presence in [33], achieving notable AUC scores for mass and calcification classification. While improvements were observed with the multitask approach, they were not statistically significant compared to single-task models. In [35], MTL was leveraged to enhance mass malignancy classification by integrating BI-RADS and biopsy information. The model achieved state-of-the-art mass detection accuracy on the DDSM and an in-house dataset.

A notable MTL approach incorporating density was presented by [32] which predicted multiple outputs, including binary malignancy and BI-RADS breast density, to enhance binary cancer classification. Although validated on a public dataset, it was limited by its use of a private dataset and did not address data imbalance or multi-class predictions. Similarly, the Man and Machine Mammography Oracle (MAMMO) [13] used a MTL CNN to classify cancer risk factors like breast density before categorizing breast cancer but baseline single-task results for auxiliary tasks were not reported.

An example of a density and cancer MTL approach was presented in [22] which employed MTL to simultaneously extract features for BI-RADS cancer classification and breast density assessment from VinDr-Mammo data. Focal loss was used to address the data imbalance, focusing training on challenging examples. The model leveraged data augmentation techniques, such as mixup and concatenation, demonstrating promising results compared to traditional ResNet variants [7]. However, the study did not provide a comparison with single-task baseline performances.

A related work [24] also leveraged MTL to predict BI-RADS density and cancer using a two-stage approach. The process involved initially extracting features from both the left and right breasts independently. These features were then averaged and processed through separate density and cancer classifiers that utilized a decision-tree-based gradient-boosting algorithm, LightGBM [10]. The final output was the highest prediction values from the two breasts. However, this approach did not learn end-to-end and also lacked a comparison with baseline single-task performance metrics.

Based on the success of these MTL approaches we explore MTL which incorporates breast density and BC information.

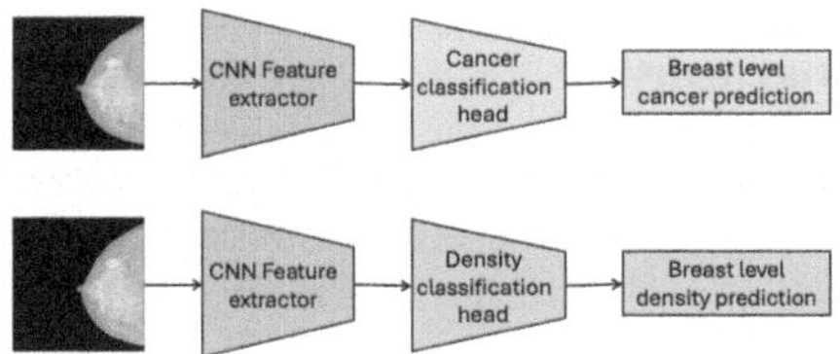

(a) Overview of the single-task framework for breast density and cancer prediction.

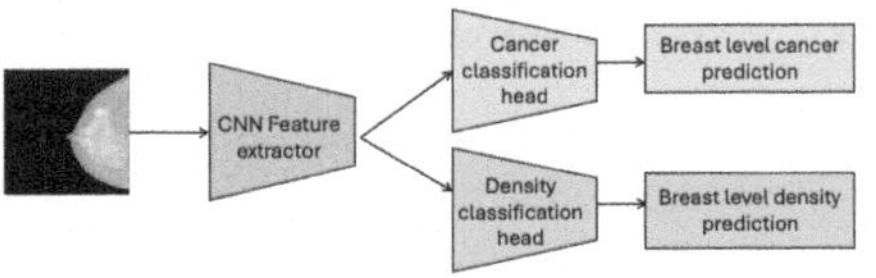

(b) Overview of the multi-task framework for breast density and cancer prediction.

Fig. 2. Comparative overview of single-task and multi-task frameworks for breast density and cancer prediction.

3 Methods

3.1 Single-task Framework

In our experimental setup, we explore both single-task, shown in Fig. 2a, and multi-task configurations, shown in Fig. 2b, for mammography classification. Let $I \in \mathbb{R}^{H \times W}$ represent an input mammogram image. In the single-task framework, each image I has one label y corresponding to cancer for the cancer classification network and density for the density classification network. The label y is defined as $y \in \{1, 2, \ldots, k\}$, where k-class classification indicates that the label y can belong to one of k classes. In the case of cancer and density, the k classes correspond to the respective BI-RADS categories.

3.2 Multi-task Framework

MTL involves optimizing a single model, characterized by parameters $\theta \in \mathbb{R}^m$, to perform well across multiple tasks, where each task i is associated with its loss function $\ell_i(\theta) : \mathbb{R}^m \rightarrow \mathbb{R}_{\geq 0}$ [37]. The typical objective in MTL is to minimize the average loss across all tasks:

$$\min_{\theta \in \mathbb{R}^m} \left\{ \ell_0(\theta) := \frac{1}{n} \sum_{i=1}^{n} \ell_i(\theta) \right\}. \tag{1}$$

Here, n refers to the number of tasks. In this work, within the multi-task framework, each image I has two label sets, one for cancer prediction and one for density classification, leading to the target vector $\mathbf{y} = [y_{\text{cancer}}, y_{\text{density}}]$. Both y_{cancer} and y_{density} in this vector are defined within their respective class set, with both tasks being multi-class classifications. The CNN classifier, $f(\cdot, \theta)$, predicts the relevant task labels (BI-RADS density and cancer) for the input image I, where θ represents the trainable parameters of the model. For each task i, the model outputs a prediction $\hat{y}_i$, and we use cross-entropy as the classification loss function ℓ_i for each task. In the multi-task setup for cancer and density classification, we consider two tasks, i.e. $n = 2$.

4 Data and Class Imbalance

4.1 Datasets

VinDr-Mammo [23] was selected as the primary dataset in our analysis as it is a public, large-scale dataset with both BI-RADS density and cancer labels. Due to these properties, VinDr-Mammo has emerged as a benchmark dataset for MTL in mammography and for mammographic density classification. For completeness, we have also leveraged the much smaller public INBreast [21] dataset for analysis of single versus multi-task. Additionally, we have simulated a MTL scenario with BI-RADS density and binary cancer classification using the CSAW-CC dataset.

The VinDr-Mammo dataset is a Vietnamese dataset of 20,000 full-field digital mammography (FFDM) images with breast-level assessments. The BI-RADS cancer categories range from BI-RADS 1 to BI-RADS 5, with no BI-RADS 6 due to the absence of biopsy results. Breast density is categorized into four levels, BI-RADS A-D. The data was split into training, test, and validation sets in a 60/20/20 ratio while ensuring that no two images from the same patient were put into different splits. One sample was excluded due to the absence of all four distinct mammographic views resulting in a total of 19,996 images.

The INbreast dataset is a much smaller-scale public Portuguese database of FFDM images [21]. It comprises 410 images and includes American College of Radiology (ACR) 4th edition BI-RADS breast density annotations (1–4), and BI-RADS cancer classification (1–6). One image was excluded from the dataset due to the absence of ACR density annotation. We split the data into training, test, and validation sets in a 80/10/10 ratio.

This study also utilized the large public CSAW-CC [3] dataset which includes screening FFDMs from Sweden. The dataset contains breast density information for 98,772 images, estimated with the LIBRA software [11]. The conversion of LIBRA's breast density percentages to BI-RADS categories A through D was performed according to the BI-RADS 4th edition guidelines. This dataset provides only binary cancer labels, lacking the multi-class BI-RADS cancer labels. A 80/10/10 ratio for the training, test, and validation sets was used.

4.2 Data Pre-processing

Data pre-processing is a key step to ensure reliable and effective training on mammography data. For all three datasets images were initially segmented to isolate breast tissue using a threshold-based method. A bounding box approach was applied to ensure precise cropping around the breast tissue with additional processing for images with possibly bifurcated backgrounds. Finally, the images were standardized in size by padding to the maximum dimensions found in the respective datasets and resized to a uniform resolution of 512×512 pixels. This dimensionality was chosen as a good trade-off in terms of both computational demands and classification accuracy.

4.3 Data Balancing Strategies

The datasets used exhibited significant class imbalances as highlighted for VinDR-Mammo in Fig. 3. As mentioned in Sect. 2, the class balance plays a crucial role in both breast density and cancer classifications. To address these imbalances and mitigate biased learning towards certain classes by adopting a balancing strategy referred to as balanced subset sampling by creating a balanced subset of data for training models. This can be tailored to the task being classified–whether it be breast density (BI-RADS A-D) or BC (BI-RADS 1–5). This strategy identifies unique labels for the current task (density or cancer), collecting samples corresponding to each label, and then randomly selecting an equal number of samples from each class to ensure uniform representation within each training batch. In the MTL setting, we investigated several strategies to address the class imbalance. Given the complexity of balancing multiple tasks, we considered the following approaches.

1. Unbalanced Sampling: Baseline approach, no balancing strategy was used.
2. Balanced Sampling for One Task Only: Balancing was performed with respect to one task only leaving the other detection task imbalanced.
3. Alternating Balancing Strategy: In this approach, the balancing was alternated between the two tasks from epoch to epoch.

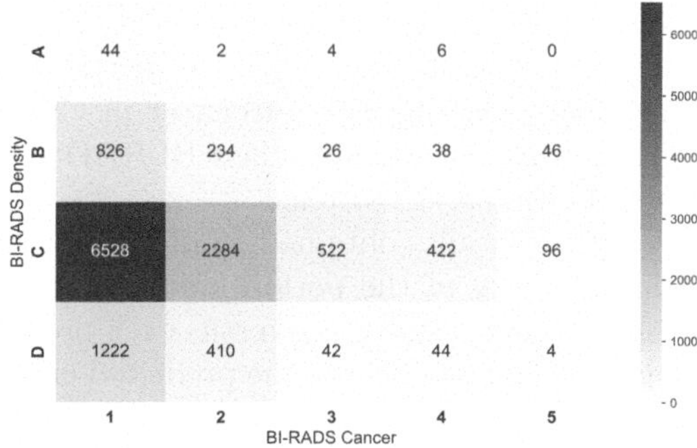

Fig. 3. Label relationship and distribution within the VinDr-Mammo traning set.

5 Experimental Setup

Similar to prior works [6,16,24], we leveraged ResNet-based CNN architectures in our experiments which were pre-trained on ImageNet with the last fully connected layers replaced and fine-tuned for the tasks considered. In our case we leveraged ResNet 18 and 34 specifically.

We leverage the classification performance, F1 metric specifically, on the validation data to determine and record the best-performing classification models during training. We train three runs of each setup and the best-performing models, with respect to the validation set, from each are then evaluated on unseen test data to obtain average performance metrics. We focus on the best-performing density models in particular from each setup and the results of these on the test set are presented in Sect. 6.

We mention that we also investigated EfficientNet B2 [31] but found ResNet 18 and 34 to perform better throughout all experiments. Thus we narrowed our investigations down to the latter two architectures. We also found learning rate annealing to not be an effective strategy in our experiments, see supplementary materials. Models were trained using NVIDIA RTX 3090, A6000 and 1080-ti GPUs, employing cross-entropy loss for optimization. We used the Adam optimizer [12]. The best models were selected with respect to density F1 scores obtained on the validation set within the first 150 epochs for VinDR-Mammo and INBreast and within the first 50 epochs for CSAW-CC.

5.1 Evaluation Metric

In all experiments conducted, either the binary F1 (in the case of CSAW-CC cancer classification) or the unweighted macro F1 score were used as an evaluation metric to assess the performance of the models. The unweighted macro F1 score is obtained by averaging these individual F1 scores across all classes, ensuring that each class contributes equally to the final metric. This metric was adopted as for all multi-class experiments as it is favourable for cases with class imbalances [22,24].

AUC was also adopted as an evaluation metric as it is often used within the context of mammography classification [6,32]. In particular binary AUC was used when performing binary classification of cancer in CSAW-CC and macro AUC was leveraged when considering multi-class BI-RADS density and cancer classification in VinDR-Mammo and INBreast.

For further analysis of the results obtained, confusion matrices were plotted which provide a visual summary of the performance of a model by displaying the number of instances for each actual class versus each predicted class. Within such a matrix, the diagonal elements, where the predicted classes match the true classes, highlight the number of correct predictions made by the model.

6 Results

6.1 Single-task vs. Multi-task

In this section, we compare performances of single and multi-task frameworks for breast density classification within the VinDr-Mammo, INBreast and CSAW-CC datasets using ResNet 18 and ResNet 34 architectures.

The results are presented in Tables 1 and 2 using F1 and AUC as performance metrics respectively with the best-recorded metrics for each setting highlighted in bold. Here, data balancing is not considered as the purpose is to ascertain the relative performance of single compared to multi-task settings using the inherent data distributions. We observe that single-task either surpasses or is on par with multi-task settings for both ResNet 18 and ResNet 34 across all three datasets considered with respect to both evaluation measures. The relative class-wise classification performances of single versus multi-task ResNet 18 models are further examined using confusion matrices in Fig. 4. In the confusion matrices, the transition from blue to yellow indicates an increasing frequency of predictions, with blue representing lower counts and yellow representing higher counts.

6.2 Investigating Class Balancing Strategies

In this section, we investigate the effects of class imbalance due to its important role when considering mammography density classification as discussed in Sect. 2. The class imbalance may have led to discrepancies in single- compared to multi-task classifications which we explored in Sect. 6.1.

We conduct a focused investigation of the impact of four different balancing strategies on the multi-task classification of breast density and BC on the VinDR-Mammo dataset. The results of density classification are presented in Tables 3 and 4. The corresponding multi-task cancer classification performances using ResNet 18 and 34 are presented in the supplementary materials.

For density classification, we observe that for the ResNet 18 model, both single-task and multi-task settings with unbalanced data achieve comparable density F1 and AUC scores. For the ResNet 34 model, the single-task setting with unbalanced data outperforms the multi-task setting in terms of density F1 and AUC. This suggests that the additional complexity introduced by MTL may not be justified even as the single-task unbalanced data strategy performs on par with or better and none of the considered balancing strategies significantly enhance MTL performance.

Table 1. Comparison of density F1 scores between single and multi-task for VinDr-Mammo, INBreast and CSAW-CC. No data balancing is used here. Results are presented here for two different architectures, ResNet 18 and ResNet 34.

Architecture	Task	VinDr-Mammo F1	INbreast F1	CSAW-CC F1
ResNet 18	Single	0.597 ± 0.017	**0.794 ± 0.038**	**0.782 ± 0.005**
ResNet 18	Multi	**0.600 ± 0.012**	0.753 ± 0.031	0.777 ± 0.009
ResNet 34	Single	**0.623 ± 0.013**	**0.688 ± 0.035**	**0.768 ± 0.012**
ResNet 34	Multi	0.598 ± 0.007	0.656 ± 0.044	0.760 ± 0.013

Table 2. Comparison of density AUC scores between single and multi-task for VinDr-Mammo, INBreast and CSAW-CC. No data balancing is used here. Results are presented here for two different architectures, ResNet 18 and ResNet 34.

Architecture	Task	VinDr-Mammo AUC	INbreast AUC	CSAW-CC AUC
ResNet 18	Single	**0.901 ± 0.005**	**0.949 ± 0.009**	**0.951 ± 0.001**
ResNet 18	Multi	0.898 ± 0.003	0.921 ± 0.007	0.950 ± 0.002
ResNet 34	Single	**0.906 ± 0.010**	**0.916 ± 0.008**	**0.954 ± 0.008**
ResNet 34	Multi	0.893 ± 0.006	0.883 ± 0.002	0.948 ± 0.003

7 Discussion and Future Work

Based on our analysis of single versus multi-task conducted over three different datasets for the task of breast density prediction, presented in Sect. 6, we found MTL to not offer substantial benefits over single task learning. This was the case for two different CNN-based architectures as well as two different evaluation metrics. This finding contrasts with the purported benefits of MTL for mammography classification, as discussed in Sect. 2, highlighting the need for further investigation into its effectiveness for these specific tasks.

To investigate the effects of class balancing we assessed four different balancing strategies for MTL within VinDr-Mammo. Our analysis did not reveal any considerable benefits towards MTL with the investigated balancing strategies.

MTL, particularly for mammography leveraging breast density, presents several unique challenges with one of the most significant being the pronounced class imbalance. In the VinDR-Mammo dataset, for instance, the multi-task setup

Table 3. Examining balancing strategy effects on density classification in single and multi-task settings on VinDr-Mammo dataset using ResNet 18. The reported results used trained models that achieved the best validation density F1.

Task	Balancing	Best Density F1	Best Density AUC
Single	Unbalanced	0.597 ± 0.017	**0.901 ± 0.005**
Single	Density balanced	0.553 ± 0.014	0.851 ± 0.009
Multi	Unbalanced	**0.600 ± 0.012**	**0.898± 0.003**
Multi	Density balanced	0.573 ± 0.007	0.870 ± 0.008
Multi	Alternate balanced	0.575 ± 0.009	0.872 ± 0.005
Multi	Cancer balanced	0.566 ± 0.008	0.882 ± 0.006

Table 4. Examining balancing strategy effects on density classification in single and multi-task settings on VinDr-Mammo dataset using ResNet 34. The reported results used trained models that achieved the best validation density F1.

Task	Balancing	Best Density F1	Best Density AUC
Single	Unbalanced	**0.623 ± 0.013**	**0.906 ± 0.010**
Single	Density balanced	0.564 ± 0.017	0.847 ± 0.003
Multi	Unbalanced	0.598 ± 0.007	0.893 ± 0.006
Multi	Density balanced	0.564 ± 0.007	0.842 ± 0.017
Multi	Alternate balanced	0.558 ± 0.019	0.870 ± 0.010
Multi	Cancer balanced	0.569 ± 0.004	0.851 ± 0.017

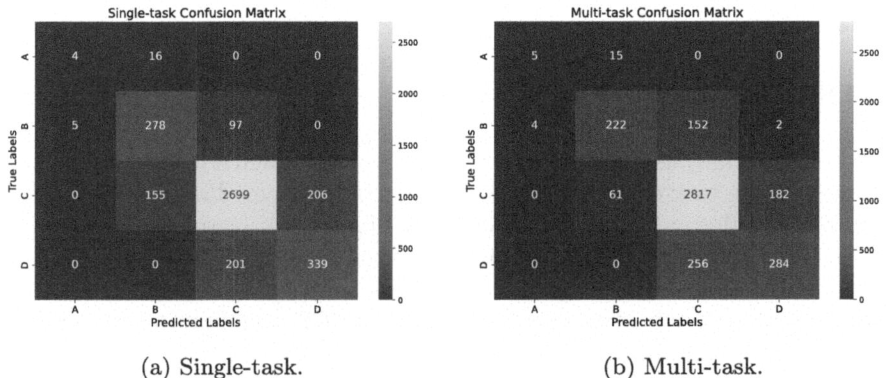

(a) Single-task.

(b) Multi-task.

Fig. 4. Confusion matrices for density labels for single and multi-task ResNet 18 models. Figure 4a corresponds to the best-performing run of the single-task model and Fig. 4b shows the best-performing run of the multi-task model, respectively. Both matrices correspond to F1 scores of 0.615.

involves classifying 20 classes compared to just 4 in the single density task. As showcased in Fig. 3, the MTL training data of VinDR-Mammo we adopted comprises of 12 out of these 20 classes having fewer than 100 images and 5 classes even having fewer than 10 samples. Despite VinDR-Mammo being a reasonably sized benchmark dataset, this imbalance complicates MTL. While we do not rule out the potential of combining density and cancer in MTL, achieving benefits may require a significantly larger dataset with a reasonable representation of each class.

Our experimental design choices, including balancing strategies, architecture size, and loss functions, follow standard practice. Other choices may potentially bring more benefit within the specific MTL setup discussed in this paper and further research is needed to corroborate or nuance our findings. Our current findings may suggest that MTL possibly faces inherent challenges such as gradient conflicts, task correlation, and varying task difficulty, which could negatively impact performance [30,36].

As discussed in [30,36], when multiple tasks are learned jointly they may not learn at equal rates and one task may dominate the other leading to conflict in task gradients. In the context of breast density and cancer, there is a discrepancy in the relative ease of the two tasks, with BC being a more difficult task. Further, while breast density is a strong risk factor for BC, high breast density is linked to masking effects on underlying tumours and inaccurate prediction of BC [28]. This creates a complex and potentially conflicting MTL scenario, where the interplay between density and BC tasks may hinder optimal learning outcomes.

Through analysis of confusion matrices for VinDR-Mammo in Fig. 4, we observed a significant class leakage between density categories B and C and C and D, particularly in the MTL setup. Notably, more true B samples were misclassified as C, and more true D samples were misclassified as C in the MTL setting compared to single-task learning. Interestingly, category C performed better in MTL. This may be due to the network capturing the relationship between density C and the cancer label BI-RADS 1, as category C is substantially represented in the dataset. The network might be exploiting shortcuts, where samples on the borderline between C and D are nudged towards C if they are associated with BI-RADS 1, which is more frequent in category C. This reliance on prevalent patterns suggests a need for refinement in model training to mitigate such biases.

Future research could explore several promising directions, including alternative strategies such as focal loss or data augmentation techniques to handle class imbalances, as suggested in [22]. Investigating gradient alignment methods to address conflicting gradients within MTL frameworks presents another valuable avenue. Further exploration could also focus on data pre-processing techniques such as enhancing contrast, attention mechanisms and employing larger architectures like ResNet 50 or Vision Transformers, which may offer improved performances. Finally, the simultaneous use of multiple views could be investigated to further enhance model accuracy and robustness.

8 Conclusion

This study critically examined the efficacy of MTL frameworks for breast density classification. Contrary to the anticipated benefits of MTL, our findings indicate that single-task models often surpass MTL approaches for density classification. This was consistent across multiple datasets, architectures, and evaluation metrics. The pronounced class imbalance in mammography datasets is amplified in the MTL training regime by increasing the number of possible classes drastically in mammography datasets. Future research could explore alternative balancing strategies, gradient alignment methods, and enhanced data pre-processing techniques to address these challenges. Additionally, leveraging larger architectures and multi-view approaches may improve MTL performance in mammography.

Acknowledgements. This work was financially supported by the Research Council of Norway (RCN), through its Centre for Research-based Innovation funding scheme (grant no. 309439), and Consortium Partners. It was further supported by RCN FRIPRO grant no. 315029 and 303514. The work was further supported by Pioneer Centre for AI (RJ), DNRF grant number P1.

References

1. Berg, W.A., Campassi, et al.: Breast imaging reporting and data system. Am. J. Roentgenol. (2023)
2. Dembrower, K., Crippa, E., Eklund, M., et al.: Artificial intelligence for breast cancer detection in screening mammography in Sweden: a prospective, population-based, paired-reader, non-inferiority study. Lancet Digit. Health **5**(10), e703–e711 (2023)
3. Dembrower, K., Lindholm, P., Strand, F.: A multi-million mammography image dataset and population-based screening cohort for the training and evaluation of deep neural networks-the cohort of screen-aged women (CSAW). J. Digit. Imaging **33**(2), 408–413 (2020)
4. Dif, N., Boudinar, M.E.A., Abdelali, M.A., et al.: FCC-FMLO and FLeft-FRight: two novel multi-view fusion techniques for breast density assessment from mammograms. Multimed. Tools Appl. 1–24 (2024)
5. Goh, C., Ho, F.: The growing problem of radiologist shortages: perspectives from Singapore. Korean J. Radiol. **24**(12), 1176 (2023)
6. Gudhe, N.R., Mazen, S., Sund, R., et al.: A multi-view deep evidential learning approach for mammogram density classification. IEEE Access **12**, 67889–67909 (2024)
7. He, K., Zhang, X., Ren, S., Sun, J.: Deep residual learning for image recognition. In: Proceedings of the IEEE Computer Society Conference on Computer Vision and Pattern Recognition (2015)
8. Jain, K., Bansal, A., Rangarajan, K., et al.: MMBCD: multimodal breast cancer detection from mammograms with clinical history. In: International Conference on Medical Image Computing and Computer-Assisted Intervention, pp. 144–154. Springer, Cham (2024)
9. Kamiri, J., Wambugu, G.M., Oirere, A.M.: Multi-task deep learning in medical image processing: a systematic review. Int. J. Comput. Sci. Res. (2025)

10. Ke, G., Meng, Q., Finley, T., et al.: LightGBM: a highly efficient gradient boosting decision tree. In: Advances in Neural Information Processing Systems, vol. 30 (2017)
11. Keller, B.M., Chen, J., Daye, D., et al.: Preliminary evaluation of the publicly available Laboratory for Breast Radiodensity Assessment (LIBRA) software tool: comparison of fully automated area and volumetric density measures in a case-control study with digital mammography. Breast Cancer Res. **17**(1) (2015)
12. Kingma, D.P., Ba, J.L.: Adam: a method for stochastic optimization. In: ICLR (2014)
13. Kyono, T., Gilbert, F.J., van der Schaar, M.: MAMMO: a deep learning solution for facilitating radiologist-machine collaboration in breast cancer diagnosis. arXiv preprint arXiv:1811.02661 (2018)
14. Lauby-Secretan, B., Scoccianti, C., Loomis, D., et al.: Breast-cancer screening – viewpoint of the IARC working group. New Engl. J. Med. (2015)
15. Lauritzen, A.D., Lillholm, M., Lynge, E., et al.: Early indicators of the impact of using AI in mammography screening for breast cancer. Radiology (2024)
16. Li, C., Xu, J., Liu, Q., et al.: Multi-view mammographic density classification by dilated and attention-guided residual learning. IEEE/ACM Trans. Comput. Biol. Bioinform. (2021)
17. Lizzi, F., Scapicchio, C., Laruina, F., et al.: Convolutional neural networks for breast density classification: performance and explanation insights. Appl. Sci. **12**, 148 (2021)
18. Lotter, W., Diab, A.R., Haslam, B., et al.: Robust breast cancer detection in mammography and digital breast tomosynthesis using an annotation-efficient deep learning approach. Nat. Med. **27**(2), 244–249 (2021)
19. Luo, L., Wang, X., Lin, Y., et al.: Deep learning in breast cancer imaging: a decade of progress and future directions. IEEE Rev. Biomed. Eng. (2024)
20. Mann, R.M., Athanasiou, A., Baltzer, P.A.T., et al.: Breast cancer screening in women with extremely dense breasts recommendations of the European Society of Breast Imaging (EUSOBI). Eur. Radiol. (2022)
21. Moreira, I.C., Amaral, I., Domingues, I., et al.: INbreast: toward a full-field digital mammographic database. Acad. Radiol. **19**(2), 236–248 (2012)
22. Nguyen, C.T., Minh, H.C., Dinh, N., et al.: Improving multi-task learning for breast cancer detection. In: ACM International Conference Proceeding Series (2024)
23. Nguyen, H.T., Nguyen, H.Q., Pham, H.H., et al.: VinDr-Mammo: a large-scale benchmark dataset for computer-aided diagnosis in full-field digital mammography. Sci. Data **10**(1), 1–8 (2023)
24. Nguyen, H.T.X., Tran, S.B., Nguyen, D.B., et al.: A novel multi-view deep learning approach for BI-RADS and density assessment of mammograms. In: 2022 44th Annual International Conference of the IEEE Engineering in Medicine & Biology Society (EMBC), pp. 2144–2148. IEEE (2022)
25. Pawar, S.D., Sharma, K.K., Sapate, S.G., et al.: Multichannel DenseNet architecture for classification of mammographic breast density for breast cancer detection. Front. Public Health (2022)
26. Rodriguez-Ruiz, A., Lång, K., Gubern-Merida, A., et al.: Stand-alone artificial intelligence for breast cancer detection in mammography: comparison with 101 radiologists. JNCI **111**(9), 916 (2019)
27. Saffari, N., Rashwan, H.A., Abdel-Nasser, M., et al.: Fully automated breast density segmentation and classification using deep learning. Diagnostics **10**(11), 988 (2020)

28. Sorkhei, M., Liu, Y., Azizpour, H., et al.: CSAW-M: an ordinal classification dataset for benchmarking mammographic masking of cancer. arXiv preprint arXiv:2112.01330 (2021)
29. Sprague, B.L., Conant, E.F., Onega, T., et al.: Variation in mammographic breast density assessments among radiologists in clinical practice: a multicenter observational study. Ann. Internal Med. (2016)
30. Standley, T., Zamir, A., Chen, D., et al.: Which tasks should be learned together in multi-task learning? In: ICML, pp. 9120–9132. PMLR (2020)
31. Tan, M., Le, Q.V.: EfficientNet: rethinking model scaling for convolutional neural networks. In: ICML, pp. 10691–10700 (2019)
32. Tardy, M., Mateus, D.: Leveraging multi-task learning to cope with poor and missing labels of mammograms. Front. Radiol. (2021)
33. Tsai, H.Y., Kao, Y.W., Wang, J.C., et al.: Multitask deep learning on mammography to predict extensive intraductal component in invasive breast cancer (2024)
34. Xu, S., Murtagh, S., Han, Y., et al.: Breast Cancer Incidence Among US Women Aged 20 to 49 Years by Race, Stage, and Hormone Receptor Status (2024)
35. Yang, Z., Cao, Z., Zhang, Y., et al.: MommiNet-v2: mammographic multi-view mass identification networks. Med. Image Anal. (2021)
36. Yu, T., Kumar, S., Gupta, A., et al.: Gradient surgery for multi-task learning. Adv. Neural. Inf. Process. Syst. **33**, 5824–5836 (2020)
37. Zhang, Y., Yang, Q.: A survey on multi-task learning. IEEE Trans. Knowl. Data Eng. **34**(12), 5586–5609 (2022)

Fine-Grained Classification of Unpigmented Skin Cancer from Paired Dermatoscopy Images

Anna Gummeson[1(✉)], Gabrielle Flood[1], Kari Nielsen[2,3,4], Carolina Nätterdahl[2,3], Fredrik Johansson[2,4], Åsa Ingvar[2,3], and Ida Arvidsson[1]

[1] Centre for Mathematical Sciences, Lund University, Lund, Sweden
{anna.gummeson,gabrielle.flood,ida.arvidsson}@math.lth.se
[2] Department of Dermatology and Venereology, Department of Clinical Sciences, Lund University, Lund, Sweden
[3] Department of Dermatology, Skåne University Hospital, Lund, Sweden
[4] Department of Dermatology, Skånes Sjukhus nordväst, Helsingborg, Sweden

Abstract. Unpigmented skin cancer is the most prevalent form of cancer, and it burdens healthcare substantially even if it is not as aggressive as the more well-known malignant melanoma. Dermatoscopy images are commonly used for diagnosis, but differentiating between the many subdiagnoses is a hard task. In this study we focus on these unpigmented cancers, performing both detection of basal cell carcinoma as well as fine-grained classification of its subclasses. We do this using a new dataset with more than 2'000 cases from a fair-skinned population. A deep learning model is specially designed for the task, handling pairs of polarised and non-polarised dermatoscopy images as input. We investigate transfer learning with different backbones as well as adding a mid-step of contrastive learning. The performance is compared to the accuracy of dermatologists on the subset of our test data where we have additional ground truth from histopathological diagnosis. This is the first study focusing on fine-grained classification of unpigmented lesions using only dermatoscopic images, and we reach a balanced accuracy of 39.4%, to be compared to 51.9% for dermatologists. This is a promising first step towards an algorithm to assist dermatologists in their work and we hope that this will open up for further studies on this interesting problem.

Keywords: Fine-grained classification · Unpigmented skin cancer · Deep learning · Dermatoscopy

1 Introduction

Deep learning (DL) models have shown to have great potential for various medical image assessments [30], including assessment of dermatoscopy images for skin

Supplementary Information The online version contains supplementary material available at https://doi.org/10.1007/978-3-031-95918-9_21.

J. Petersen and V. A. Dahl (Eds.): SCIA 2025, LNCS 15726, pp. 293–306, 2025.
https://doi.org/10.1007/978-3-031-95918-9_21

cancer detection, see e.g. [6,22,27]. Malignant melanoma are well known to the public and is the most aggressive form of skin cancer, accounting for 90% of all skin cancer mortality [14,15]. However, the unpigmented types basal cell carcinoma (BCC or basalioma) and squamous cell carcinoma (SCC), which together are almost as common as all other types of cancer in Sweden combined [19,25], have not been given as much attention.

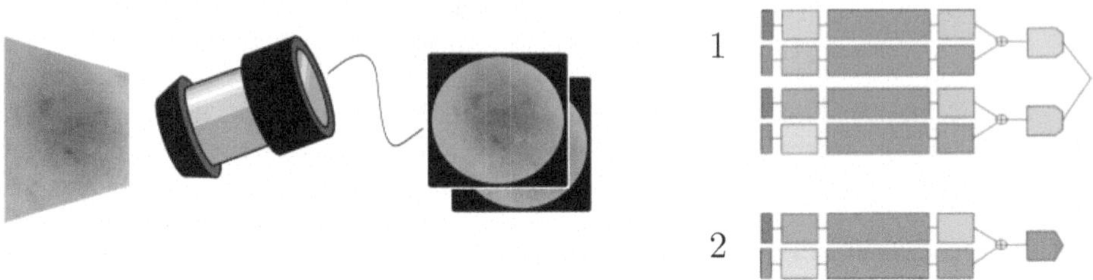

Fig. 1. The workflow schematic. We collect images using a dermatoscope both with settings for polarised and non-polarised light. The images are used to train a network using transfer learning (2). As an option we train a part of the network in a pre-step using contrastive learning (1).

To diagnose unpigmented skin cancers through teledermatoscopy is considered difficult and suspicious lesions may be sent for biopsy and histopathological analysis. This is considered the "gold standard" of diagnosing skin cancers [8,20,21]. However, biopsies have several drawbacks: it is an invasive intervention that may cause infections or scarring, it is expensive (mainly due to the labour-intensive handling), and time consuming (often leading to a two-step management of patients with two visits, interventions, and histopathological assessments). The non-pigmented BCC and SCC tumours, which commonly occur in the face, may result in mutilating surgeries [19,25]. To minimise associated morbidity and mortality of skin cancer, early detection and correct treatment are of substantial importance. The treatment plan varies for different subtypes of basalioma, which makes assistance with sub-classification an interesting application for DL.

The goal of this study is to investigate the potential of DL models for assessment of non-pigmented skin cancers in a fair-skinned population (in Sweden). Several classification algorithms, with two inputs to handle paired polarised and non-polarised images, are developed. An overview of the pipeline can be seen in Fig. 1. Consistently combining these two image modalities is to our knowledge rare in an DL setting but common practice when dermatologists perform teledermatoscopy in Sweden. Therefore, it could be a valuable and insightful addition to DL models as well. Dermatologists often have access to macro images and metadata as well, but we only use dermatoscopy images in this study. We investigate techniques like transfer learning and contrastive learning, as the amount of available data is limited, and compare models based on vision transformers and convolutional neural networks. We also compare the model performance to dermatologists' assessments, considering both binary BCC detection and fine-grained classification of BCC, SCC, and actinic keratosis (AK).

2 Related Work

There are several previous studies on classification of non-pigmented skin cancers from dermatoscopy images, for example basalioma detection or benign-malignant classification [22,27]. Use of other types of images, including optical coherence tomography, high frequency ultrasound and diffuse reflectance spectroscopy, has also been researched previously [24,27]. When developing DL solutions for medical applications, a lack of data is a common problem. Therefore, different forms of self-supervision, semi-supervision and transfer learning are commonly used in the creation of DL diagnostic models, where contrastive learning is an example [13]. Vision transformers [7] have also successfully been applied to medical data in their transfer learning modus. For example, the transformer based model EVA-02 [10] has successfully been used in the ISIC Challenge 2023 and 2024 [18]. In the specific case of sub-diagnosing BCC there is to our knowledge no other study that uses only dermatoscopy images. There are, however, studies using whole slide images from biopsies [9,28], optically-guided high frequency ultrasound [3] and a combination of diffuse reflectance spectroscopy, optical coherence tomography and high frequency ultrasound [24]. In comparison to these other methods dermatoscopy is cheaper, non-invasive and already widely used, so no specialist training is required. The extra effort to always sample the lesion with both polarised and non-polarised light is thereby in comparison a very small step.

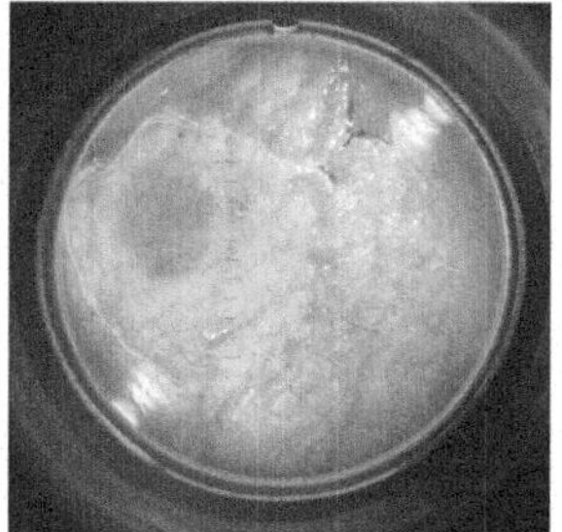

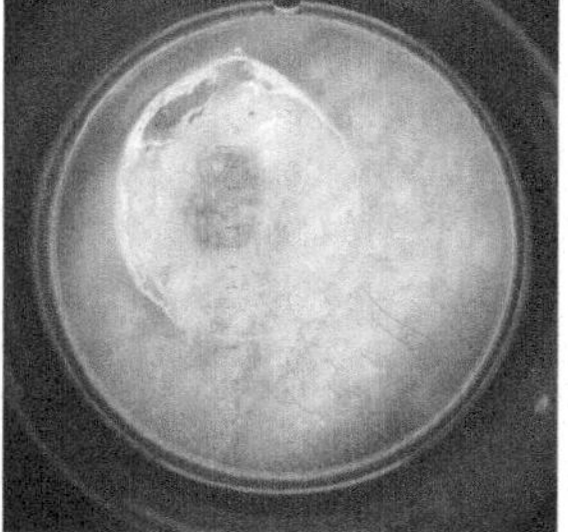

Fig. 2. Examples of the pair of non-polarised (left) and polarised (right) dermatoscopy images. This is an example of nodular BCC.

3 Data

The dataset used in this study consists of dermatoscopic images with two images per lesion: one from polarised and one from non-polarised light, see Fig. 2. The images have been collected from patients either visiting their primary health care provider or one of the skin clinics featured in this study, all located in Scania, a province in southern Sweden. The handling of patient data follows GDPR regulations. The images were taken with a mobile phone through a clip on dermatoscope. Most images are 3024×3024 pixels with a field of view of 2.7 cm. This, however, does include a fair bit of the dermatoscope, see Fig. 2. For comparison, this resolution is 10x higher than the ISIC challenge 2024 dataset [18]. In total 10'085 image pairs were collected. Images without a corresponding diagnosis or a diagnosis not relevant to this study (e.g. other diagnoses than BCC, SCC and AK) were excluded. For anonymity and ethical reasons (tattoos or other identifiable traits), three lesions were excluded, leaving 2'021 lesions for further assessments. AK was included as a benign class due to its high prevalence and its potential for misclassification as cancer.

Each lesion has been diagnosed by a trained dermatologist with full access to dermatoscope images, macro images and metadata. Some lesions have been sent for further histopathological diagnosis. We use the histopathological diagnosis as our label if it exists, and otherwise the dermatoscopical diagnosis. The subset with histopathology is not representative of the dataset in general, since only cases with uncertain dermatology diagnoses will be biopsied. For example, lesions deemed benign by a dermatologist are rarely biopsied, meaning no histopathological diagnosis is available. However, for evaluating dermatologists' performance on our data, doing so on this subset is the only possibility. For the binary classification we have in total N = 993 (49%) BCC and N = 1028 (51%) non-BCC. For the BCCs, N=633 (64%) have a subclass label into either nodular BCC, superficial BCC or infiltrative BCC. All of the non-BCCs were subclassified into SCC in situ, invasive SCC or AK. The subclasses were used for fine-grained classification.

The data was divided into training and test sets randomly, but under conditions that the same patient should not feature in both the training and test set and that the same proportions of diagnoses (BCC, SCC, AK), as well as proportions of primary health care versus clinic and different lesion placements (chest, head/neck, hand, abdomen, foot, forearm, arm, pelvis, thigh, leg, back, buttock) should be upheld in both the training and test set. We do this proportional split to be able to track bias, since there is a correlation between diagnosis and placement of the lesion as well as a difference in patients visiting a general practitioner (GP) versus those referred to a skin clinic. A summary of the resulting allocation of samples can be seen in Table 1 and a complete division in Supplementary, Table 2. Out of the 2'021 relevant lesions, 1'620 (80%) have been used for training and validation, and 401 (20%) for the hold-out test set. We use five-fold cross validation on the training set when developing our models.

4 Method

Many applications of DL in medicine faces the problem of lack of data, and in machine learning terms we have too little data to train large models from

Table 1. Overview of our dataset, note that not all BCCs have a sublabel.

	Clinic tot	GP tot	Clinic test	GP test
nodular BCC	111	65	18	5
superficial BCC	94	50	21	5
infiltrative BCC	224	89	48	20
total BCC	437	556	88	107
SCC in situ	57	272	12	53
invasive SCC	53	118	11	26
AK	174	354	37	67
total non-BCC	284	744	60	146

scratch as this most likely will result in over-fitting. Therefore, we utilise transfer learning and contrastive learning as a means to overcome this problem. We experiment with two different types of models; a direct classification (DC) network, and a supervised contrastive learning (SCL) approach. We did also try an unsupervised contrastive learning (UCL) approach, but with unsatisfactory results. The two latter approaches were trained in a two-step process, first with a feature extracting network using a contrastive loss and thereafter classification based on the trained contrastive network. All models are further described in the following subsections.

4.1 Direct Classification Network

The DC network is a transfer learning model built upon a backbone network pre-trained on ImageNet [5]. The backbone network is either an implementation of MobileNetV2 architecture [23], which is similar to MobileNetV1 [12] while also utilising inverted residual blocks with bottleneck features, or an implementation of EVA-02 [10] which is a vision transformer [7] variant with a more efficient parameter design than its predecessors. To make use of both the polarised and non-polarised images we use two copies of the backbone network followed by additional blocks of convolutional and max-pooling layers in the case of MobilenetV2, and a multi-layer perceptron (MLP) in the case of an EVA-02 backbone. The two representations are then concatenated and the model ends with an additional MLP. A schematic over the DC network can be seen in Fig. 3. During training the weights in the pre-trained backbone network are frozen.

4.2 Contrastive Learning

Contrastive learning has been widely used for medical applications due to the advantages when applied to problems with limited amount of labelled data [13].

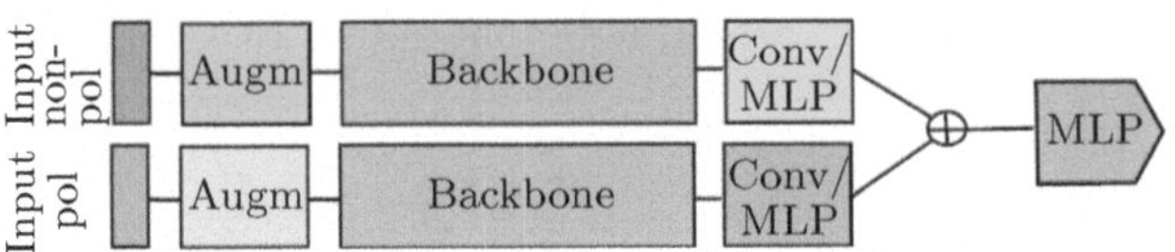

Fig. 3. Direct classification (DC) model. The input images with different light settings are independently augmented and run through the backbone network. We train two separate blocks of either convolutional layers (for MobileNetV2) or MLP (for EVA-02) and concatenate them for a final trainable MLP for classification.

Unsupervised contrastive learning (UCL), also known as self-supervised contrastive learning, has been explored in several works, including early formulations in [11], and more recent advancements such as SimCLR [4]. It can be used to cluster unlabelled data in a hyperspace of a chosen dimension. Thereafter, the space can be partitioned using a simple neural network or another classification algorithm. The loss function is constructed as to attract different augmentations of the same image in the hyperspace and repel augmentations of other images in the batch. Here, we use a supervised version of contrastive learning [16]. In this approach, negative examples are drawn from different classes, while positive examples come from the same class. This differs from the standard unsupervised contrastive learning algorithm [4] where only augmentations of the image itself are considered positive examples, as no labels are available. For the supervised contrastive loss, we need to set a temperature hyper-parameter τ. It controls how much hard negative samples are penalized, such that a small temperature puts harder emphasis on the hardest negatives while a higher temperature gives less sensitivity to hard negatives [26]. Temperatures of $\tau < 1$ are more common, however, a temperature parameter $\tau > 1$ has been discussed in several works [26,29].

We combine supervised contrastive learning (SCL) with transfer learning in the backbone network. An overview of the structure, which is the same for supervised and unsupervised contrastive learning, can be seen in Fig. 4.

4.3 Evaluation Metrics

We measure accuracy, i.e. number of correct diagnoses divided by total number of samples, and balanced accuracy, i.e. mean accuracy (sensitivity) per class. Balanced accuracy is a more fair measurement if the classes are unbalanced. For binary classification, we also present area under the receiver operating characteristics (ROC) curve (AUC), where the ROC curve is given by the sensitivity as a function of one minus the specificity.

5 Experiments

Details for all experiments are given in the following subsections. We use five-fold cross validation for most of the models, and present mean accuracies and

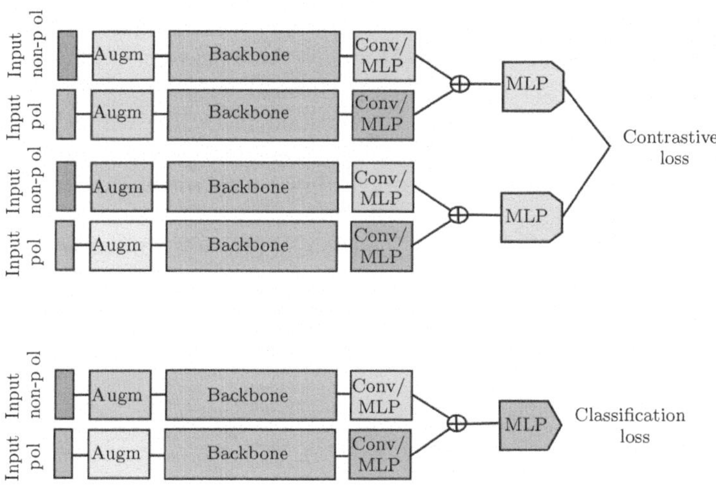

Fig. 4. Contrastive learning (CL) model. First, we train with two setups of the network and train using the contrastive loss between outputs. In the second step, we replace the final layers of the MLP and retrain using the same loss function as in the DC model.

standard deviation over the five folds. Furthermore, some additional ablation tests were also performed using the first fold only.

In addition to results on the validation and full test set, we also present evaluations on the subset of the testset where we have histopathological diagnosis, since these can be compared to the dermatologist score.

5.1 Pre-processing and Training Settings

We downscale the images a factor three and while augmenting clip them to a size of 448×448 pixels for EVA-02, as the input size for the pre-trained EVA-02 network was fixed. We did test the MobileNetV2 mostly on 666×666 pixels, but also on the smaller input size 448×448. After the downsampling the images still have approximately three times higher resolution than the ISIC 2024 Challenge dataset [18]. Inspired by the top contributions of ISIC Challenge 2023 and 2024 [18] we augment the images for training by changes in contrast and brightness, flips, rotations, zooms, translations and cutouts. All networks are trained with the Adam optimiser [17] and use batch normalisation. As classification loss we use weighted categorical cross-entropy. All code is implemented in TensorFlow and Keras, and the MobileNetV2 implementation was sourced from the original website. The EVA-02 implementation is from [2]. More specifics on the networks and training parameters can be found in Supplementary, Table 1.

5.2 Temperature Studies for Contrastive Learning

We compare the balanced accuracy and AUC for several different values of the hyper-parameter $\tau \in [0.1, 1.5]$. The results can be seen in Fig. 5. Our experiments showed better results for higher values of τ. The experiments were run on a single fold binary classification with MobileNetV2 backbone and only evaluated on the validation data. The study was performed without batch normalisation but we do not expect this to affect the optimal τ. In line with the results, we proceed with a temperature $\tau = 0.9$. For this temperature, we also re-train the network two additional times to investigate how much the results vary for different trainings of the same network. When using EVA-02 as backbone we chose the same temperature $\tau = 0.9$, as the training was unstable for low τ.

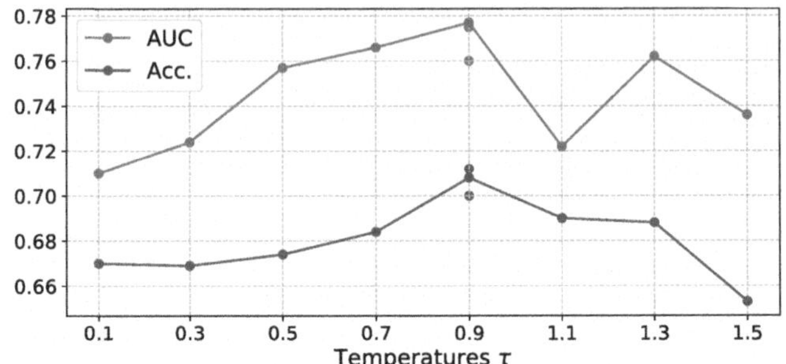

Fig. 5. Accuracies of the SCL network after training and evaluation on one fold with MobileNetV2 backbone. Each dot represents validation results for a trained network. Three different networks were trained for the best-performing value $\tau = 0.9$.

5.3 Binary Classification

For binary classification of BCC versus non-BCC we evaluate both different backbones, i.e. MobileNetV2 and EVA-02, and three training strategies: DC, UCL and SCL. The results on all folds for the best performing models are presented in Table 3. Unfortunately, UCL performed not much better than random on the test sets and the method was therefore not further evaluated and is not included in the table. To test the gain from using the pair of both polarised and non-polarised light, we also train a network using only the non-polarised light image. This was done only on the first fold, and results for this and other models on the first fold can be seen in Table 2.

5.4 Fine-Grained Classification

After experience from the binary classification, we continue to perform the multiclass, fine-grained classification on our data. The network designs are the same as for the binary classification with the modification that we get six outputs. Our best results for the different methods are presented in Table 4 with mean and standard deviations for our five-fold cross validation.

Table 2. A comparison of different architectures/models and backbones for BCC detection. Results for a single fold are presented. Non-pol refers to the case with only one, non-polarised, input image, while all other cases use paired images.

	Accuracy	Bal. acc.	AUC
Eva DC, test	0.746	0.747	**0.815**
subtest	**0.747**	0.667	0.717
Eva DC non-pol, test	0.696	0.695	0.766
subtest	0.630	0.581	0.616
Eva SCL , test	**0.751**	**0.750**	0.801
subtest	0.741	**0.724**	**0.747**
Mobilenet DC, test	0.658	0.663	0.737
subtest	0.728	0.629	0.687
Mobilenet DC non-pol, test	0.676	0.677	0.749
subtest	0.691	0.657	0.705
Mobilenet DC (EVA-02 size), test	0.643	0.651	0.777
subtest	0.753	0.575	0.709
Mobilenet SCL, test	0.611	0.616	0.676
subtest	0.698	0.529	0.563
Mobilenet SCL (EVA-02 size), test	0.586	0.595	0.710
subtest	0.728	0.515	0.612
Dermatologists	0.895	0.876	-

Table 3. Results for detection of basal cell cancer (binary problem). Mean and standard deviation are presented for each model. The result for the subtest set (marked in blue) can be compared to the accuracy of the dermatologists.

		Accuracy	Balanced accuracy	AUC
DC EVA-02	validation	**0.741** $\pm$ 0.034	**0.742** $\pm$ 0.034	**0.818** $\pm$ 0.040
	test	**0.748** $\pm$ 0.006	**0.749** $\pm$ 0.006	**0.818** $\pm$ 0.011
	subtest	**0.740** $\pm$ 0.023	0.680 $\pm$ 0.025	0.727 $\pm$ 0.022
SCL EVA-02	validation	0.683 $\pm$ 0.038	0.684 $\pm$ 0.037	0.756 $\pm$ 0.034
	test	0.720 $\pm$ 0.017	0.718 $\pm$ 0.018	0.790 $\pm$ 0.015
	subtest	0.683 $\pm$ 0.037	**0.695** $\pm$ 0.012	**0.756** $\pm$ 0.021
DC Mob	validation	0.674 $\pm$ 0.023	0.676 $\pm$ 0.022	0.749 $\pm$ 0.045
	test	0.665 $\pm$ 0.008	0.666 $\pm$ 0.006	0.733 $\pm$ 0.020
	subtest	0.674 $\pm$ 0.071	0.610 $\pm$ 0.016	0.675 $\pm$ 0.020
SCL Mob	validation	0.611 $\pm$ 0.028	0.612 $\pm$ 0.022	0.682 $\pm$ 0.048
	test	0.614 $\pm$ 0.015	0.62 $\pm$ 0.014	0.687 $\pm$ 0.021
	subtest	0.702 $\pm$ 0.015	0.545 $\pm$ 0.022	0.592 $\pm$ 0.022
Dermatologists		0.895	0.876	-

Table 4. Results for fine-grained classification, into six classes. Mean and standard deviation are presented for each model. The result on subtest set (marked in blue) can be compared to the accuracy of the dermatologists.

		Accuracy	Balanced accuracy
DC EVA-02	validation	**0.439** ± 0.041	**0.432** ± 0.036
	test	0.374 ± 0.018	**0.432** ± 0.024
	subtest	0.333 ± 0.024	0.374 ± 0.028
SCL EVA-02	validation	0.433 ± 0.041	0.399 ± 0.038
	test	**0.393** ± 0.024	0.396 ± 0.014
	subtest	**0.345** ± 0.020	**0.394** ± 0.042
SCL Mob	validation	0.213 ± 0.064	0.245 ± 0.048
	test	0.191 ± 0.040	0.259 ± 0.037
	subtest	0.261 ± 0.049	0.259 ± 0.036
Dermatologists		0.495	0.519

5.5 Comparison of Confusion Matrices to Dermatologists

On the subset of our testset that have fine-grained classification on both dermatology and histopathology diagnosis we can compare our best model (SCL with EVA-02 backbone) with the dermatologists. However, it is important to highlight that the limited number of samples results in significant variance in the findings. The resulting confusion matrices can be seen in Table 5.

Table 5. Confusion matrices for dermatologists to the left and our best model (EVA-02 backbone SCL) to the right, where the rows and columns give the true and estimated labels, respectively. The numbers are mean values over five folds. Acc. is the accuracy (sensitivity) per class.

	Dermatologists							Our best model						
	Estimated label						Acc.	Estimated label						Acc.
SCC in situ	7	5	0	1	0	0	0.54	5.6	3.2	0.6	1	1.8	0.8	0.43
Invasive SCC	1	6	0	3	0	1	0.55	2.8	2.4	0.6	2.8	1	1.4	0.22
AK	9	1	1	1	1	0	0.08	6.4	0.2	3.8	1	1.2	0.4	0.29
Nodular BCC	0	1	0	8	1	2	0.67	0.4	1	1.2	5.6	0.2	3.6	0.47
Superf. BCC	1	0	0	1	9	0	0.82	0.8	0	1.2	0.4	7.8	0.8	0.71
Infilt. BCC	2	0	0	20	2	21	0.47	7.4	3.2	4.8	13	6	11	0.24

6 Discussion

Our results are highly dependent on our dataset and the variance in our measurements is quite high since we have few examples to train on. This can be seen in Table 3 and 4 where we declare the standard deviation over the five folds. We may also see that there is high variance between training sessions in Fig. 5, where several re-runs of the same fold and temperature $\tau = 0.9$ are pictured. The dataset size has a very big impact, but the results are also affected by the fact that it is relatively uniform in technical details and patient group. For example we are limited in this study to a cohort of patients in Sweden, which sets a strong prior of what skin types are present in the dataset. We therefore refrain from drawing any conclusions of results on other datasets. It has previously been shown that large domain shifts are detrimental to the accuracy [1]. We may however draw conclusions from differences in performances, even though we do it with slight caution due to our small dataset.

Choice of Backbone network for transfer learning does affect performance. For our classification tasks EVA-02 is the better choice (accuracy 74.8% vs 66.5% for DC and 72.0% vs 61.4% for SCL, Table 3). These networks differ significantly in architecture, but a dermatology-specific consideration is that high-frequency details, such as micro-scarring and small blood vessel structures, may be more effectively preserved and propagated through a vision transformer than a more conventional network.

Using Paired Images of one polarised and one non-polarised dermatoscopy image might improve results. We did a small ablation study only using the non-polarised images, giving accuracy 74.6% to compare with 69.6% for EVA02 and accuracy 65.8% to compare 67.6% for MobileNetV2, see Table 2. From these opposing results it is hard to draw a conclusion. However, in teledermatoscopy the images are considered to contain different information (the polarised light visualizes structures further down into the skin). Our best model might utilize this extra information, but to investigate this further, we would need more images of the same lesions of either polarised or non-polarised light such that we can compare to using pairs that are both of the same setting.

Usage of Contrastive Learning is not conclusive for either MobileNetV2 or EVA-02. We must however remember that our dataset is small and the variance between training sessions and folds is large. UCL has interresting properties but we did not achieve sufficient results when applying UCL which is unfortunate. Dermatology is a field with numerous diagnoses, and unsupervised clustering could prove valuable, especially considering the relative ease of acquiring unlabelled data. More work on UCL might allow us to leverage all 10,000 examples, rather than the approximately 2,000 with appropriate labels used in this paper.

Input Size does not seem to be a very sensitive factor. For the MobileNetV2 network we tested two different input sizes, giving only small differences in performance, see Table 2.

Comparison to Dermatologists is slightly in their favour because they often have access to macro images of the lesion, patient history, and metadata such as lesion changes, patient age, and palpation of the specimen when evaluating a sample. For anonymity reasons, we only include dermatoscopy images in this study. However, it should be noted that by excluding some of the diagnostic support available to dermatologists, we are making an already challenging task even harder. More details can be found in Table 5.

Subclassifcation of Non-pigmented Skin Cancer is a hard task, and although the amount of samples per class is lower than for BCC detection we have our most interesting results when attempting to subclassify the data. It is also here we may have the most positive impact as DL for sublcassifying BCC is a medically hard and valuable task. If we could get close to or surpass dermatologist performance we could save a lot of patient and dermatologist time as well as suffering from unnecessary biopsies.

Future Work. More data is generally a very effective way to improve results in DL. In addition to the data we already have, it would be valuable to incorporate data from publicly available datasets. This could help address some of the challenges related to generalization to other data sources and make the model more inclusive of different skin types. Moreover, increasing the amount of data allows for the training of larger models, potentially even from scratch.

7 Conclusions

We have investigated the overlooked but highly impactful problem of DL-aided classification of unpigmented skin cancer, where we believe that we are the first to study fine-grained classification of BCC using only dermatoscopy images. We have a high-resolution dataset with paired images captured with and without polarised light, and design our model to combine the information from both. We use strategies such as transfer learning and contrastive learning to adjust to that our dataset unfortunately is limited in size. We see potential in using a vision transformer backbone for transfer learning and view this as a promising direction for future research. It is challenging to draw strong conclusions with limited data, but we hope that this study can be a first step for more research on the subject.

Acknowledgements. This work was supported by the Swedish Cancer Society, Vinnova and Analytic Imaging Diagnostics Arena (AIDA), Vinnova Grant 2021-01420.

Disclosure of Interests. The authors have no competing interests to declare that are relevant to the content of this article.

References

1. Aguilera Manzanera, M.D.P.: Design, implementation and evaluation of a deep learning prototype to classify non-pigmented malignant skin cancer from dermatoscopic images. Lund University: Master's Theses in Mathematical Sciences 2022, E17 (2022). ISSN: 1404-6342
2. Arse, L.D.G.: keras_cv_attention_models (2020). https://github.com/leondgarse/keras_cv_attention_models. Accessed 17 Feb 2025
3. Bozsányi, S., et al.: Optically guided high-frequency ultrasound to differentiate high-risk basal cell carcinoma subtypes: a single-centre prospective study. J. Clin. Med. **12**(21), 6910 (2023)
4. Chen, T., Kornblith, S., Norouzi, M., Hinton, G.: A simple framework for contrastive learning of visual representations. In: International Conference on Machine Learning, pp. 1597–1607. PMLR (2020)
5. Deng, J., Dong, W., Socher, R., Li, L.J., Li, K., Fei-Fei, L.: ImageNet: a large-scale hierarchical image database. In: 2009 IEEE Conference on Computer Vision and Pattern Recognition (CVPR), pp. 248–255 (2009). https://doi.org/10.1109/CVPR.2009.5206848
6. Devaraj, G.P., Ravi, R.: Advancing skin cancer diagnosis with a multi-branch ShuffleNet architecture. Int. J. Imaging Syst. Technol. **34**(2), e23051 (2024)
7. Dosovitskiy, A.: An image is worth 16×16 words: transformers for image recognition at scale. arXiv preprint arXiv:2010.11929 (2020)
8. Dreyfuss, I., Kamath, P., Frech, F., Hernandez, L., Nouri, K.: Squamous cell carcinoma: 2021 updated review of treatment. Dermatol. Ther. **35**(4) (2022)
9. Duschner, N., et al.: Applying an artificial intelligence deep learning approach to routine dermatopathological diagnosis of basal cell carcinoma. JDDG: J. Deutschen Dermatologischen Gesellschaft **21**(11), 1329–1337 (2023)
10. Fang, Y., Sun, Q., Wang, X., Huang, T., Wang, X., Cao, Y.: Eva-02: a visual representation for neon genesis. Image Vis. Comput. **149**, 105171 (2024)
11. Hadsell, R., Chopra, S., LeCun, Y.: Dimensionality reduction by learning an invariant mapping. In: 2006 IEEE Computer Society Conference on Computer Vision and Pattern Recognition (CVPR), vol. 2, pp. 1735–1742 (2006). https://doi.org/10.1109/CVPR.2006.100
12. Howard, A.G., et al.: MobileNets: efficient convolutional neural networks for mobile vision applications. arXiv preprint arXiv:1704.04861 (2017)
13. Huang, S.C., Pareek, A., Jensen, M., Lungren, M.P., Yeung, S., Chaudhari, A.S.: Self-supervised learning for medical image classification: a systematic review and implementation guidelines. NPJ Digit. Med. **6**(1), 74 (2023)
14. Isaksson, K., Katsarelias, D., Mikiver, R., Carneiro, A., Ny, L., Olofsson Bagge, R.: A population-based comparison of the AJCC 7th and AJCC 8th editions for patients diagnosed with stage III cutaneous malignant melanoma in Sweden. Ann. Surg. Oncol. **26**, 2839–2845 (2019)
15. Karimkhani, C., et al.: The global burden of melanoma: results from the global burden of disease study 2015. Br. J. Dermatol. **177**(1), 134–140 (2017)
16. Khosla, P., et al.: Supervised contrastive learning. Adv. Neural. Inf. Process. Syst. **33**, 18661–18673 (2020)
17. Kingma, D., Ba, J.: Adam: a method for stochastic optimization. In: International Conference on Learning Representations (2014)
18. Kurtansky, N., Rotemberg, V., Gillis, M., Kose, K., Reade, W., Chow, A.: ISIC 2024 - skin cancer detection with 3D-TBP (2024). https://kaggle.com/competitions/isic-2024-challenge

19. Lomas, A., Leonardi-Bee, J., Bath-Hextall, F.: A systematic review of worldwide incidence of nonmelanoma skin cancer. Br. J. Dermatol. **166**(5), 1069–1080 (2012)
20. Matsumoto, M., et al.: Estimating the cost of skin cancer detection by dermatology providers in a large health care system. J. Am. Acad. Dermatol. **78**(4), 701–709 (2018)
21. Peris, K., et al.: Diagnosis and treatment of basal cell carcinoma: European consensus-based interdisciplinary guidelines. Eur. J. Cancer **118**, 10–34 (2019)
22. Salinas, M.P., et al.: A systematic review and meta-analysis of artificial intelligence versus clinicians for skin cancer diagnosis. NPJ Digit. Med. **7**(1), 125 (2024)
23. Sandler, M., Howard, A., Zhu, M., Zhmoginov, A., Chen, L.C.: Mobilenetv2: inverted residuals and linear bottlenecks. In: Proceedings of the IEEE Conference on Computer Vision and Pattern Recognition, pp. 4510–4520 (2018)
24. Surkov, Y.I., et al.: Multimodal method for differentiating various clinical forms of basal cell carcinoma and benign neoplasms in vivo. Diagnostics **14**(2), 202 (2024)
25. Swedish Cancer Registry 2018. www.socialstyrelsen.se
26. Wang, F., Liu, H.: Understanding the behaviour of contrastive loss. In: Proceedings of the IEEE/CVF Conference on Computer Vision and Pattern Recognition, pp. 2495–2504 (2021)
27. Widaatalla, Y., et al.: The application of artificial intelligence in the detection of basal cell carcinoma: a systematic review. J. Eur. Acad. Dermatol. Venereol. **37**(6), 1160–1167 (2023)
28. Yacob, F., et al.: Weakly supervised detection and classification of basal cell carcinoma using graph-transformer on whole slide images. Sci. Rep. **13**(1), 7555 (2023)
29. Zhang, C., et al.: Dual temperature helps contrastive learning without many negative samples: towards understanding and simplifying MoCo. In: Proceedings of the IEEE/CVF Conference on Computer Vision and Pattern Recognition, pp. 14441–14450 (2022)
30. Zhou, S.K., et al.: A review of deep learning in medical imaging: imaging traits, technology trends, case studies with progress highlights, and future promises. Proc. IEEE **109**(5), 820–838 (2021)

Explainable AI for CV

A Meaningful Perturbation Metric for Evaluating Explainability Methods

Danielle Cohen(✉), Hila Chefer, and Lior Wolf

Blavatnik School of Computer Science, Tel Aviv University, Tel Aviv-Yafo, Israel
danielcohen5@mail.tau.ac.il

Abstract. Deep neural networks (DNNs) have demonstrated remarkable success, yet their wide adoption is often hindered by their opaque decision-making. To address this, attribution methods have been proposed to assign relevance values to each part of the input. However, different methods often produce entirely different relevance maps, necessitating the development of standardized metrics to evaluate them. Typically, such evaluation is performed through perturbation, wherein high- or low-relevance regions of the input image are manipulated to examine the change in prediction. In this work, we introduce a novel approach, which harnesses image generation models to perform targeted perturbation. Specifically, we focus on inpainting only the high-relevance pixels of an input image to modify the model's predictions while preserving image fidelity. This is in contrast to existing approaches, which often produce out-of-distribution modifications, leading to unreliable results. Through extensive experiments, we demonstrate the effectiveness of our approach in generating meaningful rankings across a wide range of models and attribution methods. Crucially, we establish that the ranking produced by our metric exhibits significantly higher correlation with human preferences compared to existing approaches, underscoring its potential for enhancing interpretability in DNNs.

Keywords: Interpretability · Image classification · Image generation

1 Introduction

In recent years, deep neural networks (DNNs) have revolutionized research in countless domains. Yet, unlike some predecessors *e.g.*, decision trees and linear regressors, DNN predictions remain opaque. DNNs often rely on spurious correlations or "shortcuts", leading to biases [6,22] or robustness issues [10,28,29] that hamper their widespread adoption.

To overcome this obstacle, several explainability and interpretability techniques have been proposed to clarify the decision-making process of DNNs [9, 12,24,46,50]. In this work, we focus on attribution methods, which constitute one of the most prominent approaches for producing such interpretations. Attribution methods assign a corresponding relevance value to each of the input parts (e.g., image pixels), where a higher relevance score suggests a greater impact

J. Petersen and V. A. Dahl (Eds.): SCIA 2025, LNCS 15726, pp. 309–324, 2025.
https://doi.org/10.1007/978-3-031-95918-9_22

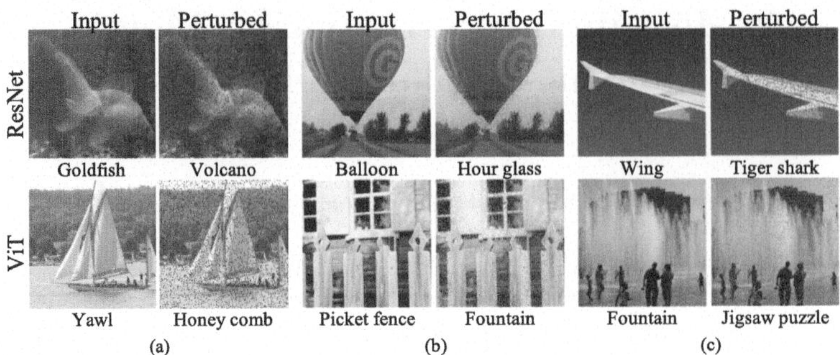

Fig. 1. Examples of OOD predictions following perturbation with (a) pixel deletion [30] (b) pixel blurring [21] and (c) per channel mean [30] on both ResNet [27] and ViT [17] backbones. For each image, we randomly draw pixels and apply the perturbation method. We enclose the original and modified prediction. As can be seen, even when the majority of the content is still visible, the model is sensitive to the OOD effect of the perturbation, causing it to modify its prediction.

on the model's prediction. Various attribution methods have been proposed to interpret the predictions by DNNs, each claiming to be the most "faithful" to the model, while often producing entirely different results than their counterparts. However, in the absence of a rigorous definition of faithfulness and a set of standard acceptable evaluation metrics, it becomes impossible to decipher which attribution method should be employed for a given use case.

In this work we adopt the dominant notion of faithfulness, where a relevance map is considered to be "faithful" to the model if it reflects the true reasoning process behind the model's prediction [9,31,35]. In other words, given a relevance map, our metric should estimate the correlation between the pixels with high relevance values and the prediction made by the model. Another important parameter for the evaluation of attribution methods is "plausibility", *i.e.*, how convincing the explanation is to humans [31]. Both faithfulness and plausibility are important when evaluating an explanation. Without faithfulness, there is no direct connection between the relevance map and the model we attempt to explain, and therefore the explanation has no utility [2]. Without plausibility, the explanation may not be clear to humans, which defies its goal of helping humans understand the reasoning behind the decision-making [19].

The predominant approach to evaluating attribution maps in the context of image classifiers is through image perturbation [15,21,30,34]. The intuition is that pixels are perturbed from most (least) relevant. If high (low) relevance marks important (unimportant) pixels, perturbing alters (keeps) the prediction.

However, existing perturbation methods make unnatural modifications to the input image, such as deleting pixels [30,35], blurring pixels [21] or per channel mean [30]. As we demonstrate in Fig. 1 these alterations often drive the input out of distribution, leading the model to produce predictions that do not correlate with the image pixels. Importantly, in Fig. 1 we apply *completely random masks* such that the content of the image is still mostly preserved, and even so, the perturbation causes the model to change its prediction. This random baseline demonstrates the significant challenge to disentangling between the OOD effect of the perturbation approach on the prediction and the relevance of the modified pixels to the prediction, making existing perturbation approaches unreliable.

Our research hypothesis is that a better approach is to use the perturbed pixels in a way that presents the classifier with a valid alternative. The easiest alternative target is likely to be the class with the second-highest classification score. We employ the capabilities of Stable Diffusion (SD) inpainting [42] with the appropriate conditioning to modify the high relevance pixels, such that they will align with the second highest prediction, and test whether the prediction changes to *the specified target class*. This allows us to filter out modifications that lead to unrelated classes, such as the ones in Fig. 1. Additionally, we add a novel weighting function that compares the relevance maps before and after the inpainting procedure. This validates that the change in prediction was indeed *caused* by the inpainting procedure, further establishing the connection between the perturbed pixels and the modification in the prediction.

We perform extensive experiments to compare our Stratified Inpainting approach to existing perturbation approaches, such as deleting pixels, blurring pixels, and per channel mean. We test different prominent attribution approaches [1,9,12,46,50], and various image backbones, such as different variants of CNNs [27,48] and Vision Transformers [17], and find that in many of the cases, existing methods struggle to provide a meaningful ranking and that the perturbation graphs do not give a meaningful distinction between methods. Surprisingly, we show that the existing perturbation methods often struggle to separate between valid attribution methods and *randomly generated relevance maps*, further indicating the inadequacy of employing these methods for evaluating explanations. Conversely, our Stratified Inpainting approach maintains decent separation between methods and produces robust results across different models and datasets. Via a user study, we find that the ranking produced by our metric is much more correlated with human preferences, demonstrating that in the case of attribution methods, faithfulness and plausibility are not mutually exclusive, and are both identified by our metric simultaneously.

2 Related Work

Explainability in Computer Vision. Many methods were proposed for generating heatmaps to explain deep networks. *Gradient based* methods use loss gradients via backpropagation [46,47,50–52]. *Attribution propagation* methods, such as LRP, recursively decompose decisions based on the Deep Taylor Decomposition framework [4,5,39]. Other methods that do not fall into these two categories include saliency-based methods [13,38,49,54,56], Activation Maximization [18], Excitation Backprop [55], concept-based methods [23,32], and Perturbation methods [12,20,21,36], which analyze network responses to input changes. With the rise of Transformers, attention explainability methods have gained interest. LRP was applied to Transformers [53], while methods like rollout [1] and Transformer attribution [8,9] extend LRP and gradients to explain attention mechanisms.

Generative Models for Explainability. Several works have investigated the use of generative models for explainability-related tasks. Agarwal et al. [3] and Chang et al. [7] proposed incorporating inpaiting with existing attribution methods [21] to improve quality. While sharing a similar intuition, we introduce a metric for evaluating existing attribution methods rather than proposing a new one. Critically, fully inpainting-based attribution can be computationally expensive necessitates at least 300 iterations [3]. In contrast, by leveraging inpainting only for evaluation, our metric reduces this overhead to just 10 iterations, greatly easing the computational burden. Finally, note that our metric also proposes a novel data filtering method, and a weighting function to directly connect the inpainted pixels to the modification in the classification.

Evaluating Explanations. While the research around explainability algorithms continues to evolve, developing methods to evaluate the provided explanations remains a less explored field of research. Most metrics of evaluating explanations are based on perturbation tests and segmentation results, which can be seen as a general case of the Pointing-Game [21,30]. More recent works [35] have suggested testing the confidence gap induced by removing features to evaluate the faithfulness of the explanation to the explained model. A different line of work [19] evaluates the explanations based on their utility to humans, i.e., the information humans can deduce from the explanations. Another related area is counterfactual explanations. Goyal et al. [24] used inpainting to generate counterfactual explanations. The primary distinction between our paper and theirs lies in the task. Our objective is to evaluate attribution methods, we use inpaint to assess the method's attribution maps while counterfactual explanations task is identifying the minimal set of pixels that will change the classification.

Text-to-Image Generation. Recent text-to-image generative models have demonstrated an unparalleled ability to generate diverse and creative imagery guided by a target text prompt [40–42,44]. In this work we leverage the powerful Stable Diffusion model [42] to produce natural perturbations to images based on text. Specifically, we leverage the Stable Diffusion inpainting model[1], which was fine-tuned on the inpainting task. Given an image, a mask, and a target text prompt, the model completes the masked areas in a manner that corresponds to the input text prompt.

3 Method

We begin by describing the problem setting. We then provide motivation for our method by highlighting the pitfalls of the widely used perturbation metrics and describe our method in detail.

[1] The SD inpainting model was trained by Runway ML, and implemented over the diffusers library: https://huggingface.co/runwayml/stable-diffusion-inpainting.

3.1 Problem Setting

Given a classifier C and an explainability method E, the goal of our method is to evaluate the faithfulness of the explanations produced by E to the predictions made by C. Formally, given a set of test images to examine, $I = \{i_1, \ldots, i_N\}$ s.t. $\forall i \in I : i \in \mathbb{R}^{h,w}$, the explainability algorithm produces a relevance map $E(C, i) = R_{E,C}(i) \in \mathbb{R}^{h,w}$. Our goal is to produce a score $f(C, E, I)$ that reflects the faithfulness of the relevance maps $R_{E,C}(i)$ to the classifier's predictions $C(i)$ on the test set, where a higher score indicates that the maps produced by E are more faithful to the classifier C.

3.2 Stratified Inpainting

The intuition behind our method is to use the powerful image generation prior of text-to-image diffusion models [42] to *replace the removed pixels by other plausible pixels* instead of unnaturally perturbing them (*e.g.*, by deletion or blurring), in order to avoid OOD inputs. Additionally, rather than measuring an arbitrary change to any other class (which may be unintuitive, as demonstrated in Fig. 1), we assess the relevance's ability to transform the prediction to the *second highest prediction.* In other words, we estimate whether the highlighted features are indeed those that lead to the top-1 prediction, by substituting them with a plausible alternative and examining the change in prediction against a *specific plausible class.* Next, we describe the steps of our method in detail.

Step 1: Creating the Evaluation Dataset I. Given a classifier C and an inpainting Stable Diffusion model SD, we aim curate a test set of classes to extract our test images from. Importantly, while Stable Diffusion is very expressive, it has been demonstrated [45] that it fails to capture some classes, especially fine-grained classes in datasets such as ImageNet [43]. Therefore, we automatically curate a set S of *classes* recognizable by SD. Our Stratified inpainting algorithm is applied over a test set I, containing images from the test set such that the top-2 prediction by the classifier is in the set S defined above, i.e., $top_2\,(C(i)) \in S$. This way, we insure SD is capable of inpainting the images given the selection of the top-2 class. Please refer to the supplementary materials for additional implementation details on the construction of the dataset.

Step 2: Evaluation on the Test Classes. The method for obtaining scores for each test image $i \in I$ is outlined in Algorithm 1. We begin by calculating the corresponding relevance map $R_{E,C}(i)$ in line 2. Next (line 3), we apply bilinear downsampling to obtain patch-wise relevance values (patch size 16, following [17]), as SD inpainting struggles with distant single pixels. This limitation is alleviated by the use of patch-wise explanations (see ablation study). We consider perturbation steps p between 10% and 50%, as inpainting over 50% often obtains the top-2 class by simply adding another object. For each step p we create a mask that captures the top pixels of that step. We use this mask to inpaint (lines 5–8) the top patches of the image by the relevance map $R_{E,C}$(i) (line 5), using

Algorithm 1. Stratified inpainting

Input: A classifier to explain C, an explainability method E, an image to explain $i \in I$, and a pre-trained inpainting Stable Diffusion model SD.
Output: Stratified inpainting scores for $p = \{10\%, \ldots, 50\%\}$ of the most relevant image pixels.

1: $scores \leftarrow \{\}$
2: $R_{E,C}(i) \leftarrow E(C, i)$
3: $R_{E,C}(i) \leftarrow bilinear(R_{E,C}(i), factor = \frac{1}{16})$
4: **for** $p \in \{10\%, \ldots, 50\%\}$ **do**
5: $\quad masked\_patches \leftarrow top_p(R_{E,C}(i))$
6: $\quad \forall patch \in i : mask[p] \leftarrow 1$
7: $\quad \forall patch \in masked\_patches : mask[p] \leftarrow 0$
8: $\quad i' \leftarrow SD(i \odot mask, top\_2\_class)$
9: $\quad$ **if** $top_1\left(C(i')\right) = top_2\left(C(i)\right)$ **then**
10: $\quad\quad w \leftarrow \frac{|top_p(R_{E,C_2}(i')) \cap (top_p(R_{E,C_2}(i)) \cup M)|}{|top_p(R_{E,C_2}(i'))|}$
11: $\quad\quad scores \leftarrow$ **concat**$(scores, \{w\})$
12: $\quad$ **else**
13: $\quad\quad scores \leftarrow$ **concat**$(scores, \{0\})$
14: $\quad$ **end if**
15: **end for**
16: **Return** $scores$

the prompt "{top-2 class name}". If $R_{E,C}(i)$ indeed faithfully captures the most relevant pixels for the prediction, then prediction of the inpainted image should change to the top-2 class.

In cases where the prediction indeed changes to be the top-2 class, we wish to verify that the change was caused by the set of top p pixels that were perturbed in the inpainting process. To this end, we compare the relevance maps obtained for the original and the inpainted images. A successful inpainting should lead to high relevance either in the inpainted areas (since those are the areas actually modified to fit the top-2 class) or in areas that were already attributed to the top-2 class. Therefore, we add a weighting function to capture this behavior. Let M be the set of top p pixels modified by the inpainting step, i be the original image, and i' be the unpainted image. Let C_1, C_2 be the top-1, and top-2 class of i, respectively. By our previous notations, $top_p\left(R_{E,C_2}(i)\right), top_p\left(R_{E,C_2}(i')\right)$ are the top pixels in the relevance maps of the original and inpainted images for the top-2 class. We apply the following weighting function (line 10): $w = \frac{|top_p(R_{E,C_2}(i')) \cap (top_p(R_{E,C_2}(i)) \cup M)|}{|top_p(R_{E,C_2}(i'))|}$. This indicates the amount of top p pixels in i''s explanation of label C_2 that are also in the inpainting mask M from i, or are in the original relevance map for class C_2 extracted from i.

We update the score for p with w if the prediction changed to the target class, which is the original top-2 class, and 0 otherwise (lines 9–14). The list of all scores (one per percentile p) is returned (line 16).

This process is repeated for all images in I that follow the selection criterion $top_2\left(C(i)\right) \in S$. The final score per step p is obtained by averaging the values for that specific step across all such images i. Kindly refer to the supplementary materials for additional implementation details.

4 Experiments

Baselines. We compare our method against the most common perturbation alternatives. First, with perhaps the most common approach, which is pixel deletion [30], and then with other prominent alternatives, such as blurring [21] and per channel mean [30]. We also consider less related methods, such as faithfulness violation [35], and saliency-based evaluation [3,13]. These alternatives are less competitive and need adaptation for comparison, so the full comparison is in the supplementary materials. Similar to Algorithm 1 (lines 9–12), each method scores explanation $R_{E,C}(i)$ differently. Importantly, existing techniques check the change of the prediction to *any other class*, thus they are vulnerable to OOD prediction changes, such as those demonstrated in Fig. 1. Conversely, our metric measures the modification to *a specific plausible class*, mitigating this OOD effect.

Models and Dataset. We experiment on widely used image backbones: ResNet-50 [27], VGG [49], AlexNet [33] and ViT-B/16 [17], to demonstrate our method's ability to provide meaningful evaluations across different selections of backbones, both CNN- and transformer-based. We construct the dataset for comparison from the subset of classes S collected as described in the previous section, over the ImageNet validation set [25] and the binary cat-dog dataset for AlexNet[2].

Explainability Methods. For each model, we use the most common explainability algorithms and *random* pixel selection as a baseline to assess the metric's robustness to OOD effects, as shown in Fig. 1. For CNNs, we consider Grad-CAM (gc) [46], FullGrad [51], SmoothGrad [50], Grad SHAP [36], inputXgrad [47], IntegratedGradients [52] and layered GradCam [16]. For ViT, we consider GradCAM (gc) [46], Transformer Attribution (transformer) [9], and partial LRP (lrp) [53], as well as the vanilla attention (attn last layer), and rollout [1].

4.1 Qualitative Results

In this section, we demonstrate our method's advantage over the baselines in generating more natural results and producing a statistically significant separation between the different explainability methods. First, we qualitatively show the corresponding graphs for each of the evaluation methods (ours and the baselines) side by side for removing $p \in \{10\%, \ldots, 50\%\}$ of the pixels (see Sect. 3.2). As can be seen in Fig. 2, our method is able to provide a meaningful ranking of the methods, where different explainability methods are *distant from each*

[2] Taken from kaggle: https://www.kaggle.com/datasets/tongpython/cat-and-dog.

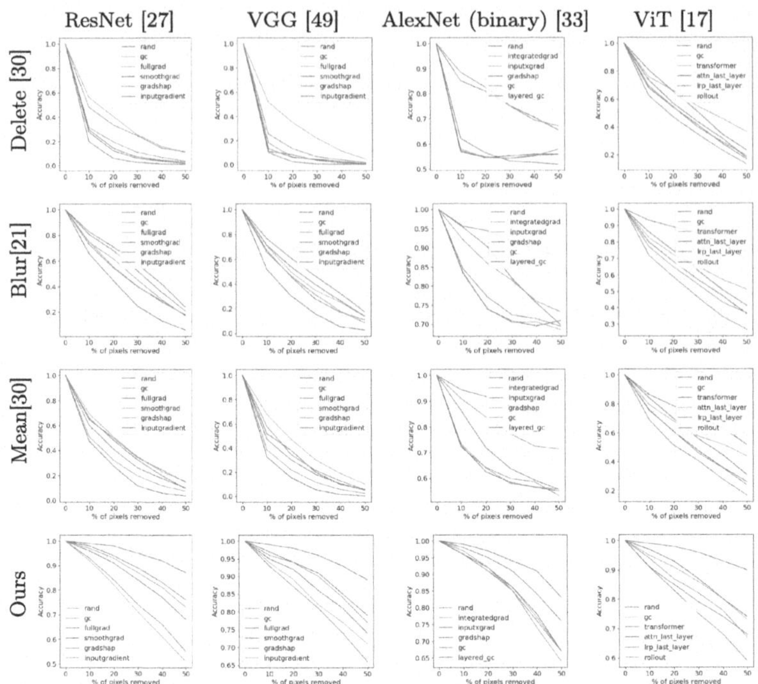

Fig. 2. Perturbation comparison against the leading baselines with ResNet-50, VGG, AlexNet-based binary classifier, and ViT-B (please zoom in to view better). For each model, we consider the most common explainability algorithms, in addition to a *random* selection of pixels (see Sect. 4 for details). As can be observed, the baselines often struggle with separating random maps from actual relevance maps (*e.g.*, delete for all models, blur for AlexNet, mean for all CNN variants) and appear to produce very similar results for all methods. Conversely, our method produces consistent ranking and meaningful distinction from the random baseline.

Fig. 3. Qualitative comparison for SmoothGrad. According to all baselines, SmoothGrad is the leading method for CNN-based networks. We demonstrate examples where the baseline metrics indicate success for SmoothGrad (*i.e.*, the prediction changed) while ours indicates failure (*i.e.*, the prediction did not change to the top-2 class). The relevance maps produced by SmoothGrad often lead to OOD effects with the baselines (similar to Fig. 1), thus their reliability for these results is questionable.

Fig. 4. Qualitative comparison of successful class changes against baselines. We showcase image examples (Input) where both our method and the baselines induced a change in prediction (predictions indicated below each image). As seen, even when our method agrees with the baselines (*i.e.*, the relevance map is faithful by all metrics), our method produces plausible pixel changes, while baselines cause OOD predictions.

other, while also *significantly separating the viable explanations from the random pixel selection baseline.* Conversely, the baseline evaluation metrics often produce graphs that are very similar for most of the attribution methods (see ResNet, VGG graphs), making it hard to tell whether a method's advantage is due to a statistical error or actual faithfulness to the classifier. The separation between different explainability algorithms is instrumental for the ability to empirically support the advantage of one explainability algorithm over the other, since the methods often produce *entirely different results* (see supplementary materials for such examples).

Additionally, the ranking of methods by the metrics often varies. For example, for all CNN variants, the baselines rank SmoothGrad as the top method, while our metric favors GradCAM or FullGrad. We hypothesize that this is due to SmoothGrad's relevance maps being scattered throughout the image, making them more susceptible to OOD effects, as shown in Fig. 1. Figure 3 qualitatively supports this claim, showing examples where the baselines indicated success with SmoothGrad (prediction changed), while our method showed the relevance was not faithful (prediction unchanged). These examples are similar to those in Fig. 1, where relevance maps appear random or scattered, and the perturbed classification does not correlate with the image content.

Next, Fig. 4 presents a qualitative comparison for cases where our method and the baselines *agree*, *i.e.*, both determine that the relevance map is faithful, as the perturbation successfully changed the prediction. Note that even in these cases, bizarre predictions are still observed when considering the baselines (*e.g.*,

Table 1. Quantitative evaluation of sensitivity to OOD effects. We evaluate the average distance from the random relevance baseline to estimate the impact of OOD inputs. A higher result indicates better separation.

Method	ResNet	VGG	AlexNet	ViT
Delete [30]	0.26	0.18	0.32	0.19
Blur [21]	0.43	0.42	0.20	**0.59**
Mean [30]	0.22	0.10	0.07	0.54
Ours	**0.49**	**0.55**	**0.69**	0.57

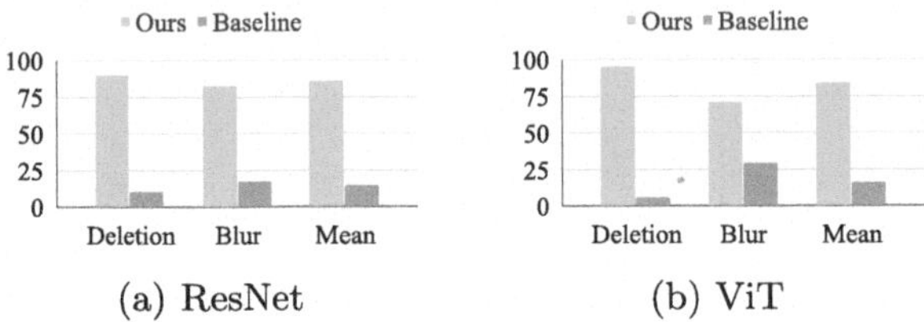

(a) ResNet (b) ViT

Fig. 5. User study evaluating the plausibility of our metric and baselines. We randomly select cases where the best relevance maps differs for (a) ResNet and (b) ViT. Users select the most plausible map based on the input and prediction. The table shows the percentage favoring our method, with users overwhelmingly preferring it.

jigsaw puzzle, fountain), while our method demonstrates actual plausible modifications of the class. This further indicates that even when our metric agrees with the baselines, it produces a much more coherent result, one that humans can understand *why the metric deemed the relevance to be faithful.*

4.2 Quantitative Results

Metrics. As mentioned, the main issue with existing methods is that the perturbed images are OOD, which causes prediction changes even for a random selection of pixels to remove. Therefore, we use the random pixel selection scores as a lower bound to estimate the impact of the OOD inputs on the method. For each evaluation method, we consider the overall score of each explainability method to be the area-under-the-curve (AUC) of the resulting graph (see graphs in Fig. 2). This results in an aggregated score for all the explainability methods by each of the evaluation methods (ours and the baselines). Next, to compare the different evaluation metrics, we wish to estimate how well each metric separates the viable explainability algorithms from the OOD results of the random baselines. We calculate the average distance of the scores from the random lower bound as follows: Let B denote the evaluation metric (ours or the baseline) that produces an AUC score for each attribution method ($\mathbf{AUC}_B(E)$). We begin by normalizing the AUC results for B, to ensure that the different scales of the evaluation metrics do not impact our calculation, *i.e.*:

$$\forall E \in \{E_1, \ldots, E_n\} : \mathbf{AUC}_B(E) = \frac{\mathbf{AUC}_B(E) - \min_E \mathbf{AUC}_B(E)}{\max_E \mathbf{AUC}_B(E) - \min_E \mathbf{AUC}_B(E)} \quad (1)$$

Next, we calculate the average distance between the viable algorithms and the random relevance baseline as follows:

$$rand\_dist(B) = |\frac{1}{n} \sum_{E \in \{E_1, \ldots, E_n\}} \mathbf{AUC}_B(E) - \mathbf{AUC}_B(rand)|, \quad (2)$$

The comparison results are presented in Table 1. As can be seen, compared to the baselines, our method produces scores significantly more remote from the random lower bound for all CNN variants. Especially for the binary classifier and VGG, where the gaps are at least 37% and 13%, respectively, in favor of our method. This shows our method's robustness to OOD inputs, rating viable attribution methods much higher than random pixel selection, with a greater ranking gap than the baselines. For ViT, our results align with pixel-blur and per-channel-mean, indicating that ViT is more robust to OOD inputs than CNNs.

User Study. As mentioned in Sect. 1, *plausibility* is key in evaluating attribution methods, reflecting human interpretability of explanations. As shown in Fig. 2, our ranking of explainability algorithms differs significantly from baselines, especially for CNNs. In addition to the improved faithfulness demonstrated above, we verify its correlation with human preferences, ensuring greater *plausibility*.

Hence, we select ResNet (CNN) and ViT (Transformer) as representatives and sample 10 images per baseline (delete, blur, mean), where our metric and the baseline favor different attributions. We use the Two-Alternative Forced Choice (2AFC) protocol, where users compare relevance maps from both methods and select the one best aligned with the input and classification. Figure 5 shows results from 22 participants and 656 responses. When our metric disagrees with baselines, users overwhelmingly prefer our top-ranked attribution, demonstrating improved *faithfulness* and *plausibility*. This shows the properties align and can be optimized together. Fleiss' Kappa (mean 0.79, std 0.18) indicates strong inter-reviewer agreement, suggesting minimal subjective bias.

Ablation Study. In Fig. 6, we consider three variants of our method to estimate the impact of each design choice. (1) *Top 10 inpainting*: Instead of inpainting to the top-2 class, we use the top-10 class. While top-2 classes share more semantic features, improving SD inpainting, inpainting to top-10 slightly reduces stability. Yet, the random baseline remains distinct, and top methods stay consistent, demonstrating robustness. (2) *no weighting*: Removing the weighting function (Algorithm 1, line 10) makes attribution methods cluster more, whereas with it, they are better distinguished. Clear differentiation is crucial for a meaningful metric, as methods often yield entirely different relevance maps for the same input (see supplementary materials). (3)*Pixel-wise inpainting*: We remove patch-wise downsampling (Algorithm 1, line 3), forcing pixel-level inpainting, which is out-of-distribution for SD. Results confirm that patch-wise inpainting is crucial, especially for CNN-based models.

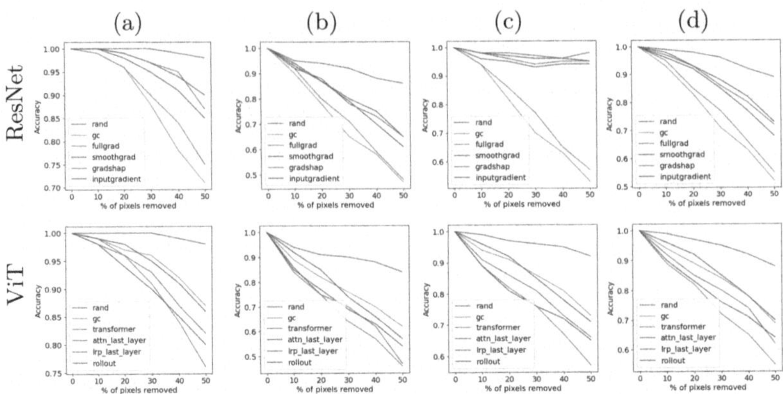

Fig. 6. Ablation study. We compare our full method with different variants of it on ResNet and ViT; (a) top-10- using the top-10 class instead of the top-2, (b) no weighting- results without applying our weighting function, (c) inpaint pixelwise- performing pixel-wise inpainting instead of patch-base inpainting, (d) full method.

5 Discussion and Limitations

As shown in our experiments, our method is a viable, stable alternative to perturbation-based evaluations. However, there are some limitations to consider. First, a limitation shared among perturbation methods is that they are only applied to the top-1 classification, not to all target classes. While some explainability methods, e.g., [9,46] can provide explanations for any class (*i.e.*, *class-specific explainability methods*), perturbation-based metrics only measure the faithfulness of the relevance map for the top-1 prediction of the classifier. A generalization to additional classes could provide deeper insight into the attribution method and indicate whether the provided class-specific signal is meaningful. Additionally, since we rely on an inpainting engine, we are limited to test images for which where the alternative (top-2) class is recognized by that engine (see step I of our method in Sect. 3.2). One solution is to fine-tune the inpainting engine for all classes of a given classifier, but that is beyond our scope. Finally, inpainting is time-consuming, making evaluation on very big datasets harder, we believe progress in diffusion model inference time [37] will soon alleviate this hurdle, allowing our metric to be used for any data size.

6 Conclusion

As deep neural networks (DNNs) permeate everyday applications, the need for explainability has become paramount. Attribution methods offer an intuitive means of explanation serving diverse purposes, from enhancing classifier robustness [10] to guiding generative models [11]. However, current evaluation metrics for attribution methods suffer from severe out-of-distribution (OOD) issues, rendering them unreliable. To address this challenge, we propose a novel metric, leveraging the potent inpainting capabilities of Stable Diffusion (SD). Our metric excels in both faithfulness and plausibility, crucial elements of an effective explanation. As generative models continue to advance, we believe that the fusion of explainability and generative capabilities is poised to play an increasingly prominent role in enhancing research in both fields.

Acknowledgements. This work was supported by the Tel Aviv University Center for AI and Data Science (TAD).

A Implementation Details

Data Construction

We create dataset S with all images recognized by the SD model, following these steps: (1) For each class $t \in 1,\ldots,T$, we generate 20 images $i_1^t,\ldots,i_{20}^t$ using the prompt "class-name". (2) We classify these images: $\forall t \in 1,\ldots,T$: $C(i_1^t),\ldots,C(i_{20}^t)$ (3) We add class t to S if at least 50% of its images are classified as t. This results in $|S| = 661$ classes for ResNet and VGG, $|S| = 704$ classes for ViT, and all classes for the binary classifier. Finally, we sample an image per class as in Step 2 of our method.

Inpainting details

We utilized the public version stable-diffusion-inpaiting [?] from Hugging Face, initialized with the Stable-Diffusion-v-1-2 weights, guidance scale parameter was set to 7.5. We used NVIDIA GeForce RTX 2080 Ti GPU with 11GB of memory. Images and their masks were resized to 512×512 for inpainting and resized back to 224×224 before classification.

B Variance in Attribution Methods

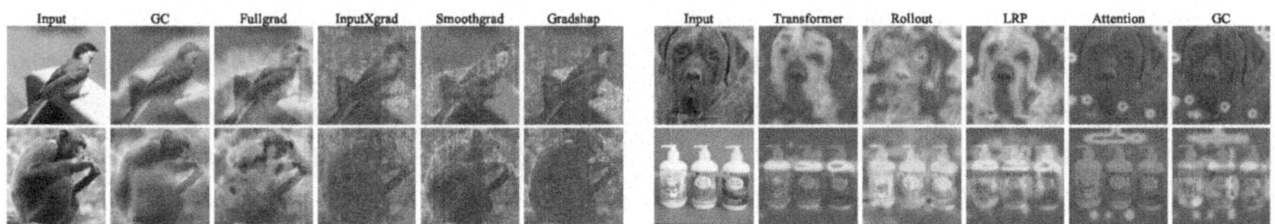

Fig. 7. Relevance maps from different attribution methods for the same input using ResNet (left) and ViT (right), highlighting the need for a metric.

As mentioned in the main paper, different attribution methods yield very different relevance maps. To empirically substantiate this point, Fig. 7 illustrates this with examples from ResNet and ViT, showing that each method highlights different pixels for the same input.

C Additional Baselines

We compare our methods to additional baselines. The first is faithfulness violation [35] which checks that the deletion of the relevant pixels decreases the confidence of the predicted class. The second baseline is a saliency-based evaluation [14] where for each percentage of perturbation pixels (10%, ..., 50%), one first extracts the smallest rectangle patch that contains the top pixels, then applies the classifier to that patch to test whether the prediction remains the same.

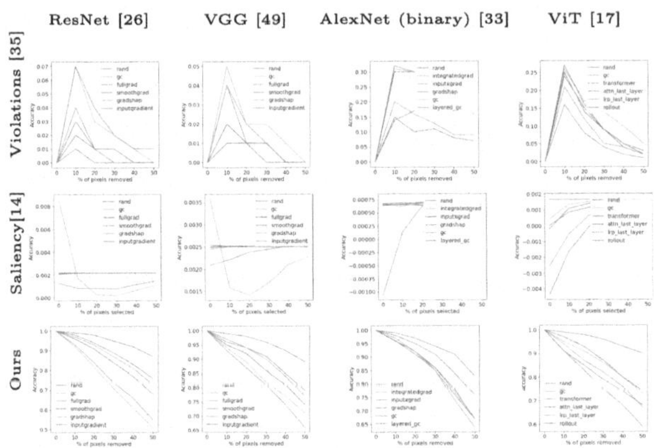

Fig. 8. Perturbation comparison against additional baselines We consider most common explainability methods and a *random* pixel selection. These baselines often fail to distinguish relevance maps from random ones [35] and yield similar results across methods [14].

Figure 8 shows that the violation method fails to separate the random baseline from real attribution methods, indicating that, like the baseline perturbation metrics, it is susceptible to OOD modifications. Additionally, the saliency-based metric often produces nearly indistinguishable outputs for various attribution methods (for all classifiers except ViT). Distinguishing these is crucial for evaluating different methods. Moreover, the random baseline cannot be clearly separated from valid attribution methods for all models except ViT.

References

1. Abnar, S., et al.: Quantifying attention flow in transformers. In: ACL (2020)
2. Adebayo, J., et al.: Sanity checks for saliency maps. In: NeurIPS (2018)
3. Agarwal, C., et al.: Explaining image classifiers by removing input features using generative models. In: ACCV (2020)
4. Bach, S., et al.: On pixel-wise explanations for non-linear classifier decisions by layer-wise relevance propagation. PloS One (2015)

5. Binder, A., et al.: Layer-wise relevance propagation for neural networks with local renormalization layers. In: ICANN (2016)
6. Buolamwini, J., et al.: Gender shades: intersectional accuracy disparities in commercial gender classification. In: FAT (2018)
7. Chang, C.H., et al.: Explaining image classifiers by adaptive dropout and generative in-filling. arXiv:1807.08024 (2018)
8. Chefer, H., et al.: Generic attention-model explainability for interpreting bi-modal and encoder-decoder transformers. In: Proceedings of the IEEE/CVF ICCV (2021)
9. Chefer, H., et al.: Transformer interpretability beyond attention visualization. In: Proceedings of the IEEE/CVF CVPR (2021)
10. Chefer, H., et al.: Optimizing relevance maps of vision transformers improves robustness. In: Advances in NeurIPS (2022)
11. Chefer, H., et al.: Attend-and-excite: attention-based semantic guidance for text-to-image diffusion models (2023)
12. Chen, J., et al.: L-shapley and C-shapley: efficient model interpretation for structured data. In: ICLR (2019)
13. Dabkowski, P., Gal, Y.: Real time image saliency for black box classifiers. In: Advances in NeurIPS (2017)
14. Dabkowski, P., Gal, Y.: Real time image saliency for black box classifiers. In: Neural Information Processing Systems (2017). https://api.semanticscholar.org/CorpusID:41766449
15. DeYoung, J., et al.: Eraser: a benchmark to evaluate rationalized NLP models. In: ACL (2019)
16. Dhamdhere, K., et al.: How important is a neuron? ArXiv abs/1805.12233 (2018)
17. Dosovitskiy, A., et al.: An image is worth 16×16 words: transformers for image recognition at scale. arXiv:2010.11929 (2020)
18. Erhan, D., et al.: Visualizing higher-layer features of a deep network. UdeM (2009)
19. Fel, T., et al.: What i cannot predict, i do not understand: a human-centered evaluation framework for explainability methods. ArXiv abs/2112.04417 (2021)
20. Fong, R., et al.: Understanding deep networks via extremal perturbations and smooth masks. In: ICCV (2019)
21. Fong, R.C., Vedaldi, A.: Interpretable explanations of black boxes by meaningful perturbation. In: ICCV (2017)
22. Fuster, A., et al.: Predictably unequal? The effects of machine learning on credit markets. Regulation Financ. Inst. eJ. (2017)
23. Ghorbani, A., Wexler, J., Zou, J.Y., Kim, B.: Towards automatic concept-based explanations. In: NeurIPS (2019)
24. Goyal, Y., et al.: Counterfactual visual explanations. In: ICML (2019)
25. Guillaumin, M., et al.: Imagenet auto-annotation with segmentation propagation. IJCV (2014)
26. He, K., Zhang, X., Ren, S., Sun, J.: Deep residual learning for image recognition. In: Proceedings of the IEEE Conference on Computer Vision and Pattern Recognition, pp. 770–778 (2016)
27. He, K., et al.: Deep residual learning for image recognition. In: CVPR (2015)
28. Hendrycks, D., Dietterich, T.G.: Benchmarking neural network robustness to common corruptions and perturbations. ArXiv abs/1903.12261 (2019)
29. Hendrycks, D., et al.: Natural adversarial examples. In: CVPR (2019)
30. Hooker, S., et al.: A benchmark for interpretability methods in deep neural networks. In: Advances in NeurIPS (2019)
31. Jacovi, A., Goldberg, Y.: Towards faithfully interpretable NLP systems: how should we define and evaluate faithfulness? In: ACL (2020)

32. Kim, B., et al.: Interpretability beyond feature attribution: quantitative testing with concept activation vectors (TCAV). In: ICML (2017)
33. Krizhevsky, A., et al.: ImageNet classification with deep convolutional neural networks. In: NeurIPS (2012)
34. Li, J., et al.: Understanding neural networks through representation erasure. ArXiv abs/1612.08220 (2016)
35. Liu, Y., et al.: Rethinking attention-model explainability through faithfulness violation test. ArXiv abs/2201.12114 (2022)
36. Lundberg, S., et al.: A unified approach to interpreting model predictions. In: NeurIPS (2017)
37. Luo, S., et al.: LCM-LoRA: a universal stable-diffusion acceleration module (2023)
38. Mahendran, A., Vedaldi, A.: Visualizing deep convolutional neural networks using natural pre-images. IJCV (2016)
39. Montavon, G., et al.: Explaining nonlinear classification decisions with deep Taylor decomposition. Pattern Recogn. (2017)
40. Nichol, A., et al.: GLIDE: towards photorealistic image generation and editing with text-guided diffusion models. arXiv:2112.10741 (2021)
41. Ramesh, A., et al.: Hierarchical text-conditional image generation with CLIP latents. ArXiv (2022). https://arxiv.org/abs/2204.06125
42. Rombach, R., et al.: High-resolution image synthesis with latent diffusion models. In: IEEE/CVF CVPR (2021)
43. Russakovsky, O., et al.: ImageNet large scale visual recognition challenge. IJCV (2015)
44. Saharia, C., et al.: Photorealistic text-to-image diffusion models with deep language understanding. ArXiv abs/2205.11487 (2022)
45. Schwartz, I., et al.: Discriminative class tokens for text-to-image diffusion models (2023)
46. Selvaraju, R.R., et al.: Grad-CAM: visual explanations from deep networks via gradient-based localization. In: ICCV (2017)
47. Shrikumar, A., Greenside, P., Kundaje, A.: Learning important features through propagating activation differences. In: ICML (2017)
48. Simonyan, K., Zisserman, A.: Very deep convolutional networks for large-scale image recognition. CoRR abs/1409.1556 (2014)
49. Simonyan, K., et al.: Deep inside convolutional networks: visualising image classification models and saliency maps. arXiv:1312.6034 (2013)
50. Smilkov, D., et al.: SmoothGrad: removing noise by adding noise. arXiv:1706.03825 (2017)
51. Srinivas, S., Fleuret, F.: Full-gradient representation for neural network visualization. In: NeurIPS (2019)
52. Sundararajan, M., et al.: Axiomatic attribution for deep networks. In: ICML (2017)
53. Voita, E., et al.: Analyzing multi-head self-attention: specialized heads do the heavy lifting, the rest can be pruned. In: ACL (2019)
54. Zeiler, M.D., et al.: Visualizing and understanding convolutional networks. In: ECCV (2014)
55. Zhang, J., et al.: Top-down neural attention by excitation backprop. IJCV (2018)
56. Zhou, B., et al.: Learning deep features for discriminative localization. In: Proceedings of CVPR (2016)

The Weighting Game: Evaluating Quality of Explainability Methods

Lassi Raatikainen and Esa Rahtu(✉)

Tampere University, Tampere, Finland
esa.rahtu@tuni.fi

Abstract. The objective of this paper is to assess the quality of explanation heatmaps for image classification tasks. To assess the quality of explainability methods, we approach the task through the lens of *accuracy* and *stability*. In this work, we make the following contributions. Firstly, we introduce the *Weighting Game*, which measures how much of a class-guided explanation is contained within the correct class' segmentation mask. Secondly, we introduce a metric for explanation stability, using zooming/panning transformations to measure differences between saliency maps with similar contents. Quantitative experiments are produced, using these new metrics, to evaluate the quality of explanations provided by commonly used CAM methods. The quality of explanations is also contrasted between different model architectures, with findings highlighting the need to consider model architecture when choosing an explainability method.

Keywords: Explainable AI · XAI · Computer Vision

1 Introduction

The rise of deep learning in the field of computer vision has seen models become increasingly powerful, however, this has been with the cost of interpretability. Due to this, the field of explainable AI has recently seen a rise into prominence. Several explainability methods have been proposed, which aim to explain what parts of an input image are most relevant to the given output of a model [3,5, 9,12,21,25], the output of an explainability method is usually called a *saliency map*. Therefore, a saliency map is an image that highlights regions, that are proposed to be most relevant to the prediction decision. These methods can be used to evaluate the quality of models' predictions, and especially can be useful when trying to understand incorrect predictions.

Previous works in evaluating saliency maps has found that many explainability methods are independent of the model parameters and labeling of training data, thus rendering them to be simple edge detectors [1]. While it is necessary for an explainability method to be independent of such factors, it does not imply a method is accurate in its explanations, as the sanity checks do not measure how explanations align with input images. Our goal in this paper is to build a

J. Petersen and V. A. Dahl (Eds.): SCIA 2025, LNCS 15726, pp. 325–338, 2025.
https://doi.org/10.1007/978-3-031-95918-9_23

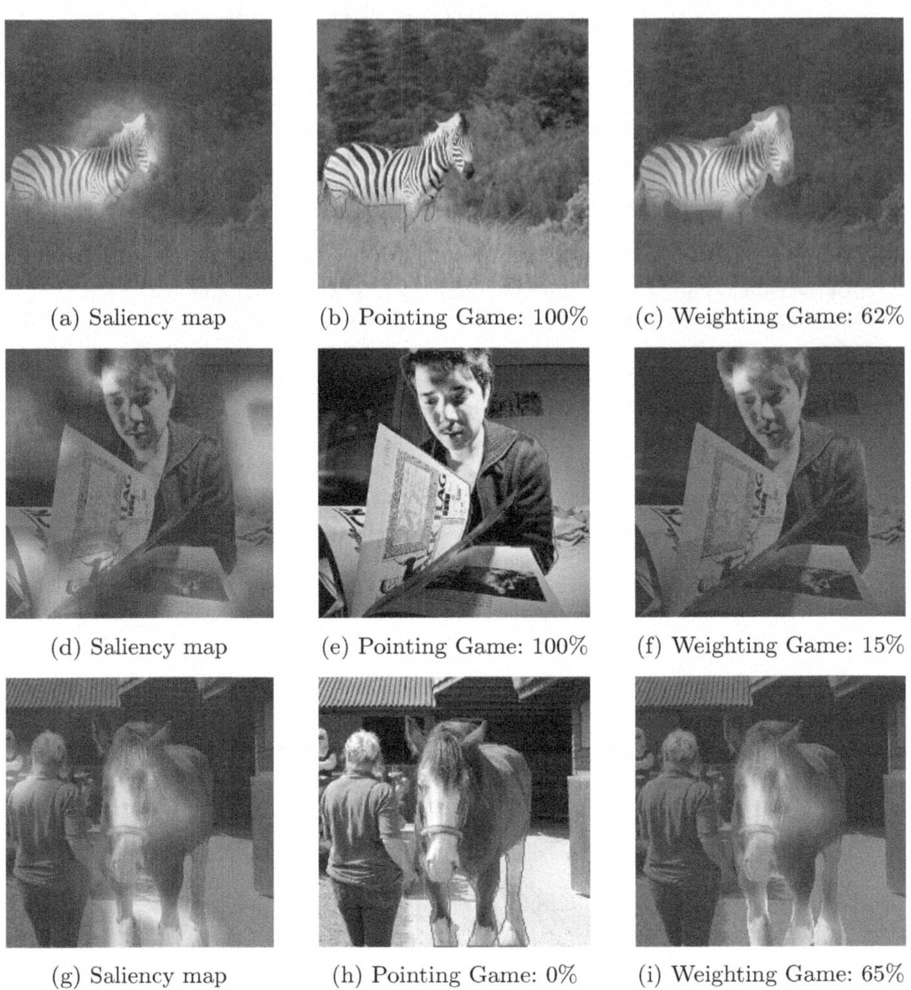
(a) Saliency map (b) Pointing Game: 100% (c) Weighting Game: 62%

(d) Saliency map (e) Pointing Game: 100% (f) Weighting Game: 15%

(g) Saliency map (h) Pointing Game: 0% (i) Weighting Game: 65%

Fig. 1. Comparison of Pointing Game [29] the proposed Weighting Game scores and visualizations for three saliency maps (one per row). The Pointing Game assesses if the location of maximum response (marked as green dot) is withing the ground-truth mask (100%) or not (0%), which results in unstable behavior as seen in last row where maximum is slightly outside the mask. Similarly, it is not able to tell difference between the samples on first two rows. In contrast, the Weighting Game measures the saliency map strength within the mask area and correctly assigns highest scores to top and bottom row examples. The Pointing Game rates the first two examples equally high since the pixel with the highest activation (green) is inside the ground-truth detection mask. The bottom example obtains zero accuracy since the maximum is slightly outside the mask. (Color figure online)

framework for measuring the accuracy and stability of explanations. By accuracy, we refer to the ability of a class-guided explanation to correctly concentrate on the correct class' objects, while with stability we refer to the consistency of explanations when minor transformations are applied to the image.

For assessment of explainability method accuracy, mainly two types of metrics have been presented. Pointing Game [29] measures the accuracy at which a saliency map's highest magnitude pixel is contained within the correct class's segmentation map, thus recording individual instances as either hits or misses. The problem with this method is that the binary nature of the metric does not allow for measuring the quality of hits and misses. An example of the limitations of Pointing Game is shown in Fig. 1, where both explanations are counted as equally good, while the one found in Fig. 1a seems more concentrated on the correct class compared to the one shown in Fig. 1d. The second type of methods for measuring the precision of explanation is usually performed by removing patches or points from an image, to record how the output changes based on the removal of different points [2,13,19]. However, removal-based methods tend to introduce unnatural patches into images.

In this paper, we extend the idea of Pointing Game and introduce the Weighting Game, which measures how much a saliency map's *mass* or *magnitude* is contained within the correct class's segmentation map and neighborhood. The measurement of the distribution of the explanation map allows for more flexibility in assessing the quality of individual explanations, as it is not binary in its nature, such as the Pointing Game. Pointing Game can slightly miss a target and define an explanation as a miss, while Weighting Game is able to see the whole picture and give a better representation of the accuracy of the explanation, such as in Fig. 1 (bottom row).

For assessing the stability of an explainability method, we introduce two separate methods. Firstly, we create 3D effect videos from an image and observe the correlation between saliency maps from consecutive frames. With sufficiently small changes between frames, this grants us the ability to assess how minor changes can affect the resulting saliency maps. Secondly, we create a predefined zoom and pan transformation on an image, and process saliency maps for the original and transformed images. Then, we measure the correlations between the saliency map for the transformed image and a correspondingly transformed saliency map for the original image. These methods provide a way to assess how consistent explainability methods are, thus giving insights on how well they can function in real-world use cases.

The main contributions of this work are: (**1**) The Weighting Game, a novel metric to measure the precision of an explainability method; (**2**) two novel metrics to evaluate the stability of an explainability method; and (**3**) comparison of several prominent CAM methods using proposed metrics with commonly used image classification architectures.

2 Related Work

Backpropagation Explainability Methods. First methods to visualize network attribution were based on gradients of a model found via backpropagation. These methods include DeConvNet [28], Network Saliency [22], and Guided Backpropagation [25]. The methods have slight difference, however, all are based in the idea of visualizing gradients of a loss function from a forward propagated input image. Later introduced backpropagation methods include SmoothGrad [24], Integrated Gradients [26], and Expected Gradients [7].

Class Activation Explainability Methods. The use of class activation maps was first introduced with CAM as a way to visualize the attribution for an output of a network [30]. This was later expanded by Grad-CAM, which used the backpropagated gradients to weight the class activation maps [21]. After Grad-CAM, multiple different methodologies have been introduced to attempt to improve upon Grad-CAM. Notable later CAM methods include, Grad-CAM++ [3], Ablation-CAM [5], Layer-CAM [12], XGrad-CAM [9], and SmoothGrad-CAM++ [17].

Evaluation of Explainability Methods. In 2018, a study into explainability methods was conducted, which determined that most methods were acting as simple edge detectors [1]. The experiments showed that produced saliency maps were mostly independent of the model parameters and labels of the training data, except for notably the saliency maps created using Grad-CAM and DeConvNet. The Pointing Game was introduced as a way to measure the class-sensitivity and accuracy of class-guided explainability methods [29]. Finally, ranking methods for pixel importance have been proposed via removal based methods [2,13,19].

3 Method

We aim to generate methods for assessing how accurate and stable explanations are. For accuracy, the method should be class-specific and concentrated within the segmentation map of an object of the correct class. Stability should be quantified by the consistency of explanations when minor changes are applied.

3.1 The Weighting Game

We defined the Weighting Game as a metric of class-guided explainability method accuracy. The prerequisite for Weighting Game is a dataset, which contains class segmentation masks for the classes that the given model is trained to classify. Thus, object detection datasets, such as Pascal Visual Object Classes 2007 [8] or MS COCO [14], function well for this task.

For calculating one instance of the Weighting Game, we begin with an image and a corresponding class segmentation mask for a class contained within the image. The class segmentation mask is the union of all corresponding class' object segmentation masks found within the image. After obtaining the class segmentation mask, it is dilated using a simple 2D convolution defined with the following equation:

$$D = M \circledast K, \tag{1}$$

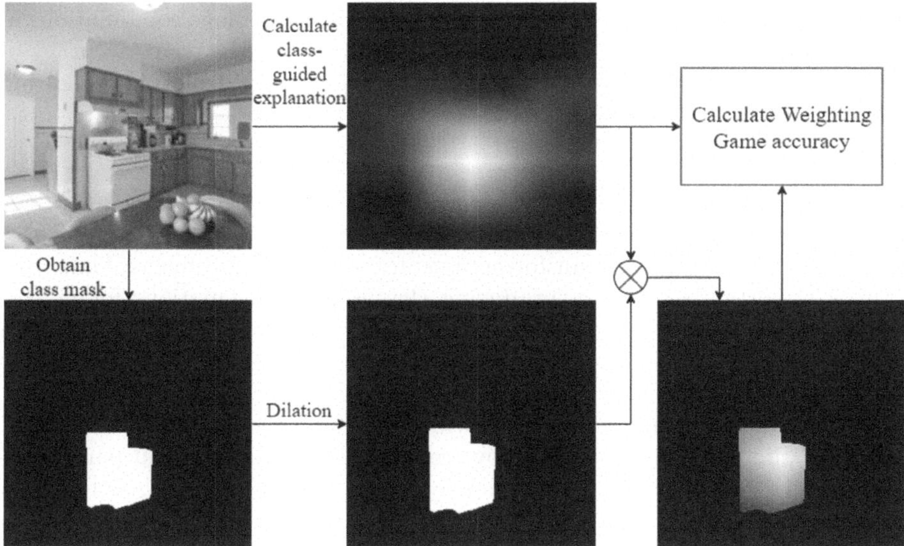

Fig. 2. Process of Weighting Game starting from an input image and a class. Firstly, we obtain the class mask for the image and calculate an explanation for the corresponding class. Then, we dilate the class segmentation mask with the produced explanation, which creates a masked explanation. Finally, the mass of the masked explanation is divided with the mass of the explanation, to obtain the accuracy of an explanation.

where M is the binary class segmentation mask to be dilated, K is a 9×9 convolution kernel, D is the resulting mask, and $\circledast$ denotes the 2D convolution operation. As the resulting mask D is binary, all parts of the mask M that overlap with the convolutional kernel K, will be included in the dilated mask D. The dilation provides the method with more robustness, as outer edges of the desired objects may also be relevant to the classification of the objects.

Then, using the class corresponding to the segmentation mask, the class-guided explanation is calculated for the image. The resulting explanation is a saliency map, which is either produced in the resolution of the input image or upscaled to it.

The dilated class segmentation mask is then used to measure, how much of the *magnitude* or *mass* of the saliency map is contained within the segmentation mask. This gives a measure of how much of the saliency map is contained within the relevant class' area in the image, thus providing a value for the accuracy of an explainability method. One class/image combination of Weighting Game can be formulated with the following equation:

$$Accuracy = \frac{\sum_{i,j} S_{i,j} D_{i,j}}{\sum_{i,j} S_{i,j}}, \tag{2}$$

where S is the class-guided saliency map and D is the corresponding class' dilated segmentation mask. As Equation. 2 provides the accuracy of one class within an image, it will be used to calculate the accuracy of all class/image combinations in the chosen dataset. The final result of Weighting Game is the mean accuracy of all class/image combinations. A diagram of the process of calculating Weighting Game accuracy is visualized in Fig. 2.

3.2 Explanation Stability

Stability of an explainability method can be assessed via the consistency of saliency maps when minor transformations are applied to an image. Thus, this factor is important when, e.g. producing class-guided explanations for videos. Stability can characterize the reliability of an explainability method, as when given similar information to an explainability method, it should produce a similar saliency map.

Stability of an explanation method will be assessed via two separate metrics. Firstly, we simulate camera motion by creating 3D effect zoom and pan videos using 3D Ken Burns effect [16]. Artificially created videos are used, as they can provide a diverse set of videos with a consistent level of motion between frames. This also provides the opportunity to stay within the same distribution of data as in training, as the validation split of the training dataset can be used to generate the videos. The created videos will have 150 frames, of which, the first half consists of a zoom in with ratio 1.5, with the second half consisting of a zoom back to the first frame. Using these videos, we calculate explanations for five pairs of consecutive frames in the video.

The explanations will be processed using class-guided methods, with the chosen class being the highest probability class for the first frame of the video. Then, the correlation between these saliency maps is calculated, to produce a measure for the stability of an explanation method. If the difference between frames is small enough, where the placement of objects is practically the same, the correlation between the saliency maps should be high. The correlations between saliency maps are calculated using Spearman's rank-order correlation, as in other literature where correlation between saliency maps is evaluated [27]. Thus, the equation used to calculate the stability of a method will be as follows:

$$Stability = corr(e(M_i, c), e(M_{i+1}, c)), \tag{3}$$

where M_i is the ith frame of a video, c is the highest probability class for frame M_0, $corr$ is the Spearman rank-order correlation, and e is some class-guided explanation function that takes in an image and a class and returns a saliency map.

The second method for assessing explanation stability is by calculating a zoom, pan, and upscaling (back to original resolution) transformation on an image. After this, saliency maps are calculated for both images. Then, the saliency map for the original image is transformed in the same manner as the transformed image, as the transformation is greater than those seen with the earlier method using consecutive frames of a video. Finally, the correlations are calculated between the saliency map for the transformed image and the transformed saliency map for the original image. Using the same transformation on

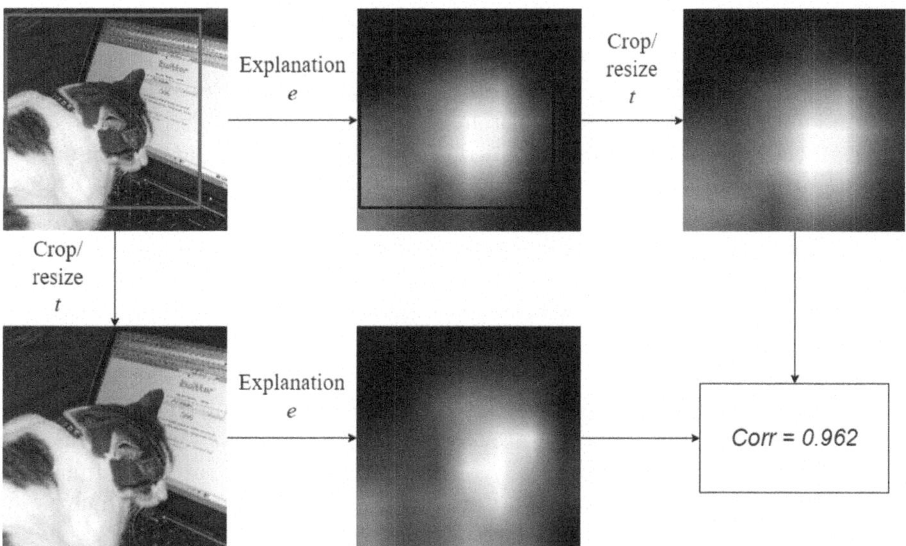

Fig. 3. Diagram of explanation stability using a zoom and pan transformation. First, a random resized crop is calculated, which is visualized as red borders in the original image. Then, a class-guided explanation is calculated for the original image, which is then cropped with the same calculated crop. After that, the original image is cropped and a class-guided explanation is calculated for the cropped image. Finally, the correlation between the resulting two saliency maps is calculated. Explanations were processed with class-guided Grad-CAM [21] and ResNet50 [11] using the class 'cat'.

the original image and the corresponding saliency map, we can align the saliency maps for the original and transformed images. Therefore, the contents are should be the same, however, with different scale in the image. The stability can be expressed with:

$$Stability = corr(t(e(M, c)), e(t(M), c)), \tag{4}$$

where t is the zoom/pan transformation function, M is the input image, c is the highest probability class for M, $corr$ is the Spearman rank-order correlation, and e is the class-guided explanation function. A diagram of the stability metric is shown in Fig. 3.

4 Experiments

In this section, we describe the evaluation of various class-guided explanation methods using our accuracy and stability frameworks.

4.1 Models and Dataset

The chosen model architectures for experiments are ResNet50 [11], VGG16-BN [23], ViT-B/32 [6], and Swin-T [15]. The models represent some of the most popular architectures from both convolutional and transformer based image classification architectures. The inference times of all models are also comparable, which is why they are desirable to compare to each other. Pre-trained versions of all models were obtained from Torchvision's available models [18].

As all the chosen models are trained on the ImageNet [4] dataset, they need to be fine-tuned in order to be used for the Weighting Game. Thus, the COCO dataset is chosen for fine-tuning the models to a multi-label classification task. The fine-tuning procedure is similar to that of the one implemented in the Pointing Game [29].

Only the weights of the last layer are adjusted in the fine-tuning procedure, using binary cross entropy loss. For the training procedure, random resized crop and horizontal flips are used for data augmentation. For validation, images are center cropped and resized to 224×224 resolution. The Adam optimizer is used for optimization, using learning rates of 0.001 and 0.002 for the convolutional and transformer models respectively. Exponential learning rate scheduler was used with 0.95 gamma. All models were trained for 50 epochs, at which point the validation losses had plateaued for all models.

For consistency in models and datasets, the COCO validation set is used in all explanation accuracy and stability evaluations. This way, the distribution of the data remains similar in training and evaluation environments, as we are not attempting to measure the out-of-distribution performance of the models or explanation methods.

4.2 Saliency Methods

The chosen saliency methods to assess in this study were Guided Backpropagation [25], Grad-CAM [21], Grad-CAM++ [3], Layer-CAM [12], XGrad-CAM [9], and Ablation-CAM [5]. All methods have implementations available in the Pytorch Grad-CAM package [10], which provides consistence in implementations. All methods except Ablation-CAM only require one pass through a model, making them computationally efficient, and interesting to determine whether the added computational cost of Ablation-CAM is worth it. Guided Backpropagation provides a baseline for comparison. All methods also satisfy the requirement of being class-guided.

The target layers for the methods are chosen to be from either the last convolutional or transformer block. For ResNet50, the chosen target layers are the layers of the last identity block, while VGG16-BN has the last max pooling layer used as the target layer. Both the ViT-B/32 and Swin-T models have the last transformer block's first layer normalization as the target layer. All saliency maps are upscaled to the input image's resolution using bilinear interpolation. Transformer models' saliency maps are also reshaped from the flattened patches to the shape of the input image.

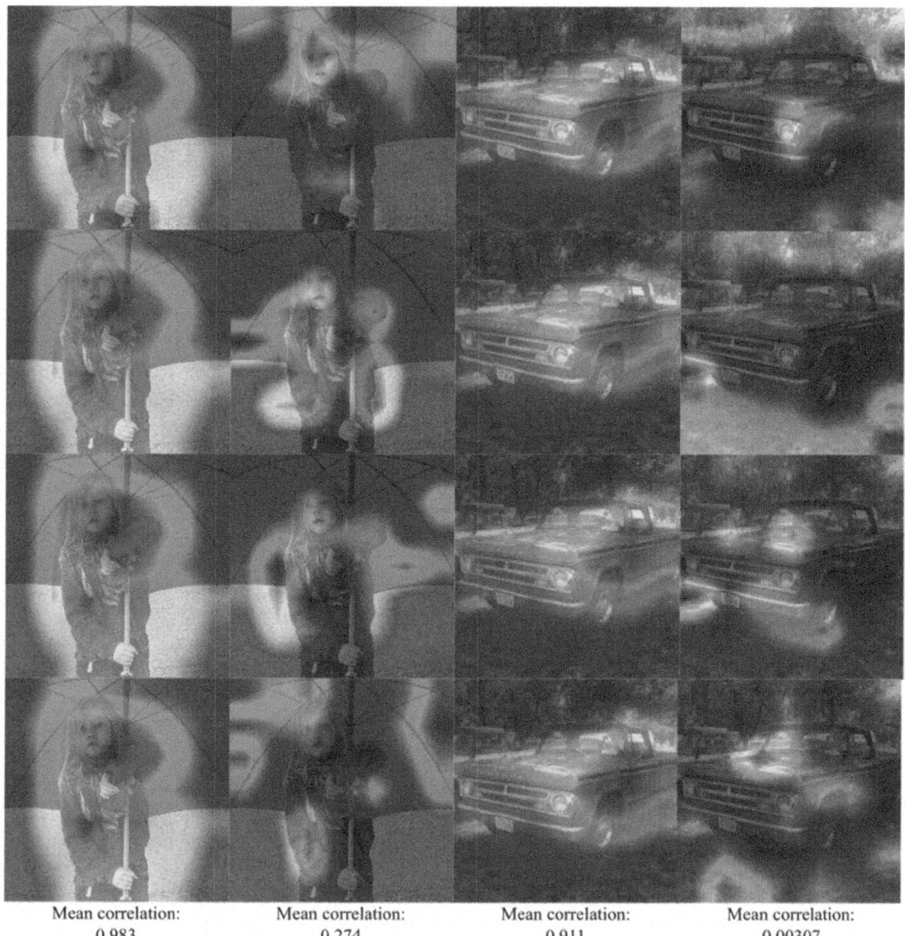

Fig. 4. Examples of consecutive frames from created 3D effect videos and their corresponding class-guided explanations. Displayed correlations values are mean correlations between consecutive frame pairs, as a measure of the explainability method's stability. First and third columns are obtained with ResNet50 [11] and Grad-CAM [21], while second and fourth columns were with ViT-B/32 [6] and Layer-CAM [12]. The first columns correspond to 'person' class and the latter ones 'car' class.

4.3 Results

Weighting Game. After calculating the results of the Weighting Game, we obtain mean accuracy results displayed in Table. 1. The table contains results for the Weighting Game using all objects and only objects that are under 10% of the size of the image. Pointing Game [29] scores are also included for comparison.

Table 1. Explanation accuracy results of different saliency methods for different model architectures. For Weighting Game, the values are the mean percentage of saliency map magnitude contained within the correct class segmentation mask. Weighting Game small is the same metric for objects under 10% of the size of the image. Results are compared to the reproduced Pointing Game [29] metric.

		Explanation accuracy (%)			
		ResNet50	VGG16-BN	ViT-B/32	Swin-T
Ablation-CAM [5]	Weighting	21.1	19.3	23.7	22.2
	Pointing	29.3	23.2	26.3	30.9
	Weighting small	4.74	4.00	5.01	4.93
Grad-CAM [21]	Weighting	30.5	29.2	26.7	14.6
	Pointing	38.0	36.6	32.4	15.6
	Weighting small	11.9	10.4	8.81	4.17
Grad-CAM++ [3]	Weighting	20.4	22.0	27.7	13.2
	Pointing	29.8	33.7	32.2	13.7
	Weighting small	5.29	6.06	9.55	4.10
Guided Backprop. [25]	Weighting	24.5	21.7	16.5	16.7
	Pointing	25.9	23.6	16.3	17.2
	Weighting small	8.40	7.04	3.79	4.49
Layer-CAM [12]	Weighting	20.3	22.3	30.6	22.5
	Pointing	28.8	34.5	34.2	30.5
	Weighting small	5.20	6.17	13.8	10.5
XGrad-CAM [9]	Weighting	30.5	29.0	16.2	16.1
	Pointing	38.0	36.4	14.6	14.3
	Weighting small	11.9	10.3	3.66	3.66

When the results of the Weighting Game are compared with those of the Pointing Game, we can immediately see that the values of the Pointing Game are higher, implying that the task of Pointing Game is generally easier. When relative values are compared, it appears that generally CNNs tend to perform better with the chosen explainability methods, when using Pointing Game as the metric. This could be due to the very low resolution of CAM methods when using late layers of CNNs. This is emphasized with for example Fig. 4, where saliency maps produced with ResNet50 are more spread out, when compared to those by ViT-B/32. Pointing Game fails to quantify this loss in fidelity of saliency maps, as the accuracy values are binary.

Pointing Game is more prone to errors in measuring the accuracy of a saliency map. In cases where the saliency map is mostly concentrated around one point, there is a possibility to narrowly miss the correct segmentation mask, such as in Fig. 1. Simultaneously, it fails to quantify cases where the saliency map is distributed all over the image, such as in Fig. 1. In these cases, the classification of an explanation as either a hit or a miss.

From observing the results of the Weighting Game, it shows that model architecture affects the results significantly, with Grad-Cam/XGrad-CAM function-

Table 2. Results of explanation stability. Consecutive frames values are mean correlations between saliency maps of consecutive frames from artificially generated 3D effect videos from still images. Zoom and pan transformation values are mean correlations calculated with Eq. 4

		Explanation stability			
		ResNet50	VGG16-BN	ViT-B/32	Swin-T
Grad-CAM [21]	3D-effect	0.961	0.938	0.966	0.878
	Crop	0.878	0.782	0.669	0.593
Grad-CAM++ [3]	3D-effect	0.987	0.984	0.943	0.747
	Crop	0.908	0.886	0.634	0.459
Guided Backprop. [25]	3D-effect	0.592	0.599	0.743	0.444
	Crop	0.473	0.572	0.132	0.107
Layer-CAM [12]	3D-effect	0.987	0.983	0.928	0.845
	Crop	0.908	0.882	0.505	0.475
XGrad-CAM [9]	3D-effect	0.961	0.938	0.0732	0.0718
	Crop	0.878	0.777	0.0238	0.0252

ing the best with CNN-based architectures, while Layer-CAM/Grad-CAM++ show great results with ViT-B/32. Swin-T shows poor results with all saliency methods, except for Ablation-CAM and Layer-CAM. Grad-CAM and Grad-CAM++ show results worse than randomly distributed saliency using Swin-T, which shows that the selection of the target layer could be poor or that the saliency methods simply do not function well with the architecture.

While CAM methods were initially designed for CNNs, they do show promise with vision transformers, as ViT-B/32 has the best results with three out of the five CAM methods. The results highlight the need to consider the model architecture in the process of choosing a saliency method, as even within the same family of architectures (CNN/ViT), the results vary significantly.

When only small objects (>10% of the size of the image) are considered, the results of the Weighting Game change significantly. Notably, the relative accuracy of Layer-CAM with transformer models improves significantly when compared to other methods. Results of Ablation-CAM can also be seen deteriorating significantly for all model architectures. CNNs perform comparatively better in the Pointing Game, which highlights the lack of the full picture of explanation accuracy when using the Pointing Game as a metric. It does not properly capture, how the explanations are distributed, which can lead to results favoring certain method/model combinations over others.

Explanation Stability Using 3D-Effect Videos. Table. 2 shows results of explanation stability using 3D-effect videos and zoom/pan transformations. Ablation-CAM was omitted from the stability experiments as a result of the high inference time of the method. The results of the video stability show that Grad-CAM++/Layer-CAM are most stable for CNNs, slightly ahead of Grad-CAM/

XGrad-CAM, which are the most accurate in the results of the Weighting Game. This shows that stability does not necessarily result in greater accuracy. However, high stability does seem like a prerequisite for good-quality explanations, as all methods with clearly worse stability show poor results in explanation accuracy. Interestingly, ViT-B/32 also shows worse stability with Layer-CAM in contrast to CNN models, while the accuracy is significantly superior with ViT-B/32.

Explanation Stability Using Zoom and Pan Transformation. Pan and zoom effect transformations were created using a random resized crop with crop scale 0.75–0.9, 1:1 aspect ratio, and bilinear interpolation. For every image in COCO validation split, a random resized crop is applied, and saliency maps for the highest probability class are processed for both images. Then, the same crop is applied for the source image's saliency map. Finally, the correlation is calculated between the cropped saliency map and the cropped image's saliency maps, as the saliency maps should align at this point. Table 2 shows the mean correlations for the method/model combinations.

The mean correlations are significantly lower than those found in the consecutive frames approach, which is sensible as the step taken in this method is far greater. However, the ranks of results also change significantly. Notably, the relative stability of transformer models drops clearly. This effect is especially seen with Layer-CAM and Guided Backpropagation, however, it is present with all other methods. The reason for this behavior could be explained via how the receptive fields operate within transformer based models. This is due to transformer receptive fields being more concentrated locally, when contrasting ViT-B/32 and ResNet50 receptive fields at later layers [20]. More concentrated receptive field could yield to results that are less stable, but perhaps more capable of correctly explaining smaller objects, as seen in results of the Weighting Game with small objects in Table 1.

5 Conclusion

We introduced a new method for assessing the accuracy of a class-guided saliency methods by evaluating the saliency mass within the correct class' segmentation mask. We also introduced two ways of assessing an explanation method's stability via transforming an image, either by a sufficiently small transformation or by a slightly larger zoom-and-pan operation. Our experiments show that, generally, sufficient stability is a prerequisite for accurate saliency maps. However, better stability does not always yield better results in explanation accuracy, as especially small objects are not always detected well with highly stable methods. We note that the observed relationship among stability and accuracy will benefit from further investigation. We also highlight the importance of considering model architecture when choosing a saliency method.

Acknowledgments. The work was supported by the Academy of Finland projects 353139 and 362409.

References

1. Adebayo, J., Gilmer, J., Muelly, M., Goodfellow, I., Hardt, M., Kim, B.: Sanity checks for saliency maps. In: Proceedings of the 32nd International Conference on Neural Information Processing Systems, NIPS 2018, pp. 9525–9536. Curran Associates Inc., Red Hook (2018)
2. Bach, S., Binder, A., Montavon, G., Klauschen, F., Müller, K.R., Samek, W.: On pixel-wise explanations for non-linear classifier decisions by layer-wise relevance propagation. PLoS One **10** (2015)
3. Chattopadhay, A., Sarkar, A., Howlader, P., Balasubramanian, V.N.: Grad-CAM++: generalized gradient-based visual explanations for deep convolutional networks. In: WACV (2018)
4. Deng, J., Dong, W., Socher, R., Li, L.J., Li, K., Fei-Fei, L.: ImageNet: a large-scale hierarchical image database. In: CVPR (2009)
5. Desai, S., Ramaswamy, H.G.: Ablation-CAM: visual explanations for deep convolutional network via gradient-free localization. In: WACV (2020)
6. Dosovitskiy, A., et al.: An image is worth 16×16 words: transformers for image recognition at scale. In: ICLR (2021)
7. Erion, G., Janizek, J.D., Sturmfels, P., Lundberg, S.M., Lee, S.I.: Improving performance of deep learning models with axiomatic attribution priors and expected gradients. Nat. Mach. Intell. **3**(7), 620–631 (2021)
8. Everingham, M., Van Gool, L., Williams, C.K.I., Winn, J., Zisserman, A.: The PASCAL visual object classes challenge 2007 (VOC2007) results. http://www.pascal-network.org/challenges/VOC/voc2007/workshop/index.html
9. Fu, R., Hu, Q., Dong, X., Guo, Y., Gao, Y., Li, B.: Axiom-based grad-CAM: towards accurate visualization and explanation of CNNs. In: BMVC (2020)
10. Gildenblat, J. (ed.): Pytorch library for CAM methods (2021). https://github.com/jacobgil/pytorch-grad-cam
11. He, K., Zhang, X., Ren, S., Sun, J.: Deep residual learning for image recognition. In: CVPR (2016)
12. Jiang, P.T., Zhang, C.B., Hou, Q., Cheng, M.M., Wei, Y.: LayerCAM: exploring hierarchical class activation maps for localization. IEEE Trans. Image Process. **30**, 5875–5888 (2021)
13. Kapishnikov, A., Bolukbasi, T., Viégas, F., Terry, M.: XRAI: better attributions through regions. In: ICCV (2019)
14. Lin, T.Y., et al.: Microsoft COCO: common objects in context. In: ECCV (2014)
15. Liu, Z., et al.: Video swin transformer. In: CVPR (2022)
16. Niklaus, S., Mai, L., Yang, J., Liu, F.: 3D ken burns effect from a single image. ACM Trans. Graph. **38**(6), 184:1–184:15 (2019)
17. Omeiza, D., Speakman, S., Cintas, C., Weldemariam, K.: Smooth grad-CAM++: an enhanced inference level visualization technique for deep convolutional neural network models. ArXiv abs/1908.01224 (2019)
18. Paszke, A., et al.: Pytorch: an imperative style, high-performance deep learning library. In: NeurIPS (2019)
19. Petsiuk, V., Das, A., Saenko, K.: RISE: randomized input sampling for explanation of black-box models. In: BMVC (2018)
20. Raghu, M., Unterthiner, T., Kornblith, S., Zhang, C., Dosovitskiy, A.: Do vision transformers see like convolutional neural networks? In: NeurIPS (2021)
21. Selvaraju, R.R., Cogswell, M., Das, A., Vedantam, R., Parikh, D., Batra, D.: Grad-CAM: visual explanations from deep networks via gradient-based localization. In: ICCV (2017)

22. Simonyan, K., Vedaldi, A., Zisserman, A.: Deep inside convolutional networks: visualising image classification models and saliency maps. CoRR abs/1312.6034 (2014)
23. Simonyan, K., Zisserman, A.: Very deep convolutional networks for large-scale image recognition. In: ICLR (2015)
24. Smilkov, D., Thorat, N., Kim, B., Viégas, F., Wattenberg, M.: SmoothGrad: removing noise by adding noise. ArXiv (2017)
25. Springenberg, J.T., Dosovitskiy, A., Brox, T., Riedmiller, M.A.: Striving for simplicity: the all convolutional net. CoRR abs/1412.6806 (2015)
26. Sundararajan, M., Taly, A., Yan, Q.: Axiomatic attribution for deep networks. In: ICML (2017)
27. Volokitin, A., Gygli, M., Boix, X.: Predicting when saliency maps are accurate and eye fixations consistent. In: CVPR (2016)
28. Zeiler, M.D., Fergus, R.: Visualizing and understanding convolutional networks. In: ECCV (2014)
29. Zhang, J., Bargal, S.A., Lin, Z., Brandt, J., Shen, X., Sclaroff, S.: Top-down neural attention by excitation backprop. Int. J. Comput. Vision **126**(10), 1084–1102 (2017)
30. Zhou, B., Khosla, A., Lapedriza, A., Oliva, A., Torralba, A.: Learning deep features for discriminative localization. In: CVPR (2016)

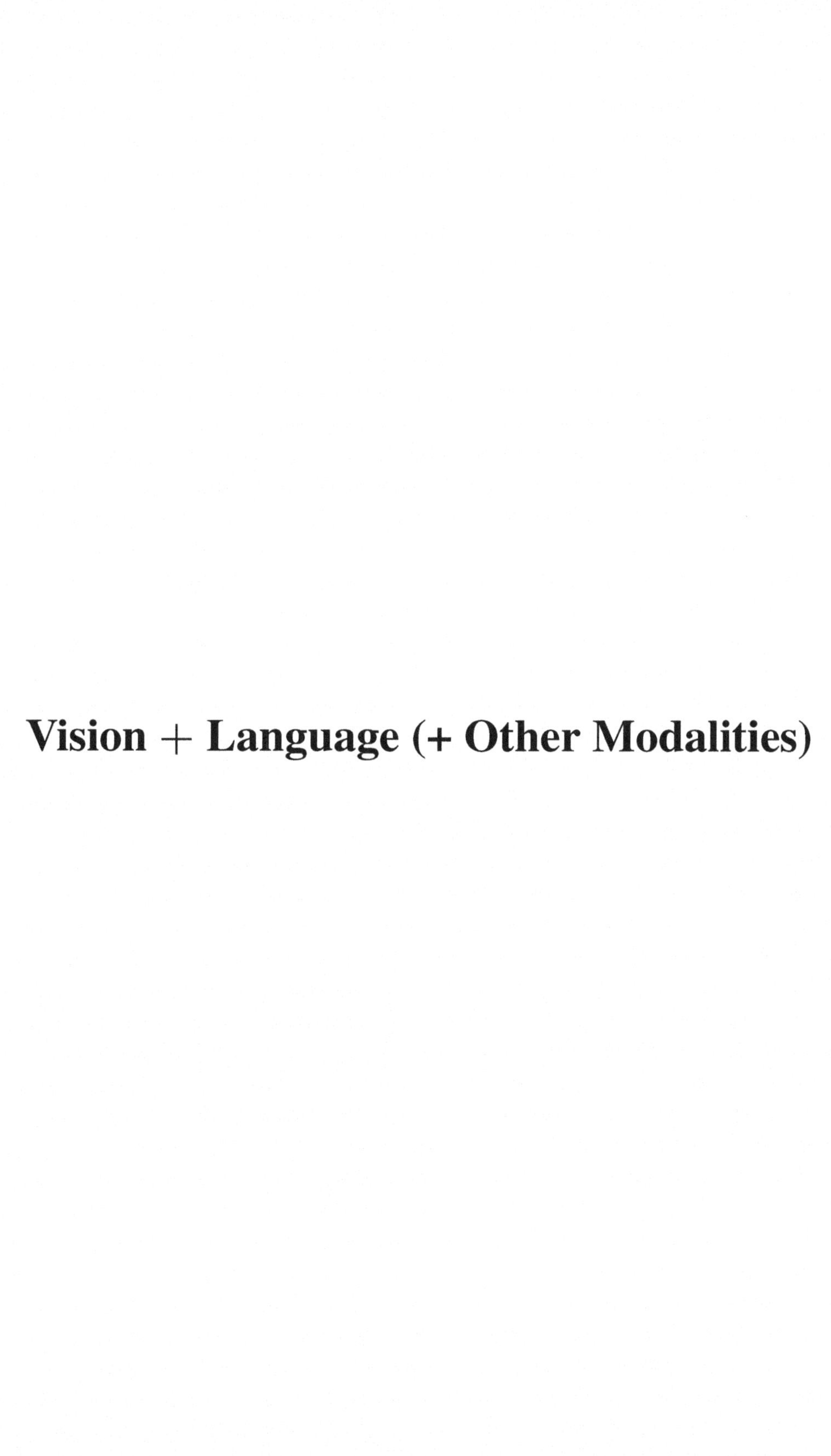

Vision + Language (+ Other Modalities)

POEM: Precise Object-Level Editing via MLLM Control

Marco Schouten[1,3(✉)], Mehmet Onurcan Kaya[1,3], Serge Belongie[2,3], and Dim P. Papadopoulos[1,3]

[1] Technical University of Denmark, Lyngby, Denmark
{marscho,monka,dimp}@dtu.dk
[2] University of Copenhagen, Copenhagen, Denmark
s.belongie@di.ku.dk
[3] Pioneer Centre for AI, Copenhagen, Denmark
https://poem.compute.dtu.dk

Abstract. Diffusion models have significantly improved text-to-image generation, producing high-quality, realistic images from textual descriptions. Beyond generation, object-level image editing remains a challenging problem, requiring precise modifications while preserving visual coherence. Existing text-based instructional editing methods struggle with localized shape and layout transformations, often introducing unintended global changes. Image interaction-based approaches offer better accuracy but require manual human effort to provide precise guidance. To reduce this manual effort while maintaining a high image editing accuracy, in this paper, we propose POEM, a framework for Precise Object-level Editing using Multimodal Large Language Models (MLLMs). POEM leverages MLLMs to analyze instructional prompts and generate precise object masks before and after transformation, enabling fine-grained control without extensive user input. This structured reasoning stage guides the diffusion-based editing process, ensuring accurate object localization and transformation. To evaluate our approach, we introduce VOCEdits, a benchmark dataset based on PASCAL VOC 2012, augmented with instructional edit prompts, ground-truth transformations, and precise object masks. Experimental results show that POEM outperforms existing text-based image editing approaches in precision and reliability while reducing manual effort compared to interaction-based methods.

Keywords: Stable Diffusion · Image Editing · LLM-Guided

1 Introduction

Recent advances in computer vision have been driven by diffusion models [34, 37], which have substantially improved high-resolution text-to-image generation, producing highly realistic and diverse images from textual descriptions. Beyond generation, image editing [22,23] has emerged as a crucial application, enabling users to modify input images according to their needs while preserving realism. A

J. Petersen and V. A. Dahl (Eds.): SCIA 2025, LNCS 15726, pp. 341–355, 2025.
https://doi.org/10.1007/978-3-031-95918-9_24

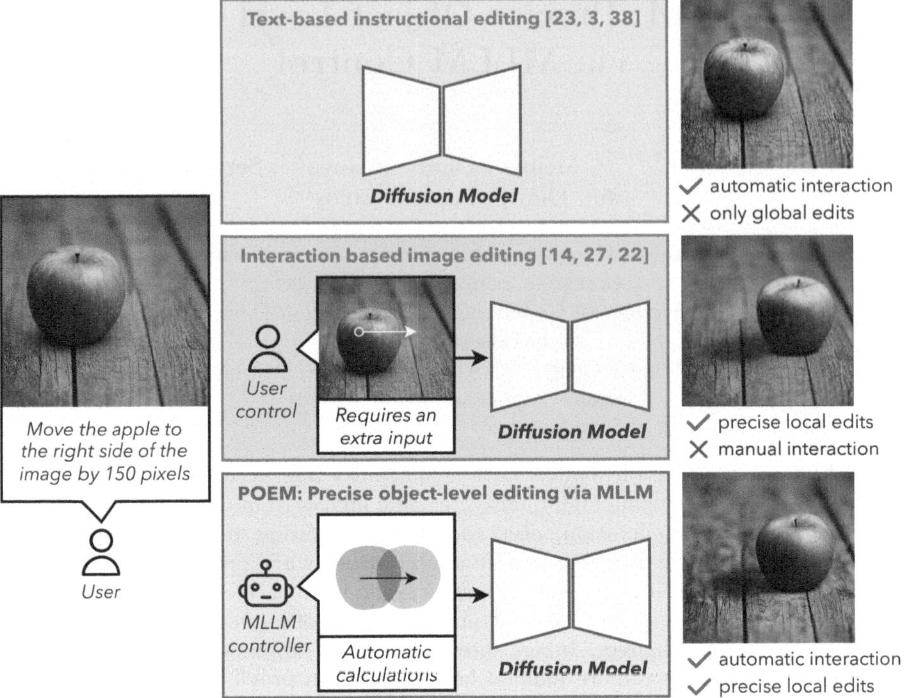

Fig. 1. POEM. Existing text-based instruction editing methods (top) struggle with precise object-level shape and layout edits. Image interaction-based approaches (middle) perform better but require significant manual user effort. Instead, we propose (bottom) leveraging MLLMs to interpret instructional prompts and automatically generate precise object masks and numerical transformations to support image editing pipelines.

challenging aspect of image editing is precise object-level modifications, such as transforming individual target objects while maintaining structural coherence. While existing techniques allow for global adjustments [3], achieving fine-grained, localized edits with high accuracy remains an open research problem [19].

Broadly, image editing methods fall into two categories: text-based instructional editing [3,23,25,38] and image interaction-based editing [5,8,14,22,26, 27,29,40,44]. The former category, exemplified by InstructPix2Pix [3], modifies input images based on a single edit prompt, making it efficient and user-friendly. Even though these methods have shown compelling results with global edits, they struggle with precise object-level shape transformations, often producing unintended global changes (Fig. 1, top). This is mainly because they purely rely on cross-attention text conditioning of a stable diffusion model [3,23]. In contrast, interaction-based approaches require users to provide additional guidance through precise object masks [22,27,29,44], specific object modification shapes [8] or click and drag [5,14,26] (Fig. 1, middle). While these methods can localize edits accurately and improve object-level editing, they demand significant manual effort, making them less scalable.

To address these limitations, we introduce **POEM** (**P**recise **O**bject-level **E**diting via **M**LLM control)(Fig. 1, bottom), a novel framework that decouples visual reasoning from the editor to achieve fine-grained object transformations. Instead of requiring users to provide precise image interactions, POEM leverages Multimodal Large Language Models (MLLMs) to interpret instructional prompts, generate precise object masks before and after transformation, and provide detailed image content descriptions. Inspired by recent advancements in large language models (LLMs) for complex reasoning [13,43] and MLLMs [10,30,46] for guiding diffusion processes, POEM ensures object localization and transformation without requiring extensive manual annotation.

Given an input image and a user edit instruction, POEM operates in two stages (Fig. 2). In the reasoning stage, MLLMs generate structured editing instructions, including precise segmentation masks that define object boundaries before and after the transformation. These masks then guide the editing stage, where we apply controlled modifications in the latent space of a pre-trained diffusion model. By constraining the generation process with explicitly defined regions, POEM ensures fine-grained control over object transformations, surpassing previous text-based approaches in precision and reliability.

Existing datasets for image editing [45,48] evaluate generic editing instructions, but they fail to capture the nuanced variations and fine details that are critical when assessing object shape edits. To address this gap and validate our method, we introduce a novel dataset, VOCEdits, by augmenting the training set of PASCAL VOC 2012 [9] with instructional edits and precise ground-truth object masks for before-and-after transformations. Our dataset enables a more rigorous evaluation of our framework's ability to handle specific edit requests, which existing datasets do not fully account for. Experimental results demonstrate that POEM achieves significantly higher edit fidelity compared to existing text-based editing approaches while requiring no additional user annotations, unlike interaction-based methods.

Our contributions are two-fold: (a) we introduce a plug-and-play reasoning block that interprets user edit instructions with high numerical precision, generating accurate object masks and transformation matrices that enhance layout modifications and mask-guided diffusion editing; (b) we present VOCEdits, a novel dataset for evaluating precise object-level edits, establishing a comprehensive benchmark for detection, transformation, and synthesis tasks.

2 Related Work

Controlling Diffusion Models. Stable Diffusion [31,37] has become a leading model for high-resolution image generation. Recent efforts have explored various approaches for controlling such models, broadly categorized into guidance [17], fine-tuning [38], textual inversion [12], and attention control [16]. Guidance methods [17] steer the generation process using auxiliary signals, such as

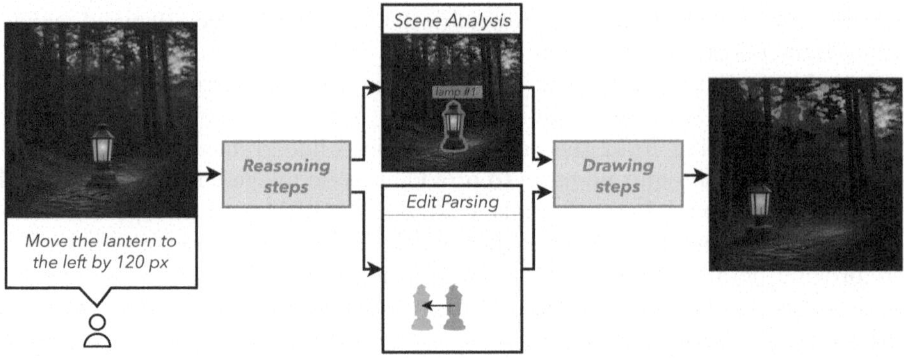

Fig. 2. Overview of our approach. An image and a user edit prompt are fed into the reasoning stage, where we analyze the scene and extract object-level masks and precise transformation parameters for appearance and shape edits. During the editing stage, we apply these edits during inference without any additional training or fine-tuning.

class labels, or text. Fine-tuning [38] modifies model weights to associate edit prompts with example images. Textual Inversion [25] optimizes concepts within the text encoder's embedding space. Finally, attention control [16] modifies spatial attention maps within diffusion layers to influence layout and geometry, enabling precise structural preservation while allowing targeted contextual edits.

Text-to-image editing extends the foundational image-guided generation approaches. Early methods [16,23] edit images by manipulating cross-attention maps. Imagic [19] finetunes the model at inference time to directly match text prompts with visual outputs while striving to preserve the image's style and structure. In contrast, InstructPix2Pix (IP2P) [3] eliminates inference-time fine-tuning by using classifier-free guidance to condition on the source image and text prompt. While IP2P enables global edits, it often over-modifies images, prompting further research [15,24,39] into more localized edits.

Editing with image interaction inputs offers control beyond text-based methods. These approaches require users to provide additional guidance through masks [22,27,29,44], or point dragging [5,14,26] and use them to optimize latent codes more precisely.

Multimodal Large Language Models (MLLMs) enhance image editing workflows [3] by interpreting context-aware user instructions [18]. They resolve ambiguities, capture the underlying user intents [46], and are adept at handling long and detailed edit prompts [20]. Another line of work focuses on layout composition and canvas-based image editing by integrating MLLMs and LLMs to enforce robust object-attribute binding and multi-subject descriptions [10,11,43,47]. For example, Ranni [11] enhances textual controllability using a semantic panel, while SceneComposer [47] enables synthesis from textual descriptions to precise 2D semantic layouts. LayoutGPT [10] acts as a visual planner for generating layouts from text, and SLD [43] iteratively refines images by employing LLMs to analyze the prompt and improved alignment. MLLMs also serve as orchestrators, decomposing complex edits into subtask tree, selecting tools, and coordinating their use [41,42]. Unlike previous methods, we utilize MLLMs to conduct visual reasoning based on edit instructions and source images, focusing on their strengths in numerical proficiency. This enables precise control, such as affine transformation parameters applied to object-level shapes.

3 Method

Given an input image I and a textual edit instruction P, our goal is to generate a modified image $\hat{I}$ that reflects precise object-level transformations specified in P. To do that, we leverage MLLM-driven reasoning to eliminate the need for additional user interaction. We propose POEM (Precise Object-level Editing via MLLM Control), an approach designed for high-precision object-level image editing. POEM decouples the visual reasoning from the image editing (drawing) to achieve fine-grained object transformations (Fig. 2).

POEM consists of five steps (Fig. 3): (a) Visual Grounding: the input image and the edit prompt are fed into an MLLM that is instructed to analyze the scene and identify and detect all objects; (b) Detection Refinement: we refine the object detection output from the MLLM to obtain more accurate object segmentation masks; (c) Edit Operation Parsing: we use an LLM that is instructed to select the target object and compute the transformation matrix; (d) Transformation: we apply the transformation to the segmented object to obtain the edited mask and (e) Edit Guided Image-to-Image Translation: given the initial input image and the masks of the target object before and after the transformation, we generate the final modified image while preserving spatial and visual coherence.

Visual Grounding. In this step, we deploy an MLLM that takes as input the image I and the prompt P. Using zero-shot prompting, we leverage the model's visual capabilities to analyze the scene and detect all objects in the image. The MLLM is directly instructed to perform object detection and detect all objects N that appears in the image. For each detected object $i \in N$, we ask the MLLM to output the detected bounding box b_i, a segmentation point on the object s_i, the object class c_i, and a unique object ID k_i.

Additionally, we instruct the MLLM to analyze the image I and user prompt P, generating four structured descriptions: the scene (S), spatial relationships (R), background prompt (P_{bg}), and generation prompt (P_g). These are not direct captions but targeted summaries capturing (1) global layout (S), (2) object relationships (R), (3) background appearance and context (P_{bg}), and (4) overall generation intent (P_g). S and R support Edit Operation Parsing to estimate the transformation matrix, while P_{bg} and P_g guide the Drawer step to maintain background consistency and apply object-specific edits.

Detection Refinement. Off-the-shelf MLLMs often struggle to produce precise object-bounding boxes when performing a visual grounding task [35]. For this reason, in this step, we refine the detection output from the previous step, and for each detected object i, we obtain a segmentation mask m_i and a refined

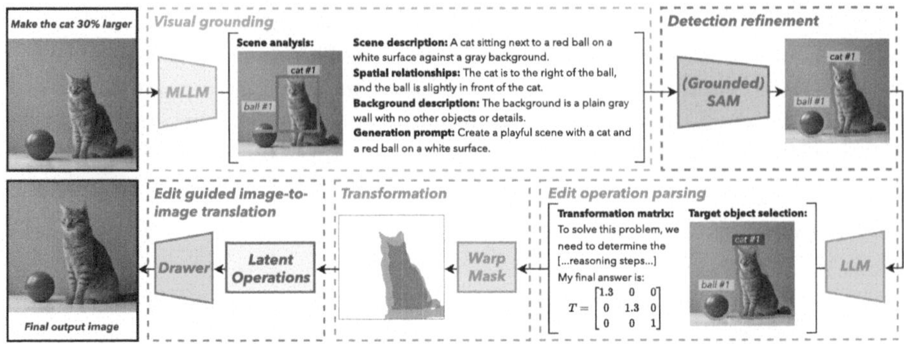

Fig. 3. Detailed pipeline of POEM. Given an image and an edit prompt, we first use an MLLM to analyze the scene and identify objects. Then, we refine the detections and enhance object masks using Grounded SAM. Next, we use a text-based LLM to predict the transformation matrix of the initial segmentation mask. Finally, we perform an image-to-image translation guided by the previous steps to generate the edited image. This structured pipeline enables precise object-level editing with high visual fidelity while preserving spatial and visual coherence.

bounding box b'_i. Without loss of generality, we use Grounded-SAM [36], and we prompt it with the predicted object classes c_i from the previous step. Grounded-SAM combines Grounding DINO [21] and Segment Anything Model (SAM) [28] to perform an open-set detection and segmentation with text prompts even for objects outside predefined categories.

Edit Operation Parsing. Given the prompt P and the set of the refined bounding boxes $B' = \{b'_i | i \in N\}$, the goal of this step is to extract a transformation matrix T and identify the ID k of the target object. Given only the prompt P, the MLLM from the first step struggles to directly infer T in a single step due to its lack of explicit scene information. For instance, if `P = 'make the cat 100px wide'`, the required transformation depends on the cat's initial dimensions in the image. If the cat is 50px wide, the scaling factor in T should be 2, whereas if the initial width is 25px, the scaling factor in T should be 4.

To address this, we use a text-based LLM optimized for mathematical reasoning to compute the transformation parameters. This separation allows for a more accurate estimation of scale, rotation, and translation transformations by explicitly incorporating object size information into the reasoning process. We use the input prompt P, the descriptive prompts S and R, and the coordinates of the detections B', and we directly instruct the LLM to predict the unique ID of the target object $i*$ and a 3×3 transformation matrix T given by:

$$T = \begin{bmatrix} a_{11} & a_{12} & a_{13} \\ a_{21} & a_{22} & a_{23} \\ 0 & 0 & 1 \end{bmatrix} \tag{1}$$

To ensure precise parsing, we employ a structured format where LLM matrices and object IDs are enclosed between the unique tokens `<MSTART>`, `<MEND>`, `<ISTART>`, and `<IEND>`. A regex-based parser extracts numerical values enclosed within the matrix tokens, ensuring the retrieval of transformation parameters.

Transformation. In this step, we select the segmentation mask m_{i*} corresponding to the selected id $i*$. Then, we perform image wrapping using T on the binary mask m_{i*} to generate the transformed mask $\hat{m}_{i*}$.

Edit Guided Image-to-Image Translation. In this step, we use the masks m_{i*} and $\hat{m}_{i*}$ of the target object, and the descriptive prompts P_{bg} and P_g from the first step to perform the image synthesis and generate the final input image $\hat{I}$. We apply these edits during the inference of pre-trained diffusion models without additional training or fine-tuning. Inspired by [43], we perform object-level shape manipulations in the latent space of diffusion models [37]. We use the region of the mask $\hat{m}_{i*}$ to define the area of interest, which is processed through backward diffusion to obtain its latent representation z_{repos}. The region of the initial maks m_{i*} is reinitialized with Gaussian noise $\mathcal{N}(0, I)$, and the new latent is blended into the image latent z as:

$$z_{\text{new}} = z \odot (1 - M_j) + z_{\text{repos}} \odot \hat{M}_j + \mathcal{N}(0, I) \odot M_j. \tag{2}$$

A forward diffusion process refines the image, enhancing realism and coherence in edited and surrounding regions.

4 Experiments

This section presents our experimental results. We introduce VOCEdits, a novel dataset to ensure rigorous evaluation of precise object-level edits in Sect. 4.1. In Sect. 4.2, 4.3, 4.4 and 4.5, we systematically explore different design choices for each step of our pipeline and evaluate their impact. We also present a qualitative comparison between POEM and state-of-the-art image editing approaches [2,3,7], while in Sect. 4.6, we discuss the limitations of our approach.

4.1 VOCEdits Dataset

We present VOCEdits, a dataset for evaluating fine-grained object-level image editing involving affine transformations: flip, scale, rotation, translation, and shear. It is built upon PASCAL VOC 2012 [9] for its high-quality instance segmentation masks, enabling precise object-centric evaluation on real-world images. We augment PASCAL VOC images with instructional prompts, ground-truth transformations, and object masks before and after editing. We use images from the PASCAL VOC 2012 trainval segmentation set, containing 2913 images and 6929 object instances. We filter out images with multiple instances of the same class, truncated objects, extreme object sizes, or masks extending beyond image boundaries, resulting in 505 unique images.

Table 1. Evaluation on VOCEdits. Methods are grouped according to different steps of our pipeline, as described in the paper. For each step, we report the Intersection over Union (IoU) (%) between the sets indicated in the right part of the table. The final section of the table presents results from other state-of-the-art image editing methods.

Method	Move	Scale	Flip	Shear	Rotate	Reason	Mix	Avg.
Visual Grounding	Estimated bounding box in the input image vs. GT							
InternVL-8B [4]	15.8	19.8	9.4	23.3	13.4	27.3	24.3	17.4
InternVL-72B [4]	46.3	47.4	43.9	49.6	49.6	43.4	46.8	47.1
QwenVL-7B [33]	54.8	**57.8**	**54.0**	**55.1**	54.0	37.8	**50.1**	**55.5**
QwenVL-72B [33]	**55.1**	56.4	53.5	54.2	**54.6**	**45.1**	40.7	54.8
Detection Refinement	Estimated segmentation mask in the input image vs. GT							
QwenVL-7B [33] + SAM [28]	22.5	34.1	22.5	20.0	24.0	31.4	27.0	27.3
QwenVL-7B [33] + G-SAM [36]	**82.6**	**86.0**	**81.3**	**81.0**	**88.5**	**51.3**	**81.3**	**84.2**
Edit Operation Parsing & Transformation	Transformed segmentation mask vs. GT							
(QwenVL-7B [33] + G-SAM [36]) + DeepSeek [6]	20.6	26.4	30.7	38.6	28.1	2.5	21.7	25.3
(QwenVL-7B [33] + G-SAM [36]) + QwenM [32]	**42.0**	**50.3**	**58.0**	**80.3**	**52.5**	**9.5**	**37.7**	**49.2**
Oracle Mask + DeepSeek [6]	23.1	30.6	40.0	56.1	30.2	2.1	21.4	29.5
Oracle Mask + QwenM [32]	**47.0**	**56.0**	**74.1**	**98.6**	**56.2**	**17.5**	**38.8**	**55.6**
Edit Guided Image-to-Image Translation	Detected segmentation mask in the output image vs. GT							
(QwenVL-7B [33] + G-SAM [36] + QwenM [32]) + SLD [43]	**32.6**	**39.7**	39.1	54.5	43.5	24.9	37.6	**38.4**
(QwenVL-7B [33] + G-SAM [36] + QwenM [32]) + SLD [43] + [31]	31.6	39.4	37.4	53.3	42.2	24.7	36.1	37.6
IP2P [3]	27.4	32.9	41.9	**72.3**	38.3	8.4	33.8	34.3
TurboEdit [7]	27.1	30.9	46.4	53.6	43.9	21.8	39.7	33.8
LEDITS++ [2]	27.4	31.8	**52.7**	56.2	**44.1**	**29.9**	**39.8**	35.0

To generate human-like edit instructions, we utilize GPT-4o-mini [1] and instruct it to paraphrase default instructions, leading to diverse descriptions. The ground-truth object segmentation masks from PASCAL VOC are then transformed via open-cv transformation for exact computation. Each image from the final set undergoes two randomly selected transformations, with three corresponding paraphrased prompts, yielding a total of 3030 unique samples.

Our pipeline processes all 3030 samples but applies an additional refinement step, excluding cases with more than five foreground objects per image. This restriction is imposed due to the limitations of [43] in handling excessive object occlusions and intersecting boxes. After this filtering, a final set of 193 images and 921 samples is retained for evaluation. Unless stated otherwise, we will use this set to evaluate our pipeline for the remainder of this section. A comprehensive summary of these results is provided in Table 1. Figure 4 provides detailed statistics on transformation distribution and object categories of the final set.

4.2 Visual Grounding

Evaluation Protocol. To assess the quality of the detected bounding box, we compute Intersection over Union (IoU) with the ground truth. If the MLLM fails to detect a bounding box, we fallback to a prediction covering the entire image. For images with multiple objects, we evaluate only the bounding box corresponding to the target object for transformation.

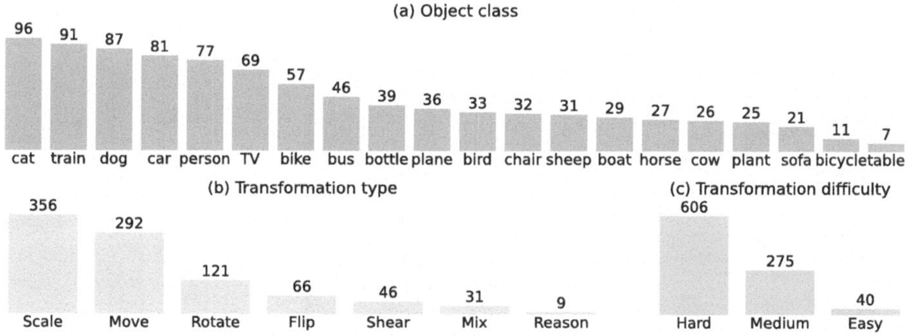

Fig. 4. VOCEdits evaluation subset statistics. Distributions of (a) object classes, (b) transformation types, and (c) transformation difficulty levels.

Comparison. We compare two MLLMs—Qwen2.5-VL [33] and Intern-VL-2.5 [4] in their 7B/8B and 72B variants. Model selection is guided by OpenCompass Open VLM leaderboard performance and dual H100 GPU compatibility.

Results. QwenVL-7B achieves an average IoU of 55.5%, outperforming Intern-VL-8B by 38.1% (Table 1). This performance advantage is evident across all transformation categories. Considering their similar model sizes, these results highlight Qwen-VL's superior effectiveness for this task. While InternVL-72B shows improved performance over its 8B variant, a similar trend is not observed for the QwenVL models. Therefore, we use QwenVL-7B for Visual Grounding in the remainder of our experiments.

4.3 Detection Refinement

Evaluation Protocol. We assess the segmentation quality by computing the IoU between the ground truth segmentation mask of the target object and the corresponding detected segmentation masks we obtain after the refinement stage. Similar to Sect. 4.2, when dealing with images containing multiple objects, we evaluate only the segmentation mask corresponding to the target object.

Comparison. We compare Grounded-SAM [36] to SAM2 [28]. Grounded-SAM is prompted with the predicted object class c_i while SAM2 is prompted with the predicted segmentation point s_i. Both c_i and s_i are obtained from the MLLM.

Results. Grounded-SAM (denoted as G-SAM in Table 1) exhibits a significant performance enhancement over SAM2 across all evaluated tasks, yielding an average IoU improvement of 56.9%. These findings underscore the superior segmentation capabilities of Grounded-SAM over SAM2, particularly in refining object detection with greater accuracy and consistency.

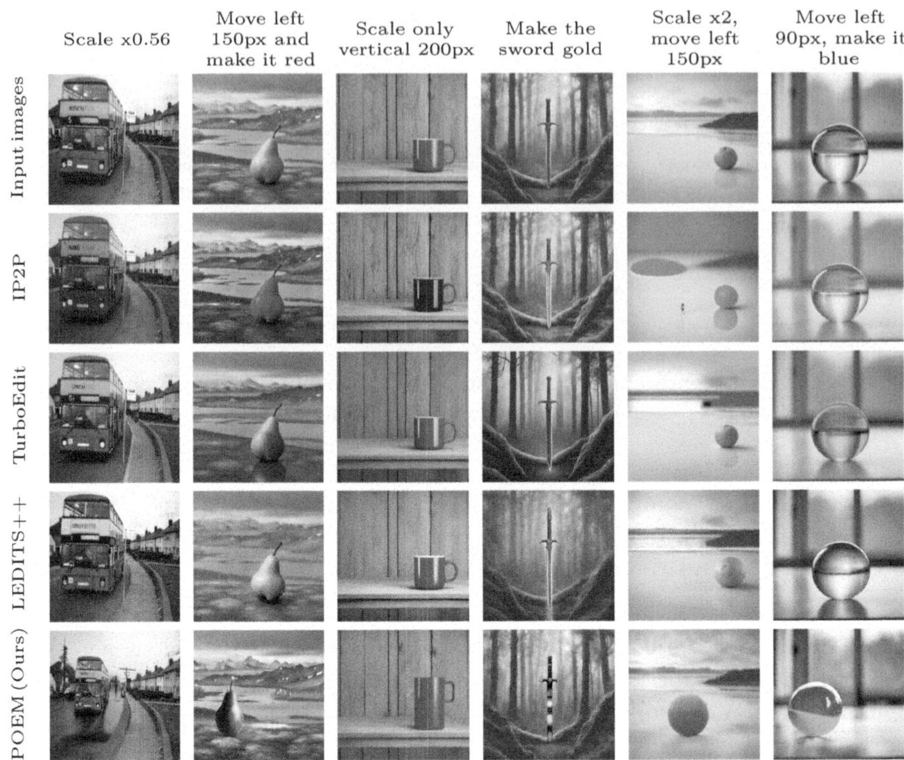

Fig. 5. Qualitative results. We compare POEM with state-of-the-art image editing models across a diverse set of edit instructions, including geometric transformations (e.g., translation, scaling), appearance changes, and combinations of both. The specific prompts used are *"Scale the bus by 0.56"*, *"Move the pear left by 150px and make it red"*, *"Scale the mug only vertically to 200px"*, *"Make the sword gold"*, *"Scale the orange by 2 and move it left by 150px"*, and *"Move the ball left by 90px and make it blue"*. (Color figure online)

4.4 Edit Operation Parsing and Transformation

Evaluation Protocol. To assess transformation accuracy, we compute the ground-truth segmentation mask of the target object after applying the ground-truth transformation matrix. We then measure the IoU between this mask and the predicted transformed mask $\hat{m}_{i*}$. This allows us to measure implicitly the error between our predicted transformation matrix T and the ground-truth one.

Comparison. We evaluate two LLMs: Qwen2.5-Math-7B-Instruct [32], which uses external tools like solvers and libraries, and DeepSeek-R1-Distill-Qwen-32B [6], relying on internal knowledge. Transformations are performed with OpenCV for geometric modifications. DeepSeek runs on a single NVIDIA H100 GPU (80 GB), using up to 74 GB of memory. We analyze two scenarios: (1) with our pipeline's best models (see Sect. 4.3) and (2) with an oracle ground-

Fig. 6. Extreme transformations. Although POEM's reasoning steps maintain robust mask quality and accurate transformation parameters, the image editing step [43] fails to generate an accurate image with an extreme edit (e.g., resizing fails at 10%, and translation errors occur when the object approaches the image boundaries).

truth mask, isolating LLM-based reasoning effects. The second scenario measures transformation errors independently, while the first evaluates cumulative error from imperfect segmentation.

Results. QwenMath (denoted as QwenM in Table 1) consistently outperforms DeepSeek across all transformation categories by a clear margin, achieving 7–41% higher IoU scores. This is likely due to QwenMath's tool-integrated-reasoning approach, which enhances matrix multiplication accuracy for complex transformations. The same trend appears for both evaluation scenarios (i.e., predicted and oracle masks). Using oracle masks improves IoU scores from 49.2% to 55.6%, suggesting that the primary source of error comes from the transformation prediction rather than segmentation inaccuracies. This highlights the importance of robust mathematical reasoning in object-level transformations.

4.5 Edit Guided Image-to-Image Translation

Evaluation Protocol. To assess the image editing quality, we go beyond standard image quality metrics (e.g., FID), and instead, we measure the alignment of the edited images with the input prompts and the transformations. Specifically, we first use Grounded SAM to estimate the segmentation mask of the transformed object in the edited image. We then compute the IoU between this mask and the segmentation mask after applying the ground-truth transformation.

Comparison. We use the Stable Diffusion v2.1 [37] as our pre-trained diffusion model and adopt the latent space operations from [43]. Additionally, we experiment with Stable Diffusion XL [31] as a refiner to improve the image quality.

Results. Comparing the two strategies for generating the final image, we observe minimal performance differences, with SDXL refinement leading to an average

IoU drop of only 0.8%. This change is statistically insignificant, but qualitatively, the refined images exhibit improved visual quality. When comparing this IoU accuracy from this step with the one from the previous section, we observe a significant 10.8% drop (from 49.2% to 38.4%). This drop is caused by the image editing process, which does not always fully adhere to the guided segmentation masks. In Sect. 4.6 and Fig. 6, we further analyze these image editing limitations, particularly in cases with extreme transformations.

Comparison to State-of-the-Art Image Editing. Figure 5 shows a qualitative comparison of POEM with state-of-the-art models, including IP2P [3], LEDITS++ [2], and TurboEdit [7]. The figure demonstrates POEM's ability to generate more faithful, targeted edits. Table 1 reports quantitative comparisons, where POEM achieves 38.4%, surpassing IP2P (34.4%), TurboEdit (33.8%), and LEDITS++ (35.0%) by about 3%. POEM excels in *translate* and *scale* operations, with improvements from 27.4% to 32.6% and 32.9% to 39.7%, respectively. These results highlight our model's superior performance, producing more precise edits and accurate transformation parameters that better align with the user's intended modifications.

4.6 Limitations

While POEM achieves precise object-level transformations, the image editing step inherited from diffusion models has certain limitations.

First, when dealing with extreme transformation, POEM can predict accurate parameters, but diffusion models struggle to generate objects that become too small relative to the image size. This issue is most pronounced when objects shrink to less than 10% of their original size or move partially outside the image boundaries (Fig. 6). To measure this effect quantitatively, we categorize the transformations of our dataset into easy, medium, and hard based on the IoU difference between the original and transformed masks. After applying the LLM-based step, we obtain an IoU of 68% for easy, 66% for medium and 40% for hard transformations. In contrast, the image editing step lowers IoU to 55%, 54%, and only 30%, respectively, highlighting challenges in handling severe modifications.

Second, our approach currently focuses on rigid-body transformations, as our editing step [43] does not support non-rigid deformations, such as altering human poses (e.g., raising an arm). A possible solution is integrating more explicit control signals, similar to Self-Guidance [8]. However, Self-Guidance is very sensitive to hyperparameters which impacts its reproducibility and generalization ability. Future work could refine its framework to ensure reliable image edits, such as preserving background integrity across transformations [8].

5 Conclusion

We proposed POEM, a novel approach that leverages MLLMs, LLMs, and segmentation models to enhance image editing capabilities through precise text-instruction-based operations. Our approach facilitates object-level editing by generating accurate masks alongside relevant contextual information derived from the input image. This feature empowers users to perform precise modifications directly from natural language instructions. Additionally, we introduced VOCEdits, a comprehensive dataset designed for evaluating object-level editing, which establishes a robust benchmark for tasks related to detection, transformation, and synthesis. By integrating MLLMs with diffusion models, POEM bridges the gap between high-level instructional reasoning and low-level spatial control, laying the foundation for future research in multimodal image editing. We believe our work will drive advancements in controllable image synthesis, making precise and intuitive editing more accessible to users.

Acknowledgments. D. Papadopoulos was supported by the DFF Sapere Aude Starting Grant "ACHILLES". This work was partly supported by the Pioneer Centre for AI, DNRF grant number P1. We would like to thank Thanos Delatolas for the insightful discussions.

References

1. Achiam, J., et al.: GPT-4 technical report. arXiv (2023)
2. Brack, M., et al.: LEDITS++: limitless image editing using text-to-image models. In: CVPR (2024)
3. Brooks, T., Holynski, A., Efros, A.A.: InstructPix2Pix: learning to follow image editing instructions. In: CVPR (2023)
4. Chen, Z., et al.: InternVL: scaling up vision foundation models and aligning for generic visual-linguistic tasks. In: CVPR (2024)
5. Cui, Y., Zhao, X., Zhang, G., Cao, S., Ma, K., Wang, L.: StableDrag: stable dragging for point-based image editing. In: ECCV (2024)
6. DeepSeek-AI: DeepSeek-V3 technical report. arXiv (2024)
7. Deutch, G., Gal, R., Garibi, D., Patashnik, O., Cohen-Or, D.: TurboEdit: text-based image editing using few-step diffusion models. In: SIGGRAPH Asia (2024)
8. Epstein, D., Jabri, A., Poole, B., Efros, A.A., Holynski, A.: Diffusion self-guidance for controllable image generation. In: NeurIPS (2023)
9. Everingham, M., et al.: The pascal visual object classes challenge: a retrospective. In: IJCV (2015)
10. Feng, W., et al.: LayoutGPT: compositional visual planning and generation with large language models. In: NeurIPS (2023)
11. Feng, Y., Gong, B., Chen, D., Shen, Y., Liu, Y., Zhou, J.: Ranni: taming text-to-image diffusion for accurate instruction following. In: CVPR (2024)
12. Gal, R., et al.: An image is worth one word: personalizing text-to-image generation using textual inversion. In: ICLR (2023)
13. Gani, H., Bhat, S.F., Naseer, M., Khan, S., Wonka, P.: LLM blueprint: enabling text-to-image generation with complex and detailed prompts. In: ICLR (2024)
14. Geng, D., Owens, A.: Motion guidance: diffusion-based image editing with differentiable motion estimators. In: ICLR (2024)
15. Guo, Q., Lin, T.: Focus on your instruction: fine-grained and multi-instruction image editing by attention modulation. In: CVPR (2024)
16. Hertz, A., Mokady, R., Tenenbaum, J., Aberman, K., Pritch, Y., Cohen-Or, D.: Prompt-to-prompt image editing with cross attention control. In: ICLR (2023)

17. Ho, J., Salimans, T.: Classifier-free diffusion guidance. arXiv (2022)
18. Huang, Y., et al.: SmartEdit: exploring complex instruction-based image editing with multimodal large language models. In: CVPR (2024)
19. Kawar, B., et al.: Imagic: text-based real image editing with diffusion models. In: CVPR (2023)
20. Liu, L., Du, C., Pang, T., Wang, Z., Li, C., Xu, D.: Improving long-text alignment for text-to-image diffusion models. In: ICLR (2025)
21. Liu, S., et al.: Grounding DINO: marrying DINO with grounded pre-training for open-set object detection. In: ECCV (2024)
22. Lugmayr, A., Danelljan, M., Romero, A., Yu, F., Timofte, R., Van Gool, L.: RePaint: inpainting using denoising diffusion probabilistic models. In: CVPR (2022)
23. Meng, C., et al.: SDEdit: guided image synthesis and editing with stochastic differential equations. In: ICLR (2022)
24. Mirzaei, A., et al.: Watch your steps: local image and scene editing by text instructions. In: ECCV (2024)
25. Mokady, R., Hertz, A., Aberman, K., Pritch, Y., Cohen-Or, D.: Null-text inversion for editing real images using guided diffusion models. In: CVPR (2023)
26. Mou, C., Wang, X., Song, J., Shan, Y., Zhang, J.: DiffEditor: boosting accuracy and flexibility on diffusion-based image editing. In: CVPR (2024)
27. Nichol, A., et al.: GLIDE: towards photorealistic image generation and editing with text-guided diffusion models. In: ICML (2022)
28. Nikhila, R., et al.: SAM 2: segment anything in images and videos. In: ICLR (2024)
29. Park, D.H., et al.: Shape-guided diffusion with inside-outside attention. In: WACV (2024)
30. Pei, Y., et al.: SOWing information: cultivating contextual coherence with MLLMs in image generation. arXiv (2024)
31. Podell, D., et al.: SDXL: improving latent diffusion models for high-resolution image synthesis. In: ICLR (2024)
32. Qwen: Qwen2.5-math technical report: toward mathematical expert model via self-improvement. arXiv (2024)
33. Qwen: Qwen2.5 technical report. arXiv (2025)
34. Ramesh, A., Dhariwal, P., Nichol, A., Chu, C., Chen, M.: Hierarchical text-conditional image generation with CLIP latents. arXiv (2022)
35. Rasheed, H., et al.: GLaAMM: pixel grounding large multimodal model. In: CVPR (2024)
36. Ren, T., et al.: Grounded SAM: assembling open-world models for diverse visual tasks. In: ICCV (2023)
37. Rombach, R., Blattmann, A., Lorenz, D., Esser, P., Ommer, B.: High-resolution image synthesis with latent diffusion models. In: CVPR (2022)
38. Ruiz, N., et al.: DreamBooth: fine tuning text-to-image diffusion models for subject-driven generation. In: CVPR (2023)
39. Sheynin, S., et al.: Emu edit: precise image editing via recognition and generation tasks. In: CVPR (2024)
40. Voynov, A., Aberman, K., Cohen-Or, D.: Sketch-guided text-to-image diffusion models. In: SIGGRAPH (2023)
41. Wang, Z., Li, A., Li, Z., Liu, X.: GenArtist: multimodal LLM as an agent for unified image generation and editing. In: NeurIPS (2024)
42. Wei, C., Xiong, Z., Ren, W., Du, X., Zhang, G., Chen, W.: OmniEdit: building image editing generalist models through specialist supervision. In: ICLR (2025)

43. Wu, T.H., Lian, L., Gonzalez, J.E., Li, B., Darrell, T.: Self-correcting LLM-controlled diffusion models. In: CVPR (2024)
44. Xie, S., Zhang, Z., Lin, Z., Hinz, T., Zhang, K.: SmartBrush: text and shape guided object inpainting with diffusion model. In: CVPR (2023)
45. Yu, Q., et al.: AnyEdit: mastering unified high-quality image editing for any idea. arXiv (2024)
46. Yu, Y., Zeng, Z., Hua, H., Fu, J., Luo, J.: PromptFix: you prompt and we fix the photo. In: NeurIPS (2024)
47. Zeng, Y., et al.: SceneComposer: any-level semantic image synthesis. In: CVPR (2023)
48. Zhang, K., Mo, L., Chen, W., Sun, H., Su, Y.: MagicBrush: a manually annotated dataset for instruction-guided image editing. In: NeurIPS (2024)

Food State Recognition from Recipes Using Multimodal Model for Task Monitoring in Autonomous Cooking Robots

Rina Tagami(✉), Hiroki Kobayashi, Shuichi Akizuki, and Manabu Hashimoto(✉)

Chukyo University, 101-2 Yagoto Honmachi, Nagoya, Aichi 466-8666, Japan
{tagami,mana}@isl.sist.chukyo-u.ac.jp

Abstract. Monitoring the cooking process with autonomous robots requires recognition of food states, such as mixing and doneness levels. A previous approach applied CLIP [1] to cooking videos but needed to exclude frames where food regions were occluded by cooking utensils to maintain recognition accuracy. However, in cooking processes that involve frequent mixing, the number of usable frames for state recognition decreases (Challenge 1). Additionally, in the training process, images were divided into two categories—before and after completion—based on a manually specified completion timing. However, the pre-completion category contained images ranging from the start of cooking to just before completion, leading to high intra-category variance and training inconsistencies, which degraded recognition accuracy (Challenge 2). To address these challenges, our proposed method reduces the learning contribution of occluded regions, allowing all frames to be used for processing (Challenge 1). Furthermore, instead of using only two categories (completed and incomplete), we also incorporate 3 to 6 intermediate categories into the training process to capture the gradual progression of cooking, mitigating training inconsistencies and enhancing recognition accuracy (Challenge 2). Experimental results showed that the model with reduced occlusion contribution achieved 30.4% recognition accuracy, a 7.4% improvement over the model trained by removing occluded frames. Additionally, the model trained with intermediate categories achieved 31.5% accuracy, a 10.8% improvement over the conventional model trained with only two categories.

Keywords: State Recognition · Multimodal Model · CLIP · Cooking Robot · Contrastive Learning

1 Introduction

In recent years, methods for generating robot actions using Large Language Models (LLMs) have been proposed, necessitating feedback on whether the robot has

J. Petersen and V. A. Dahl (Eds.): SCIA 2025, LNCS 15726, pp. 356–369, 2025.
https://doi.org/10.1007/978-3-031-95918-9_25

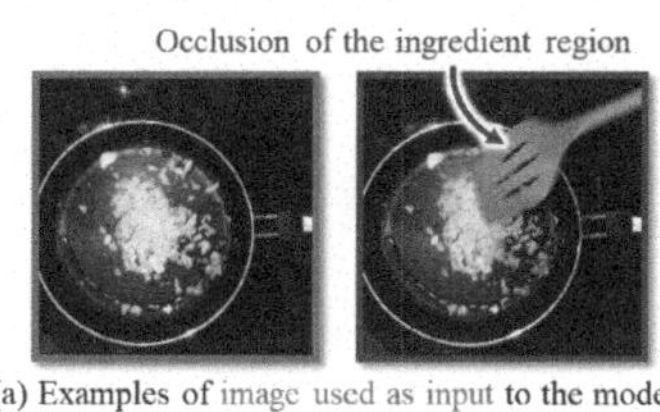

(a) Examples of image used as input to the model and image unused by the method [2]

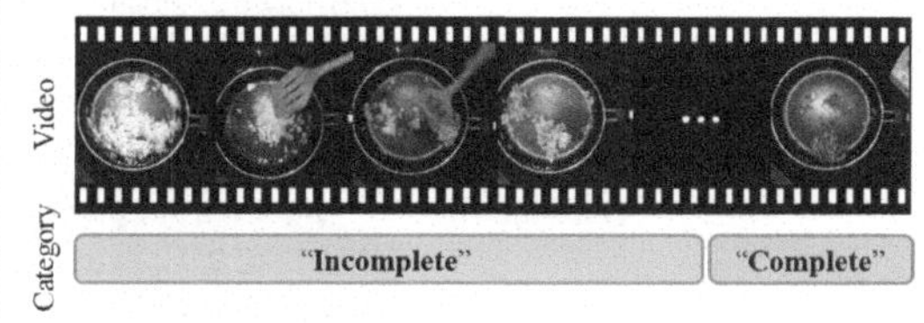

(b) Category division of the images in the method [4]

Fig. 1. Examples of images used in conventional methods and methods of category division. In (a), the red-bordered image represents an image without occlusion in the ingredient region, while the blue-bordered image represents an image with occlusion. In (b), the frame segmentation method used in conventional approaches is shown, where frames before the manually specified timing are classified as "incomplete", and frames after that point are classified as "complete". (Color figure online)

successfully executed its intended task. In the context of autonomous cooking, if the robot cannot accurately monitor the cooking state of ingredients, such as their degree of doneness or mixing level, based on visually acquired information, generating the appropriate actions for the next step becomes challenging. Therefore, food state recognition, which determines whether the ingredients have reached the condition specified in the recipe, plays a crucial role in action generation for cooking robots.

Several methods for food state recognition have been proposed [2–4] and are considered applicable to monitoring the cooking process. However, these methods achieve a recognition accuracy of only 10–20%, which remains insufficient for practical implementation. Proposed method focuses on improving the practicality of these techniques by addressing two independent challenges. The first challenge is mitigating occlusions caused by cooking utensils that obscure the ingredient regions. The second challenge involves improving recognition accuracy through better data and learning strategies.

The first challenge arises from the input images captured by the robot over time, which are fed into the CLIP (Contrastive Language Image Pretraining) model [1]. However, when food regions in the images are occluded by cooking utensils or hands, the accuracy of state recognition significantly decreases. To mitigate this issue, frames where the food region is occluded, such as the blue-framed images in Fig. 1(a), were excluded from the model's input. However, when monitoring cooking processes that involve frequent stirring, such as using a spatula, this approach significantly reduces the number of frames available for recognition. Consequently, if the food region is occluded at the moment of completion, the system fails to provide real-time feedback on the completion timing. The second challenge arises in the training process, where images were categorized into "incomplete" and "complete" classes based on the manually specified completion timing (Fig. 1(b)). However, the incomplete category contained a wide range of images, from the very beginning of the cooking process to just

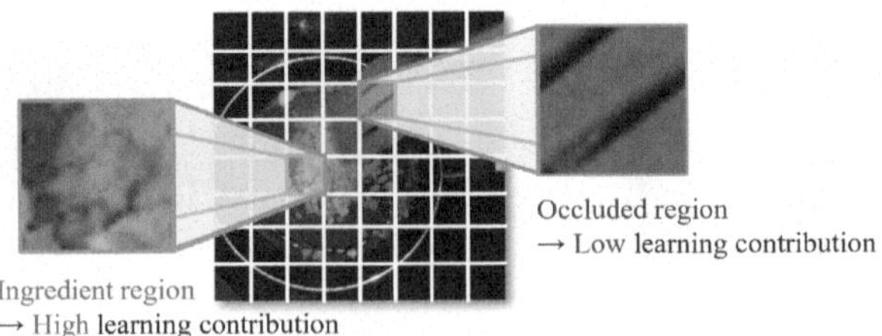

Fig. 2. The contribution of each local region to learning in the proposed method. The pink-framed images represent food regions, which are assigned a high contribution to learning, while the blue-framed images represent occluded food regions, which have their contribution reduced. (Color figure online)

before completion, resulting in high intra-category variance. This led to inconsistencies during training, ultimately degrading the accuracy of state recognition.

To address Challenge 1, our method reduces the learning contribution of highly occluded food regions, as shown in the right side of Fig. 2. Specifically, the input images are divided into multiple patches, and a mask is generated where the learning contribution decreases as the number of occluded pixels in each patch increases. This mask is fed into the image encoder of the model (Vision Transformer (ViT) [5]), where high-contribution regions receive more attention, while low-contribution regions are excluded from the learning process. As a result, even when food regions are frequently occluded, the model can utilize non-occluded regions for state recognition, enabling all frames to be used for processing. To address Challenge 2, we introduce 3 to 6 intermediate categories in addition to the two conventional categories (complete and incomplete), as shown in Fig. 1(b). The degree of completion is expected to follow a gradual increase over time, then a rapid rise to a peak, followed by a plateau at full completion, and eventually a decline. By leveraging these segmented image groups and recipe descriptions for contrastive learning in the CLIP model, training inconsistencies can be significantly reduced, leading to improved accuracy in state recognition.

The two independent ideas in the proposed method mitigate the negative effects of food region occlusion and optimize the category division of cooking data, thereby improving state recognition accuracy. If the proposed method successfully enhances recognition accuracy, it will expand the range of dishes that can be monitored in autonomous cooking, representing a significant step toward the practical implementation of food state recognition technology.

2 Related Work and Its Challenges

In recent years, multimodal models have advanced significantly, leading to the proposal of numerous methods for integrating information from different modalities [6–9]. In the field of food computing [10,11], techniques for precisely aligning text-based information (such as recipe title, ingredients, and instructions) with images have also gained attention [12,13]. For example, multimodal regularization using cross-attention [14], methods that exclude irrelevant textual expressions using self-attention [15], and hierarchical encoders that effectively learn relationships between different textual formats [16,17] have been proposed. However, these methods are primarily designed for retrieving images of completed dishes and are not suitable for recognizing cooking states in real-time. Consequently, there is a need for technology that can provide real-time feedback on ingredient states during the cooking process.

Challenge 1: Occlusion of Food Regions. The method proposed in [2] inputs overhead-view images captured by a robot along with pre-designed recipe text into the CLIP model. It determines completion of cooking when the image-text similarity, calculated over time, reaches a predefined threshold. However, this approach excludes frames where the food region is occluded, which reduces the amount of essential visual information needed for recognition. This challenge is particularly problematic in cooking processes that involve frequent stirring, as it significantly reduces the number of usable frames, leading to decreased recognition accuracy.

Challenge 2: Data Quality and Category Division. The method [3] in constructs multiple text sets and uses the CLIP model to compute image-text similarity over time, quantifying the change in similarity using a sigmoid function. However, if the text set lacks sufficient variation in state-change descriptions, recognition accuracy declines. For example, if expressions describing intermediate cooking stages are insufficient, the model fails to learn continuous state transitions properly, increasing the likelihood of misrecognition. Additionally, the method in [4] categorizes frames into "Before Change" and "After Change" based on a manually specified completion frame and incorporates a linear classifier into the model. While this method enables real-time recognition, it suffers from a significant drawback: high intra-category variance within the Before Change category. Since images belonging to the same category exhibit significantly different visual characteristics, training inconsistencies arise, leading to reduced recognition accuracy.

Although these methods [2–4] have achieved certain levels of success, they still face challenges such as the reduction of available visual information, decreased accuracy due to insufficient variation in text descriptions, and high intra-category variance in images. To address these challenges, this study proposes a method that leverages unoccluded food regions for learning and incorporates finely divided image-text data reflecting gradual cooking progress, aiming to improve the accuracy of food state recognition.

3 Proposed Method

To address the two main challenges of food region occlusion and data quality issues in conventional methods, this study proposes a method based on two key ideas.

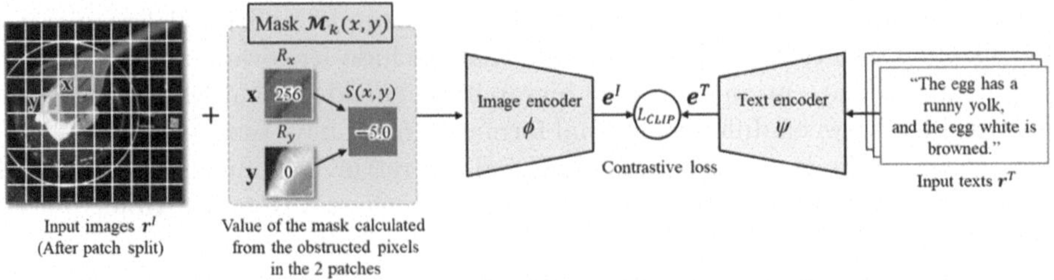

Fig. 3. The training process with images in which the food region is occluded. The fewer occluded pixels (blue-colored pixel regions) contained within a patch, the higher the learning contribution rate assigned to the mask. The output image feature e^I emphasizes only the food region, and by performing contrastive learning using this image feature e^I and the text feature e^T, reliable learning can be achieved.

The first idea is to enable food state recognition even when the food region is occluded by identifying occluded regions and applying a mask that reduces their contribution to learning in proportion to their size. Previous studies have demonstrated the effectiveness of using masks in deep learning, including methods that apply masks to input images before feeding them into an encoder [18–20] and methods that introduce masks into the decoder [21]. Conventional methods [18–20] randomly mask images. In particular, the method proposed in [18] generates masked areas to resemble the original image, but the quality of the generated images is not high, making it difficult to recognize the state of cooking using such images. The proposed masking technique enhances the extracted image features by emphasizing non-occluded food regions, thereby improving recognition accuracy. The second idea is to refine the categorization of training data. Instead of using only two categories—"complete" and "incomplete"—this method introduces multiple intermediate categories that reflect different stages of food state, incorporating them into contrastive learning. This approach reduces learning inconsistencies caused by high intra-category variance and contributes to improving recognition accuracy. By integrating these two key ideas, the proposed method mitigates the negative effects of occlusion while optimizing category division, thereby improving the accuracy of food state recognition.

3.1 Idea 1: Method for Reducing the Learning Contribution of Occluded Regions in Images

This section describes a method for identifying occluded food regions in an image and adjusting their contribution to learning based on the number of occluded pixels. The overview of Idea 1 in the proposed method is illustrated in Fig. 3. First, K pairs of images r^I and text r^T are given as input. As shown in Eq. 1, the image encoder ϕ extracts the image feature e^I, while the text encoder ψ extracts the text feature e^T. At this stage, the image encoder ϕ receives both the image r^I and a mask M, which represents the learning contribution rate for each corresponding image.

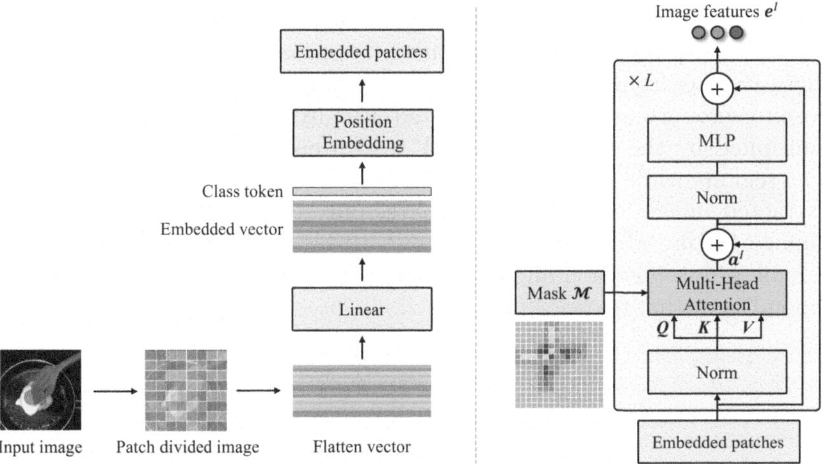

Fig. 4. Detailed structure of the proposed image encoder. First, a cooking image is input into the encoder and flattened to obtain a vector representation. A class token is then appended to this vector, and positional embeddings are added to generate the image embeddings. These embeddings are fed into the multi-head attention mechanism, where self-attention is computed internally, and the class token vector is finally output as the image feature.

The mask M is created through the following process. First, the input image is divided into 16×16 patches, and the number of occluded pixels R within each patch is calculated. In this experiment, we assumed that no food ingredients contain blue color, and therefore, blue-colored cooking utensils and gloves were used when capturing the image data. Based on this, blue pixels were counted as occluded pixels. Next, based on the number of occluded pixels R in patches x and y, the learning contribution rate is determined and assigned to the mask, as described in Eq. 2. Specifically, if both patches are entirely occluded regions, the mask value is set to $-\infty$. Otherwise, a specific value $S(x, y)$ is assigned, where values closer to 0 indicate a higher learning contribution rate. The value $S(x, y)$ represents the learning contribution rate when either of the two patches is at least partially unoccluded, and it is calculated using Eq. 3. The number of occluded pixels R of the class token is calculated as 0. In this process, the parameter α is set to 10. As a result, the higher the number of occluded pixels R in both patches, the lower the learning contribution rate assigned to the mask M.

Next, the details of the image encoder are explained (Fig. 4). First, the input image is divided into multiple patches, and each patch is converted into an embedding vector before being fed into the multi-head attention along with a mask (Eqs. 4, 5). The mask has the same size as $QK^\top$, and l represents the number of layers in the encoder. Within the multi-head attention mechanism, the relationships between different local regions of the image features are computed, and the resulting values (attention scores $QK^\top$) are adjusted by adding the mask values to control the degree of attention. The final output image feature is the feature vector of the Class token. In regions where the mask values are highly negative, the output of the softmax function approaches zero, effectively excluding those regions from learning. Conversely, in regions where the mask values are close to zero, attention is concentrated. Finally, the adjusted attention scores are multiplied by the image features V, which enhances the features of highly relevant regions while suppressing those of less relevant regions. Consequently, the extracted image feature a_l^I emphasizes only the food regions, allowing the final image feature e^I and text feature e^T to be used for contrastive learning. This enables high-precision alignment between food regions in images and their corresponding recipe texts.

$$\mathbf{e}_k^I = \phi(\mathbf{r}_k^I, \mathcal{M}_k), \quad \mathbf{e}_k^T = \psi(\mathbf{r}_k^T), \tag{1}$$

$$\mathcal{M}_k(x,y) = \begin{cases} -\infty & \text{if } R_x(x,y) = 256 \wedge R_y(x,y) = 256, \\ S(x,y) & \text{otherwise}, \end{cases} \tag{2}$$

$$S(x,y) = -\alpha \left(\frac{R_x(x,y)}{256} + \frac{R_y(x,y)}{256} \right) /2, \tag{3}$$

$$\mathbf{a}_l^I = \text{MultiHeadAttn}(\mathbf{Q}_l, \mathbf{K}_l, \mathbf{V}_l, \mathcal{M}_l) + \mathbf{a}_{l-1}^I, \tag{4}$$

$$\text{MultiHeadAttn}(\mathbf{Q}_l, \mathbf{K}_l, \mathbf{V}_l, \mathcal{M}_l) = \text{softmax}\left(\mathcal{M}_l + \mathbf{Q}_l \mathbf{K}_l^\top\right) \mathbf{V}_l, \tag{5}$$

The contrastive learning process itself remains unchanged from conventional methods. Image features e^I and text features e^T are extracted from the image encoder ϕ and text encoder ψ, respectively, and the model is trained to increase the similarity of related image-text pairs while decreasing the similarity of unrelated pairs. The loss function L_{CLIP} is computed according to Eqs. 6, 7, and 8. At this stage, the batch size is set to N, the similarity measure is cosine similarity, and the temperature parameter τ is initialized to 0.07.

$$\mathcal{L}_{\text{i2t}} = -\frac{1}{N} \sum_{i=1}^{N} \log \frac{\exp(\text{sim}(\mathbf{e}_i^{\text{I}}, \mathbf{e}_i^{\text{T}})/\tau)}{\sum_{j=1}^{N} \exp(\text{sim}(\mathbf{e}_i^{\text{I}}, \mathbf{e}_j^{\text{T}})/\tau)}, \tag{6}$$

$$\mathcal{L}_{\text{t2i}} = -\frac{1}{N} \sum_{i=1}^{N} \log \frac{\exp(\text{sim}(\mathbf{e}_i^{\text{T}}, \mathbf{e}_i^{\text{I}})/\tau)}{\sum_{j=1}^{N} \exp(\text{sim}(\mathbf{e}_i^{\text{T}}, \mathbf{e}_j^{\text{I}})/\tau)}, \tag{7}$$

$$\mathcal{L}_{\text{CLIP}} = \frac{1}{2} \left(\mathcal{L}_{\text{i2t}} + \mathcal{L}_{\text{t2i}} \right). \tag{8}$$

3.2 Idea 2: Contrastive Learning Using Images Divided into Multiple Categories

This section explains the contrastive learning method using a dataset divided into complete, intermediate, and incomplete categories. In conventional meth-

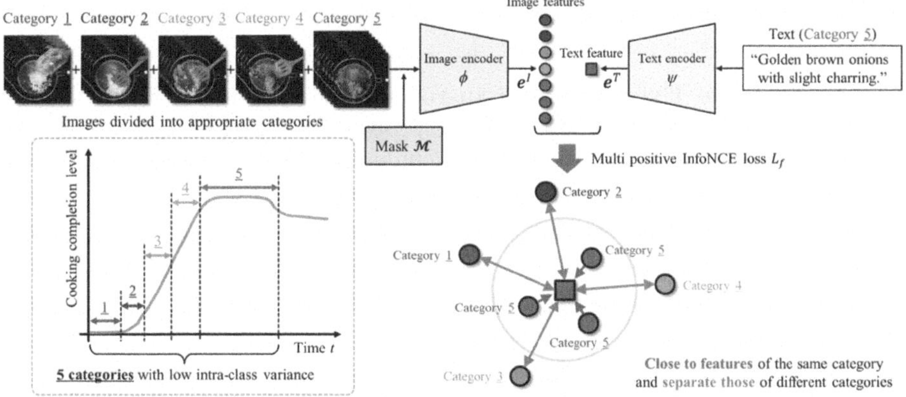

Fig. 5. The training process using multiple category-divided images. The model is trained to increase the similarity between image features and text features from the same category while decreasing the similarity with image features from different categories.

ods, dataset categorization was based on the completion timing specified by a human, where frames after the completion point were classified as complete, and frames before it were classified as incomplete. In contrast, the proposed method classifies each image according to the procedural steps of the recipe it belongs to. As shown in the Fig. 5, the example is divided into five categories c. Next, for all images in the five-category dataset, a mask is generated using the method described in Sect. 3.1, and both the image dataset and the mask are input into the image encoder ϕ to obtain image features e^I. The training process used in the proposed method differs from the contrastive learning employed in conventional CLIP models. In our method, for each recipe text (text feature) e^T, the similarity with image features e^I from the same category e^I_c (corresponding to the text category e^T_c) is increased, while the similarity with image features e^I from different categories is decreased (Eqs. 9, 10). This approach eliminates the learning inconsistencies present in previous methods [3] and enhances the alignment between the recipe text and the corresponding range of images. The final loss function is computed by summing the loss functions in Eq. 8 and Eq. 9, and then taking their average (Eq. 11).

$$\mathcal{L}_f = -\frac{1}{N}\log\left(\frac{\sum_{i\in\mathcal{S}}\exp(\mathrm{sim}(\mathbf{e}^{\mathrm{I}}_i,\mathbf{e}^{\mathrm{T}})/\tau)}{\sum_{i\in\mathcal{S}}\exp(\mathrm{sim}(\mathbf{e}^{\mathrm{I}}_i,\mathbf{e}^{\mathrm{T}})/\tau)+\sum_{i\in\overline{\mathcal{S}}}\exp(\mathrm{sim}(\mathbf{e}^{\mathrm{I}}_i,\mathbf{e}^{\mathrm{T}})/\tau)}\right), \tag{9}$$

$$\mathcal{S} = \{i | e^I_{i,c} = e^T_c\}, \tag{10}$$

$$\mathcal{L} = \frac{1}{2}\left(\mathcal{L}_{\mathrm{CLIP}} + \mathcal{L}_f\right). \tag{11}$$

4 Experimental Results and Analysis

Objective of the Experiment. Experiments were conducted to verify that the food state recognition accuracy of the proposed method is higher than that of conventional approaches that either exclude occluded frames of food regions or use data divided into only two categories ("complete" and "incomplete").

Evaluation Method. To evaluate state recognition performance, we conducted an experiment to retrieve frames from text queries. The evaluation focused on the model's ability to output the most similar image from a sequence of frames given a text input. For each video, 100 random non-duplicate images were selected, regardless of category. The results were averaged over 100 trials with different image sets and were evaluated using Rank@1, Rank@5, Rank@10, and mean Average Precision (mAP).

Dataset. The evaluation used a custom cooking dataset (Fig. 6) with 652,922 image-recipe pairs, split into 510,672 pairs for additional training and 142,250 pairs for testing. This dataset consists of four classes (Caramel sauce, Pancake, Sunny-side-up, and Fried onions) and does not contain unnecessary objects or subtitles. The annotations were performed by the cooks themselves. Unlike conventional datasets [22,23], this dataset does not fast-forward intermediate cooking frames, making it potentially suitable for challenging food state recognition tasks.

Model. Methods such as ImageBind and SigLIP, which project images and text into a common space, also exist. However, the training methods and encoders used in these methods are based on the CLIP model. Therefore, the proposed method in this study utilizes the CLIP model. The proposed method uses a pre-trained CLIP model with ViT-B/16 [5] as the image encoder and Transformer [24] as the text encoder. The maximum text length is set to 77. Input images were resized to 224×224, and the embedding dimension was set to 512. For fine-tuning the CLIP model, the AdamW optimizer [25] was used with a learning rate of 0.02. Since the proposed method aims to recognize cooking states from recipe texts, it employs fine-tuning instead of simple binary classification using linear probing, to handle intermediate expressions such as "slightly browned" or "lightly mixed".

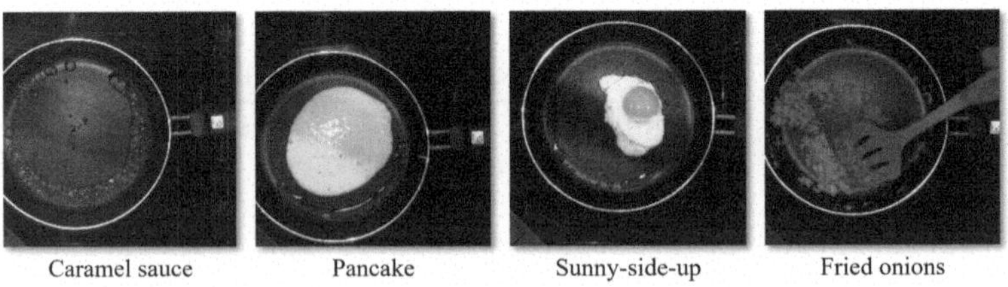

Fig. 6. Examples of images included in the created dataset.

Table 1. Recognition performance with occlusion for each method (Text-to-Frame). For each evaluation metric, the best-performing method is shown in bold red with an underline, and the second-best is indicated in plain red text.

method	fine-tuning	Method [2]	Method [4]	idea 1	idea 2	train images (occlusion)	text-to-frame			
							mAP	R@1	R@5	R@10
(1)		–	–	–	–	–	0.12	5.0	15.0	18.0
(2)	✓	–	–	–	–	✓	0.16	9.0	17.0	32.0
(3)	✓	✓		–			0.15	9.0	20.0	24.0
(4)	✓		✓		–	✓	0.20	14.0	22.0	30.0
(5)	✓		✓		–		0.19	13.0	23.0	28.0
(6)	✓	–		✓		✓	0.19	12.0	25.0	33.0
(7)	✓		–		✓	✓	0.22	12.0	28.0	45.0
(8)	✓		–		✓		0.21	12.0	27.0	42.0
(9)	✓	–	–	✓	✓	✓	**0.28**	**16.0**	**39.0**	**54.0**

Comparison Methods. The experiment included the proposed method and the following models.

(1) Pre-trained CLIP model released by OpenAI [1].
(2) Model trained with occluded images.
(3) Model trained with only non-occluded images (Method [2]).
(4) Model trained with completed and incomplete categories using occluded images (Method [4]).
(5) Model trained with completed and incomplete categories using only non-occluded images (Method [4]).
(6) Model trained to suppress the contribution of occluded regions (Idea 1 of the proposed method).
(7) Model trained with completed, intermediate, and incomplete categories using occluded images (Idea 2 of the proposed method).
(8) Model trained with completed, intermediate, and incomplete categories using only non-occluded images (Idea 2 of the proposed method).
(9) Proposed method combining Idea 1 and Idea 2.

4.1 Results of Food State Recognition Accuracy for Each Method

Table 1 and Table 2 present the state recognition accuracy of each method. The vertical axis represents each method, while the horizontal axis indicates the training conditions and the recognition accuracy.

Effect of not Contributing the Occluded Region to Learning (Idea 1). Method (1), which was not fine-tuned, exhibited lower accuracy overall. When tested on non-occluded images, its accuracy was relatively higher, indicating

Table 2. Recognition performance without occlusion for each method (Text-to-Frame). For each evaluation metric, the best-performing method is shown in bold red with an underline, and the second-best is indicated in plain red text.

method	fine-tuning	Method [2]	Method [4]	idea 1	idea 2	train images (occlusion)	text-to-frame			
							mAP	R@1	R@5	R@10
(1)		–	–	–	–	–	0.18	10.0	23.0	34.0
(2)	✓	–	–	–	–	✓	0.17	8.0	25.0	32.0
(3)	✓	✓		–			0.17	7.0	23.0	35.0
(4)	✓		✓		–	✓	0.23	18.0	26.0	30.0
(5)	✓		✓		–		0.23	18.0	27.0	33.0
(6)	✓	–		✓		✓	0.20	11.0	27.0	43.0
(7)	✓		–		✓	✓	0.31	20.0	42.0	57.0
(8)	✓		–		✓		0.33	21.0	43.0	**58.0**
(9)	✓	–	–	✓	✓	✓	**0.34**	**23.0**	**44.0**	52.0

that the occlusion of food regions is a major factor that makes state recognition more challenging. Method (3), which was trained only on non-occluded images, demonstrated improved accuracy when tested on similar non-occluded images. A comparison between Method (2) and Method (6) reveals that reducing the learning contribution of occluded regions during training resulted in higher accuracy than training directly with occluded images. This confirms the effectiveness of Idea 1 in addressing Challenge 1 of the proposed method. Additionally, Method (6) outperformed Method (3), suggesting that leveraging non-occluded food regions for training is beneficial for improving recognition performance.

Effectiveness of Intermediate Category on Learning (Idea 2). The results of Method (3) and Method (5) indicate that training with two-category division (completed and incomplete) is more effective than simply fine-tuning the model with additional images. Furthermore, comparing Method (5) and Method (8) shows that incorporating intermediate categories beyond completed and incomplete leads to improved state recognition accuracy, particularly when tested on non-occluded images. This is likely because Idea 1 was not included in Methods (5) and (8), making them less robust to occlusions. Consequently, Method (9), which integrates both Idea 1 and Idea 2, maintained high accuracy even when occlusions were present during testing. These results demonstrate that the proposed training method is effective for food state recognition.

4.2 Output Results of Images with High Similarity to Text for Each Method

Figure 7 illustrates the top-5 retrieved images for each method based on the input text. The system retrieves the most similar frames from a video in the

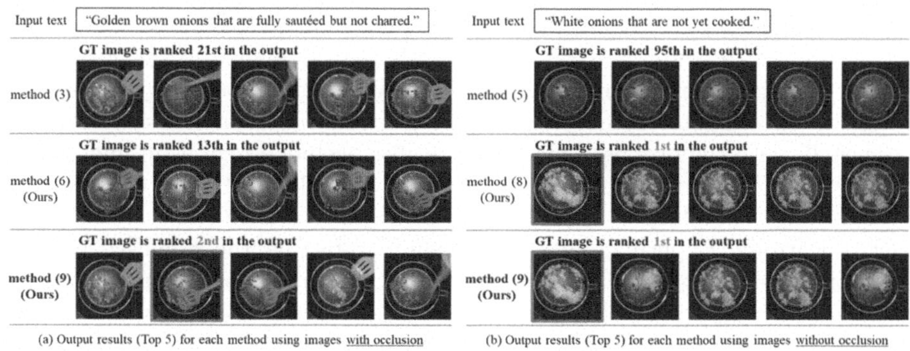

Fig. 7. Top 5 images with the highest similarity to the input text for each method. The red-bordered image is the ground truth image. (Color figure online)

custom dataset using recipe descriptions. The top row shows the input recipe text, and the red boxes indicate correct category images. When tested with occluded images, Method (9) ranked the most relevant images higher, demonstrating its effectiveness even under significant occlusion. Additionally, when tested with non-occluded images, both Methods (8) and (9) achieved high accuracy, indicating that the proposed method successfully recognizes food states from time-sequential images.

4.3 Limitations

Currently, in the proposed method, blue cooking utensils are used during data collection, and occluded regions are assumed to be detected as blue pixels. Therefore, when food is occluded by cooking utensils that are not blue, it becomes difficult to detect the occluded regions, which is a limitation of the method. By combining object detection models or segmentation models, it is expected to develop a more practical state recognition system.

Additionally, the data used for evaluation is a cooking dataset consisting of four categories, but the cooking environment is fixed, and the cooking processes are relatively simple. Therefore, it is necessary to add more diverse data for evaluation. By using datasets that include data captured from various angles and cover a wide range of dishes, we aim to investigate the limitations of the proposed method in real-world environments.

5 Conclusion

This study proposes a food state recognition method for monitoring cooking tasks in autonomous robots using the CLIP model. Conventional methods face two primary challenges. First, they struggle when food regions in captured images are occluded by hands or cooking utensils, reducing the number of usable frames and making them less effective for dishes that require frequent stirring.

Second, during training, datasets were constructed by dividing frames into only two groups—"complete" and "incomplete"—based on manually specified completion timing. However, this approach resulted in high intra-category variance within the incomplete category, leading to learning inconsistencies. To address these challenges, the proposed method identifies occluded regions in images and reduces their contribution to learning, while also introducing three to six intermediate categories in addition to the existing two and incorporating them into contrastive learning. Experimental results demonstrate that the proposed method achieves higher recognition accuracy compared to conventional approaches. By leveraging this food state recognition system, autonomous cooking robots can handle a broader range of dishes, enhancing their adaptability and potentially mitigating labor shortages in restaurants and professional kitchens.

References

1. Radford, A., et al.: Learning transferable visual models from natural language supervision. In: International Conference on Machine Learning, pp. 8748–8763, PMLR (2021)
2. Kanazawa, N., Kawaharazuka, K., Obinata, Y., Okada, K., Inaba, M.: Recognition of heat-induced food state changes by time-series use of vision-language model for cooking robot. In: International Conference on Intelligent Autonomous Systems, pp. 547–560, Springer, Cham (2023)
3. Kawaharazuka, K., Kanazawa, N., Obinata, Y., Okada, K., Inaba, M.: Continuous object state recognition for cooking robots using pre-trained vision-language models and black-box optimization. IEEE Robot. Autom. Lett. (2024)
4. Kanazawa, N., Kawaharazuka, K., Obinata, Y., Okada, K., Inaba, M.: Real-world cooking robot system from recipes based on food state recognition using foundation models and PDDL. Adv. Robot. **38**(18), 1318–1334 (2024)
5. Dosovitskiy, A.: An image is worth 16x16 words: transformers for image recognition at scale, arXiv preprint arXiv:2010.11929 (2020)
6. Jia, C., et al.: Scaling up visual and vision-language representation learning with noisy text supervision. In: International Conference on Machine Learning, pp. 4904–4916, PMLR (2021)
7. Li, J., Li, D., Savarese, S., Hoi, S.: Blip-2: bootstrapping language-image pre-training with frozen image encoders and large language models. In: International Conference on Machine Learning, pp. 19730–19742 PMLR (2023)
8. Min, W., Liu, L., Wang, Z., Luo, Z., Wei, X., Wei, X., Jiang, S.: food-500: a dataset for large-scale food recognition via stacked global-local attention network. In: Proceedings of the 28th ACM International Conference on Multimedia, pp. 393–401 (2020)
9. Zou, Z., Zhu, X., Zhu, Q., Liu, Y., Zhu, L.: CREAMY: cross-modal recipe retrieval by avoiding matching imperfectly. IEEE Access (2024)
10. Xu, F.F., Ji, L., Shi, B., Du, J., Neubig, G.B.Y., Duan, N.: A benchmark for structured procedural knowledge extraction from cooking videos, arXiv preprint arXiv:2005.00706 (2020)
11. Papadopoulos, D.P., Mora, E., Chepurko, N., Huang, K.W., Ofli, F., Torralba, A.: Learning program representations for food images and cooking recipes. In: Proceedings of the IEEE/CVF Conference on Computer Vision and Pattern Recognition, pp. 16559–16569 (2022)

12. Fu, H., Wu, R., Liu, C., Sun, J.: Mcen: bridging cross-modal gap between cooking recipes and dish images with latent variable model. In: Proceedings of the IEEE/CVF Conference on Computer Vision and Pattern Recognition, pp. 14570–14580 (2020)
13. Li, J., et al.: Hybrid fusion with intra-and cross-modality attention for image-recipe retrieval. In: Proceedings of the 44th International ACM SIGIR Conference on Research and Development in Information Retrieval, pp. 244–254 (2021)
14. Shukor, M., Couairon, G., Grechka, A., Cord, M.: Transformer decoders with multimodal regularization for cross-modal food retrieval. In: Proceedings of the IEEE/CVF Conference on Computer Vision and Pattern Recognition, pp. 4567–4578 (2022)
15. Wang, H., et al.: Cross-modal food retrieval: learning a joint embedding of food images and recipes with semantic consistency and attention mechanism. IEEE Trans. Multimedia **24**, 2515–2525 (2021)
16. Wahed, M., Zhou, X., Yu, T., Lourentzou, I.: Fine-grained alignment for cross-modal recipe retrieval. In: Proceedings of the IEEE/CVF Winter Conference on Applications of Computer Vision, pp. 5584–5593 (2024)
17. Salvador, A., Gundogdu, E., Bazzani, L., Donoser, M.: Revamping cross-modal recipe retrieval with hierarchical transformers and self-supervised learning. In: Proceedings of the IEEE/CVF Conference on Computer Vision and Pattern Recognition, pp. 15475–15484 (2021)
18. Feichtenhofer, C., Li, Y., He, K.: Masked autoencoders as spatiotemporal learners. Adv. Neural. Inf. Process. Syst. **35**, 35946–35958 (2022)
19. Woo, S., et al.: Convnext v2: co-designing and scaling convnets with masked autoencoders. In: Proceedings of the IEEE/CVF Conference on Computer Vision and Pattern Recognition, pp. 16133–16142 (2023)
20. Li, Y., Fan, H., Hu, R., Feichtenhofer, C., He, K.: Scaling language-image pre-training via masking. In: Proceedings of the IEEE/CVF Conference on Computer Vision and Pattern Recognition, pp. 23390–23400 (2023)
21. Cheng, B., Misra, I., Schwing, A.G., Kirillov, A., Girdhar, R.: Masked-attention mask transformer for universal image segmentation. In: Proceedings of the IEEE/CVF Conference on Computer Vision and Pattern Recognition, pp. 1290–1299 (2022)
22. Zhou, L., Louis, N., Corso, J.J.: Weakly-supervised video object grounding from text by loss weighting and object interaction, arXiv preprint arXiv:1805.02834 (2018)
23. Sener, F., Yao, A.: Zero-shot anticipation for instructional activities. In: Proceedings of the IEEE/CVF International Conference on Computer Vision, pp. 862–871 (2019)
24. Vaswani, A., et al.: Attention is all you need. NIPS **3**(6), 5998–6008 (2017)
25. Loshchilov, I., Hutter, F.: Decoupled weight decay regularization. In: International Conference on Learning Representations (ICLR), vol.5, no.6 (2019)

Computational Photography

Data-Efficient Limited-Angle CT Using Deep Priors and Regularization

Ilmari Vahteristo[1(✉)], Zhi-Song Liu[1], and Andreas Rupp[2]

[1] LUT University, 53850 Lappeenranta, Finland
i.Vahteristo@gmail.com
[2] Saarland University, 66123 Saarbrücken, Germany

Abstract. Reconstructing an image from its Radon transform is a fundamental computed tomography (CT) task arising in applications such as X-ray scans. In many practical scenarios, a full 180-degree scan is not feasible, or there is a desire to reduce radiation exposure. In these limited-angle settings, the problem becomes ill-posed, and methods designed for full-view data often leave significant artifacts. We propose a very low-data approach to reconstruct the original image from its Radon transform under severe angle limitations. Because the inverse problem is ill-posed, we combine multiple regularization methods, including Total Variation, a sinogram filter, Deep Image Prior, and a patch-level autoencoder. We use a differentiable implementation of the Radon transform, which allows us to use gradient-based techniques to solve the inverse problem. Our method is evaluated on a dataset from the Helsinki Tomography Challenge 2022, where the goal is to reconstruct a binary disk from its limited-angle sinogram. We only use a total of 12 data points–eight for learning a prior and four for hyperparameter selection–and achieve results comparable to the best synthetic data-driven approaches.

Keywords: Computed Tomography · Regularization · Deep Image Prior

1 Introduction

Computed tomography (CT) is a widely used imaging technique in medical diagnostics, industrial inspection, and scientific research. It involves measuring the permeability of an object from multiple angles and reconstructing the object's interior from these projections. The Radon transform is a mathematical tool that describes how an object's structure is projected onto a set of lines, forming the basis of CT reconstruction [4]. In practice, the Radon transform is approximated by taking a finite number of projections at discrete angles, forming a sinogram.

Reconstructing an object from limited-angle data is a challenging problem due to the restricted range of projection angles. This situation arises in many practical scenarios, such as industrial inspection where the imaging system cannot rotate around the object a full 180°, or in medical imaging where patient

J. Petersen and V. A. Dahl (Eds.): SCIA 2025, LNCS 15726, pp. 373–386, 2025.
https://doi.org/10.1007/978-3-031-95918-9_26

positioning might limit scanner access. Because few angles are captured, the problem is inherently ill-posed [12], leading to multiple plausible images that match the measured data.

This work focuses on the Helsinki Tomography Challenge 2022 (HTC'22) [22] which presents a difficult problem: reconstructing binary objects from limited angle data. We introduce a gradient-based optimization pipeline that integrates multiple regularization techniques. We demonstrate that this method yields competitive reconstruction results using only 8 data points to train a patch-based regularizer–a substantial reduction compared to typical machine learning approaches. To promote reproducibility and future research, we provide an open-source implementation of our pipeline.

2 Related Works

Reconstructing the interior of an object from its projections has been a long-standing challenge in CT. Traditional reconstruction techniques, such as filtered back projection (FBP) [13] and algebraic reconstruction techniques (ART) [11], often perform poorly under angle constraints due to missing information. FBP is an analytic solution for full-angle CT scans but leaves significant artifacts when working with limited angles. Current solutions to the limited-angle problem encompass a wide range of approaches, such as iterative and model-based solutions. This section will focus on iterative, deep learning, and hybrid methods.

2.1 Iterative Methods

Iterative methods have long been a cornerstone in addressing limited-angle CT challenges. These approaches, such as ART and simultaneous iterative reconstruction technique (SIRT) [10,19,20], reconstruct the image by iteratively minimizing the difference between measured and calculated projections. Since the limited-angle problem is ill-posed, regularization strategies such as total variation (TV) [7] and sparsity constraints [5] are often used to incorporate prior information, such as piecewise constant solutions or specific shapes. Increasingly, Deep Image Priors (DIP) [26] have been used to regularize the reconstructions to favor solutions with more "natural" image statistics [3,8,17,18]. These methods are computationally intensive but offer flexibility in integrating priors and physical models of the imaging system.

2.2 Deep Learning Methods

Machine learning has emerged as a tool for tackling limited-angle CT [1,9,14,33]. Neural networks have been used to reconstruct images from limited projection data by training on large datasets of paired sinograms and ground-truth images. However, the scarcity of real-world data in limited-angle scenarios poses a significant challenge for supervised learning. Sometimes creating synthetic data is possible, which allows researchers to create an arbitrary number of samples with known ground truth images enabling improved supervised models [9,30,34]. However, the synthetic data should accurately represent the true images, the imaging system, and noise to generalize to real data. Arguably, creating a synthetic dataset contains as many assumptions as regularization methods in iterative approaches.

2.3 Hybrid Approaches

Hybrid approaches combine iterative methods and machine learning. These methods aim to leverage the strengths of both approaches to improve reconstruction quality. For instance, iterative methods can provide a good initial estimate, while machine learning techniques can refine this estimate by removing artifacts or enhancing details.

One common hybrid approach is to use one reconstruction algorithm, such as FBP, ART, or SIRT, to create the first reconstruction, and then use a neural network, often a U-net, to remove artifacts or noise from the initial reconstruction [1,14,30,31]. The benefit of this approach compared to learning the end-to-end reconstruction is that the initial reconstruction can provide a good starting point for the neural network, the task is simplified, and the input space is fixed to the image size rather than varying with the sinogram size.

Additionally, some hybrid approaches use machine learning to learn regularization terms or hyperparameters that are then incorporated into the iterative reconstruction process [15,25]. For example, a neural network can be trained to estimate the best TV regularization strength based on the input sinogram, allowing for adaptive regularization that improves reconstruction quality.

3 Methodology

3.1 Problem

In the limited-angle CT problem, we are provided with projection measurements of an object from a restricted range of angles (sinogram), and our goal is to reconstruct an object's interior.

Formally, let Y be the true (unknown) image, S the measured sinogram, and $R(Y, \bar{\theta})$ the Radon transform operator with measurement angles $\bar{\theta}$. We aim to solve for Y in:

$$S = R(Y, \bar{\theta}). \tag{1}$$

To tackle the issue of ill-posedness, we introduce regularization methods that incorporate prior knowledge about the image, such as smoothness or sparsity. Regularization methods help stabilize the reconstruction process and balance between desired properties. We want to find an image $\hat{Y}$ that produces a sinogram $\hat{S}$ that is close to the measured sinogram S while also satisfying the regularization constraints. This can be formulated as:

$$\hat{Y} = \arg\min_{\bar{Y}} \left[\|R(\bar{Y}, \bar{\theta}) - S\|_1 + G(\bar{Y}) \right], \tag{2}$$

where $G(\bar{Y})$ is a regularization term that penalizes solutions that do not conform to the prior, and $\|\cdot\|_1$ is the $L1$ norm.

3.2 Overview

We propose a gradient-based reconstruction pipeline that uses a combination of regularization methods, such as DIP, TV, sinogram filtering, and patch similarity regularization. The regularization methods are discussed in Sect. 3.3. We optimize the pixel values of the reconstruction, or the weights of a neural network, to minimize the difference between the true sinogram, and the sinogram from our reconstruction. The pseudo-code is shown in Algorithm 1.

The algorithm reconstructs an image $\hat{Y}$ from a sinogram S and measurement angles $\bar{\theta}$. It iteratively refines the reconstruction using a neural network $\mathcal{N}$ with weights W. In each iteration i, the network reconstructs an image ($\hat{Y} \leftarrow \mathcal{N}_{W_i}(S)$), computes the filtered sinogram of the reconstruction using the Radon transform R and a filter F: ($\hat{S}_f \leftarrow F\big(R(\hat{Y}, \bar{\theta})\big)$), and compares $\hat{S}_f$ to the input sinogram's filtered version $S_f \leftarrow F(S)$. The loss function $L(S_f, \hat{S}_f, \hat{Y})$ is then computed, and the network weights are updated using the Adam [16] optimizer. The algorithm stops after a fixed number of iterations, N_{iter}, because in real-world scenarios the true image is unavailable, making early stopping conditions unreliable. Common ad hoc methods, such as those based on reconstruction variations [28], are thus avoided.

The loss function is defined as:

$$L(S_f, \hat{S}_f, \hat{Y}) = \|S_f - \hat{S}_f\|_1 + G(\hat{Y}), \tag{3}$$

where S_f is the filtered sinogram, $\hat{S}_f$ is the filtered sinogram from the reconstruction, $\hat{Y}$ is the reconstructed image, $G(\hat{Y})$ is a regularization term detailed in Sect. 3.3, and $\|\cdot\|_1$ is the $L1$ norm.

Algorithm 1. Reconstruction algorithm

Require: Sinogram S, measurement angles $\bar{\theta}$, initial neural network weights W_0, number of iterations N_{iter}
Ensure: Reconstructed image $\hat{Y}$

$S_f \leftarrow \mathrm{F}(S)$ ▷ Filter the sinogram
$i \leftarrow 0$
repeat
  $\hat{Y} \leftarrow \mathcal{N}_{W_i}(S)$ ▷ Reconstruct image using neural network
  $\hat{S}_f \leftarrow F\big(R(\hat{Y}, \bar{\theta})\big)$ ▷ Compute filtered sinogram from reconstruction
  $L \leftarrow \mathrm{L}(S_f, \hat{S}_f, \hat{Y})$ ▷ Compute loss as in Equation 3
  $W_{i+1} \leftarrow \mathrm{Adam}(W_i, \nabla L)$ ▷ Update weights using the Adam optimizer
  $i \leftarrow i + 1$
until $i \geq N_{\text{iter}}$
return $\hat{Y}$

Assuming that the real-world measurement operation $\hat{R}$ is close enough to the ideal Radon transform R, our underlying assumption is that if the Radon transform of our reconstruction $\hat{Y}$ is close to the measured sinogram $S \leftarrow \hat{R}(Y, \bar{\theta})$, then the reconstruction $\hat{Y}$ is also close to the true image Y:

$$R(\hat{Y}, \bar{\theta}) \approx \hat{R}(Y, \bar{\theta}) \implies \hat{Y} \approx Y. \tag{4}$$

The HTC'22 challenge also provides an image of a uniform disk with no holes. We use this disk as a mask to constrain the reconstructions to a disk area since we know the exterior is background. The test images are not necessarily centered and simply using the mask as-is might leave edges outside the mask making them impossible to reconstruct. To account for potential misalignment between the mask and the reconstructed image, we introduce two offset parameters that represent the displacement of the disk's center from the image's center. These offset parameters are optimized alongside the reconstruction, allowing the mask to shift to a better position. Instead of finding the best discrete mask position, we optimize a continuous offset to the mask's center by using bilinear interpolation to shift the mask, making gradient-based optimization possible.

3.3 Proposed Regularization

We propose a combination of regularization methods to improve the reconstruction quality. We divide the regularization methods into two categories: embedded and explicit regularization. Embedded regularization methods are integrated directly into the reconstruction pipeline, while explicit regularization methods are explicitly modeled inside the regularization term $G(\hat{Y})$ in Eq. 2. The regularization terms are combined as:

$$G(\hat{Y}) = \lambda_{\mathrm{TV}} \cdot \mathrm{TV}(\hat{Y}) + \lambda_{\mathrm{PSR}} \cdot \mathrm{PSR}(\hat{Y}) \tag{5}$$

where λ_{TV} and λ_{PSR} are hyperparameters that control the strength of the regularization terms.

Sinogram Filtering. Filtering the sinogram before solving the reconstruction problem is effective for noise suppression. Rather than directly minimizing $\|R(\hat{Y}, \bar{\theta}) - S\|_1$, we filter the sinogram with a customized ramp filter modulated by a squared sinc function [6], and minimize

$$\|F\big(R(\hat{Y}, \bar{\theta})\big) - F(S)\|_1, \tag{6}$$

where F represents the filtering operation.

The filter is parameterized by a single parameter α, which controls the amount of filtering. A smaller α approaches a pure ramp filter, while larger α progressively reduces higher-frequency content. An example of the effect of filtering is shown in Fig. 1.

Mathematically, the filter is:

$$r_\alpha(\omega) = \left|\frac{2}{\alpha}\sin\left(\frac{\alpha\omega}{2}\right)\right| \left[\frac{\sin\left(\frac{\alpha\omega}{2}\right)}{\frac{\alpha\omega}{2}}\right]^2, \tag{7}$$

where ω is the computed by taking the 1D Fourier transform of each projection. After filtering, we take the inverse Fourier transform to obtain the filtered sinogram. Thus, the sinogram filtering operation is:

$$F_\alpha(S) = \mathcal{F}^{-1}\left[r_\alpha(\mathcal{F}(S))\right], \tag{8}$$

where $\mathcal{F}$ is the Fourier transform, and $\mathcal{F}^{-1}$ is the inverse Fourier transform.

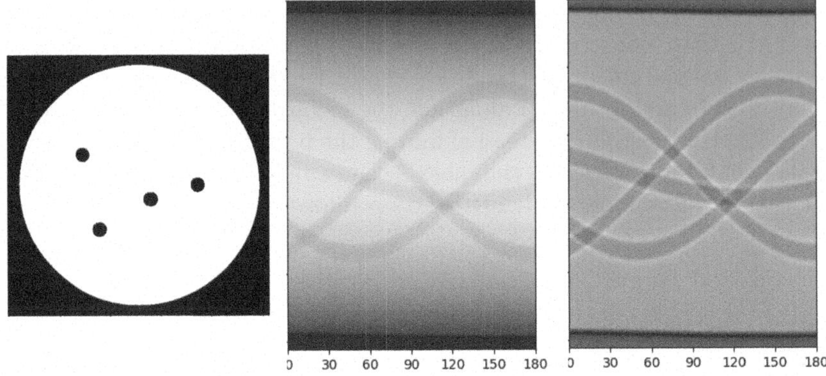

Fig. 1. The image shows the true image (left), the sinogram (middle), and the filtered sinogram ($\alpha = 5.0$) (right). The filtering reduces high-frequency content which makes the holes more prominent in the sinogram, and mitigates overfitting to noise.

Deep Image Prior. Deep Image Prior (DIP) [26] leverages the structure of convolutional neural networks as an implicit prior, favoring images with more natural statistics.

Instead of directly optimizing the pixel values of $\hat{Y}$, we instead optimize the neural network parameters so that

$$\hat{Y} = \mathcal{N}_W(S) \tag{9}$$

with

$$W = \underset{\bar{W}}{\arg\min}\left[\left|\left|F\big(R(\mathcal{N}_{\bar{W}}(S))\big) - F(S)\right|\right|_1 + G\big(\mathcal{N}_{\bar{W}}(S)\big)\right]. \tag{10}$$

where $\mathcal{N}_W$ is a convolutional neural network (CNN) with weights W, S is the input sinogram, $\hat{Y}$ is the reconstructed image, F is the sinogram filtering operator, and G is the regularization term.

By constraining the image to be a CNN output, we bias the optimization process towards solutions with natural image statistics and spatial relationships, rather than overfitting to noise in the sinogram. DIP has been successfully applied to various imaging tasks [2,3].

The backbone of our proposed DIP is based on the ConvNeXt architecture, following the winners of HTC'22 [9]. However, we avoid padding the sinogram and adding another channel to denote where the sinogram is since our use case is different and the DIP only needs to work for a single-size sinogram, thus making the input space much smaller. Using grid-search, we individually adapt the convolution parameters, such as padding and kernel size for any given sinogram.

Total Variation. Total variation (TV) regularization is widely used in CT reconstruction. It encourages piecewise constant solutions by minimizing the $L1$ norm of the image gradient [7]. We calculate the TV of an image $\hat{Y}$ using the formula:

$$TV(\hat{Y}) = \frac{\sum_{i,j} \left|(\nabla\hat{Y})_{x_{i,j}}\right| + \left|(\nabla\hat{Y})_{y_{i,j}}\right|}{N}, \tag{11}$$

where $(\nabla\hat{Y})_{x_{i,j}}$ and $(\nabla\hat{Y})_{y_{i,j}}$ are the horizontal and vertical gradients of the image $\hat{Y}$ at pixel location (i, j), respectively, and N is the number of pixels in the image.

Patch Similarity. To improve our reconstructions, we add patch similarity regularization (PSR), a patch-based regularization step that serves as an extra structural prior. We begin by training an autoencoder (AE) on image patches that match the target domain (for example, shapes similar to those in the HTC'22 challenge). This autoencoder learns a compact, low-dimensional representation of what valid patches should look like.

When optimizing, we partition the current reconstruction $\hat{Y}$ into patches and run each one through the autoencoder. We then measure the difference between the original patches and their autoencoded reconstructions. This difference is used as a penalty that nudges the overall reconstruction toward having patches that resemble the training data.

Algorithm 2. PSR penalizes patches that do not resemble the training data.

Require: Current solution $\hat{Y} \in \mathbb{R}^{n \times n}$, patch size $p \times p$, stride $s = p$
function PSR($\hat{Y}, p, s$)
  $P \leftarrow$ SplitIntoPatches($\hat{Y}, p, s$) ▷ Divide $\hat{Y}$ into patches
  $P_r \leftarrow$ EncodeDecodePatches(P) ▷ Encode and decode patches
  penalty $\leftarrow$ Mean$\Big(|P - P_r|\Big)$ ▷ Compute the mean absolute difference
  return *penalty*
end function

The autoencoder is built to capture the essential features of a patch while filtering out unusual artifacts. Figure 2 shows examples of patches and their reconstructions by the autoencoder. As can be seen, the autoencoder leaves realistic patches unchanged and removes noise and artifacts from unrealistic patches.

For training, we create eight images that mimic the HTC'22 phantoms (using the approach described in [9]) and split them into overlapping patches with stride $\lfloor \frac{\text{patch\_size}}{5} \rfloor$. We then train the autoencoder to minimize the binary cross-entropy [24] between the original and reconstructed patches. The autoencoder is trained with a batch size of 32, using the Adam optimizer with a learning rate of 0.001, for 100 epochs. The autoencoder architecture is a symmetric encoder-decoder model with three convolutional layers. We keep the latent dimension at $\lfloor \frac{\text{patch\_size}}{4} \rfloor$, which we found small enough to force the autoencoder to learn a compact representation of the patches.

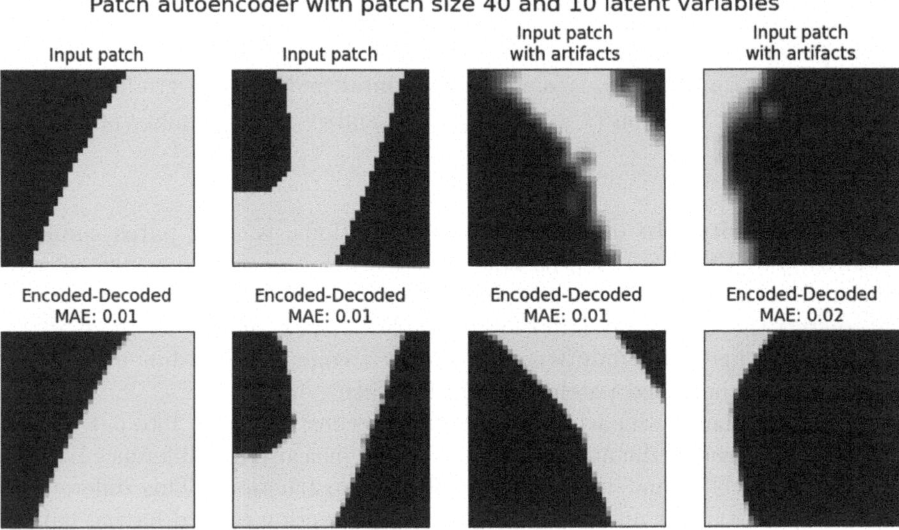

Fig. 2. Examples of patches encoded and decoded by the autoencoder. MAE is the mean absolute error between the true and reconstructed patches. Top row: Input patches. Bottom row: Reconstructed patches. The autoencoder learns to encode and decode patches that are similar to its training data. If the autoencoder is restrictive enough, it cannot encode noise or artifacts, thus providing a strong prior for the reconstruction.

4 Experiments

Our solution is implemented in Python using PyTorch, with TorchRadon [23] for differentiable Radon transforms. The entire pipeline is available in an open-source repository at https://github.com/ilmari99/data-efficient-lact.

4.1 Data

The challenge's test dataset has seven levels, each reducing the scan angle by 10°, starting from 100° (level 1) down to 30° (level 7). Three phantoms (a, b, c) were provided for each level. Figure 3 gives an overview of all 21 phantoms.

The challenge provided four training/demo images and their full sinograms to help tune algorithms. In our experiments we use these demo images with 30-degree sinograms to perform a hyperparameter search.

The evaluation metric is to find a solution $\hat{Y}$ that maximizes the Matthews correlation coefficient (MCC) [21], also known as the ϕ-coefficient, between the true image Y and the reconstructed image $\hat{Y}$.

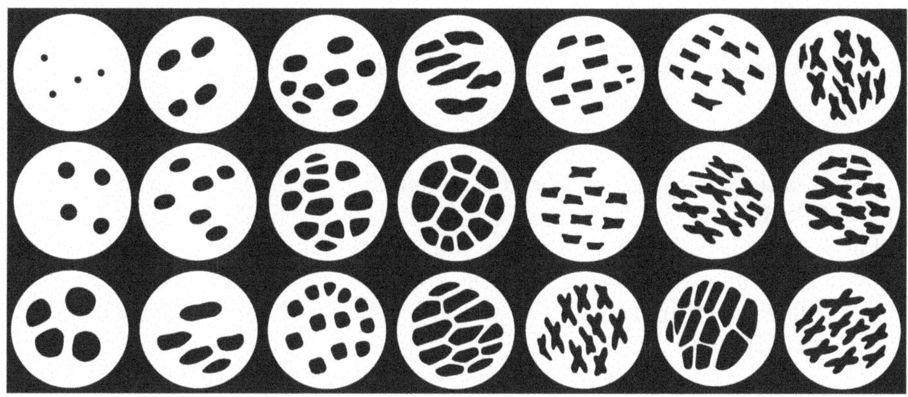

Fig. 3. All 21 phantoms used in the HTC'22 challenge. From left to right: Level 1–7. From top to bottom: Phantom a, b, c.

4.2 Tests and Validation

We conduct a hyperparameter search using the four training/demo images. To ensure an unbiased and reproducible search of the hyperparameter space, we use random sampling from a heuristically defined search space. Based on the search results, we perform an ablation study to assess the contribution of each regularization technique. For each regularization method, we identify the best-performing hyperparameter set from trials where that specific regularization was excluded.

We then compare the performance of these ablated configurations against the full model with all regularization methods included. The three best-performing hyperparameter sets are evaluated on the HTC'22 test dataset. Finally, we compare our results against the official challenge results and common baseline methods. We validate our test environment by comparing the HTC'22 winner's official results, to what our environment reported for the winner's model. We find that our environment is able to reproduce the winner's results within floating-point error, giving us confidence in our setup.

Table 1. Ablation study results. The best hyperparameter set for each ablated configuration is used. In each row, the ablated hyperparameters are in blue.

Name	DIP	F_α	λ_{TV}	λ_{PSR}	Patch size	lr	N_{iter}	MCC
No DIP	False	5.5	0.1	0.1	30	0.2	300	3.07
No Filter	True	0.0	0.5	0.1	30	0.01	1200	3.23
No TV	True	5.5	0.0	0.2	20	0.001	1200	3.44
No PSR	True	5.0	0.5	0.0	-	0.001	1200	3.44
Full	True	6.0	0.01	0.2	40	0.001	400	**3.55**

5 Results

5.1 Hyperparameter and Ablation Study

We study the importance of each regularization method by finding the best trials, where one regularization method is excluded. We then compare the MCC scores of these ablated trials against the full model that uses TV, sinogram filtering, DIP, and PSR.

The bigger the difference from the full model, the more important the regularization technique is. From the table, we can see that DIP has the biggest impact, followed by filtering, PSR, and TV. The results of the ablation study are shown in Table 1.

The HTC'22 competition allowed teams to submit as many solutions as their team had members. Thus, we also choose three of the best hyperparameter sets from our hyperparameter search and evaluate them on the test dataset. The best-performing of these three hyperparameter sets is then compared to the baseline methods and the HTC'22 results, and is referred to as "Our method" in the results.

5.2 Quantitative Results

Table 2 shows the total MCC score achieved on each level of the challenge with different methods. Our method outperforms the average HTC'22 competitors on nearly all levels and is competitive with the best challenge results. Though our method is not better than the best challenge result, it **uses only 0.006% of the data compared to the HTC'22 winner** who use 200,000 synthetic images. This makes our method a strong contender for real-world applications where data is scarce. In Table 2 "FBP" refers to filtered back projection with Otsu thresholding [32] and using a mask to constrain the reconstruction to the disk area, "No reg." refers to our method without any regularization, "Avg. in HTC'22" and "Best in HTC'22" refer to the average and best results in the HTC'22 challenge, and "Our method" refers to the best hyperparameter set in the top-3 trials.

Table 2. MCC scores of our method compared to baseline and challenge results.

Method	Level 1	Level 2	Level 3	Level 4	Level 5	Level 6	Level 7
FBP	2.79	2.75	2.52	2.38	2.44	1.82	1.69
No reg.	2.68	2.77	2.66	2.20	2.50	1.53	1.59
Avg. in HTC'22	2.79	2.69	2.62	2.57	2.61	2.28	2.07
Best in HTC'22	2.96	2.97	2.93	2.92	2.93	2.81	2.41
Our method	2.90	2.88	2.78	2.54	2.75	2.53	2.17

Figure 4 shows how our method compares to the HTC'22 submissions and FBP when accounting for the amount of data used. The x-axis shows the number of data points used in training, and the y-axis shows the MCC score on the level 7 test set. The methods we compared to are FBP, HD-DCDM [29], ConvNet [9], LPD-UNET [1], and FBP+UNET [27]. Our method is competitive with the HTC'22 submissions, despite using only 12 data points, compared to the HTC'22 winner's 200 000 synthetic images.

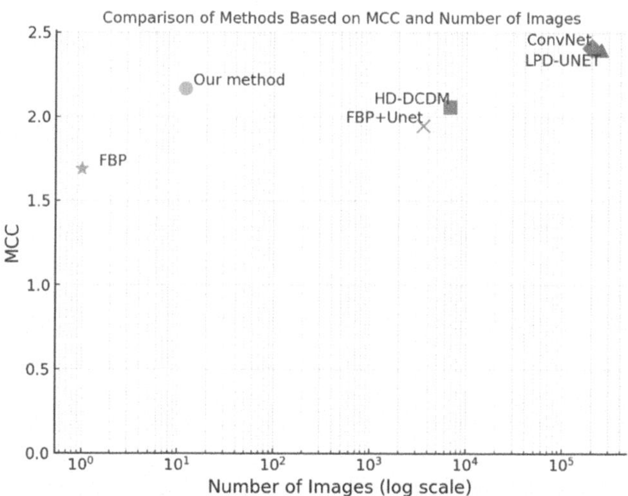

Fig. 4. Accounting for the amount of data used, our method compares favorably to HTC'22 submissions. The x-axis shows the number of data points used in training, and the y-axis shows the MCC score on the level 7 test set.

5.3 Qualitative Results

Figure 5 shows the qualitative results of our method compared to the baseline methods. As visible in the figure, FBP and our method with no regularization struggle with artifacts and noise, while our method produces clean and more accurate reconstruction.

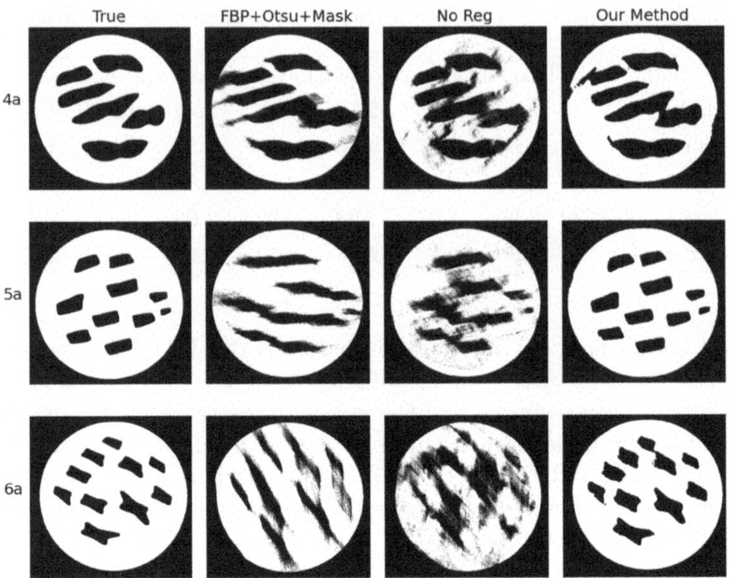

Fig. 5. Reconstruction images of the binary disk from the HTC'22 challenge. From left to right: True image, FBP + Otsu + Mask, our method with no regularization, and our full method.

5.4 Discussion

Our results demonstrate that high-quality CT reconstructions can be achieved with minimal data by using optimization-based reconstruction with good regularization. Despite using only 12 data points, our approach competes with state-of-the-art deep learning models trained on 200,000 synthetic images, making it highly applicable in data-scarce scenarios. Ablation studies highlight the importance of DIP, sinogram filtering, PSR, and TV in optimization. While our method is more computationally intensive than pre-trained deep learning models, taking approximately 30 s on an NVIDIA RTX 3060, it eliminates the need for large labeled datasets and generalizes better to real-world imaging conditions. Future work could focus on testing our method on datasets with more complicated structures and finding optimal stopping criteria for the reconstruction.

Disclosure of Interests. The authors have no competing interests to declare that are relevant to the content of this article.

References

1. Arndt, C., Denker, A., Dittmer, S., Leuschner, J., Nickel, J., Schmidt, M.: Model-based deep learning approaches to the helsinki tomography challenge 2022. Appl. Math. Modern Challenges **1**(2), 87–104 (2023)

2. Baguer, D.O., Leuschner, J., Schmidt, M.: Computed tomography reconstruction using deep image prior and learned reconstruction methods. Inverse Prob. **36**(9), 094004 (2020)
3. Barutcu, S., Aslan, S., Katsaggelos, A.K., Gürsoy, D.: Limited-angle computed tomography with deep image and physics priors. Sci. Rep. **11**(1), 17740 (2021)
4. Beylkin, G.: Discrete radon transform. IEEE Trans. Acoust. Speech Signal Process. **35**(2), 162–172 (1987)
5. Candès, E.J., Romberg, J., Tao, T.: Robust uncertainty principles: Exact signal reconstruction from highly incomplete frequency information. IEEE Trans. Inf. Theory **52**(2), 489–509 (2006)
6. csheaff: Filtered backprojection in python without explicit use of radon transform functions (2018). https://github.com/csheaff/filt-back-proj/blob/master/py/filtbackproj.py. commit OID: a174f51aa8d9c19c1c943e725f1befd83a582088
7. Estrela, V.V., Magalhaes, H.A., Saotome, O.: Total variation applications in computer vision. In: Handbook of Research on Emerging Perspectives in Intelligent Pattern Recognition, Analysis, and Image Processing, pp. 41–64. IGI Global (2016)
8. Ferreira, L.A., Beraldo, R.G., Suyama, R., Takahata, A.K., Sims, J.A.: Deep image prior with sparsity constraint for limited-angle computed tomography reconstruction. Appl. Math. Modern Challenges **1**(2), 105–125 (2023)
9. Germer, T., Robine, J., Konietzny, S., Harmeling, S., Uelwer, T.: Limited-angle tomography reconstruction via deep end-to-end learning on synthetic data. arXiv preprint arXiv:2309.06948 (2023)
10. Gilbert, P.: Iterative methods for the three-dimensional reconstruction of an object from projections. J. Theor. Biol. **36**(1), 105–117 (1972)
11. Gordon, R., Bender, R., Herman, G.T.: Algebraic reconstruction techniques (art) for three-dimensional electron microscopy and x-ray photography. J. Theor. Biol. **29**(3), 471–481 (1970)
12. Hadamard, J.: Lectures on Cauchy's problem in linear partial differential equations. Courier Corporation (2014)
13. Hounsfield, G.N.: Computerized transverse axial scanning (tomography): Part 1. description of system. Br. J. Radiol. **46**(552), 1016–1022 (1973)
14. Huang, Y., Wang, S., Guan, Y., Maier, A.: Limited angle tomography for transmission x-ray microscopy using deep learning. J. Synchrotron Radiat. **27**(2), 477–485 (2020)
15. Kim, H., Anirudh, R., Mohan, K.A., Champley, K.: Extreme few-view CT reconstruction using deep inference. arXiv preprint arXiv:1910.05375 (2019)
16. Kingma, D.P.: Adam: a method for stochastic optimization. arXiv preprint arXiv:1412.6980 (2014)
17. Liu, Z.S., Siu, W.C., Chan, Y.L.: Features guided face super-resolution via hybrid model of deep learning and random forests. IEEE Trans. Image Process. **30**, 4157–4170 (2021)
18. Liu, Z.S., Siu, W.C., Chan, Y.L.: Photo-realistic image super-resolution via variational autoencoders. IEEE Trans. Circuits Syst. Video Technol. **31**(4), 1351–1365 (2021)
19. Liu, Z.S., Wang, L.W., Li, C.T., Siu, W.C.: Hierarchical back projection network for image super-resolution. In: 2019 IEEE/CVF Conference on Computer Vision and Pattern Recognition Workshops (CVPRW), pp. 2041–2050 (2019)
20. Liu, Z.S., Wang, L.W., Li, C.T., Siu, W.C., Chan, Y.L.: Image super-resolution via attention based back projection networks. In: 2019 IEEE/CVF International Conference on Computer Vision Workshop (ICCVW), pp. 3517–3525 (2019)

21. Matthews, B.W.: Comparison of the predicted and observed secondary structure of t4 phage lysozyme. Biochimica et Biophysica Acta (BBA)-Protein Struct. **405**(2), 442–451 (1975)
22. Meaney, A., de Moura, F.S., Juvonen, M., Siltanen, S.: Helsinki tomography challenge 2022: description of the competition and dataset. Appl. Math. Mod. Challenges **1**(2), 170–201 (2023)
23. Ronchetti, M.: Torchradon: Fast differentiable routines for computed tomography. arXiv preprint arXiv:2009.14788 (2020)
24. Shannon, C.E.: A mathematical theory of communication. Bell Syst. Tech. J. **27**(3), 379–423 (1948)
25. Shen, C., Gonzalez, Y., Chen, L., Jiang, S.B., Jia, X.: Intelligent parameter tuning in optimization-based iterative CT reconstruction via deep reinforcement learning. IEEE Trans. Med. Imaging **37**(6), 1430–1439 (2018)
26. Ulyanov, D., Vedaldi, A., Lempitsky, V.: Deep image prior. In: Proceedings of the IEEE Conference on Computer Vision and Pattern Recognition, pp. 9446–9454 (2018)
27. Wang, C., Li, J.: A deep learning approach using boundary shape information for limited-angle tomography reconstruction. Appl. Math. Mod. Challenges **1**(2), 219–231 (2023)
28. Wang, H., Li, T., Zhuang, Z., Chen, T., Liang, H., Sun, J.: Early stopping for deep image prior. arXiv preprint arXiv:2112.06074 (2021)
29. Wang, J., et al.: HD-DCDM: hybrid-domain network for limited-angle computed tomography with deconvolution and conditional diffusion model. Appl. Math. Mod. Challenges **1**(2), 202–218 (2023)
30. Wang, J., Liang, J., Cheng, J., Guo, Y., Zeng, L.: Deep learning based image reconstruction algorithm for limited-angle translational computed tomography. PLoS ONE **15**(1), e0226963 (2020)
31. Xu, Y., Han, S., Wang, D., Wang, G., Maltz, J.S., Yu, H.: Hybrid u-net and swin-transformer network for limited-angle cardiac computed tomography. Phys. Med. Biol. **69**(10), 105012 (2024)
32. Yousefi, J.: Image binarization using OTSU thresholding algorithm. Ontario, Canada: University of Guelph **10** (2011)
33. Zhang, H., et al.: Image prediction for limited-angle tomography via deep learning with convolutional neural network. arXiv preprint arXiv:1607.08707 (2016)
34. Zhou, B., Lin, X., Eck, B.: Limited angle tomography reconstruction: synthetic reconstruction via unsupervised sinogram adaptation. In: Chung, A., Gee, J.C., Yushkevich, P.A., Bao, S. (eds.) IPMI 2019. LNCS, vol. 11492, pp. 141–152. Springer, Cham (2019). https://doi.org/10.1007/978-3-030-20351-1_11

Fairness

Are Generative Models Fair? A Study of Racial Bias in Dermatological Image Generation

Miguel López-Pérez[1(✉)], Søren Hauberg[2], and Aasa Feragen[2]

[1] Instituto Universitario de Investigación en Tecnología Centrada en el Ser Humano, Universitat Politècnica de València, Valencia, Spain
mlopper3@upv.es

[2] Technical University of Denmark, Kongens Lyngby, Denmark

Abstract. Racial bias in medicine, such as in dermatology, presents significant ethical and clinical challenges. This is likely to happen because there is a significant underrepresentation of darker skin tones in training datasets for machine learning models. While efforts to address bias in dermatology have focused on improving dataset diversity and mitigating disparities in discriminative models, the impact of racial bias on generative models remains underexplored. Generative models, such as Variational Autoencoders (VAEs), are increasingly used in healthcare applications, yet their fairness across diverse skin tones is currently not well understood. In this study, we evaluate the fairness of generative models in clinical dermatology with respect to racial bias. For this purpose, we first train a VAE with a perceptual loss to generate and reconstruct high-quality skin images across different skin tones. We utilize the Fitzpatrick17k dataset to examine how racial bias influences the representation and performance of these models. Our findings indicate that VAE performance is, as expected, influenced by representation, i.e. increased skin tone representation comes with increased performance on the given skin tone. However, we also observe, even independently of representation, that the VAE performs better for lighter skin tones. Additionally, the uncertainty estimates produced by the VAE are ineffective in assessing the model's fairness. These results highlight the need for more representative dermatological datasets, but also a need for better understanding the sources of bias in such model, as well as improved uncertainty quantification mechanisms to detect and address racial bias in generative models for trustworthy healthcare technologies.

Keywords: AI in medicine · Fairness · Dermatology · Racial bias

1 Introduction

Racial bias in medicine has been widely documented [4,24]. This bias is often inherited, or even amplified, by deep learning methods: First, lack of representation can lead to loss of performance and overfitting for underrepresented groups.

J. Petersen and V. A. Dahl (Eds.): SCIA 2025, LNCS 15726, pp. 389–402, 2025.
https://doi.org/10.1007/978-3-031-95918-9_27

It has been shown across tasks and datasets in medical imaging that low subgroup representation can be associated with low subgroup performance [14]. Low representation does not, however, always lead to bias [18], nor is it the only mechanism leading to bias: Differences in data quality, systematic label errors such as group-dependent over- or underdiagnosis across groups, or group-dependent prediction difficulty can also lead to biased models [5,19,25]. Such limitations can hinder the performance of machine learning algorithms in specific subpopulations, particularly among underrepresented groups. Deploying AI-driven tools in healthcare that may exhibit detrimental performance or consequences for these subgroups is not only unfair but also poses serious risks. For example, an established system that achieves high accuracy may experience an unexpected and dramatic drop in performance when deployed and tested on a different subgroup of the population. Therefore, it is crucial to detect and mitigate these biases to ensure ethical and trustworthy AI in medicine [15].

In this paper, we contribute a study of how under- and overrepresentation affects the performance of generative models, exemplified by a VAE. We show that subgroup underrepresentation is associated with subgroup underperformance, but we also obtain a general lower performance on darker skin tones. Finally, we show that the inherent uncertainty quantification built into the VAE is ineffective at capturing underrepresentation and its associated loss of performance.

2 Related Work

In dermatology, racial bias is well known, and often attributed to a significant underrepresentation of darker skin tones in clinical data [1,2,11]. Despite promising results in machine learning to accurately classify skin conditions, the existence of racial bias against darker skin tones has been reported [1,3]. This racial bias may also exacerbate accuracy disparities between light and dark skin tones among non-specialists when machine learning is deployed in the clinical practice to assist physicians [7]. Addressing this issue is essential to develop and deploy safe tools in clinical practice.

Some efforts to overcome racial bias have focused on creating large-scale dermatology image datasets that include metadata on skin tone, typically measured using the Fitzpatrick Skin Type (FST) scale. This scale goes between 1 to 6 (from the lightest to the darkest). The reference dataset to assess racial bias in dermatological images is the Fitzpatrick17k dataset, which encompasses a wide range of Fitzpatrick Skin Types (FST 1-6) [8]. While the dataset includes both skin condition and skin type labels, it remains imbalanced, with darker skin tones (FST 4-6) being underrepresented (see Fig. 2). The same study identified that existing dermatology AI models exhibit significant biases, particularly underperforming on darker skin tones. Similar observations were made on the Diverse Dermatology Images dataset [3]. Another recent example is the PASSION dataset [6], which focuses on individuals from Sub-Saharan countries and includes FST 3-6, the most common skin type in this region. This dataset aims to address the limitations of previous datasets, which have predominantly focused on lighter skin tones.

2.1 Generative Models and Algorithmic Bias

A different attempt to mitigate racial bias was to generate new synthetic samples using large generative models based on diffusion (DALLåE 2) [21]. Generative models are a type of machine learning model that aims to learn the underlying structure of data, allowing them to produce new data points from this distribution. Here, the authors achieved increased accuracy for underrepresented groups by using generative models to balance datasets. However, it remains unclear how racial bias may propagate through generative models. For instance, generative models are able to introduce racial bias even when trained on a dataset with balanced representation, as illustrated qualitatively in Fig. 1.

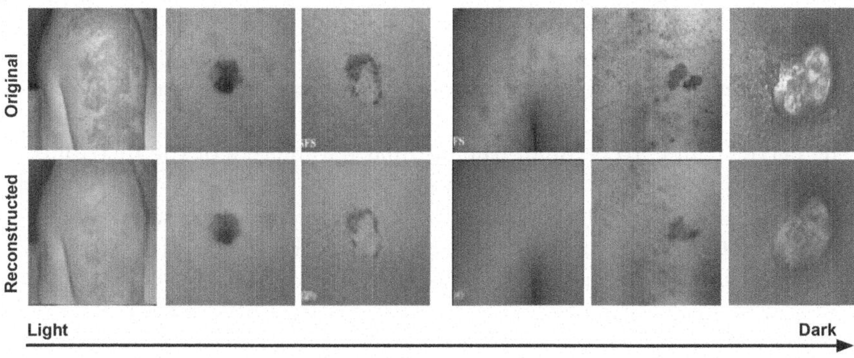

Fig. 1. Example predictions from our VAE model trained on a balanced subset of the Fitzpatrick17k dataset, comprising 50/50 light and dark skin tones. Notably, predictions for lighter tones are more accurate and better preserve the lesion compared to those for darker tones.

The focus of previous works has been on debiasing discriminative models for the classification of skin conditions. In this paper, we take a step back to address the following questions:

- *How does the performance of a deep generative model, trained to generate clinical images of dermatology, vary across different skin tones?*
- *How does this variation depend on skin tone representation?*
- *How does the inherent uncertainty quantification of the generative model perform in detecting lower subgroup representation, which is associated with lower subgroup performance?*

The way generative models encode and represent data is crucial for understanding their behavior in the context of racial bias. Although these models can generate new synthetic examples, further study is needed to determine whether they reproduce existing biases in this context. In this study, we explore the racial bias of generative models in clinical dermatology images. We utilize a Variational Autoencoder (VAE) [12,20], a popular generative model with an autoencoder architecture that comes with a probability distribution in the latent space. We use the VAE to generate and reconstruct skin images across different skin types. For this purpose, we employ the Fitzpatrick17k dataset, which offers a varied FST scale, enabling us to investigate how racial bias may affect the performance of VAEs. Finally, we propose new directions toward fair generative models.

3 Methods

3.1 Deep Learning Model

VAEs [12,20] are a popular class of deep generative models. They aim to learn the distribution of a given set of images $\mathbf{X} = \{\mathbf{x}\}$ and generate new samples from this distribution $\{\mathbf{x}_*\}$. It leverages an autoencoder-like architecture combining an *stochastic encoder* $\mathrm{q}_\phi(\mathbf{x}|\mathbf{z})$ (parameterized by ϕ), and a *stochastic decoder* $\mathrm{p}_\theta(\mathbf{x}|\mathbf{z})$ (parameterized by θ). The word *stochastic* here means that the encoder and decoder are probability distributions (in contrast to deterministic autoencoders).

The encoder aims to estimate the optimal parameters of the distribution of the latent representation $\mathbf{z}$ given an image $\mathbf{x}$. Here, we use a Gaussian distribution for this, $\mathrm{q}_\phi(\mathbf{z}|\mathbf{x}) = \mathcal{N}(\mu_\phi(\mathbf{x}), \mathrm{diag}(\sigma^2_\phi(\mathbf{x}))$. The decoder aims to reconstruct this image from the stochastic representation $\mathbf{z}$. Here, we utilize the following likelihood model for the decoder, $\mathrm{p}_\theta(\mathbf{x}|\mathbf{z}) = \mathcal{N}(\mathbf{x}|f_\theta(\mathbf{z}), \mathbf{I})$. Furthermore, we impose a prior distribution to the latent variable $\mathrm{p}(\mathbf{z}) = \mathcal{N}(\mathbf{z}|0, \mathbf{I})$. We perform variational inference to obtain the optimal variational parameters, i.e., $\{\theta, \phi\}$. For this purpose, we maximize a lower bound on the marginal log-likelihood (ELBO):

$$\mathrm{ELBO}(\theta, \phi; \mathbf{x}) = \mathbb{E}_{q_\phi(\mathbf{z}|\mathbf{x})}\left[\log p_\theta(\mathbf{x}|\mathbf{z})\right] - \mathrm{KL}(q_\phi(\mathbf{z}|\mathbf{x})||p(\mathbf{z})), \tag{1}$$

where $\mathrm{KL}(q_\phi(\mathbf{z}|\mathbf{x})||p(\mathbf{z}))$ is the Kullback-Leibler (KL) divergence between the approximate posterior $q_\phi(\mathbf{z}|\mathbf{x})$ and the prior $p(\mathbf{z})$. The first term in Eq. 1 corresponds to the negative of the reconstruction error. In this case, it is the mean square error (MSE) because of the Gaussian likelihood. The second term uses KL divergence to encourage fidelity to the prior distribution. This term can be seen as a regularizer. The training objective is to maximize the ELBO given the observational data, and the model can be optimized with respect to the variational ϕ, and generative parameters θ. To estimate the distribution $\mathrm{p}_\theta(\mathbf{z}|\mathbf{x})$, we utilize Monte Carlo sampling with the reparameterization trick to obtain samples of the latent variable $\mathbf{z}$, i.e., $\mathbf{z} = \mu(\mathbf{x}) + \epsilon \odot \sigma(\mathbf{x})$, $\epsilon \sim \mathcal{N}(0, \mathbf{I})$.

3.2 Perceptual Loss

When working with high-fidelity images, using a standard VAE may not be the best approach. The generated images are prone to be blurry [10]. This is because using a pixel-by-pixel loss that does not capture spatial correlations between two images. A typical example is the same image translated by a few pixels. To the human eyes, they are practically the same image. However, it will result in high pixel-by-pixel loss.

This drawback of VAEs working with images has been solved by feature perceptual loss [9]. This loss between two images is defined by the difference between the hidden maps in a pretrained convolutional neural network Φ. The main idea behind this approach is that the feature maps of this network have sensible information about spatial correlation in the images. Namely, it will approach better to the human vision. As a result, we can get a better visual quality output image. The loss function of the VAE with the perceptual loss used in this work is as follows,

$$\mathcal{L} = \text{ELBO} + \frac{1}{2C^l W^l H^l} \sum_{c=1}^{C^l} \sum_{w=1}^{W^l} \sum_{h=1}^{H^l} (\Phi(\mathbf{x})^l_{c,w,h} - \Phi(f_\theta(x))_{c,w,h})^2, \tag{2}$$

where C^l, H^l, W^l are the channels, height, and widht of the l-th feature map of the network, respectively.

3.3 Fitzpatrick17k Dataset

For our study, we utilize the Fitzpatrick17k dataset [8], which is a large publicly available dataset with associated information on skin tone represented using the Fitzpatrick scale. This dataset was sourced from two open-source dermatology atlases, and comprises 16,577 images with corresponding skin condition and Fitzpatrick labels. The skin conditions have associated labels at different levels of granularity. The fine-grained labels for specific conditions are grouped into 9 superclasses, which are further consolidated into 3 super-classes. The Fitzpatrick scale ranges from 1 to 7, with -1 representing missing values.

The Fitzpatrick skin-type labels were assigned by a team of human annotators from Scale AI. The labels were obtained by a consensus process from two to five annotators. They carried out a process to achieve the desired level of agreement, resulting in 72,277 annotations in total. The distribution of the Fitzpatrick labels of this dataset is depicted in Fig. 2. We observe that darker skin tones are notably underrepresented also in this dataset, and we will need to subsample to create balanced scenarios.

3.4 Experimental Details

Neural Network Architecture. The encoder and decoder of the VAE utilize residual blocks along with down-sampling and up-sampling, respectively. Each convolution block employs batch normalization and ELU activation for stable training. The resulting latent space has shape $8 \times 8 \times 64$. For the perceptual loss in Eq. (2), we used a VGG19 pretrained on ImageNet [23]. The perceptual loss was computed using the output of every convolutional layer, 16 layers in this case.

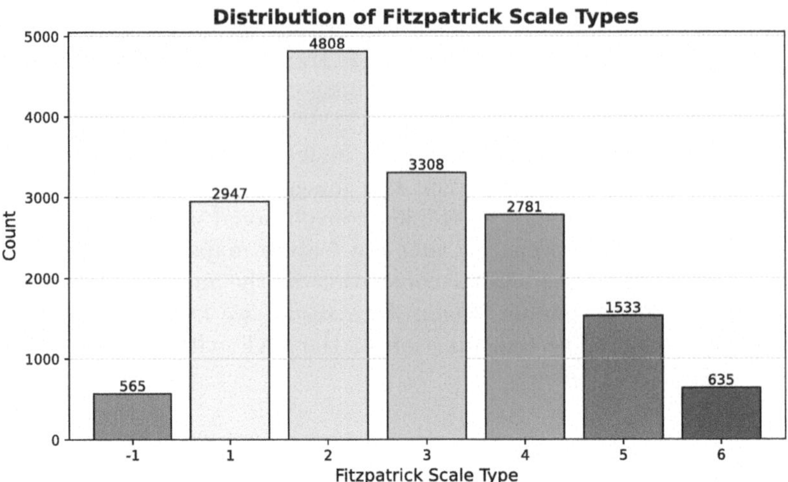

Fig. 2. Distribution of samples according to the FST in the Fitzpatrick17k dataset [8]. The label '−1' represents missing values.

Optimization Details. We train the models for 15 epochs after which we observed a convergence of the training loss. The optimizer was Adam with a learning rate of 10^{-4}. The training was performed in mini-batches of 64 samples.

Implementation. We utilize the implementation of a VAE with perceptual loss available at GitHub[1]. The code was implemented in PyTorch 2.0.1. The code was executed using one GPU NVIDIA GeForce RTX 3090 to train and evaluate the models.

Sensitive Groups. We divided the Fitzpatrick17k into two different groups. One with lighter skin, comprising FST 1-2, and one with darker skin, comprising FST 4-6. Images without a Fitzpatrick label were discarded, along with those labeled as FST 3-4, to avoid ambiguities, as these categories lie on the boundary between lighter and darker skin tones.

Experimental Setting. We imitate the experimental design from [14], originally created to showcase how subgroup performance depends on subgroup representation for image classification tasks. The experiment is repeated ten independent times. For each run, two test sets of 500 samples each are sampled from the dataset, one with lighter images (FST 1-2) and the other with darker ones (FST 5-6). Then, with the remaining images, we independently sample three training set configurations (with replacement) of 1668 samples. The 'Dataset A – Light' has 100% of lighter images, the 'Dataset B – Mixed' set has 50/50, and the 'Dataset C – Black' set has 100% of darker images.

[1] https://github.com/LukeDitria/CNN-VAE (accessed 2024-02-17).

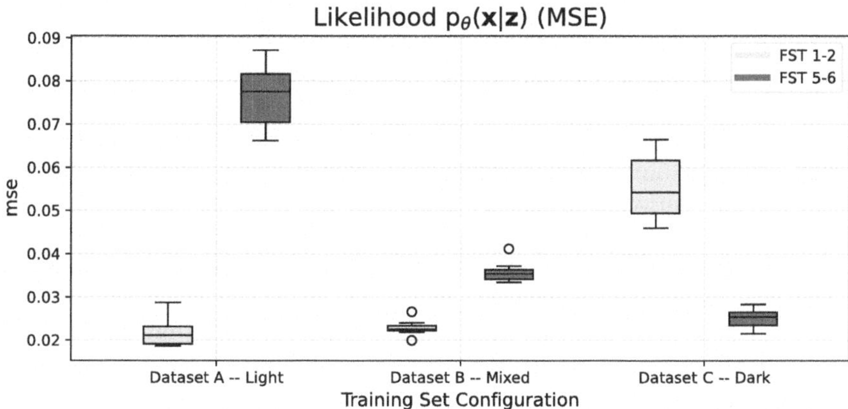

Fig. 3. Likelihood or Mean Square Error (MSE) of the VAEs in the test sets.

Metrics. To monitor how the VAE performance on subgroups depends on representation, we study two different metrics across the different test sets and training runs: To monitor reconstruction performance, we utilize the likelihood, i.e., the MSE or reconstruction error. To monitor the inherent uncertainty quantified by the model, we measure the latent standard deviation given by the stochastic encoder.

4 Results

Figure 3 shows boxplots of the likelihood (or MSE) for our VAE with perceptual loss across the three different training set configurations. For each training set configuration, we plot the results for the both test sets: light skin (FST 1-2) and dark skin (FST 5-6). The VAE generally performs well with an MSE below 0.07, which means that the neural network generally assigns a high likelihood to the data.

4.1 Subgroup Reconstruction Performance Depends on Subgroup Representation

Looking at groupwise MSE, we see the expected trend, where the MSE decreases when the representation increases: When moving towards the right, we increase the representation of dark skin, and the dark skin MSE goes down. When moving towards the left, we increase the representation of light skin, and the light skin MSE goes down.

4.2 Overall Bias Against Darker Skin

However, the boxplot also shows an overall performance gap between the two population subgroups: The MSE is generally lower for the light skin group. This is noticeable under the 'Dataset B – Mixed', when training with an equal 50/50 sample distribution from both subgroups. Here, a significant performance gap still persists between the two groups. But this is further emphasized when considering the more extreme training scenarios: Comparing the performance for dark skin (FST 5-6) using 'Dataset A - Light' (only trained with FST 1-2) with the performance for light skin (FST 1-2) using 'Dataset C – Dark' (only trained with FST 5-6), there is a performance gap showing that the worst case performance for the dark skinned group is noticeably lower than the worst case performance for the light skinned group.

Even for the best case performances, comparing the performance for dark skin (FST 5-6) using 'Dataset C – Dark' (only trained with FST 5-6) with the performance for light skin (FST 1-2) using 'Dataset A – Light' (only trained with FST 1-2), we still see a slightly better performance on the light skinned group.

Overall, our results indicate a general underperformance on the dark skinned group that does not seem to be explained only by lack of representation.

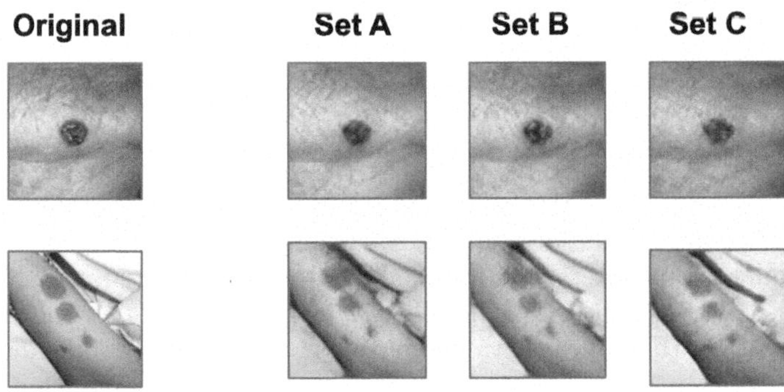

Fig. 4. Example reconstruction of lighter skin tones. Reconstructions are produced using three training set configurations: 'A', 'B', and 'C'.

To complement these results with qualitative evaluation, we include visualizations of the predicted reconstructions in Figs. 4 and 5. We include reconstructions produced by each of the three different training scenarios to illustrate the effect of the training set representation on subgroup performance. For lighter skin tones (see Fig. 4), the reconstructions do not show significant visual variation. However, for darker skin tones (see Fig. 5), the visibly highest quality reconstructions are obtained with 'Dataset C – Dark' (where 100% of the samples are dark). Notably, the color tone changes as more dark skin samples are introduced into the training set. These observations support that darker skin

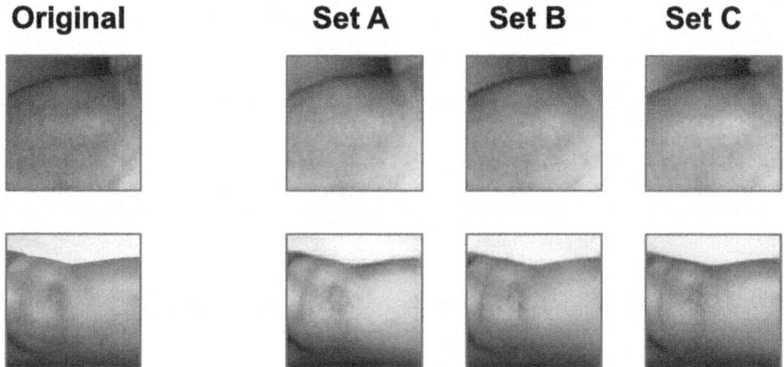

Fig. 5. Example reconstruction of darker skin tones. Reconstructions are produced using three training set configurations: 'A', 'B', and 'C'.

tones in clinical skin images are more sensitive to the training set composition, and the recovery of skin tone and details of the injury is more challenging.

4.3 Inherent VAE Uncertainty Quantification Does Not Serve as a Red Flag Warning Against Subgroup Underrepresentation and Underperformance

Figure 6 depicts the values of the averaged latent standard deviation of the latent variable $\mathbf{z}$ for our VAE with perceptual loss in the three different configurations of the training set. This figure has been made using the same process as in the previous Fig. 3. Here, while we observe a reduction in the standard deviation as we include more dark skinned images in the training set, we do not observe any differences between the estimated standard deviation across the two population subgroups at test time. Instead, the standard deviation follows a similar distribution for both test sets. This suggests that the inherent uncertainty quantification mechanism of the VAE to measure the uncertainty of the latent representation is not a proxy for subgroup representation, and therefore is unlikely to be useful to monitor the fairness of the model since the performance, assessed with the likelihood in Fig. 3, has a gap between the two test sets which is not replicated in Fig. 6.

5 Discussion and Conclusion

Fairness in AI systems for healthcare is urgently needed to ensure that trustworthy AI systems are deployed in clinical practice. Clinical images of dermatology have been observed that show racial bias, as they often fail to adequately represent the wide range of skin tones present in the population. This bias has been widely studied in the context of predictive AI for computer-aided diagnosis. In this study, we show initial results on how this bias may affect deep generative models.

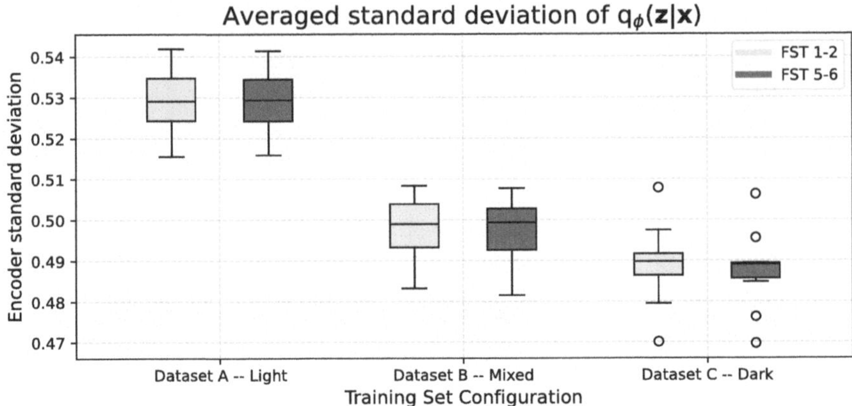

Fig. 6. Averaged standard deviation of the latent variable of the VAEs in the test sets.

5.1 Overall Findings

We have shown, as expected, that **subgroup representation affects VAE reconstruction performance**: We observe that the VAE assigns a lower likelihood to reconstructions of skin tones not seen frequently during training. We do, however, also make some additional unexpected observations:

First, we observe that **the latent standard deviation of the VAE does not provide any information about the subgroup underrepresentation or bias** in the model, and it can therefore not be used as a proxy to detect fairness problems. This raises concerns about our abilities to discover potential biases when these models are deployed to generate synthetic images of skin pathologies. This leaves an interesting future research question of whether it is possible to develop enhanced uncertainty mechanisms that can effectively capture and relate to the fairness of VAEs. This would allow those uncertainty measures to work as algorithmic bias warning signs.

Second, the VAE model proposed in this study is **generally biased against darker skin tones**. Even when the training set is balanced with a 50/50 distribution of the two subgroups, the VAE is still biased against darker skin tones. In general, lighter skin tones appear to be easier to reconstruct. This suggests that there is more at play in dermatological AI bias than just a simple representation bias.

5.2 The Reason for the Overall Performance Disparity is Unclear

Looking back at the examples from Figs. 4 and 5, one might hypothesize that a potential reason for the overall performance gap between light and dark skinned test sets comes from differences in the subgroups' diseases: The light skinned examples in Fig. 4 showcase clearly delineated lesions, whereas the dark skinned examples in Fig. 5 shows more diffuse, textured rashes whose images might be more difficult for the model to reconstruct. To assess whether this could explain the performance gap, we therefore investigated the distribution of diseases between the different skin tones.

Recall that diagnostic labels were given at three different levels of granularity, ranging from coarse, to middle, to fine-grained. At each level of granularity, we assess whether there are major differences in representation between the two groups. The scatter plot shown in Fig. 7 compares the distribution of condition prevalences for light (x-axis) versus dark (y-axis) skin tone groups, across the diagnostic hierarchy. We observe that at higher levels of the hierarchy, the distribution shows minimal differences across subgroups. However, at the finest-grained level, the distribution of conditions varies substantially between subgroups.

To see how this relates to performance across these conditions, we show the MSE for the fine-grained labels across the three different training scenarios in Fig. 8. In general, we observe that the behavior is uneven. 'Dataset B – Mixed' represents the fairest scenario, as the labels are closer to the $y = x$ line. However, performance remains biased against darker skin tones.

This leaves us with an open question: Can we disentangle the influence of skin tone versus specific skin condition on the results during image generation?

5.3 Outlook

Second, and perhaps even more importantly: How do we assess the quality of datasets for algorithmic fairness analysis? We have already shown that there may be hidden stratifications of our skin tone subgroups from the point of view of diagnoses, which opens further questions about whether these hidden groups can also work as shortcuts for models, further complicating the picture [16,17]. But skin tone is in itself a complex and controversial attribute, and recent work has focused on whether measures such as Fitzpatrick – which we rely on in this paper – are even sufficiently correct, or measuring the appropriate features for racial bias [11,13,22].

As a final note: While we have made an effort here to study the effect of how different levels of training set representation affects models, such experiments are impossible within most publicly available dermatological datasets because virtually all images are of light-skinned subjects. As such, we note that 'Dataset A – Light' (composed by only FST 1-2), which demonstrates the highest possible performance disparity in Fig. 3, is the typical scenario encountered in real life. This just highlights that there are, indeed, problems yet to be solved, and that we should expect skin tone biases in generative AI models.

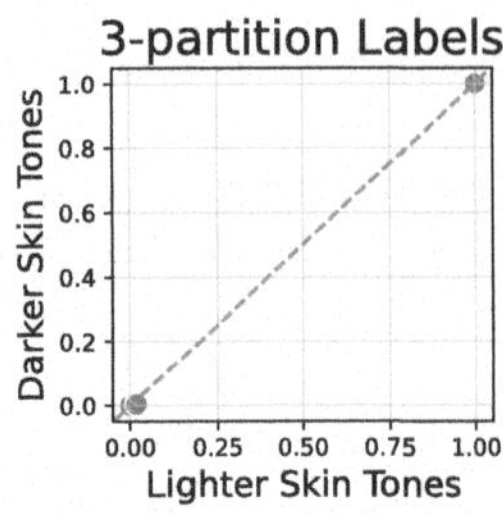

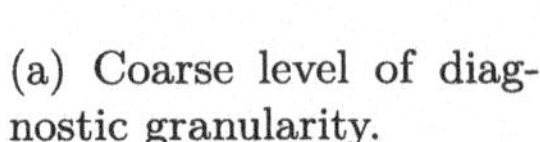
(a) Coarse level of diagnostic granularity.

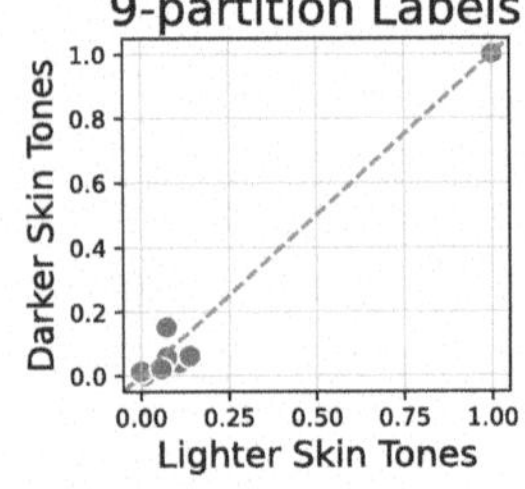

(b) Medium level of diagnostic granularity.

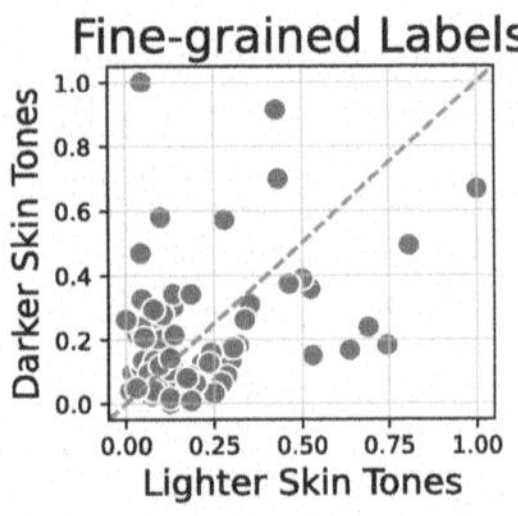

(c) Fine-grained level of diagnostic granularity.

Fig. 7. Normalized count of skin conditions categorized across varying levels of diagnostic granularity within the hierarchy of labels.

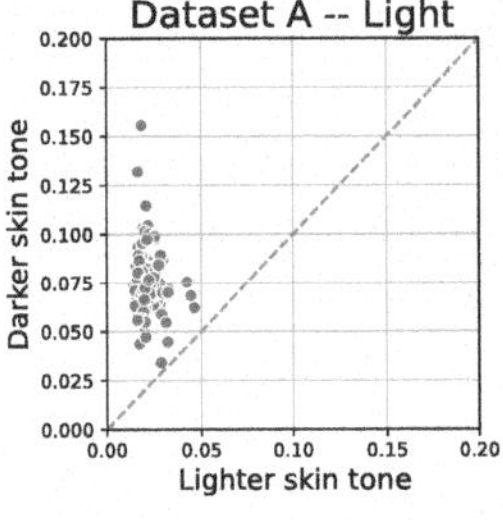

(a) Coarse level of diagnostic granularity.

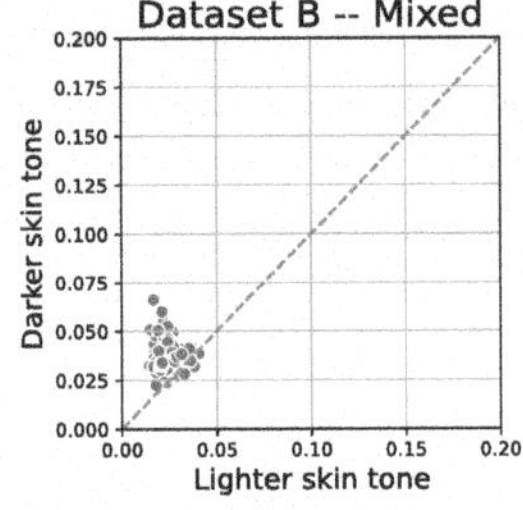

(b) Medium level of diagnostic granularity.

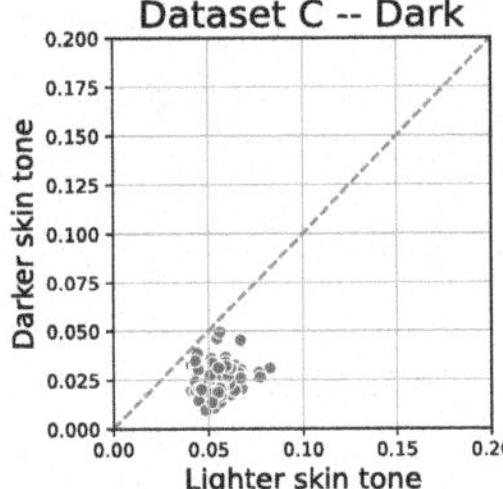

(c) Fine-grained level of diagnostic granularity.

Fig. 8. Averaged MSE of skin conditions based on fine-grained labels.

Acknowledgements. This research is partly funded by José Castillejo mobility scholar-ship (CAS23/00387) granted to MLP by the Spanish Government to visit DTU, MLP has also been supported by MICIU/AEI/10.13039/501100011033 and the European Unions "NextGenerationEU"/PRTR through grant JDC2022-048318-I and by "Ayuda a Primeros Proyectos de Investigación" (PAID-06-22), Vicerrectorado de Investigación de la Universitat Politècnica de València (UPV). The work was partly funded by the Novo Nordisk Foundation through the Center for Basic Machine Learning Research in Life Science (NNF20OC0062606). AF was supported by research grant (0087102) from the Novo Nordisk Foundation. SH was supported by a research grant (42062) from VILLUM FONDEN and from the European Research Council (ERC) under the European Unions Horizon programme (grant agreement 101125993).

Disclosure of Interests. The authors have no competing interests to declare that are relevant to the content of this article.

References

1. Almuzaini, A.A., Dendukuri, S.K., Singh, V.K.: Toward fairness across skin tones in dermatological image processing. In: 2023 IEEE 6th International Conference on Multimedia Information Processing and Retrieval (MIPR), pp. 1–7 (2023). https://doi.org/10.1109/MIPR59079.2023.00030
2. Benmalek, A., Cintas, C., Tadesse, G.A.: Impact of skin tone diversity on out-of-distribution detection methods in dermatology. In: International Conference on Medical Image Computing and Computer-Assisted Intervention (2024)
3. Daneshjou, R., et al.: Disparities in dermatology AI performance on a diverse, curated clinical image set. Sci. Adv. **8**(31), eabq6147 (2022)
4. Dehon, E., Weiss, N., Jones, J., Faulconer, W., Hinton, E., Sterling, S.: A systematic review of the impact of physician implicit racial bias on clinical decision making. Acad. Emerg. Med. **24**(8), 895–904 (2017)
5. Gichoya, J.W., et al.: Ai recognition of patient race in medical imaging: a modelling study. Lancet Digit. Health **4**(6), e406–e414 (2022)
6. Gottfrois, P., et al.: Passion for dermatology: bridging the diversity gap with pigmented skin images from sub-saharan africa. In: International Conference on Medical Image Computing and Computer-Assisted Intervention, pp. 703–712. Springer (2024)
7. Groh, M., et al.: Deep learning-aided decision support for diagnosis of skin disease across skin tones. Nat. Med. **30**(2), 573–583 (2024)
8. Groh, M., et al.: Evaluating deep neural networks trained on clinical images in dermatology with the fitzpatrick 17k dataset. In: Proceedings of the IEEE/CVF Conference on Computer Vision and Pattern Recognition, pp. 1820–1828 (2021)
9. Hou, X., Shen, L., Sun, K., Qiu, G.: Deep feature consistent variational autoencoder. In: 2017 IEEE Winter Conference on Applications of Computer Vision (WACV), pp. 1133–1141. IEEE (2017)
10. Huang, H., He, R., Sun, Z., Tan, T., et al.: Introvae: introspective variational autoencoders for photographic image synthesis. In: Advances in Neural Information Processing Systems, vol. 31 (2018)
11. Kalb, T., Kushibar, K., Cintas, C., Lekadir, K., Diaz, O., Osuala, R.: Revisiting skin tone fairness in dermatological lesion classification. In: Workshop on Clinical Image-Based Procedures, pp. 246–255. Springer (2023)
12. Kingma, D.P., Welling, M.: Auto-encoding variational bayes. In: International Conference on Learning Representations (ICLR) (2014)
13. Kinyanjui, N.M., et al.: Fairness of classifiers across skin tones in dermatology. In: International Conference on Medical Image Computing and Computer-Assisted Intervention, pp. 320–329. Springer (2020)
14. Larrazabal, A.J., Nieto, N., Peterson, V., Milone, D.H., Ferrante, E.: Gender imbalance in medical imaging datasets produces biased classifiers for computer-aided diagnosis. Proc. Natl. Acad. Sci. **117**(23), 12592–12594 (2020)
15. Li, B., et al.: Trustworthy AI: from principles to practices. ACM Comput. Surv. **55**(9), 1–46 (2023)
16. Oakden-Rayner, L., Dunnmon, J., Carneiro, G., Ré, C.: Hidden stratification causes clinically meaningful failures in machine learning for medical imaging. In: Proceedings of the ACM Conference on Health, Inference, and Learning, pp. 151–159 (2020)
17. Olesen, V., Weng, N., Feragen, A., Petersen, E.: Slicing through bias: explaining performance gaps in medical image analysis using slice discovery methods. In: MICCAI Workshop on Fairness of AI in Medical Imaging, pp. 3–13. Springer (2024)

18. Petersen, E., et al.: Feature robustness and sex differences in medical imaging: a case study in MRI-based Alzheimers disease detection. In: International Conference on Medical Image Computing and Computer-Assisted Intervention, pp. 88–98. Springer (2022)
19. Petersen, E., Holm, S., Ganz, M., Feragen, A.: The path toward equal performance in medical machine learning. Patterns **4**(7) (2023)
20. Rezende, D.J., Mohamed, S., Wierstra, D.: Stochastic backpropagation and approximate inference in deep generative models. In: Xing, E.P., Jebara, T. (eds.) Proceedings of the 31st International Conference on Machine Learning. Proceedings of Machine Learning Research, Beijing, China, vol. 32, pp. 1278–1286. PMLR (2014)
21. Sagers, L.W., Diao, J.A., Groh, M., Rajpurkar, P., Adamson, A., Manrai, A.K.: Improving dermatology classifiers across populations using images generated by large diffusion models. In: NeurIPS 2022 Workshop on Synthetic Data for Empowering ML Research (2022)
22. Schumann, C., Olanubi, F., Wright, A., Monk, E., Heldreth, C., Ricco, S.: Consensus and subjectivity of skin tone annotation for ml fairness. In: Advances in Neural Information Processing Systems, vol. 36 (2024)
23. Simonyan, K., Zisserman, A.: Very deep convolutional networks for large-scale image recognition (2015). https://arxiv.org/abs/1409.1556
24. Williams, D.R., Wyatt, R.: Racial bias in health care and health: challenges and opportunities. JAMA **314**(6), 555–556 (2015)
25. Zhou, Y., et al.: Radfusion: benchmarking performance and fairness for multimodal pulmonary embolism detection from CT and EHR. arXiv preprint arXiv:2111.11665 (2021)

Low·Level and Physics·Based Vision

Demosaicing and Neural Network Based Image Reconstruction in the Presence of Noise

Gustav Emil Mark-Hansen, Frederik Henriques Altmann, Christian Igel, and Ankit Kariryaa(✉)

Department of Computer Science, University of Copenhagen, Copenhagen, Denmark
{geha,igel,ak}@di.ku.dk, frederik@altmann.dk

Abstract. Most digital cameras use color filter array (CFA) coupled sensors to capture color images. An image signal processor (ISP) pipeline is used to convert CFA captures to visually pleasing red-green-blue (RGB) images. These images are commonly used for downstream computer vision (CV) and machine learning (ML) tasks. Despite their widespread use, the impact of different components of the ISP pipeline on ML and CV tasks remains largely unexplored. We study the impact of demosaicing (color reconstruction) and processing of images by comparing it with neural network-based reconstruction of RGB images from the CFA signal. Since noise reduction is part of the motivation for the ISP pipeline, we repeat our experiments with added noise. We find that while demosaicing has little impact, operators such as median filtering, white balance and white level adjustment significantly reduce reconstruction quality. The ML reconstruction approach that bypasses the ISP pipeline performs better, even in the presence of noise. This motivates using the raw CFA data instead of processed RGB for ML and CV tasks.

Keywords: Signal Reconstruction · Demosaicing · Color Filter Array · Deep Learning

1 Introduction

Color filter array (CFA) coupled sensors are ubiquitous in digital cameras, capturing color information by sampling the radiance of a scene in different color bands. The sampled signal undergoes demosaicing and processing to produce an RGB image optimized for human viewing, as illustrated in Fig. 1.

G. E. Mark-Hansen and F. H. Altmann—Equal contribution.

J. Petersen and V. A. Dahl (Eds.): SCIA 2025, LNCS 15726, pp. 405–418, 2025.
https://doi.org/10.1007/978-3-031-95918-9_28

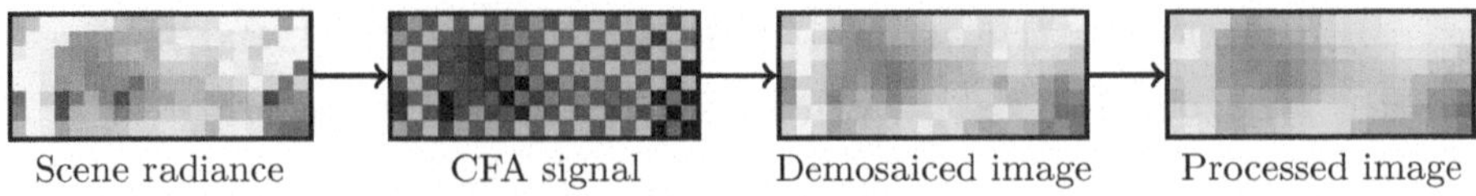

Fig. 1. A typical image signal processing pipeline where scene radiance is captured with a CFA coupled sensor. The CFA signal is processed and demosaiced to an intermediate RGB representation and processed into the final RGB image.

The processed image is often used as input for downstream tasks such as training computer vision (CV) and machine learning (ML) models, however this may not be optimal. The demosaicing and processing pipeline can introduce artifacts and noise, potentially degrading the performance of these models.

Recent advances in neural networks (NN) have led to state-of-the-art methods for denoising and reconstruction tasks [6,7,12,20,22,24,25]. A key question is, whether using the raw or processed CFA signal as input to these models can improve their efficacy and efficiency. Compared to an RGB image, a CFA signal is a better representation of the scene radiance, as it represents the raw data captured by the image sensor. Moreover, bypassing the demosaicing and processing pipeline may reduce the introduction of artifacts and noise.

To investigate this hypothesis, we conduct a systematic study comparing the performance of ML models trained on demosaiced and denoised images versus those trained directly on CFA signals. We create synthetic CFA data from various datasets (`Imagenette`, `CelebA-HQ`, and `EuroSAT`) using a simulated Bayer pattern [1]. These datasets are then converted into synthetic RGB images using four demosaicing algorithms, namely bilinear, Patterned Pixel Grouping (PPG) [19], Variable Number of Gradients (VNG) [4], and Adaptive Homogeneity-Directed (AHD) [11]. Since the conversion of RGB images to synthetic CFA data and again from synthetic CFA data to synthetic RGB image ($\overset{\text{Original}}{\text{RGB}} \rightarrow \overset{\text{Syn.}}{\text{CFA}} \rightarrow \overset{\text{Syn.}}{\text{RGB}}$) leads to information loss, we train a neural network to reconstruct the original image from the output of different demosaicing algorithms. We test three different settings: First we study the impact of demosaicing ($\overset{\text{Syn.}}{\text{RGB}} \overset{\text{U-Net}}{\rightarrow} \overset{\text{denoised}}{\text{RGB}}$). Second, we study the impact of three image processing techniques in addition to demosaicing, namely, median filtering, white balance and white level adjustment ($\overset{\text{Syn.}}{\text{RGB}} \rightarrow \overset{\text{Processed}}{\text{RGB}} \overset{\text{U-Net}}{\rightarrow} \overset{\text{denoised}}{\text{RGB}}$). Application of these image processing operators is partly motivated by noise reduction. Thus, we finally study the impact of demosaicing and image processing in presence of artificially added random noise ($\overset{\text{Syn.}}{\text{RGB}}\text{+Noise} \rightarrow \overset{\text{Processed}}{\text{RGB}} \overset{\text{U-Net}}{\rightarrow} \overset{\text{denoised}}{\text{RGB}}$). In each case, the synthetic RGB image is denoised to match the original RGB using a neural network. In contrast to these approaches, we also train a neural network to reconstruct the original RGB directly from synthetic CFA, with and without added noise ($\overset{\text{Syn.}}{\text{CFA}}$ (+Noise) $\overset{\text{U-Net}}{\rightarrow}$ $\overset{\text{denoised}}{\text{RGB}}$).

Our results show that the neural networks trained on synthetic CFA data reconstruct input images with lower mean squared reconstruction error (MSRE) compared to models trained on synthetic RGB images. In addition, we find that the processing of synthetic RGB leads to higher MSRE, with white level adjustment having the highest impact, followed by white balance and median filtering. Moreover, the image processing steps do not reduce MSRE in the presence of artificially added random noise. We find CFA data to be a better suited input source for neural networks than its corresponding demosaiced and processed RGB output. While CFA data is difficult to visualize without additional processing, utilizing it for ML methods can be beneficial as well as resource efficient.

Future use of digital photography to gather data, should account for these differences to ensure the highest data fidelity for the application, rather than focusing only on creating visually pleasant images. The code for the experiments is available at https://github.com/GustavMH/cfa-nn-image-reconstruction.

2 Related Works

This section focuses on generating synthetic CFA data and the use of synthetic CFA data in downstream ML applications.

2.1 Generating Synthetic CFA Data

There are several approaches for generating synthetic CFA from RGB images. A common approach uses approximate inverse functions to reverse an image signal processor (ISP) pipeline. Inverse functions are used to invert each step of the pipeline, for example gamma decompression is used to reverse the effect of gamma compression [3]. Another approach, known as `CycleR2R`, learns each reversal step through a deep learning framework rather than relying on predetermined steps and parameters [18]. Additionally, end-to-end deep learning methods have been developed to treat the conversion from RGB to synthetic CFA as a style transfer problem [17].

Since these approaches involve inverting the ISP pipeline, they can introduce artifacts due to the use of approximate inverse functions. Moreover, many operations in the ISP pipeline are designed to apply prior sensor information, such as gain characteristics and color correction matrices from the sensor, which standardize the image sensor output to a reference color space [14]. To avoid these limitations, we generate synthetic CFA data by directly applying the Bayer pattern to the RGB image without additional operations. It is important to note that the characteristics of the synthetic CFA data used in this study differs significantly from those of raw CFA data from camera sensors, as our synthetic CFA data generation pipeline does not invert several transformations typically applied to raw CFA data. (see Sect. 3 for details.)

2.2 Utilizing CFA Data for Computer Vision Tasks

There is growing interest in using CFA captures as input for neural networks. Li et Al. [18] demonstrated the effectiveness of using synthetic CFA data, generated through the `CycleR2R` model, to fine-tune a YOLOv3 object detection model, resulting in improved precision compared to the non-fine-tuned RGB model counterpart. Kantos et al. [15] trained VGG-13 and ResNet-34 models on real-world CFA captures and derived RGB images, and found that the CFA and RGB models achieved comparable accuracy. Brooks et al. [3] used an inverse ISP to generate data for training a CFA domain ML denoising model, which performed equally well as their RGB domain denoiser. Hansen et al. [8] investigated the impact of image processing on MobileNet performance and found a 7% improvement in accuracy when using processed RGB images compared to synthetic CFA data with a skewed lightness distribution.

However, these studies often focus on specific tasks and may not provide a comprehensive understanding of the general performance implications of using CFA signals for downstream tasks. The performance differences observed in these studies may be influenced by factors such as input source, task complexity, or noise levels. To address some of these limitations, we focus on a reconstruction task where a neural network is trained to denoise and reconstruct the original RGB image from the synthetic synthetic CFA signal as well as the original synthetic RGB image. This approach allows us to measure the differences in target and input at the level of individual reconstructed pixels, providing a more nuanced understanding of the performance implications of using CFA signals for downstream tasks.

3 Methodology

In this section, we outline our proposed methodology for evaluating differences in RGB image reconstruction using four demosaicing algorithms, namely bilinear interpolation, Patterned Pixel Grouping (PPG) [19], Variable Number of Gradients (VNG) [4], and Adaptive Homogeneity-Directed (AHD) [11]. We generate Color Filter Array (synthetic CFA) data from RGB images and then reconstruct the RGB images in three different settings. First, we study the impact of demosaicing on RGB reconstruction. Next, we study the impact of demosaicing in combination with image processing techniques, namely median filtering, white balance and white level adjustment, on image reconstruction. Finally, we study the impact of AHD demosaicing and image processing in presence of artificially added random noise. For details on the image processing methods, see Sect. 3.2.

We perform our experiments on three datasets; `Imagenette` [13], `CelebA-HQ` [16], and `EuroSAT` [10], covering a diverse set of images including ground level images, facial portraits, and satellite imagery. Additional dataset characteristics are summarized in Sect. 3.1. To create synthetic CFA data, we sample color information from images using a Bayer pattern [1] (as shown in Fig. 2). Hansen et al. [8] report that inverting the complete ISP pipeline leads to a highly skewed distribution in the synthetic CFA signal, which considerably degrades performance of ML methods in their experiments. The ISP pipeline is also designed to incorporate camera specific information such as sensor gain and color correction in the processed image. Therefore, unlike some related works [3,8], we only invert the demosaicing step and preserve other image characteristics.

To further evaluate the impact of demosaicing and image processing on ML models, we train U-Net [23] models on synthetic CFA data and synthetic RGB images to reconstruct the original RGB images. The U-Net models are trained and evaluated using mean squared reconstruction error (MSRE). See Sect. 3.3 for details on the model and training process. We use the `dcraw` program for demosaicing and image processing [5].

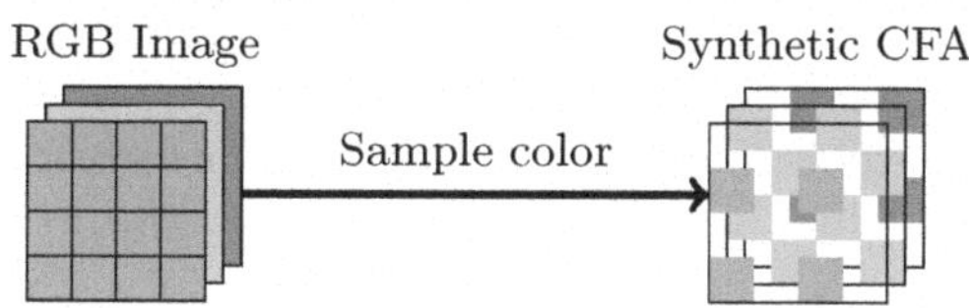

Fig. 2. Synthetic CFA generation with the Bayer pattern.

3.1 Datasets

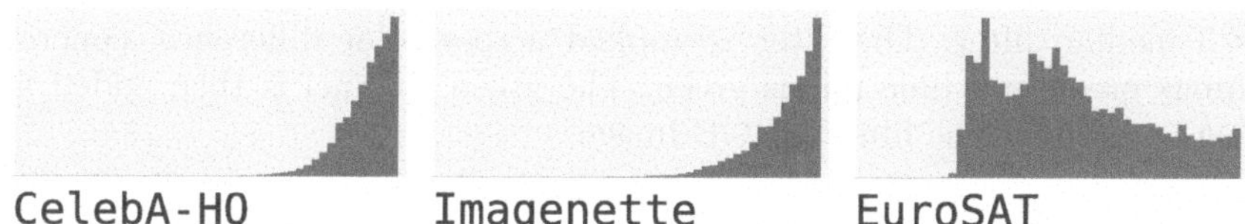

Fig. 3. Normalized histograms of 99th percentile brightness (x-axis) for images in each dataset.

The selected datasets cover a diverse range of photography conditions and subject matters. `Imagenette` contains diverse ground level images, `CelebA-HQ` human faces, an important optimization target for image processing, and `EuroSAT` satellite images. While the images in `Imagenette` and `CelebA-HQ` are brightness normalized, `EuroSAT` images are not (see Fig. 3).

Table 1. Dataset overview. Image sizes, number of samples in training and validation sets for each dataset.

Dataset	Size	Training set size	Validation set size
`Imagenette`	160 px	9469	3925
`CelebA-HQ`	256 px	20000	10000
`EuroSAT`	64 px	18500	8500

Table 1 provides an overview of the datasets, including image sizes and the number of samples in the training and validation sets. The *Size* column denotes the width of the smallest image axis in pixels. `Imagenette` contains non-square images, these are center-cropped to a 160 pixel wide square. As the datasets do not come with identifiable ICC color profiles, we use the `sRGB` color space [14].

3.2 Image Processing

Beyond demosaicing, other operations such as tone mapping, white level adjustment, white balance, gamma correction, color correction and denoising are applied in the ISP pipeline. For our study, we examine the impact of white level adjustment, white balance and median filtering operations.

White Level Adjustment: To increase the dynamic range utilized in the final image, the white level may be set automatically based on the scene capture. In `dcraw` [5], this white level is found as the 99th percentile brightness on a 512 bin histogram, the image values are then scaled to be between this level and 0.

White Balance: To make colors appear neutral, per channel scaling can be applied, making white shades have equal values in every color channel. In `dcraw`, white balance is implemented by scaling each color channel separately, such that the average channel values are equal.

Median Filtering: To remove artifacts (false color) that may occur from the demosaicing process, a median filter can be utilized. In `dcraw`, this is done with a 3×3 median filter. The filter is applied across color difference, concretely red minus green and blue minus green. The green channel is then added after filtering to produce the filtered RGB image.

3.3 Denoising Neural Network

Inspired by the image denoising approach proposed by Couturier et al. [6], we trained deep learning models based on the U-Net architecture [23]. Our U-Net model has 16.7 million parameters and we trained it optimizing for MSRE instead of the structural similarity index measure (SSIM) used by Couturier et al. We used the same model architecture for synthetic CFA and synthetic RGB data. At each pixel position, the synthetic CFA input specifies a single corresponding channel, all other channels are set to zero (see right side of Fig. 2).

The models for `Imagenette` were trained for 300 epochs, and the models for `EuroSAT` and `CelebA-HQ` were trained for 100 epochs. The training was done using the `AdamW` optimizer with learning rate 10^{-4} and batch size 32. Each experiment was repeated 5 times with weights randomly initialized with uniform He initialization [9,21].

3.4 Noise Types

In the last experimental setting, we added noise to the synthetic CFA to make the reconstruction more difficult. This also helps us to study the effectiveness of the additional image processing steps on the noisy input. We selected three types of noise to model the "real world" noise in images:

Gaussian noise models thermal noise in cameras and it is among the most common types of noise in images [2].

Speckle noise is a multiplicative noise that has similar characteristics as atmospheric speckle [2], making it useful for modeling noise in satellite imagery.

Salt-and-Pepper noise is a long-tailed noise that can be present due to dead pixels, analog-to-digital conversion errors, or bit transmission errors. Since it is often removed with median filters [2], including this noise helps to assess the effectiveness of median filtering on the reconstruction task.

4 Experiments and Results

We studied the impact of demosaicing and subsequent processing by breaking down the pipeline into its individual constituents and measuring the impact of each step on image reconstruction through the MSRE loss. Additionally, for every setting we trained a neural network to reconstruct the image directly from synthetic CFA data.

To ensure comparability in the reconstruction and denoising tasks, we used the same U-Net model architecture for both synthetic CFA and synthetic RGB data.

4.1 Effect of Demosaicing

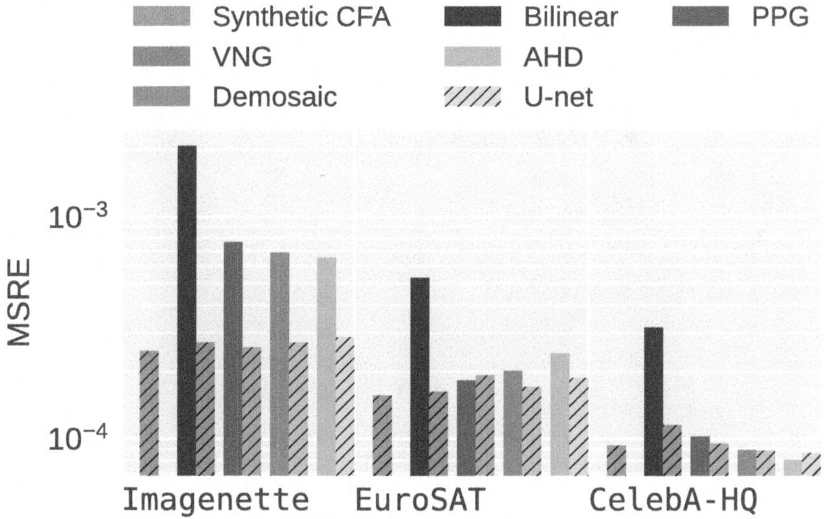

Fig. 4. Mean MSRE for different demosaicing algorithms compared with original signal, before and after U-Net denoising. The x-axis represents the demosaicing algorithm used, and the y-axis represents the mean MSRE. Solid bars indicate the MSRE before denoising, while striped bars indicate the MSRE after denoising.

We compared different approaches to generate an RGB image from the synthetic CFA data in terms of image reconstruction quality.

First, we considered the setting $\overset{\text{syn.}}{\text{CFA}} \overset{\text{demosaicing}}{\rightarrow} \overset{\text{syn.}}{\text{RGB}}$, where the RGB is generated using one of the four demosaicing algorithms VGN, PPG, AHD, and bilinear interpolation (solid bars in Fig. 4). Then we repeated the experiments, but added a U-Net trained for denoising as described in Sect. 3.3, $\overset{\text{syn.}}{\text{CFA}} \overset{\text{demosaicing}}{\rightarrow} \overset{\text{syn.}}{\text{RGB}} \overset{\text{U-Net}}{\rightarrow} \overset{\text{denoised}}{\text{RGB}}$, see striped bars next to solid bars in Fig. 4. Finally, we trained a denoising U-Net for reconstructing the image directly on the synthetic CFA data, $\overset{\text{syn.}}{\text{CFA}} \overset{\text{U-Net}}{\rightarrow} \overset{\text{denoised}}{\text{RGB}}$, see first striped bar in Fig. 4.

We found that adding a denoising U-Net to the pipeline mostly decreased the reconstruction error, except for VNG and AHD on `CelebA-HQ` and PPG on `EuroSAT`. The AHD demosaicing algorithm performed better than other approaches for the `CelebA-HQ` dataset containing high-quality close up of human faces.

The performance of applying a U-Net directly to the synthetic CFA data (i.e., without performing demosaicing) was not significantly different compared to training on synthetic RGB data, as determined by a Mann-Whitney U-test ($p > 0.05$).

4.2 Effects of Image Processing

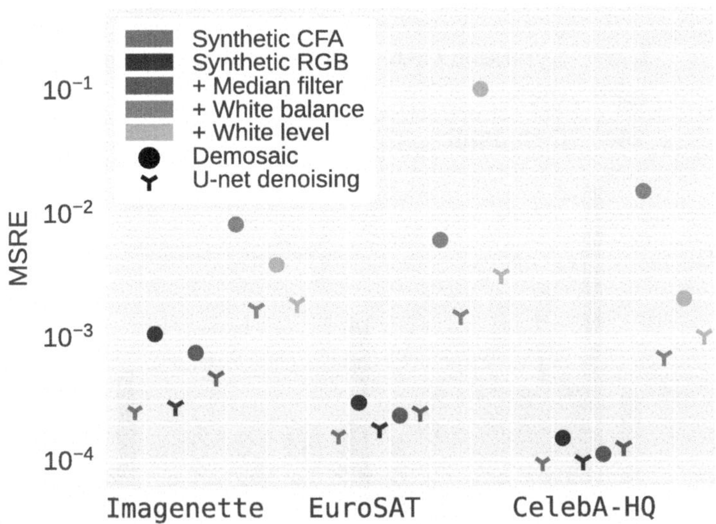

Fig. 5. Mean MSRE for synthetic RGB after demosaicing and processing with white balancing, white level adjustment, or median filtering (solid color disk) and after U-Net denoising (down spike).

In the second experiment, we applied white balance, white level, or median filtering processing to the demosaiced synthetic RGB images and measured their impact on MSRE (see Fig. 5).

We found that channel scaling image processing (white balance and white level) increased the MSRE by up to two orders of magnitude compared to demosaicing and direct U-Net reconstruction. The highest MSRE was found for `EuroSAT` with white level adjustment, likely due to the dataset's white level distribution.

Adding median filtering reduced MSRE for the synthetic RGB images, but interestingly additional U-Net denoising on top of image processing led to higher MSRE for the `EuroSAT` and `CelebA-HQ` datasets. Overall, the MSRE for all median filtered datasets remained worse compared to applying a denoising U-Net to synthetic RGB images and synthetic CFA reconstructions directly.

Notably, any additional processing after demosaicing significantly increases the MSRE (U-test, $p < 0.05$) compared to the synthetic CFA reconstruction.

Image processing operations such as median filtering are applied to deal with noise [11]. In the noise-free setting, median filtering removes information, and it is not surprising, that this deteriorates reconstruction performance. Therefore in the next experiment, we examine its effect in the presence of added noise.

4.3 Adding Noise

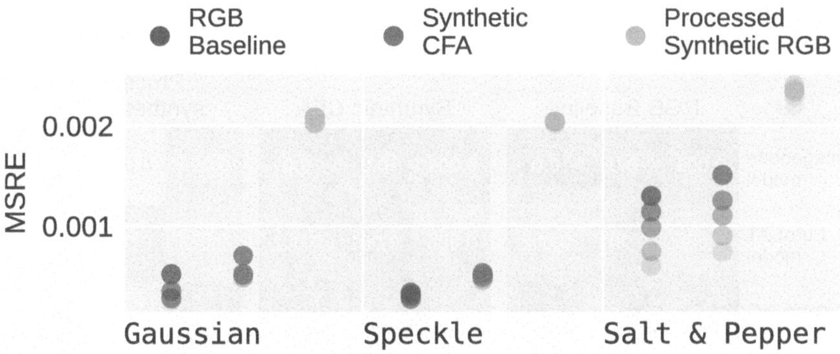

Fig. 6. MSRE for denoising original RGB images, synthetic CFA, processed synthetic RGB with 0.06%, 0.13%, 0.25%, 0.50%, 1%, and 2% noise (indicated by increasing opacity of the markers) of three types.

In this experiment, we investigated the impact of noise on our image processing pipeline. Specifically, we sought to determine how different types and levels of noise affected the reconstruction error when denoising original RGB images, synthetic CFA data, and processed synthetic RGB images. We added three types of noise to the original RGB images before applying the Bayer pattern. Then we trained denoising U-Nets [6] to remove the noise at three different stages. First, as a baseline, we applied the U-Net to the original RGB image with added noise. Then we applied the Bayer pattern to create synthetic CFA data, and denoised and reconstructed from this data. Finally, we applied our processing pipeline involving AHD demosaicing, white level adjustment, and median filtering, and then denoising the resulting synthetic RGB images. We used Gaussian, speckle, and salt & pepper noise, varying the noise level from 0.06%, 0.13%, 0.25%, 0.50%, 1% to 2%.

The results show, that the reconstruction errors were closer to the baseline models, when the processing pipeline was not applied (see Fig. 6). With processing, there was a significant increase in MSRE (U-test, $p < 0.05$). This was observed for all types of noise and at all noise intensities. This suggests, that the image processing pipeline itself introduces irreversible artifacts and that this effect is dominant in the signal.

4.4 Data Source Generalization of the Denoising Neural Networks

Our denoising neural networks have been trained and applied on each dataset separately (see Sect. 3.3). In real-world applications, we often lack denoising neural networks specifically trained on the target dataset. Therefore, we investigate cross-dataset generalization of our denoising models. We evaluate our denoising U-Net models on the RGB baseline with an additional 2% salt-and-pepper noise, on a synthetic CFA dataset derived from the RGB baseline and finally on a synthetic RGB dataset processed with automatic white level and median filtering (see Fig. 7).

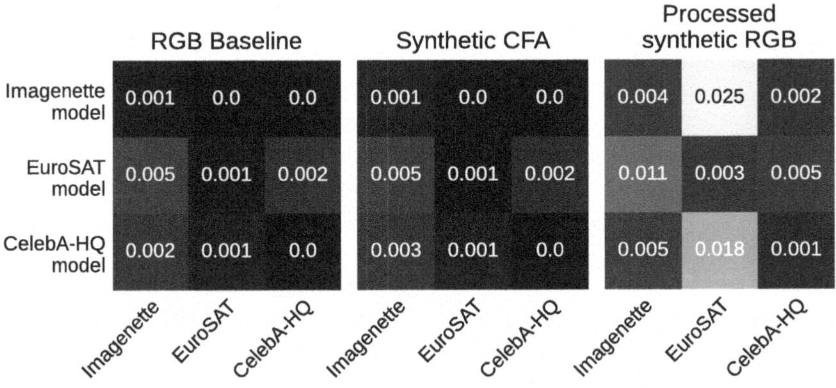

Fig. 7. MSRE by model, dataset pair for the RGB baseline in presence of additional noise (2% salt-and-pepper), synthetic CFA dataset derived from the RGB baseline and on synthetic RGB dataset processed with automatic white level and median filtering.

Training models on the processed `Imagenette` or `CelebA` datasets, yields higher MSRE when the model is evaluated on processed `EuroSAT`, which has a non-normalized level distribution before processing (see histogram in Fig. 3). Models trained on EuroSAT also have higher MSRE when denoising the Imagenette dataset. This suggests, that an accurate lightness distribution is essential to the reconstruction, and that information about this distribution is lost during preprocessing.

4.5 Computational Overhead

Evaluating the neural network model takes 23.93 ms on average for a 160 by 160 pixel CFA capture, across 13394 captures, while it takes 13.81 ms to demosaic with the AHD implementation from dcraw. Demosaicing the image before neural network inference adds 57.7% to the total latency. Both the neural network model and dcraw were run on 8 cores of an Intel(R) Xeon(R) Gold 6248 CPU @ 2.50GHz, with all 13394 image files loaded into working memory.

5 Limitations

In this study, we do not invert the complete ISP pipeline and construct CFA data through subsampling of a processed RGB image with the Bayer pattern. Consequently, the synthesized CFA data exhibits different characteristics as compared to raw CFA data. This distinction may lead to discrepancies when comparing results obtained using our synthesized CFA data versus those using raw CFA data directly.

6 Discussion and Conclusion

In this study, we conducted a systematic investigation into image reconstruction from Color Filter Array (CFA) data, exploring both traditional demosaicing algorithms with image processing steps and deep learning-based approaches.

Our experiments yielded several key findings:

1. Applying a U-Net deep neural network trained for image denoising improves the output of CFA demosaicing algorithms.
2. There is no significant difference in reconstruction error, when a U-Net is applied directly to raw CFA data versus demosaicing the CFA signal beforehand.
3. While image processing operations can support deep neural networks for classification tasks [8], common processing steps, such as white balance and white level adjustment, reduce the ability to reconstruct high-fidelity images.
4. Both in the presence and absence of added noise, processing steps such as white balance, white level adjustment, and median filtering significantly reduce reconstruction performance as compared to denoising the corresponding CFA with a U-Net deep neural network.

These findings suggest that deep neural networks applied directly to CFA data outperform demosaicing and standard image processing techniques in terms of original signal reconstruction performance measured by mean squared reconstruction error (MSRE). This is likely due to the fact, that image processing steps, particularly white level adjustment and white balance, scale all values in the image channels, reducing information about the original scene radiance. As a result, these operations make image reconstruction more challenging. These findings are in line with related literature, which show, that using CFA data as input to neural networks can achieve equivalent or even superior performance compared to traditional demosaicing and processing pipelines [3,15,18].

Notably, our results have implications for collecting imaging data for CV and ML tasks. While many types of image processing are useful for creating visually pleasing pictures, they may not be suitable for generating high-fidelity data required for downstream tasks.

In conclusion, our study highlights the importance of considering the impact of image processing steps on the quality and fidelity of imaging data. When collecting data for ML and computer vision tasks, it is essential to prioritize preserving the original information content over aesthetic considerations. By doing so, we can ensure, that the resulting data is better suited for downstream computer vision tasks.

Acknowledgements. CI and AK acknowledge support from the Danish National Research Foundation, Center for Remote Sensing and Deep Learning of Global Tree Resources (TreeSense), DNRF192. This work was supported by a research grant (VIL70006) from VILLUM FONDEN.

Disclosure of Interests. The authors have no competing interests to declare that are relevant to the content of this article.

References

1. Bayer, B.: Color imaging array. United States Patent, no. 3971065 (1976)
2. Boncelet, C.: Image noise models. In: Bovik, A. (ed.) The Essential Guide to Image Processing, chap. 7, pp. 143–167. Academic Press, Boston (2009). https://doi.org/10.1016/B978-0-12-374457-9.00007-X
3. Brooks, T., Mildenhall, B., Xue, T., Chen, J., Sharlet, D., Barron, J.T.: Unprocessing images for learned raw denoising. In: Computer Vision and Pattern Recognition Workshops (CVPRW), pp. 11028–11037 (2019). https://doi.org/10.1109/CVPR.2019.01129
4. Chang, E., Cheung, S., Pan, D.Y.: Color filter array recovery using a threshold-based variable number of gradients. In: Sampat, N., Yeh, T. (eds.) Sensors, Cameras, and Applications for Digital Photography, vol. 3650, pp. 36–43. International Society for Optics and Photonics, SPIE (1999). https://doi.org/10.1117/12.342861
5. Coffin, D.: Dcraw: decoding raw digital photos in Linux (2008). https://www.dechifro.org/dcraw/

6. Couturier, R., Perrot, G., Salomon, M.: Image denoising using a deep encoder-decoder network with skip connections. In: Cheng, L., Leung, A., Ozawa, S. (eds.) ICONIP 2018. LNCS, vol. 11306, pp. 554–565. Springer, Cham (2018). https://doi.org/10.1007/978-3-030-04224-0_48
7. Gondara, L.: Medical image denoising using convolutional denoising autoencoders. In: International Conference on Data Mining Workshops (ICDMW), pp. 241–246. IEEE (2016). https://doi.org/10.1109/ICDMW.2016.0041
8. Hansen, P., et al.: Isp4ml: the role of image signal processing in efficient deep learning vision systems. In: 2020 25th International Conference on Pattern Recognition (ICPR), pp. 2438–2445 (2021). https://doi.org/10.1109/ICPR48806.2021.9411985
9. He, K., Zhang, X., Ren, S., Sun, J.: Deep residual learning for image recognition. In: Computer Vision and Pattern Recognition Workshops (CVPRW), pp. 770–778 (2016). https://doi.org/10.1109/CVPR.2016.90
10. Helber, P., Bischke, B., Dengel, A., Borth, D.: Eurosat: a novel dataset and deep learning benchmark for land use and land cover classification. IEEE J. Sel. Top. Appl. Earth Obs. Remote Sens. **12**(7), 2217–2226 (2019). https://doi.org/10.1109/JSTARS.2019.2918242
11. Hirakawa, K., Parks, T.W.: Adaptive homogeneity-directed demosaicing algorithm. IEEE Trans. Image Process. **14**(3), 360–369 (2005). https://doi.org/10.1109/TIP.2004.838691
12. Ho, J., Jain, A., Abbeel, P.: Denoising diffusion probabilistic models. In: Advances in Neural Information Processing Systems (NeurIPS), vol. 33, pp. 6840–6851 (2020)
13. Howard, J., Gugger, S.: Fastai: a layered API for deep learning. Information **11**(2) (2020). https://doi.org/10.3390/info11020108
14. International Color Consortium: Image technology colour management – architecture, profile format, and data structure (2022), profile version 4.4.0.0
15. Kantas, C., et al.: Raw instinct: trust your classifiers and skip the conversion. In: International Conference on Pattern Recognition and Artificial Intelligence (PRAI), pp. 456–460 (2023). https://doi.org/10.1109/PRAI59366.2023.10332021
16. Karras, T., Aila, T., Laine, S., Lehtinen, J.: Progressive growing of GANs for improved quality, stability, and variation. In: International Conference on Learning Representations (ICLR) (2018). https://openreview.net/forum?id=Hk99zCeAb
17. Knl, F., Özcan, B., Kraç, F.: Reversing image signal processors by reverse style transferring (2022). https://arxiv.org/abs/2210.09074
18. Li, Z., Lu, M., Zhang, X., Feng, X., Asif, M.S., Ma, Z.: Efficient visual computing with camera RAW snapshots. IEEE Trans. Pattern Anal. Mach. Intell. 4684–4701 (2024). https://doi.org/10.1109/tpami.2024.3359326
19. Lin, C.K.: Pixel grouping for color filter array demosaicing (2003). https://web.archive.org/web/20061018012905/http://web.cecs.pdx.edu/~cklin/demosaic/
20. Parcollet, T., Morchid, M., Linarès, G.: Quaternion convolutional neural networks for heterogeneous image processing. In: IEEE International Conference on Acoustics, Speech and Signal Processing (ICASSP), pp. 8514–8518 (2019). https://doi.org/10.1109/ICASSP.2019.8682495
21. Paszke, A., et al.: Pytorch: an imperative style, high-performance deep learning library. In: Advances in Neural Information Processing Systems (NeurIPS), vol. 32 (2019)
22. Preechakul, K., Chatthee, N., Wizadwongsa, S., Suwajanakorn, S.: Diffusion autoencoders: toward a meaningful and decodable representation. In: Computer Vision and Pattern Recognition Workshops (CVPRW), pp. 10619–10629 (2022)

23. Ronneberger, O., Fischer, P., Brox, T.: U-Net: convolutional networks for biomedical image segmentation. In: Medical Image Computing and Computer-Assisted Intervention (MICCAI), pp. 234–241. Springer (2015)
24. Sohl-Dickstein, J., Weiss, E., Maheswaranathan, N., Ganguli, S.: Deep unsupervised learning using nonequilibrium thermodynamics. In: International Conference on Machine Learning (ICML), pp. 2256–2265. PMLR (2015)
25. Vincent, P., Larochelle, H., Bengio, Y., Manzagol, P.A.: Extracting and composing robust features with denoising autoencoders. In: International Conference on Machine Learning (ICML), pp. 1096–1103 (2008)

Correction to: Automated Cardiac Adipose Tissue Segmentation in Computed Tomography: A Literature Review

Andreas W. Aspe, Jonas Jalili Pedersen, Andreas Ohrt Johansen, Klaus Fuglsang Kofoed, Kristine Aavild Sørensen, Rasmus Reinhold Paulsen, and Josefine Vilsbøll Sundgaard

Correction to: Chapter 17 in: J. Petersen and V. A. Dahl (Eds.): *Image Analysis*, LNCS 15726, https://doi.org/10.1007/978-3-031-95918-9_17

The original version of Chapter 17 was mistakenly published with an incorrect author's name Andreas With Aspe instead of Andreas W. Aspe. This has now been rectified by updating it with the correct name.

The updated version of this chapter can be found at https://doi.org/10.1007/978-3-031-95918-9_17

J. Petersen and V. A. Dahl (Eds.): SCIA 2025, LNCS 15726, p. C1, 2026.
https://doi.org/10.1007/978-3-031-95918-9_29

Author Index

Galanti, Tomer I-60

J. Petersen and V. A. Dahl (Eds.): SCIA 2025, LNCS 15726, pp. 419–421, 2025.
https://doi.org/10.1007/978-3-031-95918-9

The manufacturer's authorised representative in the EU is Springer Nature Customer Service Centre GmbH, Europaplatz 3, 69115 Heidelberg, Germany. If you have any concerns regarding our products, please contact ProductSafety@springernature.com

Printed and bound by CPI Group (UK) Ltd, Croydon, CR0 4YY
15/07/2026
02167621-0007